T0302202

Unmanned Aircraft Systems Traffic Management

Unmanned Aircraft Systems Traffic Management

UTM

Michael Scott Baum

CRC Press
Taylor & Francis Group
Boca Raton London New York

CRC Press is an imprint of the
Taylor & Francis Group, an **informa** business

First edition published 2022
by CRC Press
6000 Broken Sound Parkway NW, Suite 300, Boca Raton, FL 33487-2742
and by CRC Press

2 Park Square, Milton Park, Abingdon, Oxon, OX14 4RN

CRC Press is an imprint of Taylor & Francis Group, LLC

ISBN: 978-0-367-64473-4 (hbk)
ISBN: 978-0-367-64476-5 (pbk)
ISBN: 978-1-003-12468-9 (ebk)

Typeset in Times LT Std
by KnowledgeWorks Global Ltd.

Contents

Preface

My first encounter with unmanned aviation came as a shock, following an air traffic controller's alert of an unmanned aircraft (UA) penetrating the instrument approach path I was flying to my home base airport. That vexing incident led me to fashion local mitigations and then catalyzed a deeper dive into the UA industry's nascent yet dramatically accelerating technology ecosystem, operations, standards, and regulation.

I came to appreciate that the future of manned aviation is inextricably linked to the safe, efficient, and normalized integration of UA into national airspace systems. Traffic management innovation is at such integration's core.

I also encountered a remarkably talented and passionate cadre of developers, working collaboratively with government and other aviation stakeholders. This cadre has assumed a well-deserved leadership role. Finally, I observed both challenges and a rapprochement of sorts between the manned and unmanned aviation communities whose collaboration and cooperation will chart their collective future.

Like any new technology area, as UTM has progressed from early research and testing to early deployments and on to technical standards and regulation, concepts and their implementation have been continuously evolving. And as UTM concepts are adopted globally, we are seeing diversity in their implementation. While I've done my best to describe the most current state as of the time of publication, inevitably things will continue forward. Please view this as a snapshot in time that informs the further evolution of UTM.

With both respect and fascination for the UA development community—while never losing sight of the essential interests of manned aviation—I present Unmanned Aircraft Systems Traffic Management: UTM, seeking to provide an introductory glimpse of this new, foundational system.

Acknowledgments

I'd like to thank my family, Vera, Kimberly, and Matthew for their unending patience and support during this project.

The editorial input and valuable assistance of many subject matter experts are acknowledged and greatly appreciated. Their input does not necessarily represent the opinions of their employers, and affiliations are for identification only. All reviewers do not necessarily concur with all positions taken, nor did all reviewers review the entire or final manuscript. Thanks to all, including those not listed who requested anonymity.

Alexander, Rex, Five-Alpha LLC
Ancel, Ersin, PhD, NASA
Arendt, Don, PhD, CFI, ATP, FAA
Ballal, Hrishi, PhD, Openskies Aerial Technology Ltd.
Baum, Matthew C., Independent Monitoring, LLC
Bender, Walter R., Johns Hopkins University Applied Physics Laboratory
Berchoff, Don, TruWeather Solutions
Blasgen, Alexandra, Consumer Technology Association
Bloch-Hansen, Craig, Transport Canada
Blum, Scott, PhD, Jeppesen
Blyenburgh, Peter van, UVS International
Borda, Fred, Aerial Innovation
Bruner, Tim, Kansas State Polytechnic
Carter, Andrew, Resilienx, Inc.
Chapman, Oliver, GSMA
Champagne, Robert, ASTM F38 UTM WG Member
Cook, Stephen, PhD, Northrop Grumman Corporation
Cornelius, Peter, Frequentis Comsoft GmbH
Cox, Gabriel, Intel
Curdy, Benoit, Swiss FOCA
Daniels, Jonathan, Praxis Aerospace Concepts Int'l
Davis, Mark Edwards, PhD, Ollin Aviation
Deeds, Greg, Technology Exploration Group, Inc.
Di Antonio, Giovanni, JARUS
Dietrich, Anna Mracek, Community Air Mobility Initiative (CAMI)
Eckstein, Bruce, FirebirdSE LLC
Egorov, Maxim, Airbus
Engelstad, Ken, EASA
Evans, Jonathan, KinectAir
Fisher, Brian, FAA
Flom, Nicholas T., NPUASTS
Fort, John, Frequentis
Ganjoo, Amit, ANRA

Garrity, Robin, SESAR Joint Undertaking
Giovannini, Stefano, d-Flight S.p.A.
Glasgow, Mike, Wing
Guendel, Randal, MIT Lincoln Laboratory
Gunnarson, Tom, Wisk
Hallam-Baker, Phillip, PhD, Hallam-Baker Consulting
Hallenbeck, Lane, ANSI National Accreditation Board
Hansen, Rich, AMA
Hately, Andrew, EUROCONTROL
Hayes, Stephen, Ericsson, Inc.
Hegranes, Jon, Kittyhawk.io
Hinaman, Arthur W., FAA
Homola, Jeffrey R., NASA Ames
Houston, Rory, Skyy Network
Hudson, Hunter, Northrop Grumman
Johnson, Simon, OpenStratosphere S.A.
Kannan, Suresh K., PhD, Nodein Autonomy Corp.
Kohler, Brittney, National League of Cities
Kopardekar, Parimal, PhD, NASA Ames
Kovac, Eszter, DroneTalks & Manageld
Kucera, Christopher T., OneSky Systems
Kenul, Philip M., ASTM F38
Kesterson II, Hoyt L., Avertium
Lacher, Andrew R., Noblis
Lamprecht, Andreas, PhD, AIRMAP
Larimer, Jane, JD, Nacha
Larrow, Jarrett, FAA
Lenox, Zach, JD, LLM, Wilson Sonsini Goodrich & Rosati, P.C.
Lester, Edward "Ted", REGENT Craft
Licata, James, Hidden Level, Inc.
Lubrano, Manu, involi
Lukácsy, Fanni, GUTMA
Marshall, Douglas M., JD, TrueNorth Consulting LLC
Marzac, Sergui, EUROCAE
McCarthy, Tim, PhD, Maynooth University
McDuffee, Paul, Hyundai Urban Air Mobility
Meyer, Andreas, ICAO
Moallemi, Mohammed, PhD, Concepts Beyond
Montellato, Brandon, Wing
Moore, John R., Collins Aerospace
Morrison, Adam, Streamline Designs LLC
Moskowitz, Robert, HTT Consulting
Murphy, David, ANRA
Murzilli, Lorenzo, Murzilli Consulting
Namuduri, Kamesh, PhD, University of North Texas
Negron, Reinaldo, Wing

Nuckolls, Lance, Fmr. FAA
Osantowski, Andy, Robotic Skies
Pareglio, Barbara, GSMA
Parker, Richard, Altitude Angel
Pelletier, Benjamin, Wing
Peri, Ric, Aircraft Electronics Ass'n
Plishka, Stephen, FAA
Prevot, Tom, PhD, Joby Aviation
Richards, Jeffrey, NATCA
Richter, Jennifer, JD, Akin Gump
Rios, Joseph L., PhD, NASA Ames
Robbins, Michael, AUVSI
Roth, Robert, UBER ATG
Ruff, Nathan, UASidekick
Rushton, Anthony, PhD, NATS
Sachs, Peter, FAA
Sachs, Rusty, JD, DhE, MCFI
Secen, Al, RTCA
Seely, Michael J., FlySafeUAS
Segers, Robert, FAA
Silbermann, Joshua M., Johns Hopkins U. Applied Physics Laboratory
Silversmith, Jol A., JD, KMA Zuckert LLC
Standley, Jon, L3Harris Technologies
Steinman, Don, Capt. (ret.), American Airlines
Tracy, Rick, Telos Corp.
Trickey, Graham, GSMA
Thurling, Andy, NUAIR Alliance
Voss, William, Special Advisor to ICAO
Walker, John, The Padina Group
Williams, Heidi, NBAA
Williamson, Mark R., Skyward
Wolf, Harrison, Zipline
Woods, Daniel, DPhil, University of Oxford
Wu, Stephen S., JD, Silicon Valley Law Group
Wuennenberg, Mark, ICAO

Author

Michael Scott Baum, JD, MBA, ATP, is an aviation consultant, and founder and Permanent Editorial Board member of the *Aviators Code Initiative* (ACI).

He is an active participant and contributor to various forums developing manned and unmanned aviation standards and safety best practices. Previously Baum served as vice president and executive committee member at VeriSign (VRSN), IT security law and policy consultant at Independent Monitoring, LLC, and a Network Analyst at BBN Technologies (now Raytheon). Prior thereto he practiced law in Boston. He was founder and chair of the Electronic Commerce Division and Information Security Committee of the American Bar Association's Science and Technology Law Section, an observer to the United Nations Commission on International Trade Law (UNCITRAL), chairman, of the International Chamber of Commerce (ICC Paris) ETERMS Working Party, and vice chairman of the ICC Electronic Commerce Commission. He has been board member of ANSI ASC X12, *The Jurimetrics Journal* (ABA/ASU), The PKI Forum, and the Charles Babbage Foundation, as well as various IT enterprise advisory boards.

Previous publications include *UAS Pilots Code* (ACI) (co-author), *Aviators Model Code of Conduct* (series, ACI) (co-author), *Federal Certification Authority Liability and Policy* (NIST/MITRE), *Secure Electronic Commerce* (1st & 2nd Eds., Prentice-Hall) (co-author with Ford), *Digital Signature Guidelines* (ABA) (co-author), *Electronic Contracting, Publishing and EDI Law* (Wiley) (co-author with Perritt), *Introduction to Unmanned Aircraft Systems*, Douglas Marshall, et al. (3rd ed., CRC) (contributor), and *EDI and the Law* (Blenheim) (contributor). He was a Certified Information Systems Security Professional (CISSP). Baum has been honored with awards including: ABA Committee on Cyberspace Law's *Excellence Award*; National Notary Association's *Achievement Award*; Journal of EDI, *EDI Pioneer;* and LightHawk's *President's Award.*

Baum is a graduate of Carnegie Mellon University (BA), Western New England University School of Law (JD), and is a member of the Massachusetts Bar. He received his MBA from the Wharton School at the University of Pennsylvania. He is an FAA-certificated Airline Transport Pilot in single engine aircraft, and holds Private Pilot certification with airplane single engine sea, multiengine land, rotorcraft-helicopter, and glider ratings. He is also an Advanced Ground Instructor.

Abbreviations

3GPP	3rd Generation Partnership Project
4D or 4-D	four dimensional (3D plus time)
5G	fifth generation wireless technology
5GAA	5G Automotive Association
AAM	advanced air mobility
AAO	authorized area of operation
ABOVs	area based operational volumes
ABSAA	airborne sense and avoid
ACAS	airborne collision avoidance system
ACAS sXu	ACAX Xu for smaller UAS
ACAS Xu	airborne collision avoidance system for UAS
ACJA	Aerial Connectivity Joint Activity
ADS-B	Automatic Dependent Surveillance-Broadcast
AFR	autonomous flight rules
AGL	above ground level
AI	artificial intelligence
AIA	Aerospace Industries Association
AIS	aeronautical information services
AIXM	aeronautical information exchange model
AMC	acceptable means of compliance
AM(R)S	Aeronautical Mobile (Route) Service
ANPRM	Advanced Notice of Proposed Rulemaking
ANSI	American National Standards Institute
ANSP	air navigation service provider
API	application programming interface
App.	appendix
ARC	aviation rulemaking committee
ARMD	Aeronautics Research Mission Directorate [NASA]
ASD-STAN	Aerospace and Defence Industries Ass'n of Europe—Standardization
ASSURE	Alliance for System Safety of UAS through Research Excellence
ASTM	ASTM International
ATC	air traffic control
ATCA	Air Traffic Controllers Association
ATCT	air traffic control tower
ATIS	Alliance for Telecommunications Industry Solutions
ATM	air traffic management
ATM-X	Air Traffic Management Exploration [NASA]
ATO	air traffic organization [FAA]
AWOS	Automated Weather Observing System
BARR	Block Aircraft Registry Request Program

BLOS	beyond line of sight
BVLOS	beyond visual line of sight
C2	command and control
CAA	civil aviation authority
CAMI	Community Air Mobility Initiative
CANSO	Civil Air Navigation Services Organization
CAP	CAA Publication [UK]
CAR	civil aviation registry
CASA	Civil Aviation Safety Authority [Australia]
CBRs	community-based rules
CBRS	Citizens Broadband Radio Service
CCPA	California Consumer Protection Act
C.F.R.	US Code of Federal Regulations
Ch. or Chap.	chapter
CIA	confidentiality, integrity, availability
CIF	common information function
CIS	common information service
CISA	Cybersecurity and Infrastructure Security Agency [DHS]
Comm.	Committee
ConOps	concept of operations
ConUse	concept of use
CNPC	Control and Nonpayload Communications
CNS	[Communications] or [Command], Navigation and Surveillance
CORUS	Concept of Operations for EuRopean UTM Systems
CPR	common pool resource
CRD	comment response document [EASA]
CS	control station
CSF	NIST Cybersecurity Framework
CSMAC	Commerce Spectrum Management Advisory Committee [NTIA]
C-UAS	counter-UAS
CV	conformance volume
C-V2X	cellular vehicle to everything
D&R	durability and reliability
DAA	detect and avoid
DAO	decentralized autonomous organization
DHS	U.S. Department of Homeland Security
DIWG	Digital Identity Work Group [ICAO]
DNS	Domain Name System
DoD	US Department of Defense
DOI	Declaration of Intent
DoT	Department of Transportation
DPA	data processing addendum or agreement
DR	dynamic rerouting
drip	drone remote ID protocols [IETF]
DSS	discovery and synchronization service(s)
EASA	European Union Aviation Safety Agency

EC	European Commission
EEA	European Economic Area
EO	electro-optical
ER	Exploratory Research [SESAR]
ETM	Upper Class E Traffic Management
EU	European Union
EUROCONTROL	European Organisation for the Safety of the Air Navigation
EUSCG	European UAS Standards Coordination Group
EVLOS	extended visual line of sight
eVTOL	electric vertical takeoff and landing aircraft
F38	ASTM Int'l, Comm. F38 on Unmanned Aircraft Systems
FAA	Federal Aviation Administration
FAARA	FAA Reauthorization Act of 2018
FATO	final approach and takeoff area
FCC	Federal Communications Commission
FESSA	FAA Extension, Safety, and Security Act of 2016
FF-ICE	Flight and Flow Information for a Collaborative Environment
FG	flight geography
FIMS	Flight Information Management System
FIXM	flight information exchange model
FIPS	Federal Information Processing Standard [NIST]
FIR	flight information region
FISMA	Federal Information Security Management Act of 2014
FL	flight level
FOCA	Swiss Federal Office of Civil Aviation
FRA	Free Route Airspace
ft	feet or foot
FTC	US Federal Trade Commission
FTCA	Federal Torts Claims Act
GA	general aviation
GAMA	General Aviation Manufacturers Association
GANP	Global Air Navigation Plan [ICAO]
GBSAA	ground based sense and avoid
GDPR	General Data Protection Regulation [EU]
geo-zones	geographical zones
GIS	graphical information system
GNSS	global navigation satellite system
GRAIN	Global Resilient Aviation Information Network [ICAO]
GSM	Global System for Mobile Communications
gTLD	generic Top-Level Domain
GUFI	globally-unique flight identifier
GUTMA	Global UTM Association
HALE	high-altitude long endurance
HAPS	high-altitude pseudo-satellites
HFR	high-level flight rules

IATA	International Air Transport Association
IATF	International Aviation Trust Framework [ICAO]
ICAC	International Conformity Assessment Committee
ICAO	International Civil Aviation Organization
ICATF	International Civil Aviation Trust Framework, subsequently revised to IATF
ICD	Interface Control Document
ICNS	integrated communications, navigation, and surveillance
IEC	International Electrotechnical Commission
IEEE	The Institute of Electrical and Electronics Engineers
IETF	Internet Engineering Task Force
IFALPA	International Federation of Air Line Pilots' Association
IFR	instrument flight rules
IFV	intended flight volume
IGO	intergovernmental organization
IMEI	International Mobile Equipment Identity
IMT	International Mobile Telecommunications
InterUSS	InterUSS Platform
IoT	Internet of Things
IP	Internet Protocol
IPP	UAS Integration Pilot Program
IR	infrared
ISMS	information security management system
ISO	International Organization for Standardization
ISSA	In-Time System-Wide Safety Assurance
IT	information technology
ITAR	International Traffic in Arms Regulations
ITU	International Telecommunications Union
ITU-R	Radio Communication Sector of ITU
JARUS	Joint Authorities for Rulemaking on Unmanned Systems
LAANC	Low Altitude Authorization and Notification Capability
LFR	low-level flight rules
LIDAR	light detection and ranging
LOA	letter of authorization
LOS	line of sight
LSA	light sport aircraft
LTE	Long-Term Evolution
LUN	local USS network
MASPS	minimum aviation system performance standards
MIMO	multiple input multiple output
ML	machine learning
MLAT	multilateration
MLS	mission logging system
MNO	mobile network operator(s)
MOA	military operations area or Memorandum of Agreement
MoC	[means] [method] of compliance; or Memorandum of Cooperation

MOPS	Minimum Operational Performance Standards or Measures of Performance
MOSAIC	Modernization of Special Class Airworthiness Certification [FAA]
MOU	memorandum of understanding
MSL	mean sea level
MTOM	maximum takeoff mass
NAA	national aviation authority
NARI	NASA Aeronautics Research Institute
NAS	national airspace system
NASA	National Aeronautics and Space Administration
NBAA	National Business Aviation Association
Net-RID	Network Remote ID
NextGen	Next Generation Air Transportation System
NGO	non-governmental organization
NIEC	NextGen Integration and Evaluation Capability
NIST	National Institute of Standards and Technology
NOTAM	Notice to Airmen
NPNT	no permission, no takeoff
NPRM	Notice of Proposed Rulemaking
NTIA	National Telecommunications and Information Administration
NUSTAR	National Unmanned Aerial System Standardized Performance Testing and Rating
ODA	Organization Designation Authorization
ODM	on-demand mobility
OGC	Open Geospatial Consortium
OI	operational intent
OIV	operational intent volume
O.J.	Official Journal of the European Union
OMM	operations and maintenance manual
OOP	operations over people
OSED	Operational Services and Environment Definition
OSO	operational safety objective [JARUS]
OTA	other transaction authority
OV	operation volume
OVN	opaque version number
PA	performance authorization
PANS	Procedures for Air Navigation Services [ICAO]
PAO	public aircraft operator/operations
Part 107	14 C.F.R. Part 107
PIA	Privacy ICAO Address
PIC	pilot in command
PII	personally identifiable information
PKI	public key infrastructure
PSU	provider of services for UAM
PV	planning volume
QMS	quality management system

RESTful	Representational State Transfer protocol
RF	radio frequency
RFC	Request for Comment [IETF]
RFI	request for information
RID	UAS remote ID
RNP	required navigational performance
RoW	right of way
RPA	remotely piloted aircraft
RPAS	remotely piloted aircraft systems
RTM	RPAS Traffic Management [Canada]
RTT	Research Transition Team
RTTA	Reasonable Time to Act
RWC	Remain Well Clear
SA	situational awareness
SAA	sense and avoid or special activities airspace
SAE	SAE International—Society of Automotive Engineers International
SAIL	Specific Assurance Integrity Level [SORA]
SARPS	Standards And Recommended Practices [ICAO]
SAS	spectrum access system
SBS	surveillance broadcast service
SC	Special Committee; or special condition
SCC	EU Standard Contract Clauses
SCDS	SWIM Cloud Distribution Service
SDO	standards development organization
SDR	Special Drawing Right
SDSP	Supplemental Data Service Provider
Sect.	section
SENSR	Spectrum Efficient National Surveillance Radar
SERA	Standardized European Rules of the Air
SESAR	Single European Sky ATM Research
SESAR JU	SESAR Joint Undertaking
SG	study group or subgroup
SLA	service level agreement
SLAM	simultaneous location and mapping
SMS	safety management system
SOA	service-oriented architecture
SORA	Specific Operations Risk Assessment [JARUS]
SP	Special Publication [NIST]
SRIA	Strategic Research and Innovation Agenda [SESAR]
SRM	safety risk management
SRMD	safety risk management document
STC	Supplemental Type Certificate
STM	space traffic management
sUA	small unmanned aircraft
sUAS	small unmanned aircraft system

SUSI Swiss U-space Implementation
SWaP size weight and power
SWIM System-Wide Information Management
sXu airborne collision avoidance system for sUAS
TBO trajectory-based operations
TBOV Transit-Based Operation Volume
TC type certification
TCAS traffic collision avoidance system
TCL UTM Technical Capability Level [NASA]
TFSG Trust Framework Study Group [ICAO]
TG task group
TLS Transport Layer Security
ToR terms of reference
TRON Trust Reciprocity Operational Needs [ICAO]
TSE Total System Error
TSO Technical Standard Order
UA unmanned aircraft
UAM urban air mobility [*see* AAM]
UAS unmanned aircraft system
UAS-AG UAS Advisory Group [ICAO]
UASFM UAS Facility Map
UASSC Unmanned Aircraft Systems Standardization Collaborative [ANSI]
UASSG UAS Study Group [ICAO]
UATM urban air traffic management
UAV unmanned aerial vehicle
UDDS UAS Data Delivery Service [FAA]
UFAA UAS Service Provider Framework for Authentication and Authorization
ULC Uniform Law Commission
UML UAM Maturity Level
UOE UAM Operating Environment
UOMS Civil UAS Operation Management System [China]
UPP UTM Pilot Program
UREP UA Report
U-space urban space [EC]
USP U-space service provider or UAS service provider
USS UAS Service Supplier(s)
USSP U-space Service Provider
UTM UAS traffic management; or Unmanned [Uncrewed] aircraft systems traffic management
UUID Universally Unique Identifier based on RFC 4122 (128 bit) [IETF]
UVR UAS volume reservation
V2I vehicle-to-infrastructure
V2V vehicle-to-vehicle
V2X vehicle-to-everything
VFR visual flight rules

VLD	Very Large scale Demonstrations [SESAR]
VLL	very low level altitude or operation
VLOS	visual line of sight
VoIP	Voice over Internet Protocol
WAVE	Wireless Access in Vehicular Environments
WG	work group or working group
WX	weather
ZCA	zero-conflict airspace

Selected Terminology

The following list includes terms from relevant ConOps, rules, standards, industry guidance, and recognized aviation usage. Definitions for many of these terms are evolving and inconsistent among implementations.

4D: 3D volume (latitude, longitude, altitude) plus time.

acceptable means of compliance: method satisfying CAA as a means to establish regulatory compliance.

accreditation: third party attestation (such as by a conformity assessment body) conveying formal demonstration of its competence to undertake specific tasks.

advanced air mobility (AAM): aviation that moves people and cargo in urban, rural, and interregional environments via increasingly automated systems and technically advanced vehicles; a superset of urban air mobility (UAM).

aircraft: any machine, manned or unmanned, that can derive support in the atmosphere from the reactions of the air other than the reactions of the air against the earth's surface.

authorized constraint provider: organization or individual authorized to provide and manage constraints.

autonomy: the ability of a system to plan and execute its course of action to achieve a goal delegated to it, without intervention from the delegating agent. When conditions change, it can adapt by replanning. When unable to progress towards its goal, the system can detect and report its inability to the delegating agent.

beyond visual line of sight (BVLOS): the operation when the individual responsible for controlling the flight of the unmanned aircraft (UA) cannot maintain direct unaided (other than by corrective lenses, sunglasses, or both) visual contact with the UA, other aircraft, terrain, adverse weather, or obstacles.

broadcast: to transmit data to no specific destination or recipient; data can be received by anyone within broadcast range.

broadcast UAS: a UAS equipped for and actively broadcasting remote ID data during an operation; being a broadcast UAS is not mutually exclusive with being a networked UAS. Broadcast UAS is generally distinct from ADS-B and uses Bluetooth or WiFi technologies.

C2 link (Command and Control Link): the data link between the remotely piloted aircraft and control station for the purpose of managing the flight.

civil aviation authority (CAA): a governmental entity [and competent authority] of a nation state that regulates civil aviation. For example, in Canada, Transport Canada (TC); in France, the Directorate General for Civil Aviation (DGAC), and in the United States, the Federal Aviation Administration (FAA).

collision avoidance: a function used to alert the pilot of the need to take immediate action by generating a recommended maneuver to mitigate the risk of collision with all threats.

common information service (CIS): a service consisting of the dissemination of static and dynamic data to enable the provision of services for the management of traffic of unmanned aircraft.

concept of operations: a user-oriented document that describes systems characteristics and limitations for a proposed system and its operation from a user's perspective. It may also describe the user organization, mission, and objectives from an integrated systems viewpoint and is used to communicate overall quantitative and qualitative system characteristics and operational procedures to stakeholders.

conformance: the situation where a UA is flying within its operational intent.

conformance monitoring: a USS service that determines whether a UA adheres to its operational intent and takes appropriate actions when it is not.

conformance volume (CV): a buffer encapsulating the flight geometry, sized to reflect the performance capability of the UA, and facilitating recovery actions; CV breach may trigger additional events such as notifications to other USS or execution of contingency plans.

conformity assessment: activities and processes undertaken to confirm adherence to applicable requirements; a systematic program of measures (including e.g., testing, monitoring, reporting, record keeping) applied to determine that governance requirements are satisfied.

constraint: 4D volume(s) created to denote specific time and geographically limited airspace information; used primarily to notify operators of areas requiring special awareness or avoidance, or to block access.

constraint management: a USS service providing constraint creation, modification, and deletion, and the dissemination of constraint information to other USS.

constraint processing: a USS service and role that enables the USS to ingest and communicate constraint information (such as a no-fly constraint), and, when appropriate take other action.

contingency: off-nominal/nonconforming conditions that remain uncorrected for a prescribed time period, or are declared a contingency by the operator, and result in a UA entering a contingency (or rogue) state.

control station (CS): the facilities, equipment, and/or interface that are remote from a remotely piloted aircraft and from which the aircraft is controlled and monitored. Also known as a ground control station or pilot control station.

corridor: an airspace volume serving as a route segment or passage with associated performance requirements.

detect and avoid (DAA): the capability to see, sense, or detect conflicting traffic or other hazards, and take the appropriate action.

discovery: the process of determining the relevant USS with which data exchange is required for UTM, net-RID, or other function(s). Discovery is generally undertaken via DSS.

discovery and synchronization service (DSS): a service enabling USS to discover relevant USS via a synchronized distributed database providing a consistent view (shared airspace representation) of operational intents and constraints.

DSS instance: a synchronized copy of the DSS supporting a DSS region.

DSS region: the geographic area supported by one or more DSS instances.

flight geography (FG): user-defined 4D volumes encapsulating an intended UA mission; representing the agreed-upon nominal area of containment.

Flight Information Management System (FIMS): an interface for data exchange of approvals, constraints, and related data between UTM system, participants, and ATM.

flight information region (FIR): airspace of defined dimensions within which flight information service and alerting service are provided.

flight level (FL): altitude expressed in hundreds of feet at standard air pressure (29.92 inHg or 1013.25 hPa, at sea level).

fly-away: flight outside of operational boundaries (altitude/airspeed/lateral limits) resulting from failure, interruption, or degradation of the control station or onboard systems.

free route airspace (FRA): specified airspace within which users may freely plan a route between a defined entry point and a defined exit point, with the possibility to route via intermediate (published or unpublished) waypoints, without reference to the air traffic system route network, subject to airspace availability. Within this airspace, flights remain subject to air traffic control.

general aviation (GA): any civil aircraft operation other than aerial work or commercial air transport.

geo-awareness: a function that detects a potential breach of airspace limitations and alerts the remote pilots so they can take immediate and effective corrective action.

geofence: a virtual three-dimensional perimeter around a geographic point, either fixed or moving, that can be predefined or dynamically generated and that enables software to invoke a response when a device approaches the perimeter (also referred to as geo-caging).

ground control station (GCS): *see* **control station**.

human-in-the-loop: a system requiring human interaction to perform or control actions.

key (or sync token): following a USS request for airspace data regarding an area of interest, an airspace representation and associated key are returned. The key, composed of multiple values, indicates that the USS has collected the applicable information, and is aware of that airspace's traffic, and serves to authorize a DSS ledger update and detect changes in the DSS airspace representation. *See* **opaque version number (OVN)**.

network remote ID (Net-RID) display provider: a logical entity that aggregates network remote ID data from potentially multiple Net-RID service providers and provides the data to a display application (that is, an app or website); in practice, it is expected that certain USS may be both Net-RID display providers and Net-RID service providers, but standalone Net-RID display providers are possible.

network remote ID (Net-RID) service provider: a logical entity denoting a UTM system or comparable UAS flight management system that participates in network remote ID and provides data for and about UAS it manages.

networked UAS: a UAS that during operations is in electronic communication with a Net-RID service provider (for example, via internet Wi-Fi, cellular, or satellite).

noncooperative: aircraft without an electronic means of identification due to lack of equipment, failure of equipment, or pilot action.

opaque version number (OVN): a unique value demonstrating awareness of relevant operational intents or constraints in the area of a planned operation, that serves as a prerequisite to operational approval.

operational intent: a volume-based representation of a UAS flight reflecting the airspace intended to contain a nominal flight operation. It may comprise one or more overlapping or contiguous 4D volumes where the time for each volume represents the earliest entry and the latest exit time. Operational intent reflects UAS performance to conformance requirements. Also known as an operational intent volume (OIV).

operator: individual or organization engaged in, offering to engage in, or otherwise managing a UAS operation.

performance authorization (PA): CAA authorization for an operator to conduct UTM operation(s) based upon operator proposal substantiating ability to satisfy performance capabilities in an intended area of operation.

pilot in command: the person with final authority and responsibility for the operation and safety of flight.

planning area: the area within which a USS asserts a discovery interest (where it has current or planned operations) within a finite time period.

provider of services for UAM (PSU): analogous to a USS but for UAM.

registration: the process by which an owner/operator (including contact information and other PII) and aircraft (for example, make, model) are associated with an assigned, unique identifier.

relevant USS: USS that own operational intents and/or constraints that due to their proximity to a new or updated operational intent must be evaluated by the strategic conflict detection service, or that are potentially affected by a non-conformant or contingent operational intent.

remain well clear (RWC): a temporal and/or spatial boundary around the aircraft intended to be used in a DAA system as an electronic means of avoiding conflicting traffic. Given such a boundary, it is expected that the UAS (including services/tools upon which it depends) has the ability to detect, analyze and maneuver in order to ensure that a vehicle is not being operated in such proximity to other aircraft as to create a collision hazard.

remote ID: ability of a UAS in flight to provide identification information that can be received by other parties.

remote pilot: the person controlling the flight path of an unmanned aircraft. The remote pilot may include a remote pilot in command or person supervised thereunder.

remotely piloted aircraft (RPA): an unmanned aircraft piloted from a remote pilot station.

remotely piloted aircraft system (RPAS): a remotely piloted aircraft, its associated remote pilot station(s), the required C2 link, and any other components as specified in the type design.

risk: the combination of the overall probability or frequency of occurrence of a harmful effect induced by a hazard and the severity of that effect.

safety assurance: program or processes that systematically collects, monitors, and evaluates performance and effectiveness of safety risk controls to provide adequate confidence and evidence that a product, process, or organization meets or exceeds its safety objectives and requirements.

safety management system (SMS): an organized, systematic approach to managing safety, including the necessary structures, accountabilities, policies, and procedures.

safety risk management (SRM): a process describing the system, identifying the hazards, and analyzing, assessing, and controlling safety risk.

service level agreement (SLA): a contract between a service provider, its customer and possibly other parties to define the agreed level of quality of that service.

service-oriented architecture (SOA): decomposes complex application systems into components with common characteristics and releases data/information remotely to enhance re-usability across platform, collaboration, and independence.

service provider: an organization or individual providing functions or services of ATM or UTM.

small unmanned aircraft (sUA): an unmanned aircraft (under 55lbs/25kg) on takeoff, including everything that is on board or otherwise attached to the aircraft.

small unmanned aircraft system (sUAS): one or more small unmanned aircraft with everything on board or otherwise attached, each weighing under 55lbs/25kg on takeoff, and off-board equipment as required for the safe and efficient operation of the small unmanned aircraft in a national airspace system.

Specific Operations Risk Assessment (SORA): a qualitative methodology to establish a sufficient level of confidence that a specific operation can be conducted safely.

strategic conflict detection: a USS service that ascertains operational intent conflict with other operational intents or, when applicable, constraints.

strategic coordination: a USS role constituting strategic conflict detection and conformance monitoring services.

strategic deconfliction: negotiation and coordination of intended operational volumes, routes, or trajectories for separation and collision avoidance between operations; possibly reducing the need for tactical deconfliction.

subscription service: the declared area (for operational intents) for which activity updates are requested (for RID or UTM purposes).

supplemental data service provider (SDSP): an organization that handles, manages, and disseminates decision support and other specific types of data consumed by UTM entities, including USS/USP and operators.

synchronization: the process by which discovery information is kept current. *See* **discovery and synchronization service (DSS)**.

tactical deconfliction: in-flight process by which aircraft are safely separated from collision hazards including remaining well clear of other aircraft.

total system error (TSE): the root sum of path definition error, navigation system error, and flight technical error.

track: output of a system that associates consecutive sensor plots of the same object, and may include additional data, such as identification of the object altitude, groundspeed, and velocity vector. Compared with a plot, the track of an object is smoothed in-the-loop over time, may be composed of information from many surveillance sensors, and generally include historical state information for the tracked object.

U-space airspace: a geographic zone, where UAS operations are only allowed to take place with the support of U-space.

UAS geographical zone (geo-zone): a portion of airspace established by the competent authority that facilitates, restricts, or excludes UAS operations in order to address risks pertaining to safety, privacy, protection of personal data, security or the environment, arising from UAS operations.

UAS service supplier (USS): an organization that assists UAS operators satisfy UTM operational requirements, facilitating safe and efficient airspace usage.

UAS volume reservation (UVR): a constraint or capability providing authorized USS the ability to issue notifications to UAS operators [and other stakeholders] regarding air or ground activities relevant to their safe operation. UVRs do not restrict UTM from entering the relevant airspace. *See* **constraint**.

unmanned aircraft (UA): an aircraft operated without the possibility of direct human intervention from within or on the aircraft.

unmanned aircraft system (UAS): an UA and associated equipment to control it (including, e.g., on-board subsystems, required off-board subsystems, any required launch and recovery equipment, all required crew members, and communication links).

unmanned [uncrewed] aircraft systems traffic management (UTM): federated services operating under regulatory oversight supporting safe and compliant UAS operations.

urban air mobility (UAM): a subset of **advanced air mobility (AAM)** providing passenger and cargo transport within urban environments.

USS instance: the area within which a USS provides deconfliction services.

USS network: multiple, collaborative USS operating in a region.

USS role: a grouping of one or more USS services that may be used by a competent authority to establish the granularity of authorizations that can be granted to a USS.

VFR flyway: general charted flight paths not defined as a specific course, frequently for use by pilots in planning flights into, out of, through or near complex terminal airspace.

visual line of sight (VLOS): a type of UAS operation in which the remote pilot maintains continuous unaided (except for corrective lenses, sunglasses, or both) visual contact with the unmanned aircraft, allowing the remote pilot to control the flight path of the unmanned aircraft in relation to other aircraft, people, and obstacles in order to avoid collisions.

1 Introduction

1.1 UTM DEFINED

"UTM is not a concept, it's an upcoming reality."

Reinaldo Negron
Head of UTM, Wing[1]

Unmanned aircraft systems (UAS) traffic management, or UTM, is a fundamentally new paradigm in air traffic management (ATM) in which computer systems, rather than air traffic controllers, coordinate to manage drone traffic. More formally, UTM is a highly automated and cooperative, distributed system of systems facilitating the safe, efficient, and compliant integration of UAS operations into a nation's airspace system, initially at low altitudes.[2] UTM, including related concepts such as Europe's U-space, is intended to manage UAS traffic "without overloading current air traffic control systems."[3] UTM shares certain attributes with legacy ATM systems to provide separation, flow control, and navigation services. As with ATM, airspace management, traffic flow management, and control systems are considered the "core part of UTM."[4] Yet UTM is different from ATM in important respects. While humans (e.g., pilots, air traffic controllers) play a central role in legacy ATM, UTM relies on computing infrastructure to manage several key roles and provide "an all-encompassing framework for managing multiple UAS operations."[5] "UTM is structured on principles of risk [management] and a performance-based approach, and leverages a flexibility when possible and structure where necessary principle. It is based on the following tenets: predefined protocols for cooperative data exchanges, manage by exception, third party services, and service-oriented architecture (SOA)."[6]

We can think of UTM and ATM as complementary means to provide essential traffic management services. UTM services should ideally coordinate with and could eventually merge into a universal ATM, creating a unified, seamless, and transparent infrastructure regardless of operation.[7] However it is integrated, UTM is introducing new third-party service providers, changing communications paradigms and protocols, defining new performance requirements, and spurring reexamination of the relationships, responsibilities, and capabilities of aviation actors including civil aviation authorities (CAAs), air navigation service providers (ANSPs), UAS, and other airspace users, both manned and unmanned. Depending on the jurisdiction and implementation, CAAs and ANSPs may delegate certain operational functions while CAAs continue to focus on their regulatory, safety oversight, and security roles.[8] Given the fast pace of innovation, UTM will inform applications, including urban air mobility (UAM) and associated advanced air mobility, AAM), high altitude operations, and beyond.[9] In sum, UTM will transform the architecture of air traffic management.

Unless otherwise stated, "UTM" is used generically rather than to describe a specific implementation. U-space, the European UTM initiative, is addressed primarily in Chapter 5, Section 5.3.

1.2 WHY UTM

"UTM is such a critical piece of risk mitigation."

Robert W. Brock
Dir. of Aviation & UAS, Kansas DoT[10]

Today's air traffic control (ATC) was designed over the course of the last 100 years for manned aircraft operations and relies on pilots communicating with air traffic controllers. "ATM is clearance-based because only ATC has complete awareness of the airspace operations and constraints."[11] But the growth of unmanned flight operations challenge the legacy ATM infrastructure and render its use for managing UAS traffic infeasible.[12] As the legacy system reaches its limits in scale, especially in the most congested and complex airspace, adding large numbers of UAS requires a new paradigm.

The need for UTM is driven by the rapidly growing volume of UAS operations, "potentially orders of magnitude more than current manned operations,"[13] expected to be undertaken both on- and off-airport, including from vertiports, homes, commercial building roofs, and other nontraditional aviation locations.[14] Conducting this volume of operations safety is only possible if managed with intensive automation. "[H]istorical experience suggests that increasing[ly] congested air traffic needs an appropriate level of organization [and] an organized approach to enabling [UTM] operations to balance efficiency and safety."[15]

Aside from the sheer volume of UAS traffic, fundamental characteristics of UAS challenge existing aviation conventions and protocols. For example, with no pilot on board, small UAS (sUAS) challenge the most basic requirement for a pilot to *see and avoid*[16] other aircraft. The sUAS remote pilot may have no way to see manned aircraft, and manned aircraft pilots simply cannot safely see sUAS.[17] In addition to their small size, today's sUAS are not yet generally required to broadcast their positions with cooperative surveillance transmitters, making them hard to detect, and sUAS can undertake maneuvers that are beyond the expectations and surveillance capabilities of traditional ATM.

For many operations UAS must be separated from other UAS and from manned aircraft, obstacles, and other hazards. Separation is multifaceted and multi-layered[18]—it includes and requires procedures (standards, rules, policies), processes, and technologies such as detect and avoid (DAA) systems for traffic and obstacle avoidance. To address these challenges, UTM provides a strategic awareness and deconfliction framework,[19] and its capabilities may supplement DAA systems that are intended to avert imminent collisions.[20]

Regulators and ANSPs have played a critical role in the development of UTM systems. In the United States, the Federal Aviation Administration (FAA), serving both as ANSP and regulatory body, has worked closely with other government agencies including the National Aeronautics and Space Administration (NASA), and the FAA's Next Generation Air Transportation System (NextGen) has been a core contributor to many UTM initiatives in furtherance of air traffic system modernization. Although NextGen and other ATM update initiatives will help manage UAS, they have been largely foundational.[21] They were not designed to accommodate the unforeseen rapid growth of sUAS operations.[22]

The issues driving UTM development span various UAS operation scenarios and airspace. This includes but is not limited to airspace where ATM does not provide separation services, such as at very low altitudes, below 500' above ground level (AGL), and at very high altitudes, above 60,000' mean sea level (MSL). While UTM's key benefits may be in facilitating and enhancing the safety and utility of beyond visual line of sight (BVLOS) operations or automated UAS operations, it will also benefit visual line of sight (VLOS) operations[23] and participating manned aircraft operations via traffic notifications and improved situational awareness.

1.3 DEMAND AND FORECAST

Just how large is the wave propelling UTM? The diffusion of UAS is unlike any expansion in new aircraft the aviation industry has seen. Commercial UAS are forecast to reach 2.91 million units by 2023, and operations are conservatively "forecast to nearly triple from 277,386 in 2018, to 835,211 in 2023, an average annual growth rate of 24.7 percent."[24] More than 1.6 million drones have been registered and more than 160,000 Part 107 certifications[25] have been issued in the USA, resulting in four times as many UAS as manned aircraft in the National Airspace System (NAS).[26] Additionally, the Low Altitude Authorization and Notification Capability (LAANC) (*see* Chapter 5, Section 5.2), considered the first step toward UTM, facilitated more than 170,000 auto-approvals for airspace access through 2019 and over 7,000 requests that required further coordination in 2018,[27] more than 500,000 approvals by eaerly 2021, and it continues to grow.

Europe's SESAR (the Single European Sky ATM Research project) predicts a ten-fold increase in all flight operations by 2035, with "a major part" UAS BVLOS operations.[28] Additionally, roughly 23,000 electric vertical takeoff and landing aircraft (eVTOLs) are forecasted by 2035 and another 430,000 urban air taxis by 2040[29]—most of these operations requiring some kind of automated traffic management capability.[30] One study pegs the UAM (*see* Chapter 6, Section 6.5 *Advanced Air Mobility /Urban Air Mobility*) 20-year market value at $318 billion; and Volocopter forecasts "a multi-trillion dollar market."[31] A conservative analysis of these projections underscores the pressing need for a safe, effective, and scalable traffic management capability for UAS.[32]

1.4 HISTORY

The UTM concept had its origins in NASA research that commenced in 2013, spearheaded and patented by Dr. Parimal ("PK") Kopardekar.[33] "UTM" was still not in the FAA's lexicon as late as 2014[34] although it was then the subject of a proposal by Dr. Kopardekar,[35] and NASA's plans to enable commercial drone delivery were announced that year.[36] In April 2013, Dr. Kopardekar expressly proposed "UTM" development based on his "seedling team proposal to the NASA Aeronautics Research Institute."[37] The term "UTM" and associated concepts were shortly thereafter presented at an initial stakeholder meeting at NASA,[38] and NASA commenced testing and development of its UTM Research Platform. An initial 180-person workshop in 2014 followed by a robust turnout for a 2015 NASA Ames UTM convention demonstrated to the FAA that UTM was "real" and could not be ignored.[39] Subsequently, at the International Civil Aviation Organization (ICAO), members of state research organizations proposed the UTM concept in 2016.[40] Initiatives to define and advance U-space in Europe, for example, are also historically significant.[41] *See* Chapter 5, Section 5.3 *U-space.*

More broadly, other research was already underway into how UAS could be integrated into the airspace, such as NASA's *UAS Integration in the NAS Project.*[42] Additionally in response to US Department of Defense (DoD)'s need for integrating their UAS into the NAS,[43] the DoD, in collaboration with the FAA undertook important research on sense and avoid (SAA), DAA, and associated ground-based technologies leading to the approval of certain routine BVLOS operations in the NAS.[44] Other initiatives of historical significance include the Research Transition Team (RTT) (2016, *see* Chapter 1, Section 1.5.2), the UTM Pilot Program (UPP) (2017, *see* Chapter 1, Section 1.5.4), and the LAANC (2017, *see* Chapter 5, Section 5.2).

1.5 RESEARCH AND DEVELOPMENT

1.5.1 NASA UAS Traffic Management Project

NASA has played a leading role in research, development, and testing of UTM concepts and architecture. Its work has provided a foundation and catalyst for key UTM technical developments, implementations, standards, and rulemaking.[45] NASA's "UTM research platform [is a] proof-of concept implementation … to instantiate [essential UTM] functions [and enable] NASA and its partners to conduct the research required for determining the operational characteristics for roles and responsibilities, architecture and information flows, services and performance requirements."[46]

NASA's UTM concept of operations (ConOps) has greatly influenced academic, industry, and government vision for UTM, laying out a step-by-step roadmap for developing and testing UTM technologies and capabilities. It addresses accelerating airspace access requirements, diversity of UAS operations, operating principles for UAS and manned traffic separation, and "basic mantras underlying operational characteristics," for example, flexibility where possible and structure where necessary.[47] The FAA subsequently published and continues to update a UTM ConOps.[48] *See* Chapter 2 *Concept of Operations.*

1.5.1.1 NASA TCL Trials

The NASA UTM program, which pioneered UTM research, was structured around four increasingly complex/risk-intensive[49] "Technical Capability Levels" (TCLs), each of which contributes to the understanding and development of UTM.[50] The TCL trials developed and executed multiple scenarios among participating public and private organizations, flying diverse aircraft, and seeking to meet test objectives considered essential for an operational UTM.

TCL1—Addressed initial technologies supporting, e.g., precision agriculture, firefighting, and infrastructure monitoring operations, with a focus on geofencing, altitude "rules of the road," and scheduling of vehicle trajectories. Various technical protocols were demonstrated,[51] live vehicle field testing[52] and simulated operations concluded in August 2015,[53] and the campaign completed in 2016.

TCL2—Demonstrated BVLOS operations in sparsely populated/low-density areas; tested technologies for dynamic adjustments to availability of airspace and contingency management; and developed "USS Discovery" protocols. Field testing, conducted in phases, was completed in October 2016 in Nevada,[54] and campaign was completed in 2017.[55]

TCL3—Tested technologies that maintain adequate separation between cooperative and noncooperative UAS over moderately populated areas. This included vehicle-to-vehicle (V2V) and some interaction with manned aircraft.[56] It also developed USS discovery protocols and tested airborne DAA. Field testing concluded in June 2018.

TCL4—Studied multiple BVLOS operations in higher-density urban environments for tasks such as news gathering and package delivery; and technologies for DAA, communications, and large-scale contingencies (off-nominal conditions via conformance monitoring) in both controlled and uncontrolled airspace. It included airspace authorizations, urgent/distress operations, and management of historical data. Field testing (in Nevada and Texas) concluded in Summer 2019.[57]

Going forward, NASA recognizes the need for autonomous operations: "you need highly intelligent systems on the vehicles themselves. We'd like to see the last 50 feet of operation fully autonomous."[58] As NASA UTM and the TCLs reached conclusion, NASA extended the UTM paradigm by commencing research programs to enable new concepts such as AAM.[59]

1.5.1.2 Test Sites

In 2013 the FAA selected six UAS test sites to facilitate research and gain operational experience to advance regulations and procedures for safe UAS integration into the NAS.[60] Chosen, in part, for their geographical and climate variances and connection with recognized academic and research institutions, each site participated with NASA on the TCLs beginning in 2015. Beyond initial foundational UAS testing, these sites have engaged in more sophisticated and challenging initiatives that advance UTM.[61] In 2019, three test sites also participated in the FAA's UPP.[62] Their greatest achievements have been described as "supporting NASA in their four UTM trials, and supporting FAA in their UAS Detection and Mitigation at Airports and Critical Infrastructure Pathfinder Program."[63]

The UAS Test Site Program has played an integral role in demonstrating UTM technologies. The program verifies safety, supports development of standards and regulations, coordinates with NASA and other agencies, and engages private industry partners, to test and demonstrate the latest UTM developments through national campaigns and TCL flight demonstrations.

Although not a UAS test site, the Alliance for System Safety of UAS through Research Excellence (ASSURE) provides research on diverse UAS-relevant aviation subjects via "twenty-three of the world's leading research institutions and more than a hundred leading industry/government partners."[64] The FAA Center of Excellence for Unmanned Aircraft Systems provides program oversight.[65] The FAA was mandated "to the maximum extent practicable, leverage the research and testing capacity and capabilities of the Center of Excellence … and the test ranges."[66] A final report on the Test Sites is due to Congress late 2023.

1.5.2 Research Transition Team

To begin bridging the gap from research to implementation, a research transition team (RTT) was established in 2015 between the FAA and NASA:

> to jointly identify, quantify, conduct, and effectively transfer UTM capabilities and technologies to the FAA as the implementing agency and to provide guidance and information to UTM stakeholders to facilitate an efficient implementation of UTM operations. [With the goal] to: (1) research and mature increasingly complex UTM operational scenarios and technologies; (2) demonstrate those capabilities on the NASA UTM research platform (note: TCL 1 capabilities have been demonstrated on the NASA UTM research platform prior to the formation of the UTM RTT); and (3) deliver to the FAA technology transfer packages that enable NAS service expectations for low-altitude airspace operations by providing insight and capability requirements for critical services.[67]

An RTT FAA/NASA joint UTM Research Plan documents research objectives and maps out the anticipated development of UTM, including at UAS test sites. Areas of focus include concept and use case development, data exchange and information architecture, communications and navigation, SAA, and airspace operations requirements to enable safe low-altitude VLOS and BVLOS operations.[68] The RTT is characterized as the center point of "communications and engagement with the broader public and UAS Community."[69]

The RTT reflects the FAA's stance that it will not pay for UTM, thus requiring intensive industry participation. The rationale for this stance is that industry drives innovation through their technological capabilities and resources while the FAA maintains its regulatory and operational authority. Such a private–public approach to future infrastructure underlies UTM and is being increasingly adopted by the FAA for other NextGen initiatives.[70]

1.5.3 Relevant Follow-On Efforts

The past UTM work through the TCL testing and other programs has yielded significant insights on the key areas of investment and focus needed to develop a complete UTM system. This effort has also identified new areas that require additional research. In this, the industry and government entities work in parallel: individual private UTM providers continue to develop their systems, and NASA continues to pursue research. Long-term implementations will benefit from NASA's pioneering TCL research, as well as its follow-on work including UTM-based Class E operations above FL 600,[71] the Advanced Air Mobility National Campaign with its TCL-analogous UAM Maturity Level (UML) test program,[72] the Air Traffic Management–eXploration (ATM-X) Project, and Space Traffic Management (STM).[73]

1.5.4 UTM Pilot Program

Recognizing the "limited infrastructure available to manage the widespread expansion of sUAS or drone operations"[74] within the NAS, the US Congress mandated the FAA to establish the UPP in 2017 to include "development and demonstration of enterprise services to support initial UTM operations."[75] Pursuant to the *FAA Extension, Safety, and Security Act of 2016*,[76] the UPP was tasked to develop a research plan engaging both FAA and industry to support UTM, including "an assessment of the interoperability of a UTM system with existing and potential future air traffic management systems and processes."[77] UPP Phase 1 results "provide a proof of concept for UTM capabilities currently in research and development and ... serve as the basis for initial deployment of UTM capabilities."[78] "The UTM Pilot Program demonstrated that the first round of requirements has graduated from research concepts to operational capabilities."[79] UPP Phase 2 then provided a "targeted technology demonstration showcasing [various] capabilities and services supporting high-density operations[80] including remote identification and cybersecurity mechanisms, and "help[ed] inform a cross-agency [UTM] Implementation Plan."[81] "From the beginning we [knew] NASA needed to involve industry .... We've always tested with industry."[82]

1.5.5 Other Important Research

Aside from the United States, various geographic regions and national governments are engaged heavily in UTM and in making important research contributions.[83] For example, SESAR JU, the Single European Sky ATM Research Joint Undertaking, "coordinates and concentrates all EU research and development (R&D) activities in ATM," and has been a leading force in UTM,[84] particularly through its role in U-space.[85] Its CORUS ConOps publication culminated two years of intensive collaborative research, architectural work, and demonstrations, engaging many respected organizations (*see* Chapter 5, Section 5.3 *U-space*).

1.5.6 Private Industry

In addition to the role that government has played, private industry has taken a leading role in UTM R&D.[86] There were 250–300 private companies participating in and expending significant resources to help make the NASA UTM project successful.[87] Participants included providers of UAS package delivery services, UTM, enterprise fleet services, data services, ground infrastructure, and a host of other systems and services necessary to realize UTM.[88] Dr. Kopardekar remarked: "I believe that inclusion of third party services offered by multiple UAS service suppliers (USS) will accelerate the transformation, maintain up to date technology based stacks, and will allow us to take advantage of competition and the free market."[89] NASA's UAS service supplier (USS) "has served as a reference implementation for the UTM community," and thus "is part of an ecosystem of services and capabilities designed to enable an industry."[90]

1.6 SCOPE OF OPERATIONS

A key to UTM development has been to start small and expand—*crawl, walk, run*. Indeed, each of the major UTM plans, blueprints, frameworks, and roadmaps contemplate an infrastructure providing for incremental experimentation, testing, and deployment of increasingly complex systems, airspace, aircraft, performance, and risk.[91] *Crawl, walk, run* envisions moving from separation, to accommodation, and ultimately to full UAS integration in the NAS. For example, the first limited sUAS UTM deployment in the USA has been within circumscribed very low level airspace (VLL <= 500′ AGL) near airports with near-real-time automated ATC authorization,[92] followed by accelerating enablement of (limited) BVLOS implementations.[93] Larger UAS are operating in uncontrolled airspace bounded by operating limitations under CAA authority. Table 1.1, below, depicts such an incremental

TABLE 1.1
Notional Progression of UTM Operations

Type of Operation	Considerations/Examples
VLL VLOS	*See* Chap. 5, Sect. 5.2 *LAANC* for operations within UASFM.
VLL BVLOS	Inspection, surveillance, package delivery, etc.
>500′ AGL VLOS/BVLOS	DAA and C2. *See* Chap. 3, Sect. 3.2.13 *RTCA*.
High altitude ("High E")—>FL600	Diverse aircraft—some long-durations loitering; may include high-altitude pseudo-satellites (HAPS), high-altitude long endurance (HALE) UAS, balloons, supersonics, rockets, etc. *See* Chap. 6, Sect. 6.6 *Stratospheric Operations*.
Large cargo operations	Limited to certified aircraft operating IFR; traffic management will incrementally integrate UTM.[95]
Urban air mobility (UAM)—Air taxis	Passenger carrying, thus requiring extensive risk mitigations; <=˜3,000′ AGL, typically dense environments. *See* Chap. 6, Sect. 6.5 *Advance Air Mobility / Urban Air Mobility*.
Scheduled passengers	Limited to certified aircraft operating IFR; traffic management will incrementally integrate UTM.
Space (above the Kármán line)	"Heavily influenced by [UTM]."[96]

progression toward full integration into the NAS. Each subsequently listed type of operation introduces increased (or different) UTM risks and complexity, and requires commensurate sophistication, capability, infrastructure, and risk management.[94]

The scope of UTM-based operations may eventually merge into a future, more capable ATM–or vice versa. *See infra* Chapter 6, Section 6.7 ATM-UTM Integration.

1.7 BASIC REQUIREMENTS

UTM's core objectives include safety of the NAS and efficiency.[97] Key principles and requirements supporting these goals include:[98]

- overall system scalability, effectiveness, efficiency, compatibility,[99] and fairness
- interoperability and communications compatibility with ATC systems[100]
- resilience and over-all information security[101]
- situational awareness of UAS for both unmanned and manned aircraft pilots[102]
- appropriate and functional radio frequency spectrum[103]
- support for public safety
- situational awareness of UAS for the public that balances transparency and privacy.

With this overview of UTM completed, we can delve into how UTM works—the ConOps for UTM.

2 Concept of Operations

"The most amazing thing is that it actually works."

Tom Prevot, PhD
Air Taxi Product Lead, Joby Aviation[104]

2.1 CONOPS—GENERAL

The best way to demonstrate how UTM operates is to understand its concept of operations (ConOps). The ConOps covers the scope, characteristics, operational components, and performance of the system from a user perspective and presents the requirements of each participant and how they are expected to interact.[105] As it matures, the UTM ConOps should reflect increased harmonization of principles, support for architectures, additional applications and services, interoperability, and over-all robustness.[106]

2.2 PARTICIPANTS

Multiple parties participate in UTM, and the participants will vary depending upon the implementation architecture driven by civil aviation authorities (CAAs) requirements (e.g., directing centralized versus federated architecture), and other factors such as business model and available service providers. Participant roles and responsibilities—including for information security, privacy, audit—and what roles to share (or consolidate) will vary too. The following list of participants with associated descriptions is therefore notional.[107]

2.2.1 UAS Operators

An unmanned aircraft systems (UAS) operator is an organization or individual engaged in, offering to engage in, or otherwise managing an unmanned aircraft operation, and may include remote pilots.[108] Operators exercise "overall management"[109] and provide information on operational intent, or intended flight paths, to a UAS service supplier (USS) to obtain or modify authorizations to allow UTM-enabled flight operations.[110] Among other possible requirements in UTM, operators must request activation of flight authorization requests and receive confirmation of the activation before commencing flights.[111]

2.2.2 UAS Service Suppliers

USS Service Suppliers (USS)[112] provide UTM services to UAS operators. More formally, USS provide "a collection of software services, and interfaces"[113] through which UAS operators access at least a minimum set of UTM services.[114] USS may link certain other participants and systems in UTM (i.e., supplemental data service providers (SDSPs)) to the UAS operator; and they share their respective operators' flight intents, and coordinate to resolve potentially conflicting operations.

NASA characterizes a USS as "a state appointed or third party operated system that will provide service options that are similar to what is traditionally offered to manned operations solely by a state-appointed ANSP."[115] They are new airspace management/service entities "without exact analogy in manned aviation, but are inspired by elements such as Airline Operations Centers and Flight Service Stations."[116] Although the services provided by USS are still evolving, and may vary by implementation, they can generally be summarized to provide support for the safe, efficient, and compliant use of airspace, such as via remote ID,[117] and strategic coordination (*see* Chapter 2, Section 2.3 *Scope of Services*).

More specifically, a USS:

- "acts as a communications bridge between federated UTM actors to support [o]perators' abilities to meet the regulatory and operational requirements for UAS operations,"[118] such as flight authorization,[119]
- provides the operator with actionable information, including demand forecasts and active usage, and complete visibility into current and planned usage "in and around a volume of airspace so that [o]perators can ascertain the ability to safely and efficiently conduct their mission,"[120]
- assists in managing and minimizing conflicting operational intent volumes when necessary, and
- "archives operations data in historical databases for analytics, regulatory, and [o]perator accountability purposes."[121]

Some USS may provide basic services to organizations that seek to operate a UAS, some UAS operators may operate a private USS exclusively for internal corporate operations, while other USS may service industry/vertical market-specific operators. There will likely be a core minimum set of services a USS must support to participate in a UTM system (such as strategic coordination, conformance monitoring, and constraint management).[122]

In Europe, a USS is called a U-space service provider (USSP) and required to provide at least the following four services: network ID, geo-awareness, UAS flight authorisation, and traffic information.[123] For urban air mobility (UAM), a USS has been labeled a provider of services for UAM (PSU); and "[n]otionally, a USS can expand to become a PSU and vice versa, based on services provided."[124]

2.2.3 Supplemental Data Service Providers

Supplemental Data Service Providers (SDSPs) offer ancillary data and services in support of UTM to USS, other SDSPs and operators, but do not function as UTM USS.[125] Various concepts of UTM may describe SDSP roles differently, but in general, SDSPs provide data and associated "essential or enhanced services"[126] that USS then use to provide services, or that are provided directly to operators and other stakeholders. SDSP services could include:

- Weather sensing, weather information, bespoke hazardous weather detect and avoid services, and associated custom flight and payload decision analytics information—including for specialized micro-environments[127]

- Mobile communications network services and supporting products, including depictions of coverage and signal availability (e.g., "heat maps")[128]
- Population density and pedestrian movement information
- Terrain, obstacle, and other geospatial data and mapping[129]
- Surveillance—such as ground-based, low altitude sensors (e.g., radar) for sensing noncooperative aircraft[130] and other tracking[131]
- Certain deconfliction services that do not otherwise provide USS functions[132]
- Electromagnetic interference awareness[133]
- Insurance[134]
- Operational risk management and assessment[135]
- Vertiport automation
- Noise and other environmental data
- Post-flight analysis.

Government or private sector entities may provide SDSP services, and different international standards may apply in different regions. SDSP standards are in the early stages of development, and there is a strong desire to align SDSP standards internationally. Coordination remains fragmented but generally moving in the right direction.

While USS in a federated UTM architecture will necessarily coordinate via standardized interfaces, the SDSP does not have a standard API for communication, so each service may provide a custom interface that an operator (or USS provider) would implement. For example, weather SDSPs will follow Open Geospatial Consortium (OGC) standards for aviation weather information exchange.

From an architecture point of view, distributing certain UTM roles so that USS can obtain necessary data to perform their safety function from SDSPs advances a healthy UTM ecosystem and lowers barriers to entry for USS by avoiding the need for each to become a domain specialist in all types of data. Correspondingly, it creates demand for specialist data providers who can innovate on aspects like data quality, predictive analytics, and decision support. This kind of robust and flexible ecosystem should ultimately improve UTM services and safety.

2.2.4 Air Navigation Service Providers

Air Navigation Service Providers (ANSPs) are a mixture of public and/or private entities[136] (depending on national mandate) that provide aeronautical services—typically air traffic management (ATM) and critical navigation services primarily for manned aviation. An ANSP interface with USS may support applicable UTM services, including permissions management and coordination services.[137]

Many ANSPs have been partially or wholly privatized in Europe (e.g., NATS[138] in the UK; and skyguide[139] in Switzerland) and Canada (NAV CANADA[140]) and other areas, while in China, the United States, and elsewhere they remain government entities.

> With the responsibility of managing safe flight operations within national and international airspaces, ANSPs have a unique role in the operationalization of UTM policy, infrastructure, and technologies. They must work closely with the regulatory authorities, industrial partners, and other ANSPs to develop and deploy infrastructure which meets the needs of the users of the airspace. To accommodate those needs, ANSPs

> have a responsibility to reach out to these new airspace users (e.g., UAS operators, service suppliers) in order to ensure equitable access to the airspace in line with national and international policies.[141]

Indeed, ANSP engagement in UTM can help (irrespective of their public or private roles) to ensure safe and fair airspace usage through ATM-UTM interoperability, effective data exchange, and provide public safety metrics.[142] Notwithstanding, commercial ANSPs are, by necessity, driven to monetize their services, with possible implications for the free availability of operational safety data services to affected stakeholders, particularly manned aviation.[143]

2.2.5 Civil Aviation Authorities

CAAs are the aviation regulators. The duties of CAAs have been characterized as discharged within four regulatory areas: safety, economics, airspace, and consumer protection.[144] CAAs do not necessarily play an operational role in UTM. Rather, they will typically develop UTM architecture and regulations, designate where and how UTM operations may occur, qualify equipment and systems, and provide oversight. Operational functions may be delegated to, or otherwise undertaken with logical separation and coordination[145] from ANSPs and other service providers. "CAAs also work to ensure collaboration amongst partners at the international level to support the advancement of common UTM principles. This allows for productive cooperation for international operations, as well as providing a basis to support global economies of scale."[146]

2.2.6 Other Participants and Stakeholders

Other participants and stakeholders in UTM may include cities, regional governments, private enterprises, individuals and others.[147] For example, authorized firefighting or law enforcement personnel may create airspace constraints (*see* Chapter 2, Section 2.3.7 *Constraint Services*) to notify others of the presence of UAS, or to restrict UAS from potentially dangerous airspace. Other stakeholders may include UTM service consumers (e.g., customers of UTM-enabled package delivery or "receive-only" information users), other beneficiaries (e.g., recipients of search and rescue services), or even neighborhoods and community members affected by actual or perceived UA noise, and/or visual pollution.[148] This may include those affected by or engaged with urban vertiports, urban planning, zoning, and privacy and data protection compromises. USS networks have also been characterized as participants.[149] Furthermore, asset owners/operators on the ground may also be stakeholders such as for critical or sensitive infrastructure (power plants, refineries, communications hubs, national security sites, etc.). Consumers of aggregated USS data may include national security organizations and various public safety/law enforcement entities, some of which may be engaged in airspace surveillance outside the scope of an ANSP, such as for counter-UAS applications or national air defense. Finally, manned aircraft pilots are properly UTM stakeholders.[150]

2.3 SCOPE OF SERVICES

2.3.1 General—Scope of Services

Next we turn to the services provided under UTM — what UTM does. The scope, characterization, nomenclature, and consolidation of UTM services vary among UTM implementations and concepts. Moreover, to some extent, the functional distribution of "services in terms of which stakeholder[s] provide which services is an open research question."[151] Some of the services provided as part of UTM were mentioned above in the context of presenting UTM participants (*see* Chapter 2, Section 2.2 *Participants*), and others are addressed below. Also, different UAS operations will require mission-specific USS services. A fixed-wing VTOL UAS delivering packages in an urban area or near airports may need services like LAANC or supplemental data services that a multirotor UAS inspecting infrastructure in a rural area may not need. Figure 2.1 presents one high-level notional view of a UTM scope of services, as envisioned in Transport Canada's *RPAS Traffic Management (RTM) Services Trials*.[152]

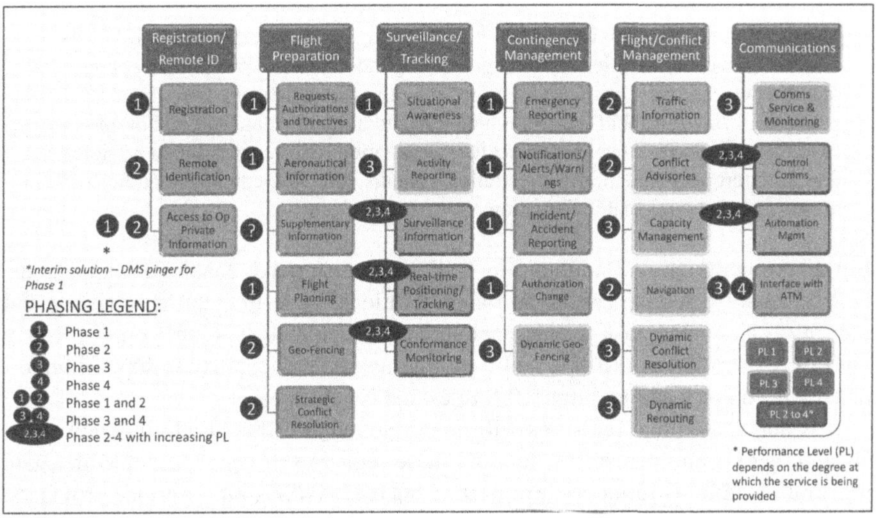

FIGURE 2.1 Scope of services example. (Reprinted with permission from Transport Canada, *RPAS Traffic Management (RTM) Services Trials, CALL FOR PROPOSALS, Phase 1 – Round 1* (May 11, 2020) Sect. 3.2, p. 3).

The following subsections survey expected UTM services and functions. Prerequisites to some of these services are addressed in Chapter 2, Section 2.4 *Architecture*. The scope and formulation of services will continue to evolve. Importantly, the variability in the formulation and boundaries of UTM services within the broader UTM community is recognized.

2.3.2 Registration

"Now is the moment for registration."

Reinaldo Negron
Co-president of GUTMA[153]

Registration creates a validated association between UAS, owner, operator, remote pilot, and state of registration (nationality), and may also "provide any required data related to their UAS,"[154] associated equipment, performance, airspace and flight authorizations, and pilot certifications.[155] Registration underlies UAS remote ID,[156] UTM generally, and is considered one "of three foundations of the U-space system."[157] It is also widely viewed as a precondition to UTM safety, security, and effective administration including enforcement.

Registration requirements are primarily within the purview of aviation authorities.[158] In some cases there may be multiple registries within a national jurisdiction, owned and operated by different administrative divisions, depending on the local administrative arrangements. Consider the following registration provisions within the European Commission's Implementing Regulation on Unmanned Aircraft:

> Member States shall establish and maintain accurate registration systems for UAS whose design is subject to certification and for UAS operators whose operation may present a risk to safety, security, privacy, and protection of personal data or environment.[159]
>
> The information about registration of certified unmanned aircraft and of operators of unmanned aircraft that are subject to a registration requirement should be stored in digital, interoperable national registration systems, allowing competent authorities to access and exchange that information.[160]

Building, operating, and maintaining a digital registry that enables instantaneous, on-line identity, authentication, and authorizations while maintaining privacy and security is a complex task for national authorities. Nonetheless, unlike most legacy aircraft registration workflows, real-time interaction is imperative to support anticipated UA volumes and facilitate USS roles and capabilities.

It is anticipated that registries will be developed at different administrative levels and there is an interest in federating these registries. Some CAAs may delegate registration administration to third parties, such as ANSPs, other service providers, or industry consortia. A global electronic aircraft registry network (including for the registration of drones) has been advocated by ICAO.[161] Additionally, a responsive "digital sandbox" open-source registry has been demonstrated by GUTMA[162] to enhance cross-border and foreign operations[163]—and to facilitate UTM.

2.3.3 Discovery

Discovery is a system-internal service that enables UTM USS to locate peers for information exchange that is primarily transient or dynamic rather than static. Once appropriate peers are discovered the USS can perform critical services such as strategic coordination, and were deployed, network remote ID. Discovery could also support services such as identifying proximate SDSPs, although its use may be inefficient for wide-area service coverage. Discovery has significant architectural implications and is addressed substantively in Chapter 2, Section 2.4.5 *Discovery.*

2.3.4 Flight Operation/Planning Management

Flight planning management is considered the input for all services that follow the creation of an operational intent[164] (analogous to a traditional flight plan in manned aviation), a recognized basic USS or SDSP function, and can be coupled with other services within UTM. The ICAO UTM Framework describes flight planning as "a service that, prior to flight, arranges and optimizes intended operational volumes, routes and trajectories for safety, dynamic airspace management, airspace restrictions and mission needs (this is not intended to refer to the existing manned aircraft flight planning services)."[165] Planning provides for USS-operator interaction to build an operational intent that is conflict free and within the performance limitations of the UAS. An operational intent is sometimes referred to as an operations plan. Such services may inform other airspace participants of new routes, trajectories or airspace reservations, and maximize the likelihood of airspace restriction conformance while permitting flight completion.[166]

2.3.5 Strategic Deconfliction

Strategic deconfliction, a core component—along with conformance monitoring—of strategic coordination, is a core essential UTM service that relies on a foundation of conflict detection.[167] *Conflict detection* is simply detecting intersections of operational intents in space and time. USS in a federated UTM system use operational intents, including those shared by other USS, to find conflict. *Deconfliction* or *coordination* is the process of resolving detected conflicts. There is benefit in separating two concepts: (1) strategic conflict detection, which should be standardized, and (2) deconfliction/planning to resolve conflicts, which need not necessarily be standardized provided the result is conflict free, thereby offering implementers flexibility to innovate planning processes (e.g., with artificial intelligence—AI—and other tools). Finally, precisely when "strategic deconfliction" ends and "tactical deconfliction" begins is subject to differing interpretations.[168]

The goal of strategic deconfliction is that a planned UAS operation "be free of 4-D intersection with all other known UTM operations prior to departure [and] a core service that will be provided by all USSs in the future UTM System"[169] to maintain awareness of current and planned traffic for an area of flight operations, and where deployed, dynamic rerouting (DR) in response to airspace constraints.[170] This will eventually apply to intersection with manned aircraft operations as well.[171] Strategic deconfliction is intent-based and may have the practical effect of reducing surveillance requirements.[172] Strategic deconfliction may be accomplished by trajectory- and/or volume-based representations of operations.[173] "Strategic deconfliction *could* take into account more advanced aspects (like demand and capacity balancing), but it won't at the beginning. It will simply try to resolve intersections through some set of algorithmic path planning processes and negotiation."[174]

Strategic deconfliction should

verify that an operator in combination with a USS is constructing operational intents that meet intended performance objectives, such as the UA being within the operational intent a percentage of flight time. These thresholds are being established through the standards process to enable strategic deconfliction to achieve/support some target level of safety – it is therefore necessary over time to verify that the thresholds are not being consistently exceeded or that a particular operator is not being consistently overly conservative which leads to airspace inefficiency. One would expect the SMS [safety management system] for a USS planning capability to include this strategic conformance monitoring.[175]

2.3.6 Conformance Monitoring

"Conformance monitoring is easy to implement I would say but it is not easy to decide what to do with the information afterwards ..."

Andreas Lamprecht, PhD
CTO, AIRMAP[176]

Once a UA is in flight, a service is needed to confirm that it follows its pre-deconflicted flight intent. Conformance monitoring may provide "real-time monitoring and alerting of non-conformance to intended operational volumes, routes or trajectories for a UAS operator,"[177] and may serve as a trigger for additional strategic coordination. In addition to initial (preflight) airspace deconfliction, mechanisms are necessary to alert other nearby USS, operators, and UTM participants that a flight operation is outside of its operational intent volumes.[178] UTM systems may trigger notification where "authorization intent thresholds are to be violated,"[179] or upon violation.[180] In such cases, the USS must communicate timely alert(s) of nonconformance and, as applicable, subsequent contingency or rogue status to other relevant USS. NASA has provided that

[c]onformance requirements include, nonexclusively, 1. Staying within conformance volumes. 2. Submitting position reports at required rate. 3. Responding to required information requests from USS. and 4. Meeting agreed requirements for operation.[181]

It has been urged that "[a]bsent surveillance to ensure conformance, the separation distance will be much larger, just like non-radar separation of conventional aircraft today."[182] However, because conformance monitoring lacks controller-administered separation minima, such as those that increase based on reporting interval in oceanic airspace, DAA may play an important conformance monitoring role as the technology evolves.[183]

Conformance monitoring can verify that an operator in combination with a USS is constructing operational intents that meet intended performance objectives. It is important for successful UTM that thresholds are not consistently exceeded or that a particular operator is not consistently overly conservative, precipitating airspace inefficiency. One would expect the SMS for a USS planning capability to include this strategic conformance monitoring. Aggregated over multiple flights, such data can provide a feedback loop function,[184] helping an operator identify critical UTM performance data for safety management, system analysis or improvement, and enforcement purposes.[185]

2.3.7 Constraint Services

Constraints represent 4D areas that may pose a hazard to nominal operations or are otherwise the subject of airspace restrictions (such as no-fly zones). USS take constraints into account when deconflicting operational intents. There may be multiple types of constraints, such as those that provide increased SA and enhance risk mitigation but may not necessarily prohibit operations; and others that prohibit operations. The ASTM UTM specification does not enforce constraints. Rather, it is up to the operator to meet regulatory obligations associated with a constraint. FAA and NASA have referred to what we now call constraints as UAS volume reservations (UVRs), "a constraint in the UTM airspace that is typically short-lived and dynamically announced."[186] Awareness of constraints by affected operators (and other relevant stakeholders) is essential to flight safety, so dedicated services manage them. For example, the ASTM UTM specification presents *constraint management services* to support the creation, modification or deletion of constraints by authorized constraint providers, and a *constraint processing service* to support consumer/user awareness of such constraints and to support deconfliction with constraints.[187] Other implementation issues include whether to limit constraints to public safety and national security or to also include commercial entities.[188]

2.3.8 Tactical Deconfliction

> "Tactical stuff is a bit of a zoo."
>
> **Maxim Egorov**
> *Research Scientist, Airbus*[189]

While strategic coordination services refer primarily to preflight deconfliction, a tactical separation service helps aircraft avoid near-term threats in flight.[190] More broadly, it may help aircraft remain well clear of other aircraft, and ensure other types of unexpected (and exigent) hazards do not interfere with the operation, such as adverse weather events or airspace contingencies. This service may, for example, take the form of real-time SA and alerts of proximate aircraft in the airspace, leaving it to the UAS or the operator to decide the best course to remain well clear, or it may automatically recalculate new flight plans based on current trajectories and well clear criteria. Tactical deconfliction can be undertaken over the network, on the aircraft, or using a combination of both. While tactical deconfliction is not among the first set of UTM services envisioned for deployment, it is supported by some USS services,[191] mentioned for future U-space advanced services,[192] and has been characterized as "part of the 'UTM umbrella'."[193]

> Tactical deconfliction will affect all manner of UTM services. Unexpected maneuvers are going to affect airspace reservations, flow control, restricted zones and airspace

> structure, terrain or obstacle avoidance, etc. that are all commonly thought of as UTM services. The problem is that the timeliness and complexity of notifications and negotiations likely needed for tactical deconfliction will challenge federated inter-USS-type architecture.[194]

Such problems may be primarily associated with ground-based tactical deconfliction. Alternatively, many experts are contemplating and addressing on-board tactical deconfliction.[195]

While route assignment or airspace reservation is done before takeoff (i.e., strategically), the need may arise to change the route (operational volumes) in-flight (i.e., tactically). A tactical or DR service[196] responds to changed circumstances where deviation from a flight plan is necessary after take-off in a safe and coordinated way. It can respond to an emergency,[197] new or changing operational criteria, or hazard mitigation, such as for conflicting aircraft. Tactical deconfliction will find increasing importance—and challenges[198]—as traffic density climbs. For example, optimization research could not:

> improve performance once we moved beyond 40 vehicles/km^2 at the 100 m separation standard. To me, what this is saying is that there is a fundamental limit on how well the airspace can be managed as a whole while every vehicle is making reactive decisions that allow it to be safe in short-time horizons but don't consider the implications of those decisions on all the other traffic in the airspace. The right approach would be to combine efficient tactical decision-making systems like DAA, and UTM services like strategic deconfliction, demand-capacity management, etc. to enable even higher densities.[199]

2.3.9 UAS Remote Identification

> "Everything should have the capacity to be remotely identified."
>
> **Tom Prevot, PhD**
> *Air Taxi Product Lead, Joby Aviation*[200]

UAS remote ID (RID or remote ID) is viewed widely as being the first UTM services needed for early deployment, due to the demands of security and public safety stakeholders. RID "allows governmental and civil identification of UAS for safety, security, and compliance purposes, [increasing] UAS Remote Pilot accountability by removing anonymity while preserving operational privacy …. "[201] Remote ID is "fundamental to connect[ing] a drone with its operator [and] a basic foundation of airspace security,"[202] and "a crucial component of UAS traffic management."[203] Tracking and location functions allow authorized third parties to remotely identify UAs and their operators in real time.[204] RID position data could also be used in collision avoidance systems.[205] RID standards specify two methods for UAS to communicate their ID: (1) radio broadcast or (2) network transmission to an ID management server,[206] although the FAA RID rules are, for now, limited to the broadcast method. The network method may rely on UTM infrastructure to receive requests to identify nearby UAS, validate the requests, correlate the requests with known nearby UAS, and report back ID information about them. RID position data could also be used in collision avoidance systems.[207] In either case, RID allows people on the ground to use a smartphone to identify the drones they see overhead. *See* Chapter 2, Section 2.4.8 *UAS Remote ID.*

2.3.10 Conflict Advisory and Alert

A conflict advisory and alert service "provides UAS operators with real-time alerting through suggestive or directive information on UA proximity to other airspace users (manned and/or unmanned)."[208] As standards for SA and DAA systems evolve, it is important to recognize the functional difference between air traffic separation alerts and advisories versus those provided by an SA or DAA system. "The intent of SA/DAA is not to preempt air traffic control separation but rather to provide advisories and alerts to UAS operators that an unsafe condition may exist."[209] *See* Chapter 2, Section 2.4.6 *Detect and Avoid.*

2.3.11 Contingency Management

Off-nominal/nonconforming conditions, such as a UA departing operational intent volumes or violating constraints, if not corrected may require prescribed, increasingly exigent actions. Contingency services become indispensable upon system or other anomalies that place the operation at risk, including for example, a public safety (e.g., emergency responder) operation that conflicts with an activated UTM-enabled flight,[210] or degraded navigation performance, including lost link or mechanical failure jeopardizing flight completion.[211] Contingency services could be triggered by notification from the operator or automatically by the USS, such as upon a finite number of continuous seconds following nonconformance with operational intent, "where the operator and USS share a standard means to detect off-nominal situations."[212] Multiple categories of increasing severity may be implemented, such as *urgency* and *distress*—with associated response protocols.

In a federated UTM, when a flight enters a contingency condition, its USS would notify other USS operating nearby. In this regard, "the operator is responsible for notifying affected airspace users."[213] Notification of a contingency condition could be communicated via network and local radio frequency (RF) broadcast and, if available, include anticipated path and UA state data.[214] Notification contributes to awareness and can improve separation by supporting revised, including enlarged, operational volumes.[215]

For flights in contingency conditions, rules may preclude replanning (i.e., to return to a nominal operation) and instead require transition to a terminal/ended state (generally invoking an emergency landing or immediate transition to an alternate landing site).[216] Finally, consider the impact of UTM infrastructure failure, such as "contingency or emergency operations of the UTM system itself vs. [as] an actor in the system [and what] needs to be considered to mitigate risks should the system degrade or have a catastrophic failure."[217]

2.3.12 Simplified Phases of Flight and Operational States

To see how UTM functions to support UAS operations, we can illustrate the phases of a flight and associated UTM (USS) actions. Figure 2.2 depicts selected, highly simplified phases of flight and operational intent states within a notional UTM.[218]

States are abstractions to enable the automated system to logically function and keep track of the UAS and ecosystem. It begins with flight planning (lower left), proceeds clockwise thru various operational states and associated actions, and concludes with fight termination in the ended state (lower right).

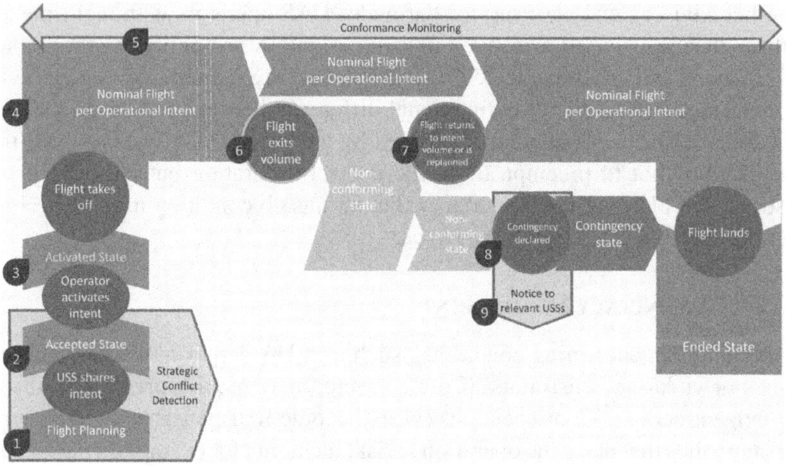

FIGURE 2.2 Simplified phases and states of flight.

The following ten paragraphs describe the numerically identified states and services presented in Figure 2.2.

1. **Creating a Nonconflicting Operational Intent**—An operator and USS may collaborate in flight planning to find an acceptable conflict-free operational intent. Where permitted, relevant USS may agree that certain identified intersections are acceptable. The USS uses a DSS (or other authorized mechanism) in this process (*see* Chapter 4, Section 2.4.5 *Discovery*), and applicable prioritization and fairness schemas (*see* Chapter 4, Section 4.4.4 *Fair Access*).
2. **Accepted State**—Once a nonconflicting OI is resolved and communicated to relevant USS, the USS sets/invokes the accepted state indicating that the operator may initiate the flight.
3. **Activated State**—Operator communicates to the USS intent to commence flight operations within the approved operational intent volume(s) and receives acknowledgement from the USS accordingly.
4. **Nominal Flight (per operational intent)**—The UA commences the flight and operates within the designated operational intent volume(s). *See* Chapter 4, Section 4.4.2 *UTM Volumes*.
5. **Conformance Monitoring**—Conformance monitoring is provided by the USS throughout the flight operation to determine whether the UA adheres to its operational intent volume(s) and provides appropriate notice upon nonconformance.
6. **Nonconforming State**—Should the UA exit such operational intent volume(s), notification(s) to the operator (and possibly to relevant USS) may

occur. The operational intent volumes may be enlarged or otherwise revised to include volume(s) encapsulating the UA's projected path.

7. **Return to Current Operational Intent or Replan**—Operational intent may transition back to active when the vehicle is again conforming to the existing OI. That is, if a vehicle goes nonconforming (exiting the existing active OI), but is able to move back to the existing OI before becoming contingent, then the operation can simply change its state back to active.
8. **Contingency State**—Should the nonconformance continue beyond a designated timeout period, or upon notification by operator, the operation transitions to the contingency state. The flight operation must thereafter transition to the ended state, i.e., it cannot return to nominal operations and activated state even if the nonconformance is thereafter corrected.
9. **Notice**—Upon contingency, relevant USS are notified with available position data and expected flight path of the UA.
10. **Ended State**—The fight operation is terminated by the operator or USS at the completion of the operational intent or from the contingency state.[219]

2.3.13 Additional Services

UTM systems could provide other services as well. Some operational possibilities include (whether integrated with other services, or provided separately), interations of the above-mentioned services, advanced deconfliction services, geo-awareness,[220] and demand management. Additional possible value-added services may include some of those presented in Chapter 2, Section 2.2.3 *Supplemental Data Service Providers (SDSPs).* Nonetheless, consistent with applicable rules, a CAA may limit the scope of UTM services, including, for example, to "safety-critical" services, or to those that do not: financially burden a CAA or ATM system, shift costs to manned aircraft pilots or operators, or conflict with core rights of stakeholders. Other services may be particularly relevant to Advanced Air Mobility.[221]

2.4 ARCHITECTURE

2.4.1 General—Architecture

A UTM notional architecture as envisioned by NASA, is presented in Figure 2.3.[222] Widely embraced, it presents one high-level view of UTM structures, functions, and relationships. This architecture has many features we associate with any UTM concept:

- envisioning UTM as being a system of systems
- providing scalability for multiple USS and other relevant stakeholders
- accommodating diverse supporting services with a focus on safe operations
- describing interface/support for its various second-level components and services
- conceptually designed to be highly automated.[223]

Responsive to its complexity, this architecture places interoperability as a critical requirement for smooth, effective operations.

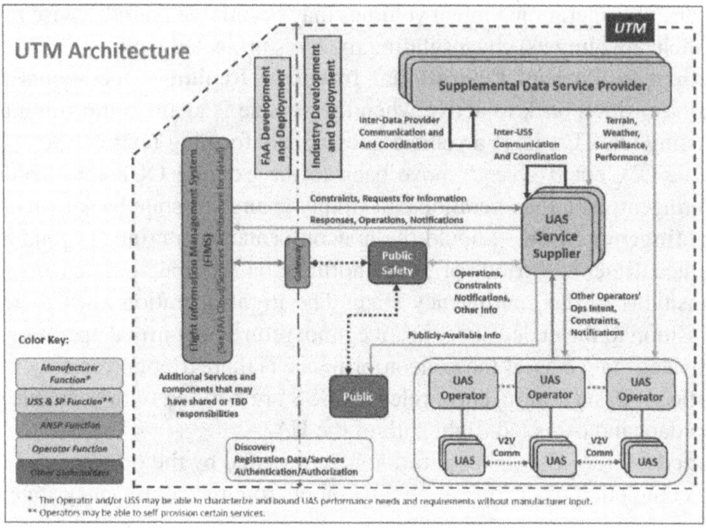

FIGURE 2.3 UTM notional architecture. (Courtesy of NASA.)

The primary ATM-UTM interface is the Flight Information Management System (FIMS) that connects a CAA/ANSP to USS and the breadth of UTM components, services, and stakeholders. *See* Chapter 2, Section 2.4.3 *Flight Information Management System*. As depicted above, the UTM architecture supports key components of safe, scalable flight operations: submission and modifications of flight plans, constraints, notifications, and communication of other information. The architecture depicted in Figure 2.3 includes registration, discovery, and authentication/authorization that support diverse UTM services. Each of these components and functions is discussed below. The CORUS ConOps (*see* Chapter 5, Section 5.3 *U-space*) represents the UTM vision in Europe, or U-space, in part, addressing both centralized and federated models.[224]

2.4.2 Centralized vs. Federated Architecture

"The FAA has chosen the federated approach."

Ali Bahrami
Assoc. Administrator for Aviation Safety, FAA[225]

"The debate has been somehow solved . . . both options are possible."

Lorenzo Murzilli,
Murzilli Consulting[226]

Although UTM can be implemented as a centralized system, its most promising capabilities—and general direction of development—are toward decentralized, federated systems. In principle, a federated system would consist of multiple USS and SDSPs to which a civil aviation authority grants authority[227] to participate in the defined UTM ecosystem. In Canada, for example, "[t]his authority would also be negotiated with the national ANSP, and, in cases where the ANSP is a non-governmental entity, be subject to the ANSP supporting a federated approach."[228]

There are pros and cons to implementing a centralized versus federated architecture within aviation's safety-critical environment, and the consequences "are profound and need careful consideration.[229] Centralized ATM systems are well-understood, widely deployed, well-supported, and formally planned. Yet they may be brittle,[230] with inadequate redundancy or network resiliency; and unable to meet the economic, political, and technical challenges of developing UTM. This was underscored by Lorenzo Murzilli, formerly of Swiss FOCA:

> We learned yet another important lesson: the need [for] a radical shift from a centralized, planned, top-down approach to UTM to a decentralized, agile, bottom-up approach. We had to abandon the idea that the government could centrally plan, build, and later deploy the UTM.[231]

Federation may offer improved redundancy, scalability,[232] availability, service customization and support for a broader diversity of UAS, specialization, flexibility, innovation, and competition. Federation may also offer CAAs the possibility of avoiding significant development, implementation, and operational costs of traffic management, shifting them to the private sector. Dale Sheridan of the Australian Government has urged, "[u]nless there is need to centralize something, it probably should be federated."[233]

Nonetheless, federation may add complexity to an already complex system, requiring additional management and co-ordination functions. For example, ceding "control of airspace access, management, and flight information to private entities may create additional complexities around access to safety related flight information data, operational data, and data of national or public interest."[234] Complexity has been demonstrated to be the root cause of many failures in software and information technology.[235] "Aviation can also learn from other industries"[236] regarding complex systems architecture.[237] Federated architectures may also create greater security risks, identity and authentication administration, and application-specific and/or interoperability challenges. Notwithstanding, federated systems boast many compelling successes, and proven mitigations, such as in the banking system, cellular communications, and the internet. Standards are key to managing the complexity of federated systems.[238]

The primary means of communication between the participants in federated UTM is enabled via application programming interface (API) and harmonized protocols.[239] For example, operational intents and constraints are shared via API (*see* Chapter 2, Section 2.4.9 *UTM Data Exchange*). Reflecting the technology focused development history of UAS flight, APIs can reduce complexity and accelerate interoperability and system performance. NASA has taken pioneering steps in defining requirements for the APIs, with subsequent critical contributions by standards development organizations.[240]

Airspace planning and optimization may present further challenges in a federated infrastructure.

> For example, centralized planning is not possible in the NASA UTM architecture unless a single USS is designated as the sole planner ["source of truth"] for the airspace in a given region. Folding the [flight] planning function into a single Air Navigation Service Provider (ANSP) would also enable centralized planning.... The intent of the NASA UTM architecture is to be distributed in nature, and that leads to a number of trade-offs from a planning perspective. At a high level, the advantage of a distributed architecture is that it enables multiple entities to independently manage their UAS operations, while adhering to coordination when needed to ensure safety.[241]

It has been suggested that federated architecture "leads to sub-optimal airspace utilization in regions where multiple USSs are operating simultaneously and can lead to cascading effects that cause severe airspace congestion in highly used regions [in particular] in free-flight operations where multiple USSs may need to plan around each other without any structural constraints in the airspace to organize operations. While negotiation schemes have been discussed in the context of the distributed UTM architecture, the implementation and the impact of the negotiation on sub-optimal use of airspace is not clear."[242] "What may be needed in some airspace is some structure that enables more optimal use of the airspace ... even if the implementation is centralized.[243]

Many federated systems challenges have been mitigated through intensive research, development, and standards.[244] The deployment of a hybrid architecture may exploit beneficial attributes of both centralize and federated systems.[245] Notably, the U-space regulation offers member states a choice, accommodating either architecture (*see* Chapter 5, Section 5.3 *U-space*). The following subsections describe key architectural components of UTM systems, with emphasis on federated architecture.

2.4.3 Flight Information Management System

A FIMS coordinates air traffic information between UTM systems and the ATM system, and serves as "a central component of the ecosystem providing the CAA/ANSP with access to UTM data."[246] (*See* Figure 2.3 – ***UTM Notional Architecture***). The term FIMS originated in NASA's UTM architecture and has been adopted by various UTM ConOps (with the same or different name, such as Common Information Function (CIF)) to embrace similar or additional functions.[247]

The FIMS has been described as an "interface ... for data exchange"[248] facilitating "automated exchange of airspace constraint data,"[249] a collection of capabilities, and as "a central, cloud-based component that acts both as a bridge to the NAS and "a broker of information between stakeholders."[250] It avoids need for all operators to connect directly to the CAA. "Connections to the FIMS are allowed from [authorized] USSs that meet minimum requirements related to functionality, quality of service, and reliability."[251] FIMS has further been described as:

- real-time (possibly with a "look back" option)
- a gateway to ANSPs
- coordinating diverse services (e.g., identity and authentication—at least to distribute credentials)

- a set of public APIs that interact with external parties (e.g., FIMS–USS)[252]
- a "trust anchor" and single source of truth[253]

The way FIMS is designed plays a critical role in the nature of the UTM architecture. Tom Prevot of Joby Aviation put it this way: "[s]ome entities want FIMS to be a more powerful central tool to deconflict within the airspace where we don't use air traffic services; others want to keep the FIMS as tiny as needed. It doesn't have a defined scope; countries want to drive it. The US wants a decentralized UTM architecture. If FIMS authority grows it will limit the effectiveness of its users and we'll get back some of the inefficiencies of ATM today."[254] FIMS architecture can be flexibly defined and deployed.[255] For example, a country may decide to monopolize the FIMS architecture and require that all weather information come from its national weather service, while others may designate a sole provider for deconfliction. These are administrative decisions that may be implemented in and reflected by certain FIMS, CIS, or other centralized information services. "There is not a one size fits all FIMS solution."[256] In the future, the FIMS could come to be seen as a specialized SDSP.[257]

FIMS is always a centralized function (i.e., there can only be one FIMS per region), but the services it provides are what may vary by implementation (*see below*). "Centralizing services" essentially adds them into the FIMS block of the architecture, versus the federated, decentralized portion of the architecture.[258]

Figure 2.4[259] depicts a noteworthy model—the Swiss U-space FIMS implemented "as the source of trust [and] initial [qualified] distributed system before future revision."[260]

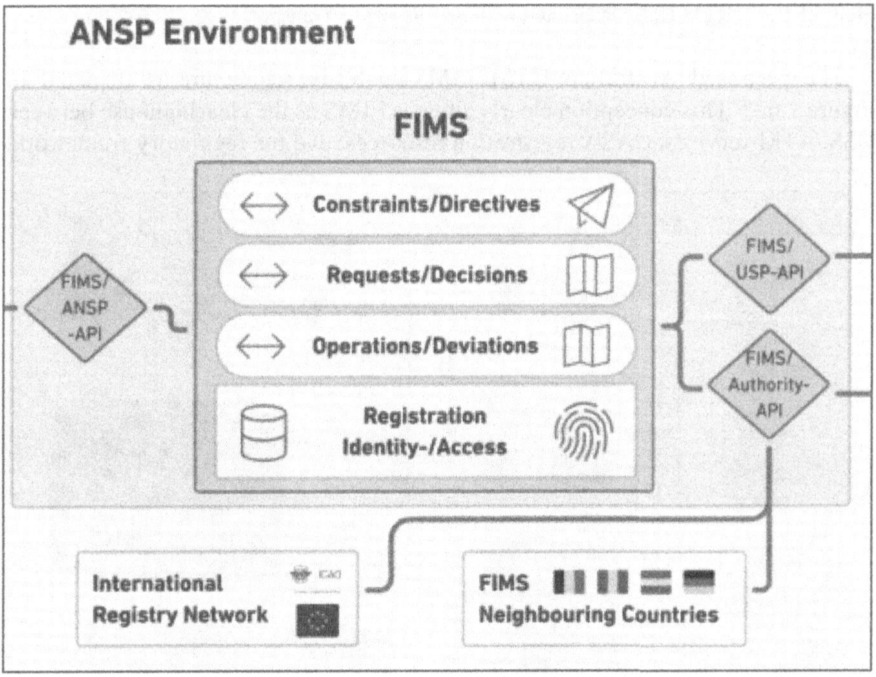

FIGURE 2.4 SWISS U-Space FIMS. (Courtesy of Swiss FOCA (2019).)

FIMS services sit between ANSP services and federated USP services, as well as managing interactions with neighboring countries and external registration resources.

Additionally, Transport Canada has proposed an "RTM Hub" within its proposed RTM (RPAS Traffic Management) System, as presented in Figure 2.5.[261] The RTM Hub is envisioned to eventually connect Transport Canada services, NAV Canada ANSP services, USPs, and UAS operators.

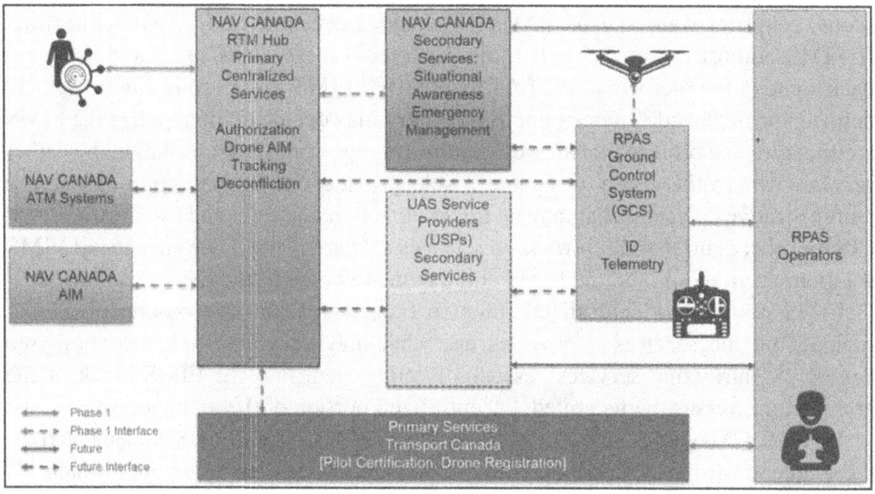

FIGURE 2.5 RTM Hub. (Reprinted with permission of Transport Canada (2020).)

The proposed Australian UTM FIMS-enabled architecture is depicted in Figure 2.6.[262] This conception clearly situates FIMS as the clearinghouse between USS, ATM services, CASA registration resources, and the regulatory framework.

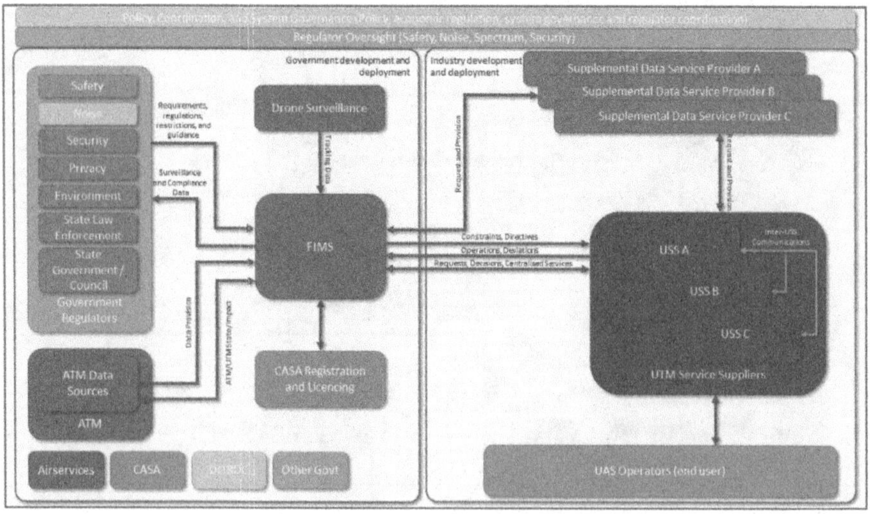

FIGURE 2.6 Proposed Australian UTM architecture. (Source: Australian Government (2020).)

The proposed Indian UTM ecosystem presented in Figure 2.7 includes the DigitalSky Platform (central block of the figure) depicting, in part, the DigitalSky-UTM Service Provider (DS-UTMSP) and the DigitalSky-Engine (DS-ENG) as "the core of India's UAS Ecosystem [that] connects with all other blocks of the UTM architecture."[263]

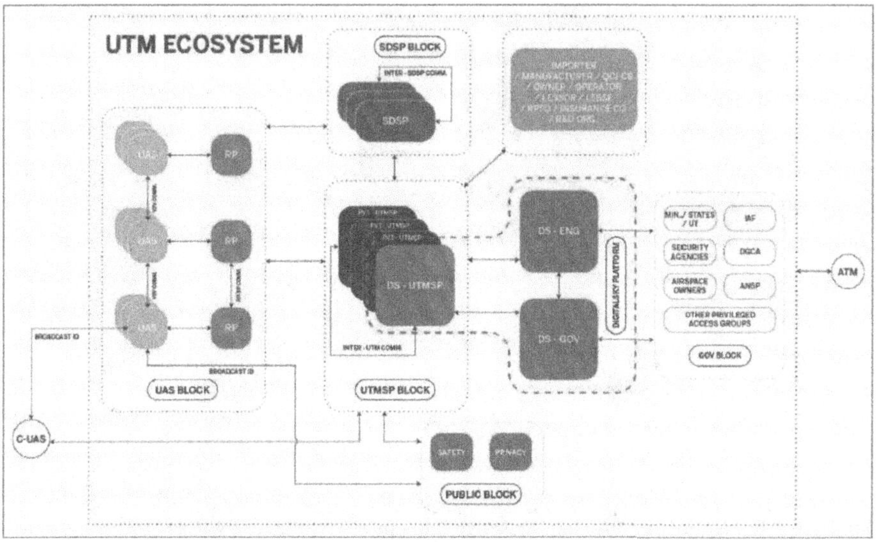

FIGURE 2.7 Proposed Indian UTM Ecosystem. (Source: Government of India (2020).)

Viewed in a slightly different context, Figure 2.8 illustrates how FIMS services could fit into the FAA's cloud-based architecture, in general. This view envisions FIMS information services using common data to support cloud applications, connected to FAA and industry/public users of those services through gateway-like information services that manage the connections.[264] NASA Ames has completed a technology transfer of initial FIMS code to the FAA where it has been evaluated for its fit into the existing and future operational architecture.[265]

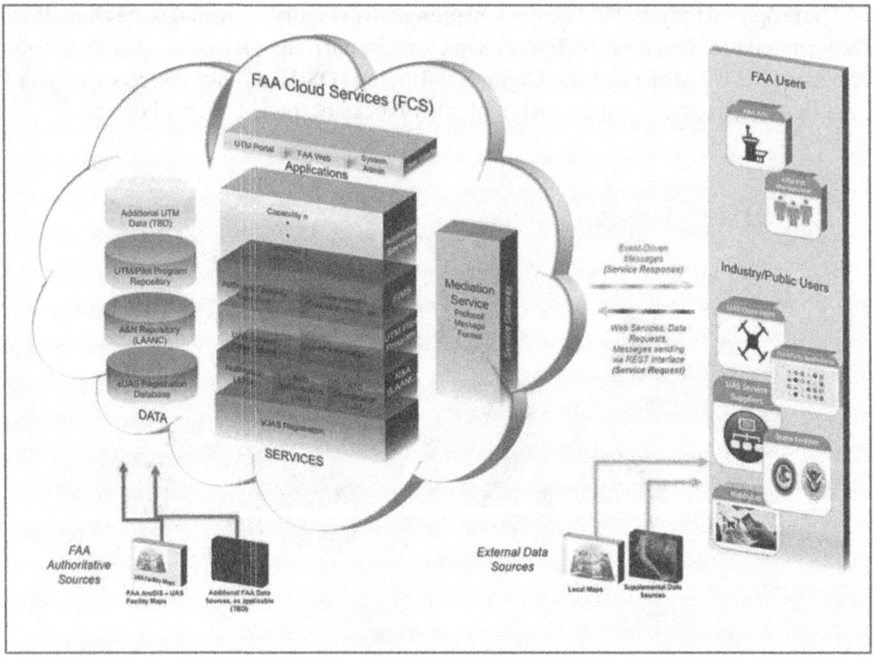

FIGURE 2.8 Notional FAA cloud-based architecture.[266] (Source: FAA.)

Globally there is a high degree of variability in UTM implementation based on airspace, regularity, and existing infrastructure complexity, and how services are provided.[267]

However, the FIMS core function of acting as a data exchange mechanism between the CAA/ANSP and UAS operation via the USS remains unchanged regardless of the specific implementation. Ultimately, each CAA/ANSP will decide what approach best suits their needs. However, global variation in the approach to FIMS is itself a potentially significant hindrance to global interoperability. And yet, FIMS could also serve as a bridge for cross-jurisdictional interoperability—by allowing different UTM ecosystems to interoperate so long as there are standards.[268]

2.4.4 System-Wide Information Management

"SWIM principles hold for all operations."

Steve Bradford
Chief Scientist for Architecture and NextGen Development, FAA[269]

Before UTM was even a concept, System-Wide Information Management (SWIM) emerged within the ATM domain as a means to share diverse operational and strategic planning information among a range of stakeholders and systems, both inside and outside CAA/ANSP boundaries.[270] "The aim of SWIM is to provide information to users with access to relevant and mutually understood information in an interoperable manner."[271] Services provided include aeronautical navigation information, meteorological, flight and flow, surveillance/position, and certain ATM strategic coordination information exchange between qualified parties.

SWIM is based on a one-to-many architecture instead of point-to-point connections between stakeholders. The scope of SWIM includes information exchange and governance,[272] standards, and the infrastructure required to exchange information between SWIM-enabled applications. SWIM-enabled applications consume or provide SWIM information services using SWIM standards.

> The emerging nature of new entrants such as UAS, UAM and others have substantially different operational and business models than those of traditional aviation stakeholders. The variability in the way these vehicles will operate will necessitate new ways for the FAA to communicate and conduct business with these stakeholders to enable their operations and maintain safety and efficiency in the NAS, while retaining the existing infrastructure and data distribution model established under NextGen with commercial aviation stakeholders that operate today.
>
> The long-term vision and implementation of data exchange and distribution methods to serve new entrant stakeholders, whether it be directly between the ANSP/CAA and the operator, or through a third party service (USS, SDSP) will ultimately be captured in some future expansion of SWIM as governed by the ICAO Aviation System Block Upgrades or, in the case of the FAA, the NAS 2035 vision.[273]

UTM will interact with the ANSP data exchange portal, which in some cases is the direct SWIM interface. In others, interaction is via intermediary exchange service, such as FIMS within the FAA's trusted domain rather than via public SWIM portal.[274]

ICAO offers that "[SWIM] principles should be applied to support information exchanges between UTM and ATM. For this to occur … current aviation connections, through SWIM, will need to be extended to new airspace users … ."[275] FIMS implementations can draw on SWIM data, delegate some FIMS data sharing interfaces directly to a SWIM implementation, or even delegate the FIMS functions as a SWIM service depending on the particular CAA/ANSP SWIM implementation.[276] "SWIM will be one of the things FIMS plugs into and supports,"[277] and should play an important role for UTM.[278] In U-space, "[t]o prepare the flight, the drone operator uses information-sharing services connected to ATM via SWIM (e.g., NOTAMs, meteorological conditions and forecasts at the nearest aerodrome)."[279] SWIM data services supporting UTM may include terrain, obstacles, and specialized airspace information.[280]

The rapid growth of cloud-based services and technologies has eclipsed conventional, burdensome methods of consuming SWIM services—catalyzing transition from the SWIM core, although at a modest pace.[281] In 2017, the FAA conceived of the SWIM Cloud Distribution Service (SCDS), to provide a publicly accessible cloud-based infrastructure dedicated to providing near real-time SWIM data—with a simplified, quick method to access FAA SWIM data in preexisting formats.[282] SCDS or comparable migration should enable UTM to scale operations, and instantiate services for new users while allowing ANSPs to better manage data distribution from the exchange portal.

Further engagement between the SWIM and UTM development communities (including for education, API vetting, standards collaboration, and governance harmonization) should contribute to UTM integration in the NAS and offer new options.[283]

2.4.5 Discovery

"It's really in the middle of everything."

Mike Glasgow
Technical Standards and UTM System Architect, Wing[284]

Discovery is the essential process required in a federated system by which a USS determines what other USS it must contact to deconflict operational intents, constraints, and obtain other relevant information. This supports strategic separation to enable aircraft "to stay clear of each other."[285]

Like any distributed system, each independent UTM USS requires a service to facilitate federation. However, given the number of expected USS instances in a region, a single group of all instances is likely to have significant data exchange overhead. Instead, forming multiple small collections of USS instances by connecting only those that have potentially overlapping flights helps ensure a minimum of unnecessary communication. The Discovery and Synchronization Service (DSS) enables each USS to identify peers that meet these criteria, allowing independent UTM USS cohorts to be formed. A data exchange within each cohort—independent from the DSS—enables operators to strategically separate their flight operations from those of their peers. Discovery services can be implemented as a CAA/ANSP managed service in the FIMS or by private industry.[286] Conceptually, discovery is easy to understand, but it can be challenging to implement.

For a network remote ID service, similarly, the discovery service "determine[s] the required USS data exchanges to successfully complete Net-RID services."[287] *See* Chapter 2, Section 2.4.8 *UAS Remote ID.*

2.4.5.1 Discovery Methods

"Just call it a phone book."

Amit Ganjoo
Founder & CEO, ANRA Technologies[288]

UTM architectures implement varying methods to detect potential conflicts, represent the location where UAS operations may intersect, and determine which USS must communicate and resolve conflicts,[289] and each architecture drives needs for discovery services.

As a highly-simplified illustration of why discovery is needed, Figure 2.9 depicts three USS and their maximum expected area of flight operations: polygons A, B, and C, with A and B intersecting at D, thereby invoking a requirement to provide notification to each other if they plan new flights in area D. How USS A and B discover and share information is a function of the particular UTM implementation.[290]

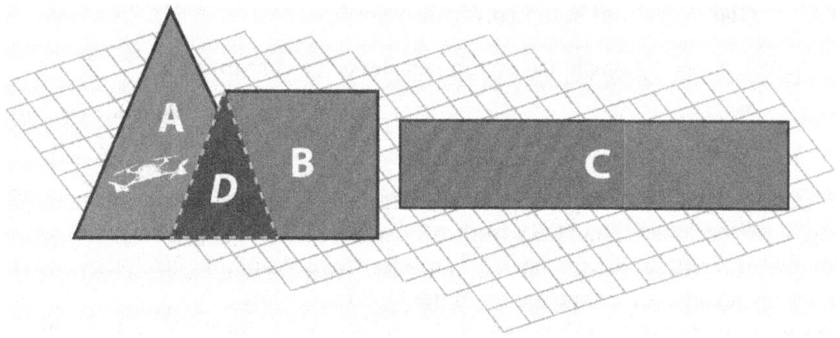

FIGURE 2.9 Discovery.

Discovery in UTM systems can be complex due to policy, regulatory, and administrative factors. For example, USS and their customers may wish to minimize the sharing of proprietary data such as company-specific details of a planned flight operation, position reports, or personally identifiable information (PII).[291] (*See* Chapter 4, Section 4.7.6 *Data Rights and Privacy Management*). It is easy to see how a discovery service could become a centralized data store of flight information and at the same time a critical point of failure in the system. This risk has led naturally to UTM developers favoring a discovery paradigm that facilitates minimal sharing of flight data, except in exigent circumstances such as nonconforming, contingency, or intersecting operations.[292]

The paradigm that has emerged gives individual USS a way to ascertain which other "peer" USS they must contact in the area in which they are planning flights to determine what other operations exist, the state of airspace, coverage areas, and dynamic no-fly zones or other constraints. Once a USS has discovered its peers it can then communicate with them directly to exchange 4D airspace volumes, identify potentially conflicting operations, and perform required functions such as networked remote ID, strategic coordination,[293] and constraint management.[294] Discovery implementations can be characterized by the extent to which they are centralized or distributed, pull or push data, require the sharing of transactional details, synchronize,[295] scale, and create (or mitigate) *race conditions*.[296]

Today the primary discovery mechanism for UTM has been standardized by ASTM—the DSS (*see* Chapter 2, Section 2.4.5.2). In fact, ASTM not only defined the DSS, but "an interoperability paradigm that consists of two parts: the DSS to resolve who to talk to (and making sure you did), and peer-to-peer interfaces that are how you talk to them. Within that context, the DSS and peer-to-peer interfaces for *every* UTM service that uses the interoperability paradigm *will* vary, to accommodate service-specific data and nuances."[297]

2.4.5.2 Discovery and Synchronization Service

"The DSS is weirdly simple and weirdly complex at the same time."

Reinaldo Negron
Head of UTM, Wing[298]

The standardized discovery service for federated UTM in the form of the DSS is being deployed today. The "DSS from ASTM is the leading approach with support from industry, NASA, Switzerland,"[299] and elsewhere. As of mid-2021, there was one leading open source implementation of the DSS (*see below).*

As described above, DSS operates as a *lookup table* for USS to determine the other USS with flights in a particular area. It accepts data to be written (e.g., "use these 4D volumes to represent my flight operation"), or responds to requests (e.g., "get me other UAS operations in this area"). The DSS response to a USS request for nearby operations points the requestor to the appropriate USS with operations in the area, rather than identifying the 4D volumes in use. The DSS does not pass messages between USS or provide coordination.[300]

Figure 2.10 below illustrates the DSS discovery process with its associated data exchanges. Its key attributes are the: (1) USS query of the DSS for conflicting operational intents, constraints, or other information, (2) return of the relevant USS data query endpoints, (3) and (4) subsequent responsive communications among relevant USS peers independent of the DSS.[301]

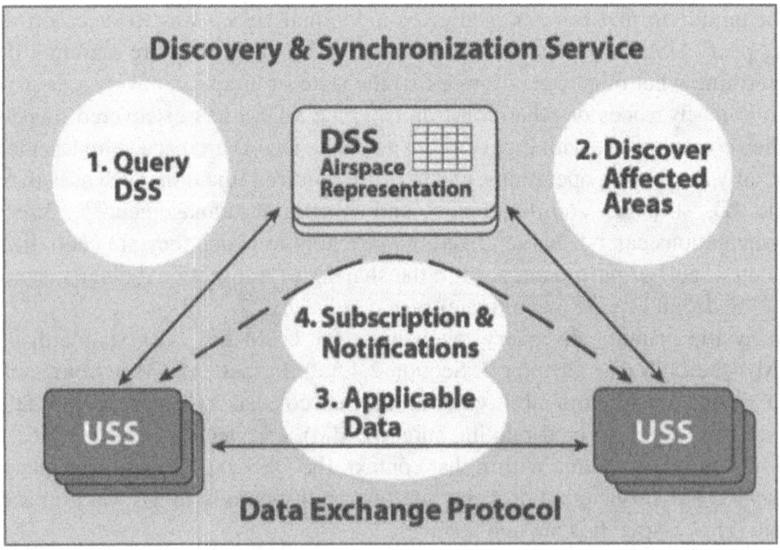

FIGURE 2.10 DSS process. (Derived from and reprinted, with permission, from ASTM F3411-19 Standard Specification for Remote ID and Tracking, copyright ASTM International, 100 Barr Harbor Drive, West Conshohocken, PA 19428. A copy of the complete standard may be obtained from ASTM International, www.astm.org.)

An example of discovery using DSS follows.[302]

a. When a USS is planning a new UAS operation, it queries the DSS to discover other operations or constraints in the area. If the operation must fly a specific route, the geographic scope of the query corresponds to that specific route. If the operation has flexibility to vary the route to avoid other operations and constraints, the query corresponds to the area within which the operator is willing to fly (referred to as the *planning area*).

   How a USS interacts with the DSS can depend on the frequency of operations within an area. For a one-time flight, a USS may perform a one-time query to the DSS and then *pull* the applicable planning information from the relevant USS. Alternatively, when a USS has multiple flights in an area, and therefore will be planning flights on a persistent basis in that area, it is helpful to maintain awareness of other operations on an ongoing basis. In this case, the USS can establish a subscription for information about operations. In response, the DSS notifies relevant USS when their operations or constraints intersect a subscription, and the relevant USS then *push* the applicable planning information to the subscribing USS.
b. When the DSS receives a query, which includes a 4D volume, it maps the query onto the airspace representation and finds any intersecting operations and/or constraints.
c. Once a set of operations and/or constraints are discovered, the DSS returns to the planning USS a list of IDs for the operations and/or constraints and the URL(s) for the owning USS(s). It does not share terrain or obstruction data.[303]
d. The planning USS then contacts directly the owning USS for each operation discovered by the DSS and obtains the complete details of the operation necessary for deconfliction. Part of the information returned by the owning USS is the "opaque version number" (OVN) for an operation.[304] The OVN ensures that changes to airspace are known to clients before they act. Comparing version numbers is an efficient way of detecting changes rather than reading the complete data structures.
e. The planning USS uses the information about other operations in the area to plan a conflict-free route. When this process is complete, the USS attempts to write its newly planned operation to the DSS. The USS provides the OVNs for all other operations and constraints in the area. The DSS verifies that the USS provided a complete and current set of OVNs. If the set is current and complete, it demonstrates that the USS planned the flight with awareness of all other operations and constraints; and the DSS then allows the new operation to be added. If the OVN set is not current and complete, it indicates the USS was either not aware or did not have current data for one or more operations or constraints. In this case, the planning DSS does not allow the operation to be added and informs the USS of the operations for which additional information is needed.

f. Absent intervening (conflicting) updates from other USS for a higher priority operation (e.g., public safety) or a new constraint, and with proper authorization, an operator may conduct the flight.
g. Following completion of the operation, the owning USS deletes the operation from the DSS.

APIs—To participate in discovery, for both USS-DSS and USS-USS data exchange described above, each USS must support a standard set of APIs.[305] For interactions with the DSS, APIs facilitate storing operations in the DSS and querying the DSS for operations in an area of interest. For interactions between USS, APIs facilitate the *pull* and *push* methods for obtaining information about operations from the owning USS.[306]

The DSS is built on distributed database technology that does its own communication outside the scope of the UTM/DSS standards.[307] The DSS can be viewed as a (mesh) network of interconnected nodes.[308] Requirements for what information must be synchronized between nodes is provided in the UTM standard, but the API is not specified, enabling use of varying technology. For each USS region "multiple geographically dispersed DSS are desirable"[309] and should therefore have two or more connected DSS nodes to avoid a single point of failure, handle anticipated traffic density, support associated load balancing, and provide the required service level.[310] In practice, it is anticipated that there will be many nodes operated by various providers. At the boundary of each flight information region (FIR) national boundary, each country must decide how to structure DSS.

Administration—DSS could be operated by industry association,[311] a large-scale USS,[312] another USS, government, or its delegee, such as a specialized ANSP. The applicable regulator will specify minimum performance/service level requirements[313] and authorize DSS. Such requirements should facilitate interoperability[314] via international standards. Bilateral or multilateral agreements among CAAs to stand-up and operate DSS are anticipated.

To facilitate USS finding the applicable DSS, a DSS static endpoint list could be issued by either the applicable CAA, an industry-based consortium, or perhaps ICAO.[315] Such a list of available DSS should be available as a fallback for USS should the primary DSS become unavailable. Other procedures will facilitate various operations and maintenance activities, such as for priority (e.g., public safety), "cold starts," "heartbeats" (distinct protocols to ascertain USS and optionally, DSS availability), DSS failure, and other contingencies.

Preventing Abuse—A USS that requests data from the DSS under the pretext of seeking to plan flights and manage airspace, when in fact, the intent is solely to discern competitors' operations and proprietary data for economic advantage should be deemed unacceptable, in breach of UTM rules, with responsive action/enforcement.[316] As an illustration, one could imagine a progression where "a first offense is likely detected and reported (to the regulator), USS must explain, potentially given a warning. Repeat or chronic abuse behavior should result in a USS's authorization to operate being revoked by the regulator."[317] In addition, a DSS or other designated entity could maintain a *Bad Actor List* to help monitor such conduct and support remedial action.[318]

Discovery will continue to evolve, but today industry leadership believes "we are now coalescing given consensus standards-based development and broad industry participation."[319]

2.4.5.3 InterUSS Platform

A noteworthy implementation of the DSS is the *InterUSS Platform™* ("InterUSS"), which preceded standardization of DSS. Used in several well-known demonstrations that informed development of the DSS in the ASTM Remote ID and UTM specifications,[320] InterUSS is now a compliant implementation of the DSS standard maintained by the Linux Foundation's open source InterUSS Project.[321] Recognizing the InterUSS Platform's impact on UTM developments, it is mentioned here for its historical role and perspective.

Initiated by Wing,[322] InterUSS was also used in multiple NASA TCL demonstrations[323] and in a portion of the FAA's UPP. Its objectives were similar to those of the DSS, but the DSS added capabilities based upon lessons learned from the early demonstrations. "The DSS only uses a portion of the InterUSS concepts and is really a hybrid of the InterUSS model (of synchronization) and the geographic-based shared operating area model of the NASA LUN [local USS network]."[324]

2.4.6 Detect And Avoid

> "Everyone has an obligation and right to sense and avoid."
>
> **Hon. Michael Huerta**
> *Former Administrator, FAA*[325]

> "You will never solve uncooperatives."
>
> **Jay Merkle**
> *Exec. Dir., UAS Integration Office, FAA*[326]

A foundation of aviation safety is that pilots must be able to "see and avoid" other aircraft.[327] UTM systems can aid UAS in avoiding other participating aircraft by deconflicting flight paths before takeoff, but noncooperative traffic not participating in the system remains a separation challenge.[328] This is where tactical (in-flight) deconfliction becomes critical. For BVLOS operations, where the remote pilot is unable to visually surveil the airspace near the UAS, a suitably robust suite of technologies must serve several discrete functions of DAA. DAA is a critical set of capabilities meant to stand in for the see and avoid requirement, and provides the following key functions:

1. Detection of manned aircraft, whether cooperative or noncooperative.
2. Identifying intruder aircraft that pose a threat to ownship and alerting the pilot (or the UAS) of that threat.
3. Deciding how to maneuver to safely avoid the intruder aircraft—whether manually by the human pilot, via algorithm such as ACAS X (*see below*), or by some other means.
4. Providing guidance to the pilot (or the UAS) and then executing the vehicle (UA) maneuvers (either via pilot commands or automatic response

capabilities—recognizing variability in the system boundary) to avoid the conflict.

5. Continuing to assess the state of the intruder aircraft in relation to the UAS' new trajectory in a feedback loop until the intruder no longer poses a threat.

DAA systems, as tactical risk mitigations, complement UTM's strategic risk mitigation and are typically not a part or requirement of a UTM system, but provide a critical safety capability generally expected in conjunction with UTM.[329]

There are many ways in which the above steps can be performed. For example, a human pilot (possibly with the help of additional visual observers) can do all of them without additional tools or UTM services if they are physically present near the UA. But this is infeasible for scaled, cost-effective BVLOS and most anticipated UTM-based operations.

Most interestingly from a UTM architecture standpoint is that aspects of a DAA solution can possibly be offloaded to various parts of the UTM system—either residing as services within the operator's USS, or as a set of interconnected SDSPs that have knowledge of the UA's position, maneuvering abilities, and other information. Moreover, there are many ways to delegate the essential DAA functions between the operator, UAS, USS, and other UTM services (such as via *sXu—see below*).[330] Although there is little consensus among industry, standards development organizations or regulators as to which of these approaches is best in terms of meeting performance requirements, cost-effectiveness, scalability, or other metrics, DAA is essential to BVLOS approvals and key to many business models and scaling. This is an area of active research, testing, and rulemaking.

The key performance metric of a given DAA solution is its robustness expressed as a risk ratio. This is the probability of a midair collision (or loss of well clear) happening with the DAA system operational, divided by the probability of that occurrence happening without a DAA system. The risk ratio to be achieved is informed and validated by studies and set by industry standard or regulation. A given solution requires testing and validating many variables, such as vehicle maneuverability characteristics, surveillance uncertainty, timing, communications or decision latency, and algorithmic efficiency, to determine whether the risk ratio is achieved.

"Safe and effective BVLOS airspace integration will require an approved UTM service that provides acceptable surveillance performance to mitigate the hazards to manned aircraft as well as meet the requirements of [operating near other aircraft and giving right-of-way]."[331] DAA "is a method of meeting the requirement but not the only means."[332] Nonetheless, to complement the strategic deconfliction via UTM, UAS "in areas with high density or heterogeneous traffic may be required to equip with DAA technologies,"[333] that preferably should be effective for both cooperative *and* noncooperative aircraft.

DAA equipage can be airborne-based, ground-based, or both, and exploit diverse technologies such as RF spectrum, acoustics, and electro-optics.[334] The performance level required remains an active area of inquiry and development (although some required performance metrics have been standardized), as expressed by Terrence Martin, PhD of Nova Systems:

> The fielding of DAA for collision avoidance continues to pose challenges. I think the answer lies in acknowledging that you don't need the gold plated solution that has

historically been imposed on large military platforms, particularly if you can show that for low traffic environments, you can still meet regulatory expectations for Target Levels of Safety with DAA system that perform the detect part of the "*detect-decide-move*" feedback loop at say 70% rather than 99%. What I will say is that concession needs to be supplemented by quantitative evidence demonstrating what the actual traffic is, through airspace characterization or a qualitative decomposition accepted by regulatory bodies, in conjunction with containment and an acceptable response time horizon.[335]

A challenge to using ground surveillance for tactical deconfliction is the sufficiency of ground infrastructure. As recognized by the UK Civil Aviation Authority, "the maturity of technological and operational mitigations is not yet sufficient to authorise BVLOS operations in non-segregated airspace."[336] Consider, for example, that surveillance infrastructure in very low-level airspace (VLL) is practically nonexistent; radar and ADS-B towers were built to handle traffic at higher altitudes—not looking for manned traffic lower than 400 feet, and risk sUAS traffic saturating these surveillance systems; and UAS remote ID is neither primarily designed nor deployed for deconfliction purposes.[337] Another big challenge is terrain/obstacle avoidance and how that function interacts with DAA—"this is more of an issue now since we are operating at such low altitudes and don't want a DAA maneuver to steer into a mountain."[338]

Standards are emerging for sUAS DAA. ASTM F38 has developed a DAA specification, *Standard Specification for Detect and Avoid System Performance Requirements*, providing minimum safety performance thresholds and associated methods for "smaller UAS BVLOS operations for the protection of manned aircraft in lower altitude airspace."[339] EUROCAE and other standards development organizations also are engaged in sUAS DAA standards development.[340] A promising collision avoidance capability particularly suited for sUAS and poised for future DAA integration has been advanced by the FAA, research, and standards communities as an extension of the Airborne Collision Avoidance System X (ACAS Xu – for larger UAS), called ACAS sXu.[341]

Recognizing the compelling need for sUAS surveillance services to support UTM, standards are also in development to stand-up surveillance SDSPs.[342] Surveillance SDSPs may also enhance counter-UAS systems by providing strategic deconfliction data to help validate authorized UAS and improve threat assessment.[343] In sum, DAA will be a key factor in enabling BVLOS operations, and UTM in general, and is an intensive focus of standards development.

2.4.7 Cybersecurity

"Security should be a first-class citizen in this discussion."

Joseph Rios, PhD
ATM-X Project Chief Engineer, NASA[344]

2.4.7.1 General

By moving from a largely closed and proprietary human-in-the-loop ATM infrastructure to a set of interconnected, federated systems and publicly published open APIs to communicate and co-ordinate actors, UTM systems reflect advances in modern communications infrastructure. UTM also introduces certain Internet and open security standards in the aviation world. Correspondingly, UTM inherits certain well-recognized and researched security risks of open, distributed systems.[345] "We face a new era of asymmetric attacks, since the malicious actor can be anywhere, mask their identity and location, and bring down huge swaths of infrastructure and/or intercept and distort underlying data exchanges to inject false information."[346] In fact, the attack surface includes the entire confidentiality, integrity, and availability "CIA" triad, presenting "inconceivable threats" compared to the pre-internet era.[347] Therefore, although many UTM/U-space threats have been identified,[348] there is urgent need for comprehensive aviation-focused security threat modeling, testing, and mitigation research.[349] These security risks are introduced to most levels of the UTM technology stack, including, but not limited to, unauthorized access to USS that are the gateway/service point for operators into UTM, attacks on USS data at rest or in transit, and denial of service attacks.

Given the connected and network-centric operations that UTM represents, "the risk level for the UTM increases with a new breed of advanced cyberattacks."[350] The complexity of the UTM system and the number of actors involved mean that there is an expanded attack surface that presents new challenges at each interface of this system of systems.[351] Even "a single, poorly-implemented USS will diminish trust and security in the overall system."[352] At a high level, a UTM system can look like a series of interconnected microservices managed by different entities and locations, and therefore the security risks to each of these systems can mimic those of microservice-oriented application systems.[353] Moreover, UTM will become *critical infrastructure*; as such, disruptions could potentially create "a debilitating effect on security, national economic security, national public health or safety, or any combination of those matters."[354]

"Moving forward, we simply have to spend efforts on ensuring cyber-physical security of advanced systems early on, and keep in mind that cyber security is not *one and done* type activity, on-going focus is needed on all enterprise level systems, much like our personal computing systems."[355] "It has to be secure from day one."[356]

As a matter of level-setting and perspective, many UTM experts view UTM security as able to piggyback on the advances in technologies upon which the modern Internet is built. Under this view, users and UAs are authenticated using many of the same technologies trusted for banking transactions, where communications between systems are encrypted in transit using proven security technologies.[357] Indeed, some security experts assert that UTM presents recognized security threats that are well-known and well-addressed by commercial IT systems,[358] but emphasize that at some point, UTM will require a cybersecurity culture and law enforcement relationships to supplement the technology.[359] There are other security experts who urge that neither current Internet nor aviation standards are sufficient: "I suspect that the answer will be that the new infrastructure will have to outmatch both."[360]

The development of adequate UTM platform security best practices are expected to be undertaken by UTM industry consortia[361] with consultation from government and ICAO. ICAO has championed the importance of cybersecurity in aviation for more than 15 years.[362] Its 13th Air Navigation Conference recognized the breadth of security risks associated with UTM, "encourag[ing] UTM providers to implement *the highest level of cyber security standards* that are consistent with aviation community expectations and guidelines for very low altitude airspace operations."[363] Additionally, its Global Air Navigation Plan (GANP) recognizes the need to facilitate aviation system block upgrades that include UAS and UAS management via UTM. More recently, ICAO launched development of the *Global Aviation Trust Framework,* intended (nonexclusively) to support UTM. *See* Chapter 2, Section 2.4.7.2 *International Aviation Trust Framework.* Moreover, NASA has issued a *UAS Service Supplier Framework for Authentication and Authorization.*[364]

The U.S. Congress directed the FAA in the *FAA Extension, Safety, and Security Act of 2016* (FESSA), to develop a "comprehensive and strategic framework of principles and policies to reduce cybersecurity risks to the national airspace system."[365] The Act further requires the FAA to use "a total systems approach that takes into consideration the interactions and interdependence of different components of aircraft systems and the national airspace system"[366] of which the developing UTM will become an integral component. Such a total system approach must transcend computer communications to include the computing base. Nonetheless, to date the "FAA has not completed a comprehensive and strategic cybersecurity framework of policies designed to identify and mitigate cybersecurity risks."[367]

A secure and sustainable UTM requires a set of tools to provide the range of security services assuring its resilience and performance across the varied aviation environments. These include tools for authentication, integrity, non-repudiation, and privacy of sensitive communications and operations. Identity trust and network trust are complementary, and both demand a proper information security management system (ISMS).[368] Although many of these tools have been demonstrated or are otherwise available in proprietary and open source systems, most are not yet productized in an aviation/UTM context. In part, this is because few UTM system and component security requirements are fully specified, and they continue to evolve.[369] More formal ISMS and associated practices specific to UTM are needed.

As UTM services respond to safety regulations, enterprise security management and interoperable system requirements, heightened security mechanisms are needed, including to support approvals, continuing compliance and safety assurance (*see* Chapter 4, Section 4.2 *USS Qualification*). Until security requirements are further specified, a look at certain standards relevant to UTM building blocks and service providers is instructive, including, but not limited to the following:

- **International Standards Organization ISO/IEC 27000** series of standards address ISMS,[370] and include ISO/IEC 27001, *Information technology—Security techniques—Information security management systems—Requirements.* The ASTM UTM specification references ISO/IEC 27001, and the U-space regulation also anticipates its use.[371] Other ISO standards such as *ISO 9000 Family - Quality Management* are also relevant to the UTM ecosystem.[372]
- **NIST, Federal Information Processing Standards (FIPS), and Risk Management Framework (RMF).**[373] For example, FIPS Publication 200, *Minimum Security Requirements for Federal Information and Information Systems* ("FIPS PUB 200")[374] requires the categorization of information systems per FIPS PUB 199, *Standards for Security Categorization of Federal Information and Information Systems,*[375] and then application of commensurate controls per Special Publication 800-53, *Security and Privacy*

Controls for Federal Information Systems and Organizations.[376] Its audience transcends government to include, in part, "[i]ndividuals with system development responsibilities" as well as "business owners, system owners, information owners or stewards, system administrators, system security or privacy officers."[377] FIPS Pub 200 was also designated in the LAANC onboarding MOA.[378] Its applicability to non-Federal systems and personnel is well-supported. NIST's *Framework for Improving Critical Infrastructure Cybersecurity* is also instructive.[379] Additionally, because of the complexity of UTM, "[a]chieving security objectives … requires system security activities and considerations to be tightly integrated into the technical and nontechnical process of an engineering effort—that is, *institutionalizing* and *operationalizing* systems security engineering as a proactive contributor and informing aspect to the engineering effort" as described in NIST Special Publication 800-160, *System Security Engineering: Considerations for a Multidisciplinary Approach in the Engineering of Trustworthy Secure Systems*,[380] and its Volume 2, *Developing Cyber Resilient Systems: A Systems Engineering Approach*.[381] NIST's Special Publication 800-63-3, *Digital Identity Guidelines*[382] is also noteworthy, in part, because NASA asserted that:

> USS identities in the UTM System will be assured according to the following levels in [SP 800-63-3]:
>
> 1. Identity Assurance Level 3 (IAL3)
> 2. Authenticator Assurance Level 3 (AAL3)
> 3. Federation Assurance Level 3 (FAL3).[383]

Other NIST publications also deserve attention.[384]

- **ATM-Specific Standards.** Standards addressing ATM security might inform UTM security.[385] For example, *EUROCAE ED-205—Process Standard for Security Certification and Declaration of ATM ANS Ground Systems*, provides a process to assess ATM/ANS ground systems security and other commercial concerns. It is "a resource for certification or declaration of conformity with applicable security requirements."[386]
- **Wireless Access in Vehicular Environments (WAVE) Standards.** For example, the IEEE 1609.2-2016—*Standard for Wireless Access in Vehicular Environments—Security Services for Applications and Management Messages*.[387] This and associated standards accommodate intermittent, large volume, mobile communications.

As is true with any new system(s), development of security tools and mitigations for UTM require a formal *scoping* of the security risks in its operations. Generally, this is undertaken in private industry, since industry must deliver these systems as a part of business operations. However, given the complexity of UTM,[388] relying exclusively on sectorial development may be inadequate. "Without agreed-upon scoping, there is no network trust."[389] As mentioned above, the implications of an increasing, diverse, and evolving threat surface deserve further in-depth consideration.[390] Absent such consideration, downstream assessment and assertions of assurances may be deemed untrustworthy or otherwise fail to serve their intended purposes. To the extent a USS is akin to a delegated ANSP, existing ANSP requirements in the context of data and information security also become relevant.

The foundational blocks for UTM security such as authentication technology and methods, identity mechanisms and authorization, and mobile network security services, will contribute to an effective systems-level vision of security. UTM architecture must support diverse and evolving security and organizational requirements as a core requirement. Building these specifications providing for an effective and recognized management system that includes information security for USS and other UTM components requires further work.

The development of safe, secure UTM systems is an ongoing process, and several important cybersecurity questions remain unanswered, including the following:

- Given the many stakeholders and interested parties, how are roles and access to the system defined?
- Are any existing protocol/security systems used in the public Internet sufficient, or to what extent must they be "aviation grade"?[391]
- What are the performance boundaries of these systems and would these be considered safe/practical for UAS operations?
- What are the system vulnerabilities and attack vectors, and are they considered an acceptable level of risk for flights over urban environments or populated areas?
- To what extent will nation state actors be addressed in responsive threat analysis?
- What level of autonomy in urban environments is acceptable? In non-urban?

The development of these systems requires more testing, analysis, and the sharing of such results to develop appropriate UTM-specific risk frameworks and mitigation strategies.[392]

2.4.7.2 International Aviation Trust Framework

"You don't federate with people you don't trust."

Richard Parker
Founder and CEO, Altitude Angel[393]

The historical underpinnings of ICAO providing a level of trust for digital aeronautical communications were built on the Chicago Convention of 1944 that allowed a world that had been devastated by attacks from the air, to trust the aircraft that would be used to rebuild the global economy. ICAO is now applying that

> *system of trust* to the digital age .... If you don't have a robust system of identity and trust you don't have a way to communicate digitally. That digital system of trust

> should be built on the same legal principles that have guided the development of the global aviation system. If not, we end up with one system of trust that applies to the old system and another system of trust that applies to new digital entrants. That fragmentation can't be allowed to happen. The power of aviation is that it connects the world under a single sky. Our predecessors have managed to hold this precious sky together in spite of geopolitical crises and technological revolutions. We must not be the first generation to fail in this task. The solution is simple. Use the regulator who certifies you, and knows you, to serve as the anchor for digital trust. This simple step ties the new digital entrants into the established global framework used by conventional aviation. A common source of trust was invented in 1944 and was the mechanism that created a single unified sky out of dozens of war-torn regions. By using that same source of trust, new digital entrants will be able to become part of that powerful single sky.[394]

Since UTM will interoperate among disparate systems, organizations, and to some extent between nation states, the challenges of ascertaining trustworthiness increase considerably. *See* Chapter 2, Section 2.4.7.1 *Cybersecurity—General.* Information systems underlying UTM require a secure environment that provides, among other things, assurances of each party's identity, authorization to operate in the NAS, and privacy protections.

In response to these risks and requirements, ICAO's Air Navigation Commission recommended, in part, that ICAO:

> (f) establish a formal project involving States, international organizations and relevant stakeholders for the urgent and transparent development of a globally harmonized aviation trust framework through a group of experts. Priority should be given to governance principles; [and]
> (i) develop, as a matter of priority, and promote high-level policies and management frameworks for cyber resilience to help mitigate cyber threats and risks to civil aviation based on international industry standards and preferably aligned or integrated with existing management systems;[395]

The primary tangible response has been ICAO's efforts to develop a *trust framework* that bridges the various national infrastructures to facilitate a range of aviation-related services and transactions, including global UTM.[396] ICAO created a Working Group on Current and Future Operational Needs of the Trust Framework Study Group (TFSG)[397] that is developing use cases and responsive mitigations to support "the federated decision-making model of air traffic management," and "various interconnection methods over IP" in an "environment [that] include[s] multiple stakeholders in disparate network private domains."[398] This undertaking is influencing UTM standards and implementations globally.

The envisioned Trust Framework serves as a *digital aviation ecosystem* and leverages public key infrastructure (PKI) and alternative technologies,[399] the mutual recognition and interoperability of digital certificates, a generic Top-Level Domain (gTLD),[400] and a private address block to logically isolate the infrastructure from the Internet, thus improving security.[401] It envisions an "aviation Internet" that is similar to the public Internet except all the parties would have an aviation role (much like the .aero domain[402]) and would encompass all functions of aviation including airports, airlines, USS, etc.

Since UTM communications will require trusted security services, the proposed Trust Framework initiative should help prioritize resources, engage experts, and provide coordination for UTM implementations. In the longer term, it would offer trust-assured UTM infrastructure. A proof of concept is being planned between EUROCONTROL and the FAA and will be expanded incrementally to other regions.[403] Network remote ID can be an early beneficiary of the Trust Framework. Notwithstanding, security architecture and standards continue to evolve globally, and the outcome and adoption of this work will take time to ascertain.

Designed to ensure that an unmanned aircraft (UA) digital identity is tied uniquely to a nation state registration, the Trust Framework digital identity may also contribute to an aircraft's "nationality" under international law, such as the *Chicago Convention*, Annex 13 Aircraft Accident and Incident Investigation,[404] to mitigate complex and politically burdened determinations of responsibly for an accident, and ascertain who may conduct an accident investigation. The Trust Framework might also support prosecution of cybersecurity attacks against UTM infrastructure under the *Convention on the Suppression of Unlawful Acts Relating to International Civil Aviation,*[405] since it too requires aircraft nationality. Cyberattacks are notoriously complex to prosecute, often involving actors from multiple states, and presenting jurisdictional, cross-border evidentiary, and extradition issues.[406]

2.4.8 UAS Remote ID

> "I have 193 countries that need this [RID] answer and they need it now. Aviation needs a Single Sky."
>
> **Saulo da Silva**
> *Chief, Global Interoperable Systems Section, ICAO*[407]

> "Remote identification is a crucial first step in the development of these UTM services."
> **FAA**[408]

UAS remote ID (RID) refers to the electronic identification of a UAS in flight to provide ID, location, and performance information. Security and law enforcement have made its deployment a precondition for expanding day-to-day UAS operations. RID is widely considered "vital for safety"[409] and security,[410] "an important building block in the unmanned traffic management ecosystem [and] a critical element for building unmanned traffic management capabilities,"[411] that "will facilitate more advanced operations for UAS."[412] Additionally, RID tracking capability is expected to one day be useful in collision avoidance.[413]

In sum, RID is viewed as important to:

- Enhance UAS operator accountability and behavior for public safety purposes,
- Provide a means to assist law enforcement determine UAS compliance, including as a precondition to invoking countermeasures where a sUAS presents a credible threat,[414]
- Garner public acceptance, and

- In the future, enhance pilot situational awareness by the underlying ability to track position via separation services.[415]

There is substantial government and industry support for standing-up RID, including incentives for early U.S. adoption via "voluntary compliance" pending completion of rulemaking,[416] and for example, voluntary implementation of RID in Switzerland. Nonetheless, widespread RID implementation may take years to complete.[417]

In June 2019, the European Commission issued a regulation prescribing "direct remote identification" for certain operations that:

1. allows the upload of the UAS operator registration … and exclusively following the process provided by the registration system;
2. has a physical serial number … affixed to the add-on and its packaging or its user's manual in a legible manner;
3. ensures, in real time during the whole duration of the flight, the direct periodic broadcast from the UA using an open and documented transmission protocol, of the following data, in a way that they can be received directly by existing mobile devices within the broadcasting range:
   i. the UAS operator registration number;
   ii. the unique physical serial number of the UA …;
   iii. the geographical position of the UA and its height above the surface or take-off point;
   iv. the route course measured clockwise from true north and ground speed of the UA; and
   v. the geographical position of the remote pilot or, if not available, the take-off point;
4. ensures that the user cannot modify the data …;
5. is placed on the market with a user's manual providing the reference of the transmission protocol used for the direct remote identification emission and the instruction to:
   i. install the module on the UA;
   ii. upload the UAS operator registration number.[418]

Subsequently, the U-space regulation mandated network identification services (*see* Chapter 5, Section 5.3 *U-space*). RID also "form[s] an integral part of the FAA's regulatory framework"—as evidenced by the December 2020 release of its *Remote Identification of Unmanned Aircraft* final rule ("Remote ID Rule").[419] The rule provides three compliance options for all UA required to register:[420] (1) Standard Remote ID UA (integrated RID capability), (2) UA with Remote ID Broadcast Module (attached to or contained in the UA), and (3) FAA-Recognized Identification Areas (FRIA, preapproved operation areas).[421] The Standard Remote ID UA requirements include:

§ 89.110 Operation of standard remote identification unmanned aircraft.

Unless otherwise authorized by the Administrator, a person may comply with the remote identification requirement of § 89.105 by operating a standard remote identification unmanned aircraft under the following conditions:

a. *Operational Requirements.* A person may operate a standard remote identification unmanned aircraft only if the person operating the standard remote identification unmanned aircraft ensures that all of the following conditions are met:
   1. From takeoff to shutdown, the standard remote identification unmanned aircraft must broadcast the message elements of § 89.305.
   2. The person manipulating the flight controls of the unmanned aircraft system must land the unmanned aircraft as soon as practicable if the standard remote identification unmanned aircraft is no longer broadcasting the message elements of § 89.305.

b. *Standard remote identification unmanned aircraft requirements.* A person may operate a standard remote identification unmanned aircraft only if the unmanned aircraft meets all of the following requirements:
   1. Its serial number is listed on an FAA-accepted declaration of compliance, or the standard remote identification unmanned aircraft is covered by a design approval or production approval issued under part 21 of this chapter and meets the requirements of subpart F.
   2. Its remote identification equipment is functional and complies with the requirements of this part from takeoff to shutdown.
   3. Its remote identification equipment and functionality have not been disabled.
   4. The Certificate of Aircraft Registration of the unmanned aircraft used in the operation must include the serial number of the unmanned aircraft, as per applicable requirements of parts 47 and 48 of this chapter, or the serial number of the unmanned aircraft must be provided to the FAA in a notice of identification pursuant to § 89.130 prior to the operation.

Associated minimum message elements broadcast requirements appear in Remote ID Rule, Section 89.305:[422]

A standard remote identification unmanned aircraft must be capable of broadcasting the following remote identification message elements:

a. The identity of the unmanned aircraft, consisting of:
   1. A serial number assigned to the unmanned aircraft by the person responsible for the production of the standard remote identification unmanned aircraft; or
   2. A session ID.

b. An indication of the latitude and longitude of the control station.

c. An indication of the geometric altitude of the control station.

d. An indication of the latitude and longitude of the unmanned aircraft.

e. An indication of the geometric altitude of the unmanned aircraft.

f. An indication of the velocity of the unmanned aircraft.

g. A time mark identifying the Coordinated Universal Time (UTC) time of applicability of a position source output.

h. An indication of the emergency status of the unmanned aircraft.

Performance requirements appear in Section 89.310, respectively. The rule also presents RID design and production requirements. Significantly, Network-RID capability proposed in the [RID NPRM 2019] (providing RID communication via network) was not included in the Remote ID Rule (*see below*).

The leading remote ID and tracking standard was published in February 2020 by ASTM Committee F38 on Unmanned and Remotely Piloted Aircraft Systems, *Standard Specification for Remote ID and Tracking.*[423] Its core architecture, features, and attributes are summarized as follows.

Two methods are specified in F3411-19 to facilitate remote ID: *broadcast* and *network based.* Each permits a user, via mobile phone app or other devices/interfaces, to ascertain the identity of nearby UAS.

Broadcast: The remote ID messages are broadcast directly from a transmitter on the UAS using unlicensed spectrum. The broadcast method requires no network or ground infrastructure. The receiver must be within radio range to ID the aircraft. It could also serve as a back-up for networked UAS that lose connectivity. Broadcast is viewed as essential in remote areas without mobile network service.[424]

Network: Network-based remote ID uses an internet connection to send RID data from the UAS (including ground station) to a networked server that in turn responds to queries for ID information about UAS. The internet connection will generally be provided by existing cellular/mobile communications infrastructure, but could be provided by another means as well (such as a local private wireless or satellite network). The networked ID servers that respond directly to user queries are able to gather UAS ID information from other nearby ID servers via a discovery mechanism. *See* Chapter 2, Section 2.4.5.2 *Discovery and Synchronization Service.*

Broadcast and network-based RID can complement each other. In higher population density areas network-based RID generally has a range and interference handling advantage. However, in more rural areas, broadcast RID can have an advantage in that it does not rely on any ground infrastructure.[425] Network-based RID is considered advantageous for certain types of operations, such as BVLOS, and public safety officials can identify aircraft in areas within which such officials are not physically present. The specification also defines a mechanism by which non-equipped UAS such as model aircraft can participate in networked RID by reporting the location and time of an operation in advance.

Although current U.S. RID rules have not yet adopted network-based RID (due, in part to complexity and public opposition),[426] network-based remote ID informs UTM architecture, services, standards, and anticipated future rulemaking. Nevertheless, "[t]he FAA notes that this rule does not preclude industry from establishing Remote ID USS-like networks."[427] Additionally, "the FAA notes that some aspects of ASTM F3411-19 may need to be revised or updated as a result of the requirements of th[e] final rule [before evaluation] as a potential means of compliance for [RID]."[428] The ASTM RID WG chair offered the following perspective on the Remote ID Rule's approach:

> It seems the FAA prefers the "baby step" to meet the RID requirements rather than eating the whole elephant initially (as suggested in the NPRM). This seems to be aligned with Europe and other regions as well. Given the current landscape, I think the FAA made the right decision as to their duty for RID. Rather than shoehorning UTM as a component of a mandated RID, I think they will allow UTM to prove itself as a (non-mandated) industry necessity. Then, we'll revisit it for the purpose it is intended: planned traffic management. This will allow UTM (and requisite federation of network services) to evolve at a natural/healthy pace and allow RID to expedite with fewer ground dependencies.[429]

ASTM RID mechanisms in the RID specification serve to remove anonymity while preserving privacy. They provide for the public to identify and locate proximate UAS without public disclosure of the operators' PII, while at the same time making full name, address and other PII of operators and pilots available to law enforcement through privileged database access. Figure 2.11 depicts the parties and associated RID interfaces as structured by ASTM F38:[430]

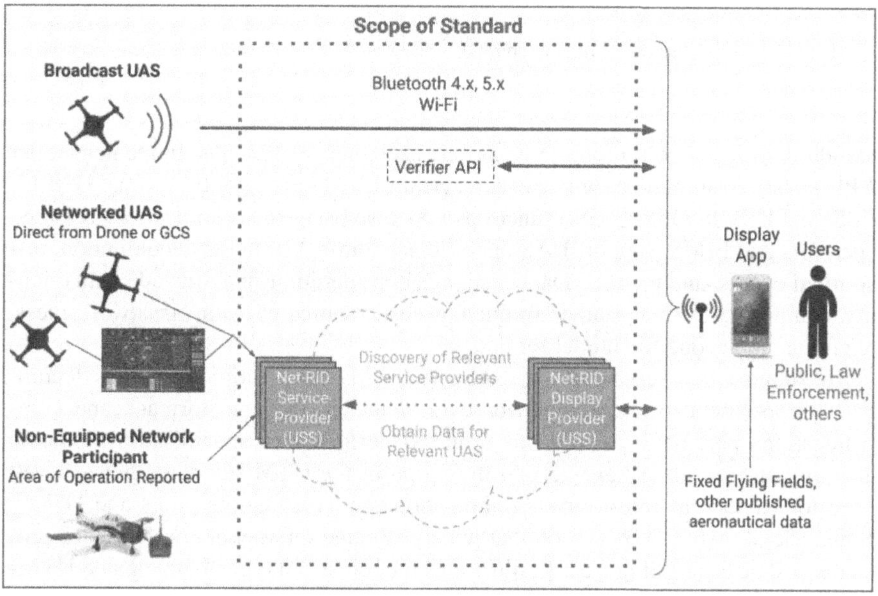

FIGURE 2.11 ASTM Remote ID and Tracking–Scope of Standard. (Reprinted, with permission, from ASTM F3411-19 Standard Specification for Remote ID and Tracking, copyright ASTM International, 100 Barr Harbor Drive, West Conshohocken, PA 19428.)

The "Scope of Standard" dotted rectangle (above figure) delineates the protocols and relationships addressed by the standard whereas the protocols from the UAS, including associated control stations, and display apps are outside of the standard. The standard specifies protocols for broadcast UAS (Bluetooth and Wi-Fi), as well as the IP-based networked communications between USS, DSS, and other specified providers. As noted above, networked RID relies on a discovery processes for initial communications with the DSS to find other aircraft in the area followed by peer-to-peer communications among Net-RID Display Providers and Net-RID Service Providers for relevant RID data.

Finally, the ASTM RID standard advances authentication evolution and interoperability by providing a pluggable authentication transport mechanism that is flexible enough to allow various authentication techniques such as asymmetric (public key) and symmetric key methods as well as methods that have not yet been developed. This allows the receiver to validate the UAS unique ID association with UAS and also allows the manager (or CAA) for a given airspace to specify the verifier and therefore the authentication requirements for that region.[431] Such methods should be harmonized with UTM, Internet, and developing International Aviation Trust Framework standards and guidance. *See* Chapter 2, Section 2.4.7.2 *International Aviation Trust Framework*.

2.4.9 UTM Data Exchange

"UTM is predicated on … data exchange."

FAA[432]

Standardized protocols, represented as data exchange formats, and implemented per APIs, enable communications between certain UTM participants, such as USS-to-USS, and USS-to-DSS.[433] USS functions may also bridge communications between operators and FIMS, support UAS operations planning, strategic deconfliction, off-nominal events, and more.[434] Data exchange considerations may include data quantity, complexity, format, update frequency, cost, "source of truth," trustworthiness, retention, and economic model for data use.

To support data exchange, the ASTM UTM specification describes technical interoperability requirements for data elements, message formats, and communications parameters as well as required performance specifications.[435] Some data exchange protocols are beyond the scope of the UTM standards being developed today, such as those for communications between remote pilots and USS.[436] Nonetheless, there is expectation that widely adopted internet and web service protocols will be used.[437] Figure 2.12 illustrates the scope and type of anticipated UTM-relevant data exchanges except for discovery services (*see* Chapter 2, Section 2.4.5.2 *Discovery and Synchronization Service*).[438]

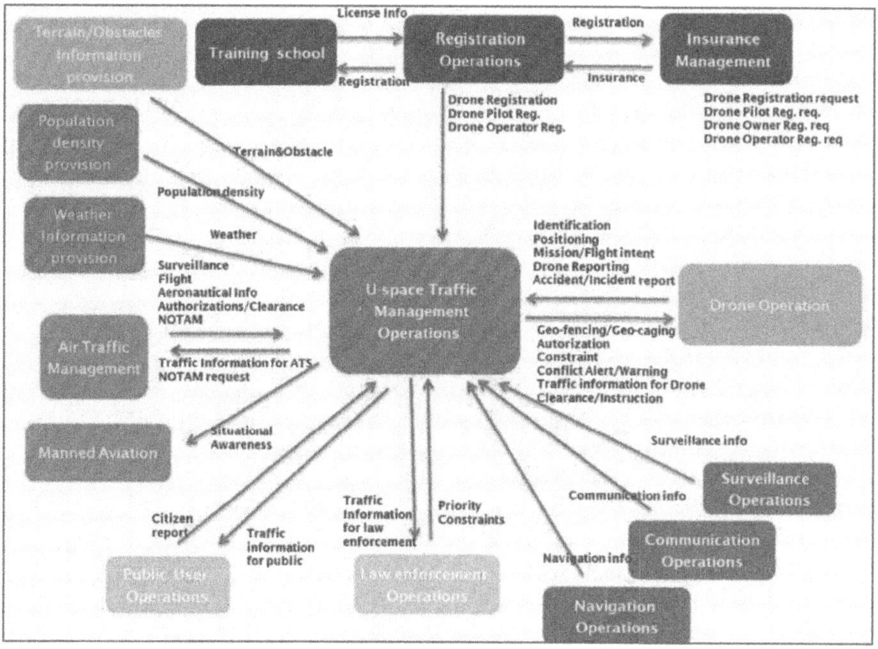

FIGURE 2.12 Notional U-space data exchange. (Reprinted with permission of SESAR JU, CORUS ConOps (2019): 82.)

SDSP data exchange standards are in development including for weather, surveillance, and mobile network operators (MNOs) coverage/availability (e.g., for preflight route network health maps).[439] Additionally, for example, the National Academies of Sciences, Engineering, Medicine recommended that "organizations should prioritize research on the protocols, data formats, and *data exchange standards* that support advanced aerial mobility vehicles in a geospatial real-time system supporting safety-critical operations across the National Airspace System.'"[440] Since AAM is dependent on many data resources from the same or similar entities (such as weather and terrain), such research should inform and advance the state of UTM. The sheer volume of the potential data exchange will require careful management, particularly in its interface with ATM.[441] The technical interactions between ANSPs and USS are sometimes presented in interface control documents (ICD) and APIs. SWIM may facilitate certain UTM data exchanges. *See* Chapter 2, Section 2.4.4 *System-Wide Information Management*. Finally, NASA has launched the Air Traffic Management-eXploration (ATM-X) project, Digital Information Platform (DIP) sub-project to develop an "easily accessible cloud-based Digital Information Platform, and ICAO has commenced an exploratory AIM for UTM project with expectation to initially develop guidance.[442]

3 Standards-Making for UTM

"UTM interoperability is all about the standards."

Lorenzo Murzilli
Murzilli Consulting[443]

"The most important thing is the validation of the standards."

Maria Algar Ruiz
EASA Drone Programme Mgr.[444]

3.1 GENERAL—STANDARDS-MAKING FOR UTM

As a complex system with safety a key variable, UTM requires standards, and standards organizations have engaged increasingly in UTM-related initiatives. The range of technologies, systems, and protocols that must cooperate, communicate, integrate, or otherwise interoperate to stand-up a UTM is complex and cross-cutting. For the purposes of this chapter, "standards-making" is viewed broadly to include the development of frameworks, guidelines, recommended practices, information reports, minimum operational performance standards (MOPS), minimum aviation system performance standards (MASPS), roadmaps, and specifications.

Aviation regulations rely increasingly on standards approved or recognized by the regulator, and CAAs increasingly engage in their formation via acceptable means of compliance–AMC–or similar methodologies.[445] Nonetheless, it is widely understood that "only those standards that are acknowledged as absolutely necessary to secure safety, efficiency and airspace integration should be developed,"[446] and "sufficient time should be given to [countries] and regions to test and validate UTM concepts and solutions before developing standards."[447]

The trend is toward industry consensus performance-based (rather than prescriptive) standards both in UTM and aviation generally.[448] "Performance-based standards are technology agnostic, enabling industry to continue to innovate and iterate current technologies to improve equipment."[449] However, as Reinaldo Negron, Head of UTM for Wing put it, "writing performance-based standards is hard."[450] The following organizations and initiatives demonstrate UTM-relevant standardization. Invariably other organizations have or will become involved.

3.1.1 ANSI

Responding to the need for coordination amid the proliferation of new UAS standards-making globally, and with the support of the FAA and industry, the American National Standards Institute (ANSI), the sole US member of the

International Organization for Standardization (ISO), convened the Unmanned Aircraft Systems Standardization Collaborative (UASSC). The UASSC has developed the *Standardization Roadmap for Unmanned Aircraft Systems* ("ANSI Roadmap").[451]

ANSI is not itself a standards development organization (SDO). Rather, its role is as "a neutral facilitator" to address coordination in areas of emerging technologies.[452] The UASSC, with the input and cooperation of the broader UAS standards-making community, surveyed UAS standards activities including UTM standards and associated standards initiatives, considered what standards are needed to facilitate integration of UAS into the airspace, and helped to inform industry resource allocation decisions related to participation in standards development.[453] The UASSC's roadmap describes published standards as well as those in development and seeks to identify gaps. The UASSC identified sixty gap areas and forty high-priority gaps (including, e.g., C2 link, DAA, cybersecurity, and operational risk assessment), and identified a gap in UTM standardization specifically as follows:

Gap O7: UTM Services Performance Standards. UTM service performance standards are needed.

R&D Needed: Yes. Considerable work remains to develop the various USS services listed as well as testing to quantify the level of mitigation they provide. Only after some level of flight testing to define the "realm of the possible" can the community of interest write performance-based standards that are both achievable and effective in mitigating operational risk.

Recommendation: There is quite a lot of work for any one SDO. A significant challenge is finding individuals with the technical competence and flight experience needed to fully address the subject. What is needed is direction to adopt the performance standards evolving from the research/flight demonstrations being performed by the research community (e.g., NASA/FAA RTT, FAA UTM Pilot Project, UAS Test Sites, GUTMA, etc.). Given a draft standard developed by the experts in the field (i.e., the ones actively engaged in doing the research), SDOs can apply their expertise in defining testable and relevant interoperability and performance-based requirements and thus quickly converge to published standards.

Priority: High (Tier 2)

Organization(s): NASA, FAA, ASTM, ISO, IEEE, EUROCAE, JARUS[454]

3.1.2 EUSCG

The European UAS Standards Coordination Group (EUSCG) is a joint coordination and advisory group established to coordinate UAS-related standardization activities across Europe, precipitating from EU regulation and EASA rulemaking initiatives. It provides a bridge from European to international standardization activities,[455] and develops, monitors, and maintains "an overarching European UAS standardization rolling development plan" that addresses UTM.[456] EUROCAE chairs the EUSCG and ensures its secretariat.

3.2 STANDARDS DEVELOPMENT ORGANIZATIONS

The following is a selected list of organizations that develop, maintain, advance, or otherwise directly contribute to UTM standards development.

3.2.1 ASD-STAN

ASD-STAN, an associated technical body to CEN (European Committee for Standardization) for Aerospace Standards,[457] is "involved in U-Space as a result of the Delegated Act (EU 2019/945) being a single market (CE marking) regulation. By law, standards related to CE marking must be handled by CEN/CENELE, ETSI or an affiliated body. EUROCAE 105/33 for instance can be used directly by EASA for establishing geo-zones (Implementing Act 2019/947, Art. 15) but everything related to the Delegated Act must still process through ASD-STAN."[458] ASD-STAN initiatives relevant to UTM include, for example:

- **Domain DO5 Autonomous Flying**—the working group addresses both "entirely autonomously or remotely controlled" UAS,[459] and
- **DO5/WG08, UAS Unmanned Aircraft Systems**—includes, but is not limited to, "classification, design, manufacture, operation (including maintenance) and safety management of UAS operations."[460] The WG has produced the Direct Remote Identification series of standards.[461]

Finally, ASD-STAN and ASTM International signed a MoU (October 2020) for close collaboration on Remote ID, information exchange, and broader technical standards alignment.[462]

3.2.2 ASTM International

> "We're pulling the brain trust together to work on the interoperability issues and are chipping away at the problem."
>
> **Philip M. Kenul**
> *Chair, ASTM Int'l, F38*[463]

ASTM International's F38 Committee on Unmanned and Remotely Piloted Aircraft Systems ("F38") develops wide-ranging UAS standards within its three substantive subcommittees: F38.01 Airworthiness, F38.02 Flight Operations, and F38.03 Personnel Training, Qualification and Certification.[464] These F38 subcommittees are hardware, procedure, and people oriented respectively, and address "issues related to design, performance, quality acceptance tests, and safety monitoring,"[465] much of which contributes to the UTM ecosystem. Through collaboration between F38, other ASTM aviation committees,[466] and the global community, ASTM has become the leading UTM standards-making organization, having "seen its relevance and influence in the ecosystem skyrocket …. "[467] Its initiatives that touch on UTM include, nonexclusively, the following:

- **UTM**—*New Specification for UAS Traffic Management (UTM) UAS Service Supplier (USS) Interoperability,*[468] *and New Specification for Surveillance UTM Supplemental Data Service Provider (SDSP) Performance*[469]
- **RID**—*Standard Specification for UAS Remote ID and Tracking*[470]
- **OOP**—*New Specification for Operation Over People*[471]
- **BVLOS**—*Standard Practice for Seeking Approval for Beyond Visual Line of Sight (BVLOS) Small Unmanned Aircraft System (sUAS) Operations, and New Specification for Positioning Assurance, Navigation, and Time Synchronization for Unmanned Aircraft Systems.*[472]
- **DAA**—*Standard Specification for Detect and Avoid Performance Requirements,*[473] *and New Test Method for Detect and Avoid*[474]
- **WX**—*New Specification for Weather Supplemental Data Service Provider (SDSP) Performance*[475]

Additionally, other F38 standards and collaborative initiatives within ASTM also contribute to UTM.[476] Finally, because UTM is first an aviation system that also intersects with IT, telecommunications, and other areas, SDOs with no aviation background have been learning how aviation treats safety priorities, having to "adjust to the requirements of aviation rather than aviation adjusting to them. We don't want to fracture a well-established safety culture." observes Philip Kenul, Chair, ASTM F38.[477]

3.2.3 Cellular Industry

"There really is no other alternative technology supporting the scale needed."

Jay Merkle
Exec. Dir., UAS Integration Office, FAA[478]

Aviation communication has traditionally used protected spectrum and communication technologies designed to provide voice communication between humans—the pilots in the cockpit, UA operators, and traffic controllers on the ground. But this communications paradigm doesn't work for UTM operations' machine-to-machine communications and increasing automation. Moreover, available spectrum for manned aviation is limited and cannot feasibly accommodate the anticipated low altitude, high volume operations, creating a potential impediment to expansion of UTM operations. However, cellular networks offer UTM operations an attractive alternative. In fact cellular connectivity is already a key technology in the drone industry for support of coming identification and tracking services, command and control communications, and payload communications. "BVLOS and wide area connectivity are a natural fit, and although cellular is not 100% ubiquitous, it does match population density well, which is also highly correlated to the most valuable airspace."[479]

While mobile cellular networks are typically considered a means of communication, they can also have characteristics as an interconnected, distributed network of physical sites with radio antennas and local computing resources for UTM. Seen this way, broader aviation uses become apparent. Among these are positioning (including GPS backup), supplementary data services, monitoring of local conditions (such as traffic, RF environment, and weather), nesting/charging sites, and edge computing for both flight and mission tasks.[480]

Over the past 30 years, the mobile industry has demonstrated its ability to transform society through 2G, 3G, and 4G, and today the industry supports a staggering 5 billion unique subscribers with 9 billion connections globally.[481] Since the introduction of LTE (which is fully IP-based), mobile networks have offered diverse services, not just for consumer handsets, but also for smart city infrastructure, automotive applications, public safety, and the industrial sector. "The LTE Network is critical for both commanding the UAV and for conducting conformance modeling (surveillance) once the operation moves beyond RF [l]ine of [s]ight."[482]

The now-emerging network technology standard, 5G, builds on these successes and allows mobile network operators (MNOs) to deliver the networks and platforms that support existing and new services, with new business models and use cases.[483] 5G has been described as more than just a generational step, representing a fundamental transformation of the role that mobile technology can play in society. 5G promises to mark an inflection point in the future of communications, bringing instantaneous high-powered connectivity to billions of devices communicating without human intervention, also known as Internet of Things (IoT).[484] Consider the following "eight currencies" of 5G, as Verizon calls them,[485] that can create an agile, purpose-built network tailored to the different needs of citizens, businesses, and the economy, including UTM:

1. Throughput—eventually expecting 10 Gbps
2. Service Deployment—via software-based network virtualization
3. Mobility—accommodating fast-moving vehicle operation and hand-off
4. Connected Devices—up to 100× more connections per km^2 than 4G
5. Energy Efficiency—10% of the energy requirements
6. Data Volume—up to 10 $TB/s/km^2$
7. Latency—supporting, e.g., autonomous driving and robotics
8. Reliability—an essential attribute.

MNOs have focused near-term 5G deployments primarily on enhancing performance in areas of high demand, such as highly dense urban areas, campuses, and stadia, etc. Because it will be a phased rollout, 4G will coexist with 5G services for the foreseeable future.[486] One noteworthy capability of 5G, the network exposure function, will enable MNOs to have applications like UTM interact with the network core for the resources needed specific to that application. While an emerging network capability with associated challenges,[487] 5G provides secure, dynamically changeable quality of service and network optimization that should enable UTM.[488]

Diverse stakeholders across the globe have been working since 2012 to define and shape what 5G should be. For example, the International Telecommunications Union Radiocommunications Sector (ITU-R) (*see* Chapter 3, Section 3.2.10 *ITU*), has defined 5G design goals, as summarized in Table 3.1. Such design goals represent

TABLE 3.1
—5G Design Goals[490]

Requirement		Value
Data rate	Peak	Downlink: 20 Gb/s Uplink: 10 Gb/s
	User experience	Downlink: 100 Mb/s Uplink: 50 Mb/s
Spectral efficiency	Peak	Downlink: 30 bit/s/Hz Uplink: 15 bit/s/Hz
	5th percentile user	Downlink: 0.12~0.3 bit/s/Hz Uplink: 0.045~0.21 bit/s/Hz
	Average	Downlink: 3.3~9 bit/s/Hz Uplink: 1.6~6.75 bit/s/Hz
Area traffic capacity		10 Mbit/s/m2
Latency	User plane	1 ms~4 ms
	Control plane	20 ms
Connection density		1,000,000 devices per km^2
Energy efficiency		Loaded: see average spectral efficiency No data: Sleep ratio
Reliability		This is 1–10^(–5) success probability of transmitting a layer 2 PDU (protocol data unit) of 32 bytes within 1 ms
Mobility		0 km/hr~500 km/hr
Mobility interruption time		0 ms
Bandwidth		100 MHz

(*Source:* Reprinted with permission of the ITU.)

an ideal situation. Performances of deployed systems may vary. Additionally, the implications of geopolitics on 5G cannot be ignored.[489]

Standards collaboration between the UTM and the cellular industry is recognized as essential, and has been formalized[491] by 3GPP, GSMA, and the Global UTM Association (GUTMA),[492] as described below (*see* Chapter 3, Section 3.2.3.3 *ACJA*).

3.2.3.1 3GPP

Founded in 1998, "[t]he 3rd Generation Partnership Project (3GPP) unites (Seven) telecommunications standard development organizations … to produce the Reports and Specifications that define 3GPP [mobile network] technologies [and] providing complete system description."[493] At the time of 3GPP's creation, there were several non-interoperable regional standards, however the mobile network community understood the need for interoperable technology. The consolidation of a single authoritative worldwide standards body drove harmonized global standards development and success of the mobile industry. Perhaps this history could be a lesson for the patchwork of standards bodies in UTM, and aviation in general today.

3GPP's UTM-relevant initiatives include work completed in Release 15 such as the study of UAV support in LTE and several general supporting functions developed as part of 5G.[494] Some additional requirements work, including those for UAS remote ID (RID) was done as part of Release 16.[495] Architectural and protocol work to support RID is expected as part of Release 17[496] and should reflect upcoming regulations in many countries.

However, "many of these services have, until now, been developed in isolation,"[497] where the cellular industry tried to satisfy the perceived needs of aviation, but due to the complete separation of the two industries from a standards perspective, could not fully meet those needs.[498] Traditionally 3GPP's membership came from the telecommunications field and lacked participation from the various vertical markets, and in particular, the aviation community. The challenge is for 3GPP, as a community of more than 1,000 experts in cellular networks capable of providing the most suitable technical solution, to understand the needs from the aviation industry. "Some of these [technical solutions] represent opportunities (or even requirements) to optimize cellular for aviation applications."[499]

The need for "better coordination" between the aviation and UTM communities and 3GPP is now recognized.[500] The sheer scale and importance of cellular communications for BVLOS, other UTM-enabled operations and network intelligence[501] have prompted mobile cellular industry collaboration,[502] as described below.

3.2.3.2 GSMA

> "[M]obile Network Operators can and will play a key role in the emerging UAS and UTM ecosystem."
>
> **GSMA[503]**

GSMA[504] represents the interests of MNOs worldwide. It is a non-profit membership association uniting more than 750 operators with almost 400 companies in the broader mobile ecosystem. "GSMA is a key organization in addressing the business issues associated with any network build out required to better support large scale deployment of UAS."[505] Founded in 1995 to support mobile operators using GSM as the first recognized global standard for cellular communication, GSMA has been the custodian of interoperability, roaming, and settlement. It also manages the Type Allocation Codes (TAC) that form the first part of the device identification IMEI (International Mobile Equipment Identity) ranges to all 3GPP-compliant device manufacturers, in addition to other related services.[506] Graham Trickey, formerly Head of GSMA IoT Programme, describes their interest in drones:[507]

> The GSMA Internet of Things Programme[508] established a Drone Interest Group in 2016, leveraging the decennial experience in IoT with the vision of everybody and everything connected for a better sustainable future. We will see safe and secure connectivity between all these items. We already see connected cities and automobiles and in the near future we will see a new dimension with drones.

With engagement in the GUTMA Annual Conference 2019, and the formation of ACJA jointly with GUTMA, GSMA has enhanced collaboration with the UTM community.[509]

3.2.3.3 ACJA

GUTMA and GSMA jointly hosted a conference entitled *Connected Skies* in June 2019 that focused on closing the gap between the aviation and mobile network communities. This led to a joint cooperation agreement between these organizations signed in late 2019, and the formation of the first joint activity commencing 2020 called Aerial Connectivity Joint Activity (ACJA) to help align UTM and mobile industry standardization.[510]

ACJA's technical program of work is robust, seeking to advance specific methods and metrics through its initial work tasks that include:

WT1: 3GPP coordination with aviation stakeholders
WT2: Interface for data exchange between MNOs and UTM ecosystem
WT3: Standard Aerial Service Profile
WT4: Translating Cellular to Aviation Standards: MOPS/MASPS.[511]

3.2.4 CTA

The Consumer Technology Association (CTA) produced the widely adopted ANSI/CTA 2063-A, *Small Unmanned Aerial Systems Serial Numbers* standard.[512] This standard is implemented in remote ID standards supporting UTM such as the ASTM *Standard Specification for Remote ID and Tracking*.[513] This serial number standard is referenced by EASA, and by the FAA in its Remote ID Rule. Drawing upon CTA's initiative to develop a *Baseline Cybersecurity Standard for Devices and Device Systems* (ANSI/CTA-2088),[514] it is also developing a supplement to address sUAS cybersecurity.[515] "The standard will build upon the baseline cybersecurity requirements in CTA-2088 to address … the unique capabilities, uses, and applications of sUAS."[516]

3.2.5 EUROCAE

The European Organisation for Civil Aviation Equipment (EUROCAE), a non-profit European aviation standardization forum, is developing diverse and extensive UAS-related standards within six focus areas.[517] Its Work Group 105 (WG-105), Unmanned Aircraft Systems (UAS), addresses "safe operation of UAS in all airspace, at all times and for all types of operations," and its primary UTM initiatives are undertaken within the following WG-105 sub-groups (SG):[518]

- **WG-105 SG-13** DAA for VLL, with the development of an Operational Services and Environment Description (OSED) and a MOPS for the DAA function in very low level (VLL) (ED-267),[520]
- **WG-105 SG-31**/UTM - General, with the development of a Workplan, identifying next standardization activities needed in support of U-space increments 2 to 4,

- **WG-105 SG-32**/UTM E-Identification, with the development of a MOPS for UAS e-identification (ED-282),
- **WG-105 SG-33**/UTM Geo-Fencing, with the development of a MOPS for geo-fencing (ED-269), and a MOPS for geo-caging (ED-270),
- **WG-105 SG-61** SORA Support, with the development of a Workplan, identifying next standardization activities needed to support the implementation of the SORA methodology and future Standard Scenarios,[519] and
- **WG-14** Environment, SG-1, Developing recommended practices for testing of UAS ground segment equipment to environmental conditions, in close collaboration with WG-105.

3.2.6 ICAO

The International Civil Aviation Organization (ICAO),[521] a specialized agency of the United Nations, is the world's governing body for international aviation. ICAO has provided RPAS[522] and general UAS leadership for many years to ensure safety and uniformity in international aviation operations[523] through various initiatives including studies, development of guidance materials, standards, development of the UAS Toolkit, and publication of Standards and Recommended Practices (SARPs). Recognizing that sUAS and UTM developments are proceeding at a breathtaking pace, affecting international flight operations, and that most sUAS are unable to meet the basic tenets of the Chicago Convention,[524] ICAO found its traditional approach to standards development was inadequate. In response to this finding, it established the Unmanned Aircraft Systems Advisory Group (UAS-AG) in 2015 to support the RPAS Secretariat in developing guidance materials intended to assist countries with regulating unmanned aircraft systems (UAS).[525]

To help ICAO member states establish "a common global framework for, and core boundaries of … UTM, in order to allow further UTM developments to focus on better defined issues, whether technical, operational or legal,"[526] the UAS-AG began work on a framework for UTM document in 2017, and released *UTM—A Common Framework with Core Boundaries for Global Harmonization*, Edition 1, in April 2019.[527] The second edition of the ICAO UTM Framework, issued in November 2019, added two appendices addressing ATM and UTM systems information exchange: *E. UTM-ATM Boundaries and Transition*, and *F. Essential Information Exchange between UTM and ATM systems*, and the third edition issued in September 2020 provided further updated content.[528] The UAS-AG is developing a UTM Toolkit. ICAO also organizes annual DRONE ENABLE symposia to inform further ICAO action and catalyze nation state and private-public UTM development and harmonization.[529] The FAA has seen a "significant pivot at ICAO, recognizing there is a larger role to play than its traditional focus on cross-border issues."[530]

Separately, ICAO's Air Navigation Commission recommended development of SARPs, guidance or best practices related to UTM, "including autonomous operations, after states and regions have had sufficient time to test and validate concepts."[531] To the extent that ICAO acts consistently with its typical SARP development, it will wait until after there has been ample time to do so, including with regard to its Annexes.[532]

ICAO is also engaged in cybersecurity initiatives that support UTM, undertaken within its Trust Framework Study Group's three working groups: Digital Identity Working Group (DIWG), Trust Reciprocity Operational Needs (TRON), and Global Resilient Aviation Interoperable Network (GRAIN), and the future governance of cybersecurity matters within ICAO is an active area of inquiry.[533] *See* Chapter 2, Section 2.4.7.2 *International Aviation Trust Framework*. Additionally, a presentation at ICAO's 40th Session Plenary (Sept. 2019) informed member states of the contribution of mobile cellular networks for future aviation ground infrastructure, with a focus on UTM.[534]

Certain work of the UAS-AG was proposed for transition into ICAO's Global Air Navigation Plan (GANP),[535] but this proposal was rejected. Finally, ICAO's RPAS Panel should remain engaged[536] regarding potential UTM implications, and other Air Navigation Commission technical panels will likely find a nexus to UTM as the UTM capability starts overlapping with ATM capabilities/services.[537] Both the RPAS Panel and the various ICAO technical panels are expected to benefit from the work being conducted by the UAS-AG.

3.2.7 IEEE

The Institute of Electrical and Electronics Engineers (IEEE) is a leading developer of industry standards,[538] with considerable focus on the device or sub-device level. It engages in UTM-relevant standards-making, including IEEE SA P1939.1—*Standard for a Framework for Structuring Low Altitude Airspace for Unmanned Aerial Vehicle (UAV) Operations,* defining "UAV capabilities and related infrastructure for UAVs to operate in and comply with low altitude air space regulations."[539] The chairman (of WG IEEE SA P1939.1) described his working group as "working on definition and construction of low altitude airspace structuring with pre-planned UAV public air routes through GIS technology etc., which indeed includes UTM management and operation."[540] Collaboration among this work group (P1939.1) with those of P1936.1—*Standard for Drone Applications Framework*, and P1937.1—*Standard Interface Requirements and Performance Characteristics for Payload Devices in Drones,* may further support UTM.[541]

Other IEEE initiatives that may benefit UTM include: P1920.1—*Aerial Communications and Network Standards,* "applicable to manned and unmanned, small and large, and civil and commercial aircraft systems," P120.2—*Standard for Vehicle to Vehicle Communications for Unmanned Aircraft Systems*,[542] and an approved initiative to develop power line inspection guidance.[543] IEEE UAS work is undertaken within its Unmanned Aerial Systems and Unmanned Aerial Vehicles Standards Committee (AES/UAS/UAV/SC), and Traffic Enforcement Technologies working group (IM/TET TC41).[544]

3.2.8 IETF

The Internet Engineering Task Force (IETF) produces essential, recognized technical standards in the form of RFCs, which underlie the design, use, and management of the Internet.[545] As such, its work product is relevant in the developing UTM, including various security supporting protocols, e.g., OAuth (RFCs 6752, 8252), and public key infrastructure (RFCs 4210, 4211, etc.). One "Applicability Statement for various IETF Technical Specifications" urges that:

> IETF can help by providing expertise as well as mature and evolving standards. Host Identity Protocol (HIPv2) [RFC 7401] and the Domain Name System (DNS) [RFC 2929] can complement emerging external standards for UAS RID, to facilitate utilization of existing and provision of enhanced network services, and to enable verification that UAS RID information is trustworthy (to some extent, even in the absence of Internet connectivity at the receiving node).[546]

The Drone Remote ID Protocol (drip) working group was formed in the IETF Internet Area with the goal to "specify how RID can be made trustworthy and available in both Internet and local-only connected scenarios, especially in emergency situations."[547] Responsive draft standards are in development.[548] It is envisioned that IETF's role in UAS, including UTM will continue to evolve.

3.2.9 ISO

The International Organization for Standardization (ISO), Technical Committee (TC) Special Committee (SC) 16 (ISO/TC 20/SC 16) addresses "[s]tandardization in the field of unmanned aircraft systems (UAS) including, but not limited to, classification, design, manufacture, operation (including maintenance) and safety management of UAS operations."[549] SC 16's Working Group 4 on *UAS Traffic Management* has focused on UTM standards and guidelines that are "to be aligned with the rules and guidance provided by aviation authorities."[550] Its UTM work program includes:[551]

ISO 21384-3, *Unmanned aircraft systems—Part 3: Operational Procedures*[552]
ISO/TR 23629-1, *UAS traffic management (UTM)—Part 1: General requirements for UTM—Survey results on UTM*[553]
ISO/WD 23629-5, *UAS traffic management (UTM)—Part 5: UTM functional structure*[554]
ISO/AWI 23629-7, *UAS traffic management (UTM)—Part 7: Data model for spatial data* (under development)[555]
ISO/WD 23629-12, *UAS traffic management (UTM)—Part 12: Requirements for UTM services and service providers.*[556]

Additionally, ISO/IEC JTC1/SC17/WG12, *Cards and security devices for personal identification* has undertaken a project ISO/IEC AWI 22460-1 2017-05-03, *ISO License and Drone Identity Module for Drone,*[557] and work within ISO/IEC JTC 1/SC 27, *Information security, cybersecurity and privacy protection* develops diverse InfoSec management standards and materials supporting distributed information systems[558] that may benefit UAS remote ID and UTM. ISO standards may eventually support some nation states' deployment of UTM, leveraging ASTM and EUROCAE standards that have attracted the engagement of leading CAAs and ANSPs. John Walker, Chair of ISO/TC 20/SC 16, envisions its future UTM standards to include: "General requirements for UTM, UTM architecture, UAS air traffic management, UTM data and information, and requirements for UTM service providers, aligned with the guidance provided by authorities."[559] ISO has also released UAS operational procedures standards.[560]

3.2.10 ITU

The International Telecommunications Union (ITU) is the United Nations organization responsible for the coordination and allocation of radio frequency spectrum bands. Its work is structured around its World Radio Conference (WRC) held every three to four years.[561] UTM is dependent on reliable, if not persistent, low latency radio communications. And yet, because spectrum is a finite resource under increasing demand, and "no single frequency range satisfies all the criteria required to deploy" mobile capabilities,[562] the ITU's work is essential to UTM. Much of ITU's relevant work is undertaken within its Radiocommunications Sector (ITU-R), and its International Mobile Telecommunications (IMT) for 2020 initiatives. *See* Chapter 4, Section 4.5 *Spectrum.*

3.2.11 JARUS

The Joint Authorities for Rulemaking on Unmanned Systems (JARUS) "is a group of experts from national aviation authorities (NAAs) and regional aviation safety organizations [developing recommendations for] a single set of technical, safety and operational requirements for the certification and safe integration of Unmanned Aircraft Systems (UAS) into airspace and at aerodromes."[563] JARUS was created in 2008 as a group of National Aviation Authorities, with the aim to share their experience in the nascent sector of the civil UAS certification and operations, and to develop common guidelines. An important evolution of JARUS has occurred with the involvement of the industry stakeholders organized under the Stakeholders Consultation Body (SCB), which supports the JARUS work by providing advice on activities and deliverables, and by providing subject matter experts as necessary. JARUS has developed its work for years within its seven working groups.[564] A general structural change to the JARUS framework, together with the 2021 election of a new Chair,[565] includes a new vision and mission, more flexible working organization,[566] an informal interface with ICAO's Air Navigation Bureau, support for UTM planning, and a new work programme addressing, among others, autonomy, and flight rules. Although JARUS does not presently have a specific UTM working group, various JARUS initiatives and publications inform UTM development and deployment.

JARUS's WG *Safety and Risk Management* - SRM (former WG6) has developed the Specific Operations Risk Assessment (SORA).[567] (*See* Chapter 4, Section 4.2.7 *Specific Operations Risk Assessment*). The SORA methodology has been adopted by several authorities, as well as EASA, as a means to assess and approve UAS operations in CAT B ("Specific" category) based on the results of a holistic and operation-centric risk assessment, that balances design and operational mitigations.[568] Its Guidelines on Specific Operations Risk Assessment (SORA) - Annex H, *Unmanned Traffic Management (UTM) implications for SORA*[569] is of particular relevance to operationalizing UTM – for both operators and service providers that may provide services to support operators to comply with safety requirements coming from the risk analysis. WG SRM has also created a cybersecurity subgroup charged with looking at SORA "through a cyber lens" and recommending risk-based and proportionate mitigations to cyber risks.[570] Other UTM-relevant initiatives within JARUS include WG 4's (DAA) input to WG 6 in defining UTM contributions to Air Risk Class treatment in SORA, the Operational Safety Objective #13 Cybersecurity Considerations Related to External Services (e.g., UTM and U-space) within Annex E, *Integrity and assurance levels for the Operational Safety Objectives (OSO)*,[571] finally the upcoming WG 5's White Paper on Use of Mobile Networks to Support UAS Operations, which is expected to be published in 2021.

3.2.12 NATO

While the North Atlantic Treaty Organization (NATO) is not directly active in civil aviation, its influence extends to UTM. NATO plays an important role in the harmonization of standards relating to military aviation, has done extensive standards development addressing unmanned systems, and may be a key stakeholder in discussions surrounding development and deployment of UTM systems.[572]

3.2.13 RTCA

RTCA (formerly the Radio Technical Commission for Aeronautics) provides a key venue for public-private development of minimum operational performance standards (MOPS) in aviation and avionics affecting the NAS.[573] It has played a pioneering role in enabling UAS, particularly via DAA and command and control (C2) standards, focused largely on civil UAS transitioning to and operating IFR in Class A airspace.

Formally, RTCA reports that it is *not* addressing UTM.[574] Nonetheless, "although not specifying performance standards in the UTM environment, the interoperation [with UTM] has to be considered."[575] Moreover, UTM standards development appears to be at least influenced by various RTCA initiatives and standards, particularly by work undertaken in two of its SCs:

SC-228—*Minimum Performance Standards for Unmanned Aircraft Systems.*[576] SC-228 is addressing DAA equipment and C2 data link MOPS within its two working groups (WG1 DAA, and WG2 C2). Its work includes authoritative definitions for *well clear.*[577]

SC-147—*Traffic Alert & Collision Avoidance System (TCAS).*[578]

Among RTCA's current standards that may influence UTM:

DO-200B—*Standards for Processing Aeronautical Data.*[579] This standard may influence the development of future assessment regimes for UTM.[580]

DO-178C—*Software Considerations in Airborne Systems and Equipment Certification.*[581]

DO-278A—*Software Integrity Assurance Considerations for Communication, Navigation, Surveillance and Air Traffic Management (CNS/ATM) Systems.*[582]

DO-362A—Updated C2 Link MOPS.

DO-365—*Minimum Operational Performance Standards (MOPS) for Detect and Avoid (DAA) Systems,*[583] and updates:

DO-365A—Reflects Phase 2 of DAA MOPS development. At least the ground-based noncooperative radar MOPS including architectural considerations and operational concepts appear relevant to UTM.[584]

DO-365B—Incorporates ACAS Xu (aligned with SC-147) to address noncooperative well clear and alerting, air-to-air radar MOPS, and certain sensors.[585]

DO-377—*Minimum Aviation System Performance Standards for C2 Link Systems Supporting Operations of Unmanned Aircraft Systems in U.S. Airspace.*[586]

DO-381—*Minimum Operational Performance Standards (MOPS) for Ground Based Surveillance Systems (GBSS) for Traffic Surveillance* - used for air traffic surveillance in support of DAA operations for unmanned aircraft. The primary applications will be used in terminal, transit, or extended operational areas in the NAS as defined in **DO-365A**.

DO-385—*Minimum Operational Performance Standards (MOPS) for Airborne Collision Avoidance System X (ACAS X) (ACAS Xa and ACAS Xo).* While not directly supporting UTM, these MOPS inform the development of *ACAS sUx* (ACAS for sUAS).[587]

DO-386—Vol 1, *Minimum Operational Performance Standards for Airborne Collision Avoidance System Xu (ACAS Xu),* and DO-386, Vol II, *Minimum Operational Performance Standards for Airborne Collision Avoidance System Xu (ACAS Xu).*[588]

While again, RTCA has not been directly involved in UTM, its SC 228 is developing DAA capabilities that address more specialized UAS operations that require more tailored performance or constrained guidance. These operations are expected to take place in all classes of airspace with the exception of surface operations and Class E above A which remain out of scope. Such specialized operations are expected to address the following use cases but will be prioritized according to community needs and support.[589]

- Smaller UAS operations that occur at slower speeds and closer to terrain and obstacles. The expectation is that the guidance may need to be constrained by airspace restrictions and terrain and obstacle concerns.
- High Altitude Pseudo-Satellite (HAPS) launch and recovery operations. This functionality will be limited to the transition to/from Class E above A. It is expected that there will be a separate layer of separation automation employed in Class E above A that will be developed outside of the scope of SC-228. *See* Chapter 6, Section 6.6 *Stratospheric Operations.*

- VTOL operations including the AAM use case. These aircraft are capable of different maneuvers and make approaches to different environments than addressed by the Phase 2 activity. Such guidance may need to be tailored for the approach and departure phase of these vehicles.
- Part 135 cargo operations. It is expected that the existing functionality will support this use case, however, detailed operations were not investigated during Phase 2 OSED development. Phase 3 OSED development will further develop the concept and capture any changes needed.

Additionally, the committee will be creating standards for use of LTE commercial networks for C2 links used for type certificated UAS, creating Guidance Material for Lost Link Behavior of UAS and developing Navigation Standards for UAS. Finally, RTCA's Al Secen stated, "we continue to work with other Standards Organizations to ensure no duplicate efforts are undertaken."[590]

3.2.14 SAE

SAE International is engaged in diverse initiatives affecting UTM. For example, SAE's international joint committee G-34/WG-114 Artificial Intelligence in Aviation (includes AI in DAA), G-32 (in collaboration with EUROCAE WG-72) Cyber Physical Systems Security Committee, G-31 Electronic Transactions for Aerospace Committee, S-18UAS/WG-63 Autonomy Working Group, and associated publications, including for quality management deserve attention.[591]

3.3 STANDARDS-SUPPORTING ORGANIZATIONS

Among the organizations supporting or influencing UTM standards development, the following have been particularly influential.

3.3.1 EASA

The European Union Aviation Safety Agency (EASA), an agency of the European Union, is a key U-space resource. Its U-space initiatives include publishing a regulatory framework for U-space foundational services,[592] collaborating on technical specifications for various U-space demonstrators,[593] and catalyzing a cybersecurity response.[594] EASA recognizes that "[i]f each EU Member State starts implementing U-space services on the basis of not commonly agreed nor validated standards this will create dis-harmonisation, inefficiency and will have an impact on safety in the long run when more autonomous operations are foreseen."[595]

3.3.2 FAA NextGen

The integration of UAS into the NAS is an essential component of the FAA's Next Generation Air Transportation System (NextGen) initiative, which seeks to unify and manage all UAS R&D execution, including support for standards.[596] "The foundational work laid by NextGen in UAS R&D and prototyping has paved the way for early integration of UAS operations and standards. This has set the stage for the FAA to transform as a regulator and service provider as they enable the NAS 2035 vision."[597] Some observers have characterized the role of UAS and UTM with NextGen as having stretched capabilities, broken traditional R&D paradigms, and catalyzed fundamental changes—spawning proposals for a new FAA Office of Innovation to advance the public-private partnership model upon which UTM has progressed.[598]

3.3.3 GUTMA

The Global UTM Association (GUTMA) supports and accelerates the transparent implementation of globally interoperable UTM. While not a SDO, it plays a key role in UTM standards, and is the preeminent "non-profit consortium of worldwide UTM stakeholder [fostering] the safe, secure and efficient integration of drones in national airspace systems."[599] Its membership includes participants from the entire UTM ecosystem: ANSPs, aviation regulators, UAS operators, and UTM/U-space suppliers, and has "extended its membership to include several members from the telecommunication industry developing systems or operating platforms for connected drone operations."[600]

With its focus on UTM advancement, interoperability, and global harmonization, GUTMA has coordinated open source protocols including: Flight Declaration (future flight scheduling), Flight Logging (bifurcating public and private data), and Air Traffic Data (combining diverse sensor and air traffic data feeds) protocols, as well as a Drone Registry Brokerage Database Schema, and Database API specification.[601] GUTMA has also helped facilitate creation of the ACJA (*see* Chapter 3, Section 3.2.3.3 *ACJA*).

Underscoring its focus on UTM interoperability, GUTMA contributes directly and indirectly through its partnership with diverse UTM standards-making and coordinating expert groups. GUTMA's mission has evolved to include having established MOUs with ASTM and GSMA among other SDOs,[602] reassessing its direction as a software development organization, and acting as a global UTM knowledge repository,[603] as well as undertaking more traditional roles of education and thought leadership.[604]

3.3.4 NASA

NASA's Aeronautics Research Mission Directorate (ARMD) is influential in UTM standards development, including, for example, in the ASTM F38 UTM, Surveillance SDSP, DAA, and other working groups. NASA has a recognized history of supporting industry standards, and where needed, development of NASA technical standards,[605] including management of multiple JARUS "working groups with industry for quite some time as part of the RTT."[606] NASA's many pioneering and leading roles in UTM are introduced in Chapter 1 *Introduction* and annotated throughout this book.

3.3.5 SESAR JU

The Single European Sky ATM Research Joint Undertaking (SESAR JU) is a public-private partnership, constituting the technology pillar of the European Union Single European Sky (SES) initiative.[607] As noted above, SESAR JU "coordinates and concentrates all EU research and development (R&D) activities … to develop the new generation of ATM."[608] Its roles and initiatives support U-space significantly (*see* Chapter 5, Section 5.3 *U-space*).

3.4 FURTHER STANDARDS COORDINATION

> "Without the standards bodies we really wouldn't be able to move beyond VLOS today."
>
> **Steve Bradford**
> *Chief Scientist for Architecture and NextGen Development, FAA*[609]

Efforts to coordinate UTM standards development have progressed globally, and, as presented above, major UTM initiatives increasingly cooperate[610] in various fora. Coordination between NextGen and SESAR is a noteworthy example.

> NextGen and SESAR are working together on the integration of UAS [including UTM/U-space] to initiate, coordinate, and prioritize the activities necessary to support the evolution of all UAS categories as fully-integrated airspace users [and includes] the development and integration of the rapidly-evolving smaller drone environment, where potentially thousands of drones will be enabled to operate through the implementation of an entirely new management concept: UAS traffic management (UTM)/U-space.[611]

The scope of relevant standards that are candidates for coordination is extensive as highlighted above in the ANSI Roadmap, and by SESAR:

- Objective-based design standards for open and specific drone categories
- e-Registration
- Geofencing
- Data Exchange Protocols [weather …]
- Data quality [obstacles …]
- Security for open and specific drone categories
- Tracking
- Deconfliction
- Ground-Ground service interoperability
- Vehicle-to-Infrastructure communication [V2I]
- Vehicle-to-Vehicle communication [V2V]
- CNS Performance for open and specific drone categories
- Detect & Avoid for open and specific drone categories
- Emergency Recovery for open and specific drone categories

- Command, Control Datalink [Terrestrial, Satcom] for open and specific drone categories
- Guidance on Spectrum Access, Use, and Management.[612]

Further standards coordination is expected also to address gaps in: common measurement such as reference points and equipment providing varying "accuracy and performance in the measurement of altitude, navigation or time,"[613] failure alerting and management, other UTM systems translation between disparate reference systems to facilitate safe flight,[614] as well as in high altitude ("High E"),[615] and AAM/UAM operations.[616]

As presented above, coordination initiatives, including by ANSI (*see* Chapter 3, Section 3.1.1), EUSCG (*see* Chapter 3, Section 3.1.2), and ICAO (*see* Chapter 3, Section 3.2.6), are committed to advancing effective UTM standardization coordination.[617] Coordination is expanding rapidly, characterized by John Walker as a "Kitty Hawk moment now", and informal, where "[t]he stitching is happening behind the scenes. The constant theme is that standards organizations are talking to each other in atypical ways that are bringing very positive results."[618]

4 UTM Governance

"Responsibilities under regulation may need to evolve to consider [an] increased level of automation, use of UTM service providers and novel business models."

Jarrett Larrow
UTM Program Manager, FAA[619]

4.1 GENERAL—UTM GOVERNANCE

Governance profoundly[620] affects the architecture,[621] code,[622] services,[623] participant roles and responsibilities,[624] security,[625] interoperability, integration, business models,[626] and fairness[627] of UTM.

Federated UTM introduces new stakeholders and presents novel demands that cry out for thoughtful governance.[628] Lorenzo Murzilli observed:

> We have a system that is managed safely today. If you want to change something in there [for UTM], it is not easy because all rules are built to manage traditionally .... At the moment, I don't think we have an answer. If we want to make the transition, then we need to have rules that allow for the transition.[629]

Subsequent developments, including introduction of new standards, the U-space regulation, and accelerating government / industry collaboration offer promising contribution to UTM governance.[630]

Certain governmental roles, particularly those regarding safety oversight and enforcement, are generally non-delegable.[631] Indeed, many unmanned systems domains challenge traditional aviation governance.[632] While aircraft certification for UAS is distinct from qualifying UTM services, it echoes governance changes no less vexing. Aviation information infrastructure governance initiatives also inform UTM.[633]

4.2 USS QUALIFICATION

"We're calling it qualification."

Michele Merkle
Dir. of ATO Operations Planning and Integration, FAA[634]

4.2.1 General—USS Qualification

"How do we approve U-Space service providers?
It's hard work but it is not a hard problem."

Benoit Curdy
Digital Transformation, Swiss FOCA[635]

This discussion adopts the term *qualification* to broadly denote the process of satisfying applicable requirements within a governance regime for UTM.[636] Governance bodies worldwide recognize a need for UTM qualification. ICAO acknowledges the need for "development of policies to address means of compliance or system approval for UTM systems,"[637] and the need to address "methodologies and processes to ensure that UTM Service Suppliers (USS) are duly authorized,"[638] further stating:

> Given the nature of the planned/projected initial capability of UTM, UTM systems may have to demonstrate and achieve a level of confidence normally found in certified aviation systems. However, it is not necessarily the case that this needs to be done using the existing, established industry standards which may be viewed as excessive or unnecessary for the intended function of UTM.[639]

SESAR sees need for "review [of] safety assessment and certification requirements considering the change of the role of human[s] in the overall system."[640] The intensive information systems platform, immutable aviation safety constructs, advanced automation,[641] new stakeholders, federation, and the key role of standards[642] also factor into UTM governance.

Gur Kimchi, former VP, Amazon, urged "we should agree on a common approach to what it means to be credentialed."[643] This will take time as regulators gain experience, and industry and stakeholders collaborate to standardize UTM practices, performance metrics and assessment mechanisms. The challenges parallel some of those in the broader UAS ecosystem, as raised by Dirk Hoke, CEO, Airbus Defense and Space:

> What is the policy framework to bring these … products to market [with] certainty? … For us it's important that the great certification bodies, the big certification bodies get their act together, agree on way forward, and that we define the standards, even if they are high standards, very high burdens, high barriers, that enable us to understand what we need to do.[644]

One recognized UTM regulator responds, underscoring that a passive approach is ineffective:

> Certification bodies are at the very end of the process and even standardization bodies are only the reflection of an act from industry stakeholders. It denotes a passive attitude from the industry which is not the approach favored by authorities. In addition, the implicit acceptance of 'high standards' sounds like a call from an incumbent to establish high barrier to entry, which is definitely not welcome.[645]

Regulators require a defined allocation of responsibilities among UTM participants (such as between USS and operator) to assure safety, accountability, and enforcement. One regulator opined, "it is a goal of mine to get some approval of the USS .... I just [first] need a clear delineation of responsibility between a USS and operator."[646] Such a delineation should reflect the respective capabilities of the various participants in the developing ecosystem. Consequently, in the short term, for example, "[because] not all operators will have the same capability to come up with [the correct] operational intent volume," some USS may unavoidably undertake certain primary responsibilities.[647]

Even at this early stage in the development of UTM governance, UTM has been the subject of legislation. In the U.S., the FAA Reauthorization Act of 2018 ("FAARA") advances UTM research and development, pilot projects,[648] testing, and implementation, including "full operational capability."[649] FAARA requirements for the FAA to develop a UTM implementation plan include:

> (c)(3)(G)(B) setting the standards for independent private sector validation and verification ...
>
> (1) enabling operations beyond visual line of sight ... and
> [(d)(2)] consider, as appropriate—
> (J) *qualifications*, if any, necessary to operate UTM services;[650]

In the U.S., the predominant UAS governance regime is the Small Unmanned Aircraft Systems rule (14 C.F.R. Part 107), limited to prescribed VLOS operations without waiver. Traditional qualification via air carrier certification for operating UAS is instructive, but arguably a stop-gap measure only.[651] The Remote ID Rule also informs UTM governance including use of industry consensus standards as a MoC (*see* Chapter 2, Section 2.4.8 *UAS Remote ID*); and the FAA may find it efficacious to promulgate a new 14 C.F.R. part specific to UTM. In Europe, the U-space regulation (*see* Chapter 5, Section 5.3 *U-space*) requires service provider demonstration of competences to "hold a certificate issued by the [applicable] competent authority"[652] and the competent authority to "establish a certification and continuous risk-based oversight programme."[653]

As UTM qualification regimes emerge, they may address four key aspects: the requirements themselves, who assesses that a UTM service conforms to requirements, how a service demonstrates that it conforms, and some way to ensure the service continues to meet the requirements going forward.

Within these categories, future UTM qualification processes may include, for example, designated means of compliance (MoC) providing an acceptable level of safety, third-party assessment, and self-declaration. Certain qualification processes may, in part, be facilitated by semi-automated *checkout* and *onboarding*. We can expect the processes to reflect identified risks, regulatory requirements, practical operational and administrative considerations—and to be data driven.[654] Recognizing the developing primacy of a risk-based approach to performance requirements for UTM governance, the Specific Operations Risk Assessment (SORA) may contribute to such governance, and is addressed below.[655]

4.2.2 Who Assesses Conformity

"Data and rules and testing equal approved provider."

Reinaldo Negron
Head of UTM, Wing[656]

Conformity assessment refers broadly to the systematic program of measures (such as testing, monitoring, reporting, and record keeping) applied to determine that governance requirements are satisfied. Qualification regimes that confirm a UTM service meets requirements are a form of conformity assessment. Such requirements may reflect rules, standards, practices, and agreements. Conformity assessment requirements for UTM are evolving in step with formal UTM safety objectives and regulation and there is increasing clarity to a broader conformity assessment vision.[657]

Who confirms that a UTM service complies with requirements? Regulators or certifiers themselves may issue some form of certification of conformity, or they could allow a third-party to certify, or permit service providers to self-certify (more formally, self-declare) conformity to certain requirements.

The term self-certification neither means that a competent authority relinquishes its roles in oversight or approval, nor implies a particular level of self-certification.[658] Under one approach, self-certification may be used as a self-declaration of belief in the completion of preconditions associated with a formal qualification process.[659] In such cases it does not alone represent governmental acceptance, approval, or certification. Arguably, the extent to which self-certification alone is permitted should be limited to lower-risk assertions,[660] otherwise CAA or designated third-party validation is indicated. Self-certification also requires implementation clarity—to distinguish from or avoid the appearance of undertaking an act of a CAA designee. In some regimes, use of self-certification alone may preclude liability exculpation.

While aircraft certification processes may not necessarily directly impact UTM qualification, the FAA's use of *level of involvement* of designees has served to expedite administration and improve utilization of FAA limited resources on safety-critical items[661] and may inform UTM governance. "[I]t is critical to the success and effectiveness of the certification process. Under this program, the FAA may delegate a matter related to aircraft certification to a qualified private person. *This is not self-certification*; the FAA retains strict oversight authority .... During the past few years, Congress has endorsed FAA's delegation authority, including in the FAA Reauthorization Act of 2018, which directed the FAA to delegate more certification tasks to the designees we oversee."[662] Limitations to level of involvement regarding new and novel technologies,[663] as highlighted by the Boeing 737 MAX accidents,[664] may also inform UTM governance.

4.2.3 Bases of Conformity Assessment

Conformity assessment processes rely on a common interpretation of requirements that systems must meet. Traditionally, aviation regulators generally set standards based on technical aspects of systems. The newer practice is to generate performance-based requirements in conjunction with industry consensus standards.[665] Such standards are a key component to governance of UTM. EUROCAE has observed:

> Here as well as in other areas standards are envisaged to be used to define detailed objectives. It is envisioned to not call [consensus] standards directly by regulations but to use them as acceptable means of compliance. This eases the introduction of alternative ways to show compliance to the high-level objectives without being blocked by regulations.[666]

A FAA Drone Advisory Committee (DAC) Task Group has recommended: "The FAA should partner with industry to create performance-based requirements/standards for UTM … to approve or certify these systems."[667] Doing so is consistent with a trend toward more quantitative, data-driven risk assessment, driven in part by the shift from *prescriptive* to *performance-based* regulations,[668] or at least eliminating prescriptive language that precludes current or future technologies. Jay Merkle, Executive Director, UAS Integration Office, FAA stated,

> industry standards will become a means of compliance. ASTM will be the first …. What is the role of the regulator in UTM—it's about describing the performance requirements. If you think of it in light of how legacy uses RNP … you can conceptualize where we're heading in encoding them in regulation.[669]

Performance-based regulations encourage and accommodate innovation. They depend on the ability to define objective, relevant, and effective performance expectations. In situations where the undesirable events are infrequent or even rare, service providers may not be required to act until occurrence of the negative events that the rule is intended to avoid. Thus, performance-based regulations depend on robust risk assessments (*see* Chapter 4, Section 4.2.6 *Risk Assessment*). Andy Thurling, CTO of NUAIR has urged, "[w]e need performance requirements for the 'ecosystem', i.e., flight planning, surveillance, weather, C2, and other services—appropriate for the airspace in which operations are being approved—beyond *What* and *How* to the *How Well*?"[670]

As for what the requirements may contain, ICAO has summarized the scope of safety-significant factors that a state might assess for "issuance of an operational approval for a UTM system," including:

- types of UA and their performance characteristics (including navigation capabilities and performance)
- adequacy and complexity of the existing airspace structure

- nature of the operation
- type and density of existing and anticipated traffic
- regulatory structure
- environmental considerations
- the requirement for all UA in the UTM airspace volume to be cooperative
- detection/separation of noncooperative UA
- management of aeronautical information service (AIS) data
- geographic information systems (GIS) data applicable to the UTM airspace.[671]

The FAA has described notionally the qualification of UTM services, taking a "modular and discrete" approach, and accommodating "tailored oversight" as follows:

1. Services that are required to be used by Operators due to FAA regulation and/or have a direct connection to FAA systems. These services must be qualified by the FAA against a *specified set of performance rules.*
2. Services that may be used by an Operator to meet all or part of a FAA regulation. These services must meet an *acceptable means of compliance* and may be individually qualified by the FAA.
3. Services that provide value-added assistance to an Operator, but are not used for regulatory compliance. These services may *meet an industry standard*, but will not be qualified by the FAA.[672]

The EU's structured use of "conformity assessment procedures" for UAS may inform UTM governance to the extent it "include[s] procedures from the least to the most stringent, in proportion to the level of risk involved and the level of safety required," and that "[i]n order to ensure inter-sectorial coherence and to avoid ad hoc variants of conformity assessment, conformity assessment procedures should be chosen from among those modules."[673] Additionally, an "assurance process consisting of type certification, site-specific commissioning and annual revalidation … ." deserves consideration.[674] As an industry veteran puts it,

> we've forgotten that there is a whole network of ground systems that get us into the airspace. I think this is the answer for a lot of the UTM issues. We need and must treat them as navigation aids—it's critical and irresponsible if we don't. As such, the certifying and commissioning process for Part 87 radars or Automated Weather Observing Systems (AWOS) for non-Federal applications provides a path forward.[675]

Other assessment vehicles for demonstrating acceptable MoC to requirements may also contribute to UTM governance. For example, meeting the FAA's government contracting requirements may contribute to a USS' system of governance:

> Each Remote ID USS would be required to establish a contractual relationship with the FAA through a Memorandum of Agreement (MOA) entered into under the FAA's "other transaction authority" under 49 U.S.C. 106(l) and (m), and to comply with a series of terms, conditions, limitations, and technical requirements

> [and] would also be contractually required to meet quality-of-service metrics that would establish the minimum requirements for providing remote identification services, including availability of the service and what happens when various failures occur.[676]

For accepted aeronautical data processes and associated databases—common in distributed/federated systems—a letter of authorization (LOA) can acknowledge that an applicant's processes provide an acceptable means of demonstrating means of compliance with the objectives of regulations and standards.[677]

Finally, certain design assurance principles and processes that underlie software and hardware built for aviation systems may have application in qualifying UTM. Historically, aviation relied on some level of regulator-prescribed design assurance used as a basis of qualification or certification.

> Standards and regulations seem to have largely either ignored this element of qualification or kicked the can down the road to some future endeavor using terminology such as "an appropriate level of design assurance ..." with no guidance or methodology for determining the appropriateness of the level. Further, the UTM ecosystem has been designed around non-deterministic technology, which limits the design assurance level these services could realistically achieve using current aviation processes such as DO-278 (for software).[678]

This lack of guidance or methodology may impede UTM qualification to the extent it precludes rendering UTM services adequate for safety critical aviation applications, or demands new methods of design assurance derived from non-aviation industries.[679] One supporting method for UTM qualification may draw from standard practices used to qualify complex function(s) in UAS "constrained through a run-time assurance (RTA) architecture to maintain an acceptable level of flight safety."[680] This remains an area of active development.[681]

4.2.4 Demonstration of Conformity

UTM qualification regimes can be expected to stipulate how a service demonstrates that it complies with requirements for demonstrating conformity. As with the question of who certifies, there are three possible nonexclusive options. The regulator could test and audit the service against requirements, third-party testing entities could test and audit the service against requirements or service providers themselves could test their own systems. There is FAA interest for the UTM industry to set requirements (via standards) for auditing and logging, to proactively stand-up and apply safety management (SMS) processes and practices to advance self-sufficiency, to promote a systematic approach to safety, advance self-sufficiency, and to take responsibility to drive its own reliability. As indicated in FAA's "modular and discreet" approach cited above (Chapter 4, Section 4.2.3 *Basis of Conformity Assessment*), the means of demonstrating conformity may vary by individual UTM services, and operating area.

4.2.4.1 Checkout and Onboarding

CAA checkout and onboarding may become semi-automated processes for demonstrating conformity for UTM qualification. An example would be a process matured from NASA UTM initiatives and the FAA's Low Altitude Authorization and Notification Capability (LAANC), respectively. For example, NASA's *UAS Service Supplier Specification* proposed early instructive UTM baseline requirements for a "USS Checkout" process in its Technical Capability Level (TCL) initiative, with emphasis on interoperability.

> An organization interested in offering services as a USS within the UTM System needs to complete a checkout process. The requirements that are checked during this process are those that are included in this specification. Since each requirement is not necessarily a software-specific requirement, an entity can expect a combination of software testing of their USS implementation along with required supporting documentation and other artifacts. Upon successful completion of this checkout process, the organization will be recognized as a valid USS .... That identity will be managed within FIMS. The checkout process is managed by the ANSP, but might be executed by an entity other than the ANSP, at the discretion of the ANSP.
>
> A potential flow to complete the checkout process may include the following steps for the interested entity:
>
> 1. Review USS documentation.
> 2. Implement USS per USS Specification.
> 3. Test implementation using an existing "sandbox" environment ...
> 4. Apply for checkout process.
> 5. Software checkout.
> 6. Obtain identity information from ANSP/FIMS.[682]

Another example is *onboarding*, whereby USS applicants (to LAANC) "demonstrate the capability to provide LAANC services on all available product configuration and platforms ... to verify compliance of all LAANC Performance Rules."[683] To demonstrate satisfaction of minimum performance metrics via automated testing and simulation, "[a] common simulation environment is needed against which a USS could test changes for validation. It would also lower manual testing overhead between USS and the FAA."[684] Testing automation for MoC may challenge quick iteration development and implementation cycles of certain USS/UTM protocols and "require [] additional development to automate."[685] In any event, they should reflect the associated level of risk. Automated testing without independent validation of certain results (typically higher-risk items) in a general UTM onboarding process may raise safety and legal issues and fail to serve its intended purpose. The ASTM remote ID and UTM specifications describe test methods and associated performance metrics for interoperability validation that should contribute to future UTM qualification automation.[686] UTM standards may test for interoperability with other UTM implementations, latency, periodicity, protocol compliance, and reliability. Onboarding to ensure compatibility with existing NAS systems and services like the System-Wide Information Management (SWIM) may also be required. *See* Chapter 2, Section 2.4.4 *System-Wide Information Management.*

A body of prior work (including research, tools, and implementations) in other industries[687] requiring safety-critical systems has advanced the integration of automated methods to complement conventional qualification in standing-up distributed trustworthy information systems. For example, such systems may support implementing public key infrastructure (PKI) certification authorities, where organizational due diligence (e.g., via credential validation, identity, creditworthiness, attachments/liens and adverse judgements, intellectual property registration, and other data sources) may evidence trust and propriety underlying semi-automated qualification processes.[688]

4.2.4.2 Third-Party Assessment

> "The role of 'approved testing providers' to verify interoperability and compliance should be defined."
>
> **Gur Kimchi**
> *Former VP, Amazon*[689]

Assessment of UTM service providers such as through third-party testing[690] and audit[691] may support CAA determination of qualification and continuing oversight. Diverse recognized third-party assessment entities in the aviation, IT, and security domains offer, or are developing, specialized mechanisms and programs that could provide assurances of UTM ecosystem quality and conformity.[692] ASTM has developed standards for both compliance audits on UAS and independent audit programs for operators[693] that are asserted to be extensible and effective for UTM assessment.[694] Standardization of qualifications and practices, plus assurance that third-party organizations providing oversight will remain independent, underly effective assessment programs.[695]

For International Aviation Trust Framework purposes, compliance with "[a]ccreditation, [c]ertification & [a]udit [s]ervices enabl[ing] federation members to validate that their governance and system/infrastructure components adhere to federation requirements either through self-assertion or a formal third party audit process" has been proposed.[696] To "complement" NASA's UTM initiatives, Dr. Koparkedar proposed the *National Unmanned Aerial System Standardized Performance Testing and Rating (NUSTAR),* seeking to "offer credible and comprehensive self-regulatory structure for small unmanned aerial systems [via] standardized tests and scenario conditions to assess performance of the UAS."[697] An iteration of that proposal has been advanced by NUAIR (the New York test site) that includes industry consensus standards and provides "third party statements of compliance"[698] for UAS, USS, and SDSPs "… as an independent third party to provide safety declarations."[699]

4.2.4.3 Self-Testing

CAAs in their assessment of conformity may rely on self-testing (and possibly self-declaration) in the demonstration of conformity for some parts of the process, while they test for others.[700] Security is a good example where self-declaration has contributed to compliance and is poised for use in UTM qualification. For instance, Swiss FOCA plans to require U-space service provider "self-declaration of compliance before requiring certification [to] ISO 27001 [Information Security Management]."[701] In other aviation domains such as light sport aircraft, self-declaration, or self-certification is used for aircraft assessment by manufacturer, providing attestation of compliance with standards, regulations, best practices, and SLAs.

4.2.5 Continuing Compliance and Safety Assurance

> "[E]stablish a certification and continuous risk-based oversight programme"
>
> **European Commission**[702]

Once qualified, we can expect UTM service providers to be required to demonstrate that they continue to comply with applicable requirements. For example, the U-space regulation's oversight regime recognizes the role of compliance, requiring the competent authority to:

> (j) establish a ... *continuous risk-based oversight programme*, including the monitoring of the operational and financial performance, commensurate with the risk associated with the services being provided by the U-space service providers and single common information service providers under their oversight responsibility[703]

Management systems supporting safety objectives (presumably responsive to its safety policy), such as safety management systems (SMS)[704] are generally indispensable (or required) and must address the dynamics of UTM to ensure the safety of the NAS.[705] Whether by CAA mandate for a formal SMS or otherwise, a means is necessary to assure that any person or entity undertaking a safety-critical or safety supporting role understands the attendant risks of their operational functions and takes steps to manage them. Safety assurance and oversight are prioritized in UTM governance and "must be considered across the lifecycle of UTM in its entirety"[706] including for software updates.[707]

The U-space regulation also requires competent authorities to "carry out audits, assessments, investigations and inspections of the U-space service providers and single common information service providers as established in the oversight programme."[708] In federated UTM, multiple USS may offer the same service based on different data sets, and "it is therefore necessary to apply and monitor performance specifications of services and their associated data to verify these federated services meet the safety requirements for their supported UAS operations."[709]

The development of In-Time System-Wide Safety Assurance (ISSA) may reduce "the safety assurance cycle time until real-time safety assurance is achieved at the system-of-systems level."[710] System-wide safety assurance can be broken down into three steps: monitor, assess, and mitigate. Mitigation of detected risks, such as faults, failures or adverse conditions demand contingency management capabilities including fault tolerance, graceful degradation, isolation, localization and remediation.[711] Standards and service level agreements (SLAs) can specify USS "health check" endpoints to accommodate real-time continuous compliance assurance by the competent technical authority.

Looking forward, where UTM-inspired traffic management services support advanced air mobility (AAM) operations, and noting, for example, that accidents in the for-hire market "underscore … the urgency of improving the safety of charter flights,"[712] UTM safety regimes may undergo heightened scrutiny. As a matter of policy, there is firm expectation that safety assurance for AAM and other manned operations supported by UTM will be no less than that for ATM given an equivalent amount of risk. Indeed, "ATM has variable levels of safety depending on classification of the platform, [and] there is no difference for UTM systems."[713] It can be expected that ISSA will be integrated into UTM.

What is logged and timestamped, for how long, and the robustness of the associated processes should reflect governance requirements, developing UTM standards and practices, and business/organizational needs.[714] Logging may demonstrate performance and conformance, support event reconstruction, and provide metrics for operational improvement, fairness, and safety enhancements, and provide evidence to help prevent or resolve disputes. Logging may also impact and support data rights and privacy management (*see* Chapter 4, Section 4.7.6 *Data Rights and Privacy Management*), audit, and continuing compliance.

Generally, logging should be undertaken by each service provider and certain other UTM stakeholders, and may include applicable state changes, subscriptions, and operational data reflecting execution of operational intents, flight or airspace authorizations, constraints, nonconformance, contingency, and accidents.[715] For UTM purposes, raw data to be logged has been characterized to include:

VPN traffic: All interactions (including requests with content and timestamps, responses with content and timestamps or no-response indication) with both DSS instances and other USSs. Basically, all the stuff covered by the API, plus timestamps and non-responses.

Op details: The details about an operational intent that might not traverse the VPN. This includes, importantly, the precise 4D volumes of each operational intent for each version of the operational intent.

Aircraft positions: The telemetry data used to support the conformance requirements ([and linking] each of these positions to an operational intent ID, and perhaps a Boolean flag indicating whether the aircraft is in flight).[716]

Finally, UTM standards may provide certain common data schemas to facilitate logged data export and exchange. Separately, in addition to logging, it can be expected that CAAs may recommend or require self-reporting. For example, LAANC USS Performance Rules state that "USSs are encouraged to proactively self-report detected violations to avoid FAA action."

4.2.6 Risk Assessment

"Understanding the different elements of the system
is key to effective risk assessment and management."

Jarret Larrow
UTM Program Manager, FAA[717]

UTM infrastructure risk assessment, which shares attributes of distributed information systems risk assessment, should be informed by relevant information system and control methods, and associated risk assessment tools.[718] These methods and tools can help understand relevant data, including who monitors the system at what level(s), and human factors—especially when and how a human must intervene upon system automation failure,[719] and "particularly if you are linking-in hundreds of aircraft in the system."[720] "The UTM risk assessment has a holistic approach to analyze multiple UA operations and UTM services."[721]

The FAA reports that "Safety Risk Management assessments have not yet been conducted by the FAA regarding UTM systems, infrastructure, integration, and/or implementation. NASA is currently working with the FAA to identify what a future UTM system would look like. As FAA and NASA research is completed, the FAA will conduct and develop SRMDs [safety risk management documents] as part of the transfer of UTM technology to operational use."[722] Separately, SRMDs have been produced for LAANC[723] and applications relevant to UTM, and a SRMD for UAS Remote ID is under development. These documents should inform development of the broader UTM safety case and responsive mitigations underlying an effective governance regime. Additionally, the FAA is developing a UTM risk model and risk message capability to inform governance.[723]

Data-driven risk analysis for a new process like UTM has inherent limitations in contrast to that for manned flight, with more than 100 years of experience—and data—to draw upon.[724] Some important data parameters may not be quantifiable, however qualitative observations are often expressed quantitatively when they are actually categorical data. Similarly, caution is warranted regarding quantitative human reliability protocols because describing human error often exceeds our ability to estimate it in reliable quantitative terms. Notwithstanding, highly automated UAS operations, currently scaling around the world, will allow for new data-driven approaches. This will be dependent on valid, reliable performance metrics and assessment methodologies.

Caution should be exercised in coupling "risk based" and "performance based" where risk data are historical (accident/incident rates—lagging indicators) and performance reflects the absence of negative outcomes. Observable events are in the past while risk is in the future. Of course, past event rates are not necessarily valid indicators of future performance.[725] Where accidents are relatively infrequent and technological implementation is new, quiescent periods may lull us into complacency that is assessed as successful risk management where, in fact, a period of accumulating but non-salient latent conditions are being experienced.[726]

4.2.7 Specific Operations Risk Assessment

"How can UTM implement some of the mitigations that are already foreseen in SORA?"

Andreas Lamprecht, PhD
CTO, AIRMAP[727]

The *Specific Operations Risk Assessment* (SORA) developed within JARUS is one significant tool that should contribute to regulators' and operators' understanding of how UTM mitigates risk and thus promotes UTM's scaling.[728] Although part of a larger risk mitigation structure, SORA may soon influence UTM service provider governance as well. Initiated by regulators, and increasingly embraced globally with participation of many companies and academic and research institutions, SORA is a risk framework that "provides a methodology to guide both the applicant and the competent authority in determining whether an operation can be conducted in a safe manner."[729] It is a "pre-digested process to identify the inherent risk of the operation [and brings] everyone up to a basic level of safety and competence [identifying] safety risks, or threats, resulting from equipment failure, operator error,[730] adverse operating conditions, etc., and then defines appropriate means for risk mitigation."[731] In principal, SORA is not tied to any particular regulatory scheme, though increasingly, competent authorities expect applications for UAS operations will follow the SORA framework for determining and mitigating air and ground risk in the "Specific" category.

SORA v. 2.0 reserves a place for UTM to provide external support services to an operator within its designated *ANNEX H—Unmanned Traffic Management (UTM)*,[732] which is currently being drafted. Annex H will attempt to link USS operational requirements to existing SORA mitigations. Thus, an operator could subscribe to a pre-approved UTM service that would fulfill portions of the operator's safety case. The first aspects will relate to evaluation and mitigation of air and ground risk (dubbed operations planning); and surveillance, a critical part of detect and avoid (DAA).[733] Annex H is also expected to provide guidance on identifying credible data sources for airspace and population density information.[734]

The discrete service mitigations identified in Annex H may be used to stack UTM services into a logical grouping of robustness and service levels that can be used to reduce the overall risk of an operation and help individual operators obtain operations approvals. Indeed,

> Civil Aviation Authorities (CAAs) around the world want to enable operators to use UTM/U-Space services as mitigations to the risks inherent in UAS operations. But, without requirements and standards that define the level to which these services are effective, it is impossible to quantify the amount of risk mitigation an Operator can claim when using a UTM/U-space service.[735]

Annex H is focused on assessing *operational* risk, not USS or other UTM component risk, or general UTM capability. Effective risk mitigation will derive from implementing robust and reliable UTM services. Notwithstanding, Annex H might help: advance a common understanding of risk mitigation effectiveness for UTM services;[736] normalize associated SLAs;[737] articulate responsibilities among operators,

UTM service providers and CAAs;[738] and contribute to qualifying UTM infrastructure.[739] UTM service level performance development will engage standards groups with the expectation that industry, rather than regulators, is better able to identify appropriate requirements. In practice, it has been the role of standards groups in conventional aviation to resolve difficult details and performance requirements, which, pending validation, can be incorporated into regulation or into acceptable means of compliance with the performance requirements.[740]

Early USS approvals should help to break down fundamental responsibilities and their oversight, and provide the basis for more complicated safety requirements based, in part, upon the SORA "because [o]bviously, in this domain, [we're] moving a little bit more into services that are actually used for mitigating safety risk. Meaning then if they don't work there might be people at harm instead of a LAANC service, which really does not have those requirements [and if] the system is down no one is at harm."[741]

SORA is not yet scalable for certain operations, is in early stages of automation, and thus requires considerable human input. Digitizing the SORA process will need further research, coordination between CAAs and industry, more investment, and even a quantitative dataset for a common risk framework. Future automation should streamline implementation of UTM services into the SORA process.[742]

4.3 UAS FLIGHT RULES

"UTM is key to the effective application of UAS Flight Rules."[743] Correspondingly, effective flight rules are essential for UTM-enabled flight operations, and full integration of UAS into a NAS. The FAA UTM ConOps recognizes that "[r]ight-of-way rules, established procedures, and safe operating rules enable harmonized interaction when aircraft encounter one another."[744] Current rules, exemplified by ICAO Annex 2 *Rules of the air*,[745] have been viewed as "very flexible … and should inspire us to first [try to] use the existing rules,"[746] and ICAO SARPs and Procedures for Air Navigation Services (PANS) developed for Remotely Piloted Aircraft Systems (RPAS) have been viewed as contributing to a path forward.[747]

Nonetheless, the need for rule reform is widely acknowledged, yet consensus on its form and scope remains in play. For example, the ICAO UTM Framework asserts:

> Rules of the Air which specify flight rules, right-of-way, altitude above people and obstructions, distance from obstacles and types of flight rules, all of which, as written, are *incompatible with the intended operations within UTM systems*.[748]

EUROCAE urges that "[o]perating the UA in compliance with the Right of Way (RoW) rules need to be reviewed, and possibly adapted to the VLL [very low level] environment, especially [for] urban areas,"[749] EUROCONTROL suggests that low-level flight rules (LFR) and high-level flight rules (HFR) are needed to supplement IFR and VFR rules,[750] and the Indian government "recommends that the flight rules for unmanned aircraft may be different (say UFR)."[751] NASA's Dr. Kopardekar proposes "digital flight rules" (DFR) and urges "the whole digital way of operating requires a different way of thinking."[752] "As flight rules evolve, it is expected that manned aircraft [o]perators will share some responsibility for separation when operating in the UTM ecosystem."[753] Additionally, specific rules to accommodate autonomous operations have been urged.[754]

Since nation states and regional authorities may have varying (and complex) flight rules addressing operational priorities and conflicts, UTM standards demand both flexibility/modularity and specificity.[755] Prioritization rules for conflicting operational intents and constraints are further considered in Chapter 4, Section 4.4.4 *Fair Access*.

Rules responsive to managing airspace at the boundaries of UTM and ATM have also been advocated, and "[s]tates will need to decide how to apply the flight rules" there.[756] As a practical matter, building and optimizing responsive UTM flight rules will require further experience, time, and stakeholder engagement.[757]

4.4 AIRSPACE

4.4.1 General—Airspace

"We don't need to reclassify airspace."

Steve Bradford
Chief Scientist for Architecture and NextGen Development, FAA[758]

"[I]ntegrating the technology may require some large changes of the National Airspace System and the rules governing it."

Cameron R. Cloar, Esq.[759]

Despite notable support for airspace reform to facilitate UAS integration, it remains a work in progress. ICAO has stated:

> The current airspace classification scheme as developed for manned aviation may not effectively support visual line-of-sight (VLOS) or BVLOS operations. This gap includes the potential [need for] modification of current classes of airspace or potentially creating new classes of airspace to accommodate the range of needs brought by UAS operations.[760]

Some UTM initiatives urge airspace reform arguing that "[t]he existing rules cannot be tweaked to adapt them to the digital nature of drone operations [and need their] own flight rules and airspace classification,"[761] such as for "flights not conducted in accordance with IFR or visual flight rules .... "[762] Nonetheless, "the difficulty to develop an airspace classification" is acknowledged.[763] For example, within the manned aviation community, one traditional view holds that new airspace entrants must conform to the existing airspace structure and rules, and "[i]ntegration must not require the transference of safety burdens, intentionally or unknowingly, to other airspace users."[764]

The FAA Reauthorization Act of 2018 ("FAARA") advances initiatives for UAS integration into the NAS,[765] and various roadmaps and research initiatives consider airspace reform to facilitate UAS integration.[766] The FAA established Aviation Rulemaking Committees (ARCs), such as the Airspace Access Priorities ARC that "recognized a need to develop an improved framework that will allow the FAA to balance the respective needs of its wide-variety of controlled airspace users, including ... *operators of Unmanned Aircraft Systems* in positive control airspace ...";[767] and the Unmanned Aircraft Systems (UAS) in Controlled Airspace ARC,[768] that did not issue public findings. In 2016 the FAA's Small Unmanned Aircraft Systems rule

(14 C.F.R. Part 107, Subpart B – Operating Rules), opened certain airspace below 400 ft. AGL and formalized regulations subject to waiver. Yet the rule contained significant restrictions, including prohibiting BVLOS operations, and operations in controlled airspace without prior ATC authorization. Subsequently, the UAS Remote ID rule (*see* Chapter 2, Section 2.4.8 *UAS Remote ID*) provided "the next incremental step" in further integrating UAS in the NAS, and serves as a "critical element to enable" operations that raise safety and security concerns such as operations at night and over people. Nonetheless, the RID rule's preamble states that it does "not alone enable routine expanded operations, affect the frequency of UAS operations in the airspace of the United States, or authorize additional UAS operations. Nor does the rule by itself open up new areas of airspace to UAS."

Despite continuing challenges, consensus may be developing: that the manned aviation community does not exert exclusive control of the airspace, that airspace is a finite resource, and that collaboration is needed to resolve separation issues that may arise in transitions between classes of airspace.[769] Such a consensus may coalesce around the realization that the regulatory reform process may delay UTM deployment. Correspondingly, "we can't underestimate the challenge if we can't operate within the current rules," urged Tom Prevot.[770]

The U-space regulation takes an affirmative approach, authorizing designation of U-space airspace by Member States.[771] Such volumes do not represent additional classes of airspace in the ICAO context, but instead service volumes where different levels of U-space service are required to support unmanned and manned operations collectively, and the airspace retains the original ICAO class.[772] Furthermore, the U-space Implementing Regulation designates ANSP and U-space service provider responsibilities in controlled and uncontrolled airspace. Furthermore, "[w]here a Member State designates a U-space airspace within controlled airspace, it shall ensure that manned aircraft which are provided with an air traffic control service and UAS remain segregated through the dynamic reconfiguration of the airspace within that U-space . . ."[773] Within this scheme, the SESAR CORUS ConOps defines three U-space volume types of VLL airspace, each reflecting incrementally greater attendant risks, with associated conflict resolution services and other requirements:

X—no conflict resolution services available,
Y—strategic conflict resolution services only, and
Z—strategic and in-flight tactical conflict resolution services.[774]

The U-space volume types are depicted in Figure 4.1.[775]

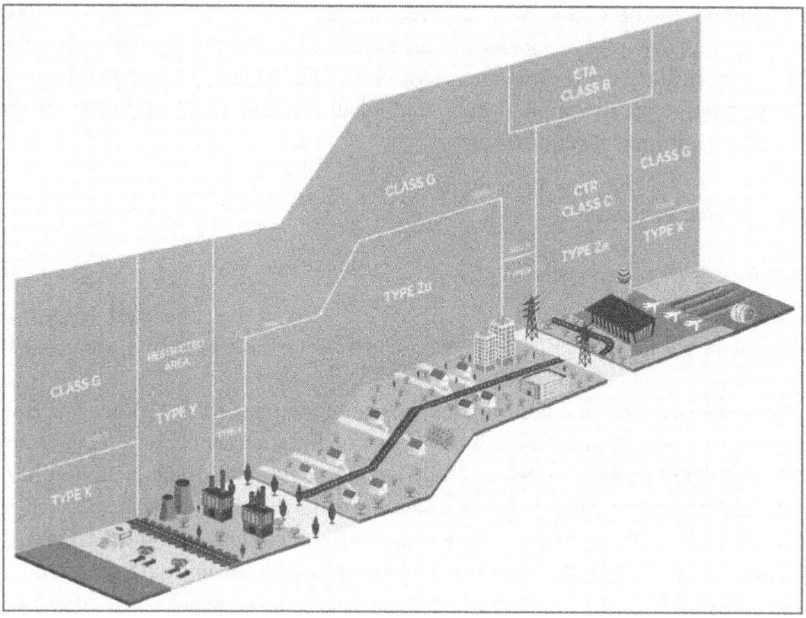

FIGURE 4.1 U-space VLL volume types. (Reprinted with permission (SESAR JU 2019), 33.)

Various proposals for airspace reform affect UTM, exploiting varying degrees of dynamic airspace management and configuration.[776] For example, Embraer proposes a "*layered approach* enabl[ing] the ANSP to increase urban airspace capacity and provide equitable airspace access for new and legacy aircraft" in airspace between VLL for UAS, and higher altitudes for ATM-controlled airspace.[777] Another proposal notionally included in the FAA's UAM ConOps depicts use of corridors (*see* Figure 4.6 *UAM Notional Corridor*). As urged by NASA's Dr. Kopardekar, "the airspace system should be ready when the vehicles are ready."[778] Finally, for context and general reference, Figure 4.2 presents the structure of U.S. airspace in which there are no separate/exclusive classes of airspace designated for UAS/UTM.[779]

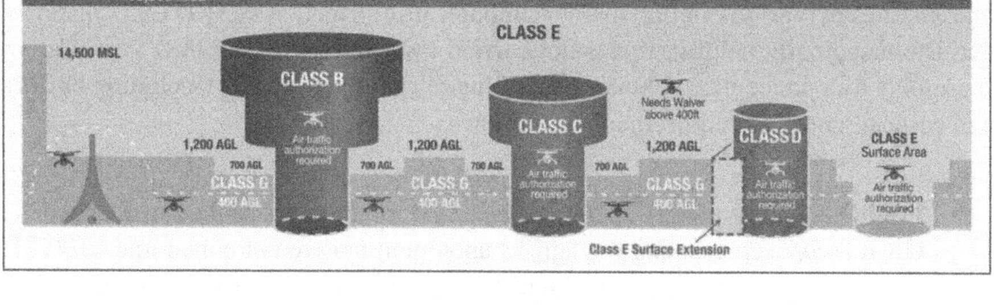

FIGURE 4.2 U.S. airspace volumes. (Source FAA.)

In addition to the Class A-G airspace depicted there are various recognized types of special use airspace including restricted areas, temporary flight restrictions (TFRs), and military operations areas (MOAs).[780] Each class of airspace serves particular purposes, has specified altitude and locations, and associated pilot, communication, and equipment requirements.

4.4.2 UTM Volumes

"Airspace never contemplated this."

Paul McDuffee
Exec./Ops. Analyst, Hyundai Urban Air Mobility[781]

A system that manages and coordinates the use of airspace for UTM needs a framework to resolve which airspace volumes a UA operation may occupy. That airspace can be abstracted as layers (and segments) demarcated with flight excursion hazard boundaries/thresholds. Such volumes can vary widely in their attributes.

For example, subject to the UTM's supported ConOps, four-dimensional (4D) volumes of airspace (3D plus time) may be designated to enable safe separation. 4D volumes may include discs, tubes, or complex shapes and segments. The airspace volume for a UA doing an inspection operation might be concentric spheres or polygons around an inspection structure, whereas the airspace volumes for a UA doing package delivery might be concentric tubes along the intended flight route.[782] The primary drivers behind volume-based representations are twofold: to allow operators to capture the uncertainties associated with the diverse operations UTM can support – requiring less precise information, and help preserve privacy for operations bound for private homes and other sensitive locations.[783]

Such volumes may accommodate strategic separation of operations and be ordered sequentially to reflect the intended flight path. "The critical piece here is which volume will be used for deconfliction and what will be the requirements associated with it? This has huge safety implications."[784] The size,[785] performance requirements,[786] traffic management actions,[787] type, and nomenclature for these volumes continue to evolve. Representing an operation using volumes in UTM is a departure from more traditional approaches common in ATM such as trajectory-based constructs. As UTM advances, trajectory-based operations (TBO) and other approaches may also be accommodated.[788]

Figure 4.3 depicts a notional approach to operation volumes developed by NASA and iterated by various initiatives.[789] Although it informed ASTM UTM standards and is conceptually helpful, it was not carried over to its work product verbatim.[790] It presents four concentric volumes describing relevant parameters (including buffer) and certain reporting requirements, as follows:

- **Flight Geography (FG)**: operator-defined 4D volumes encapsulate an intended UA mission, representing the agreed-upon nominal area of containment. FG breach provides notification to the operator (but not to other UTM participants) with the expectation that the operator performs recovery actions.[791]

- **Conformance Volume (CV)**: buffer encapsulates the FG, sized to reflect the performance capability of the UA and facilitate recovery actions; CV breach may trigger additional events such as notifications to other USS or execution of contingency plans.[792]
- **Operational Intent Volume (OIV)**: area of intended operation shared with the USS network and sized to reflect total system errors. It encloses the CV and serves as a buffered expansion of the CV to mitigate flight anomalies (e.g., unanticipated wind gusts). Additional buffer may be designated by the operator, and OIVs can abut one other.
- **Authorized Area of Operation (AAO)**: maximum airspace available in a geographic area with discretely defined boundaries for an operator to conduct its mission under its performance authorization (PA) while satisfying applicable limitations. An AAO may have "more than one AAO under a single PA [with] different levels of performance based on the underlying infrastructure."[793] For operations adjacent to controlled airspace, breaches of the AAO could be shared with ATC.

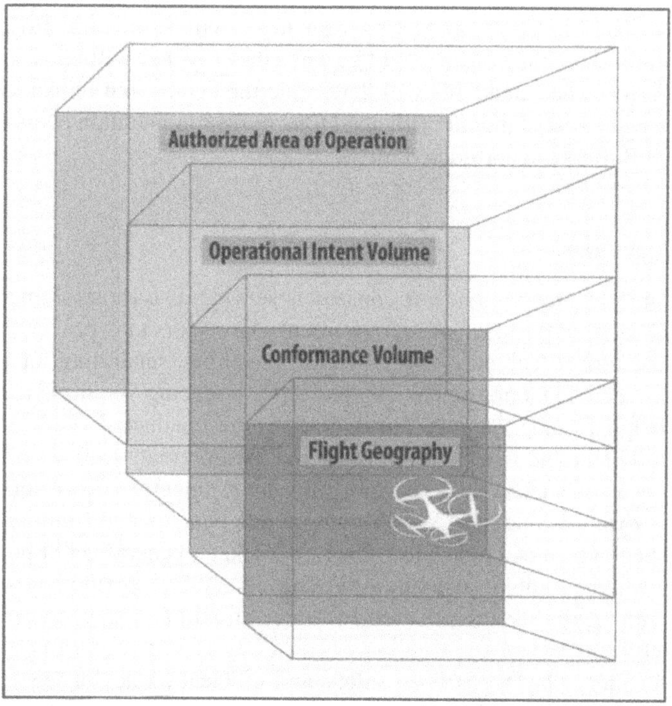

FIGURE 4.3 Notional operation volumes.

The relationship between the OIV and CV deserves particular emphasis. Depending upon applicable rules, the OIV is the 4D volume shared (i.e., communicated) with UTM participants who potentially share the same airspace, and with relevant USS. The CV is the 4D volume within which the operator *trusts* that the UA will operate, based on the accuracy of on-board and/or ground-based sensors, UA performance, and operating conditions.

- exceeding the CV should ultimately result in the UAS operator determining the cause (and hopefully preventing reoccurrence), but **not** notifying any other UAS operator (or regulator) of the incident. Exceeding conformance should be considered an *internal* issue.
- exceeding the OIV should result in *immediate* notification to other UA operators—and the declaration of the UAS as "nonconformant" [or contingent].

"Conformance" requires both: (a) staying inside the CV, and (b) staying inside the OIV. The actions taken when exceeding conformance of either are significantly different.[794] (*See* Chapter 2, Section 2.3.6 *Conformance Monitoring*).

OIVs are temporal "and this is likely to become important with increasingly autonomous, small-sized vehicles operating in a strategically-separated environment. With many air vehicles planning and re-planning routes that may intersect, and that may be of short duration in a limited altitude band, the time element of planning will be complex and dynamic. Consequently, it will be important to be assured that drones are where they say they will be when they say they will be, with a margin for error. Performance standards will determine the details, but implementation in accordance with those standards will need *time buffers* to maintain overall strategic integrity."[795] Buffers are considered further below.

4.4.3 Corridors

Corridors are airspace volumes or constructs serving as a route segment or passage with associated performance requirements. Corridors may provide additional airspace structure to enhance safety by distinguishing, separating, or bolstering traffic flow, reducing operational complexity,[796] reducing required routes,[797] or generally helping guide where UTM operations are conducted. They are viewed as particularly suitable where free routing risk is too great, such as in high-density areas, near obstacles, or around airport environments.[798] Corridors come in two types—*dynamic* and *static*, each with unique benefits and limitations. Their respective deployment should follow Dr. Kopardekar's general UTM paradigm of "flexibility where possible, structure where necessary."[799] Corridors may be static but used with flexibility (as affected by time, direction, speed or performance requirements.[800])

Static corridors play recognized safety and efficiency roles in legacy aviation for certain busy airspace. Such corridors are deployed and depicted in aeronautical charts as flyways, helicopter routes, related constructs, and have more recently informed notional development of UAM-specific corridors, described below.[801] Flyways in the complex San Francisco Bay area airspace are depicted in Figure 4.4 as a series of segmented arrows, or corridors.[802]

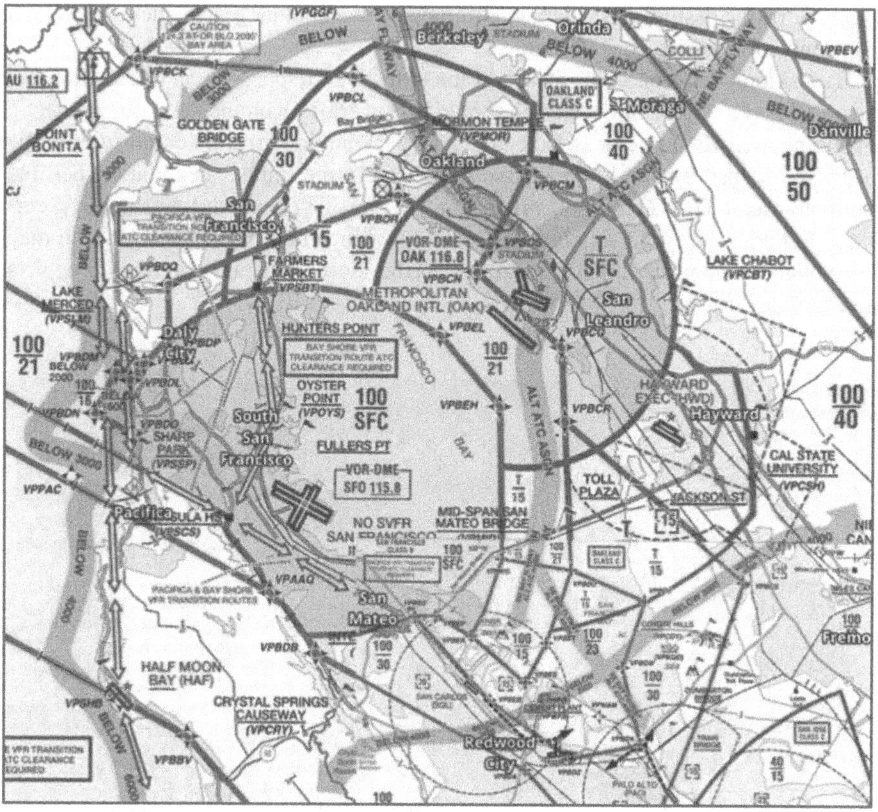

FIGURE 4.4 Flyways. (Source FAA.)

For UTM, static corridors are generally considered transitional—of possible utility until traffic management becomes further advanced/integrated.[803] They offer important benefits such as unparalleled situational awareness particularly for non-participating traffic, e.g., manned aircraft. Looking forward, the envisioned role of dynamic corridors is introduced by Tom Prevot:

> Building a dynamic structured network of routes is going to be key to enabling any type of scalable operation. Structured route systems are needed that can dynamically shift. Corridors are more "static" than the routes within them. Route networks within a corridor can shift just like on a bridge when you have the number of lanes per direction depend on traffic conditions.[804]

UTM corridors have been proposed for use within diverse airspace classes, including Class B, "to operate class G-like."[805] Corridors may terminate at or near the planned destination, or connect to other corridors or airspace to form a route, and may be created per flight by the USS across areas where defined routes do not exist.[806] Corridor design should incorporate safe distances to account for error and accommodate operational constraints such as geofences, obstacles, and conflicting flight paths.[807]

Static corridor design must include community stakeholder engagement as well as environmental justice considerations.[808]

Transitions between controlled and uncontrolled airspace, special use airspace, airports, and corridor intersections should reflect traffic volumes and other factors, including their dynamic or static character.[809] Dynamic corridors could initially be restricted to use by single participants, one at a time, unless negotiation permits multiple users. The UTM itself might only be capable of scheduling time in a corridor, not creating new corridors. The extent to which non-participating traffic may fly though or intersect corridors requires further development.[810] Figure 4.5 depicts an example corridor.[811]

FIGURE 4.5 Example corridor. (Derived from and reprinted with permission from OneSky Systems.)

Corridors have been proposed within the FAA's UAM ConOps. "The FAA will define, maintain, and make publicly available UAM corridor definitions and will manage the[ir] performance requirements,"[812] and further describes them as:

> performance-based airspace structures with defined dimensions . . . . known to airspace users and governed by a set of rules which prescribe access and operations. Where supporting infrastructure and support services meet performance requirements within a UAM Corridor, operators whose aircraft also meet performance requirements will be able to operate within the UAM Corridor.[813]

Figure 4.6 depicts a notional corridor for UAM as envisioned by the FAA.[814]

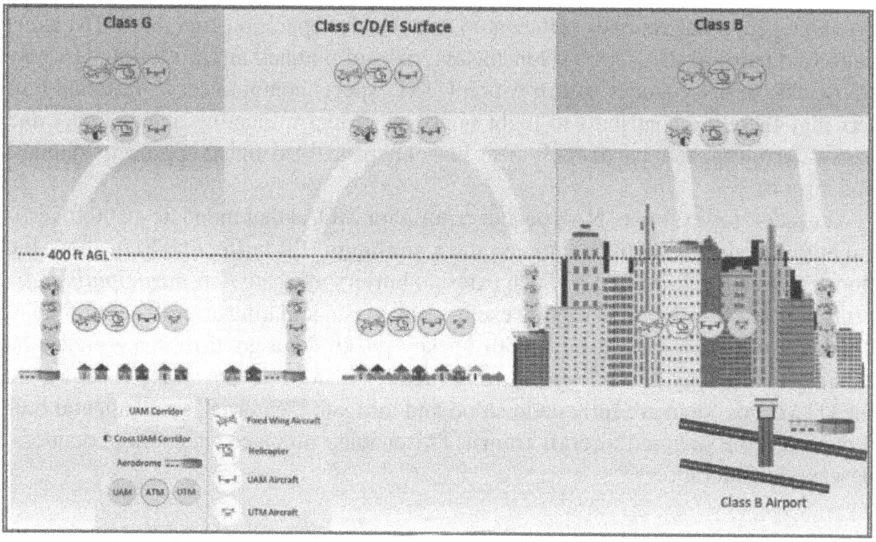

FIGURE 4.6 UAM notional corridor. (Courtesy FAA.)

Other proposals have described "specific aerial corridors [for sUAS flight] within a predefined aerial network,"[815] and "drone safety tubes [to] cocoon[] the drone."[816] Whatever the nomenclature, the "need for predictable repeatable navigable routes is recognized."[817]

Jeff Homola of NASA suggests, "[i]n the early days … corridors probably need to be more static structures, more predicable in those types of airspaces [near airports]. Away from airports we can look to more dynamic corridors."[818] Airport environments are often complex and highly structured, with choreographed arrivals, departures, and overflights. Corridors can be created as part of airspace design and reused to control traffic flow into commonly used locations such as airports, heliports, and vertiports.[819] The use of static corridors may also be particularly suitable[820] because "[w]e don't have a terminal area that is capable of [use of dynamic corridors]. We need a broad look at the infrastructure before this is done."[821] UAS corridors connecting and proximate to airports should be clearly defined, charted, and made available to both manned and unmanned pilots to enhance situational awareness and flight safety.[822]

There are important voices in the UTM development community that recognize corridor limitations, particularly in the "longer-term." Robert Champagne of the ASTM F38 UTM Working Group explains:

> We have tried to stay away from the concept of corridors. I can see the attraction (easily managed, etc.), but they are also incredibly limiting in a world where there is no physical infrastructure to limit travel. We have wide open spaces, so why should we limit air traffic? Look at automotive roads today—you only move as fast as the slowest vehicle, travel is restricted to existing routes, etc. Sure, in the short-term it makes things easier, but we need to at least be *thinking* longer-term on UTM/UAM …. I don't want to overload "constraints," but yes, a static volume that indicates noncooperative traffic (i.e., not equipped to deal with UTM) makes sense to me here.[823]

External Buffers—External buffers are generally outside of the UTM ecosystem. A UTM design premise holds that required navigation performance (RNP) for UAS may provide operational volumes sufficient to ensure safe separation, and that UTM alone can avoid harm to the NAS.[824] Nonetheless, risks of manned aircraft hitting UA may transcend the 4D volumes within which UTM expects containment.[825] External buffers may therefore contribute to flight safety by further mitigating manned pilot and aircraft error hazards particularly near busy airspace, instrument approach volumes, and airports.[826]

Consider, for example, NAS design criteria for MOAs that mandate a 500 ft vertical buffer from IFR flight operations, and a minimum IFR buffer of 300 ft. above the floor of controlled airspace.[827] Such external buffers mitigate *non-participating aircraft pilot error* rather than relying exclusively on participating aircraft containment. "Today, we have a system built with hockey pucks for a good reason—under the existing technology, *those buffers are needed*."[828] Until relevant safety risks within the UTM ecosystem are fully understood and mitigated, external, supplemental buffers separating manned aircraft from UTM-enabled airspace (including corridors) deserve consideration.[829]

4.4.4 Fair Access

"We can't let this become like ticket master."

Steve Bradford
Chief Scientist for Architecture and NextGen Development, FAA[830]

"Quantifying fairness is inherently difficult."

Airbus UTM[831]

Some airspace users could monopolize airspace to the disadvantage of other users, particularly as access becomes reliant on automated systems. Fair access to airspace addresses this issue. "Equitable airspace access is an essential concept that needs to be understood and embraced by the various users of the NAS to maximize value to the United States"[832] and the "guarantee[d] equitable and fair access to airspace for all users" is a key principle of U-space.[833] Fairness issues involve both air and ground (such as access to aerodrome and vertiport facilities). Expectations for fair airspace access extend increasingly to all altitudes for both manned-unmanned, and unmanned-unmanned usage, and include route prioritization, demand-capacity management, and deconfliction.[834] "[T]he inability to resolve conflicting requests for resources will eventually be an impediment to integration of new entrants."[835]

Certain operations are understood to require priority, including emergencies and public safety.[836] Additionally, for efficiency, certain routes may have distinct, justifiable prioritization rules and follow multiple paths with different prioritization schemes,[837] and "some prioritization paradigms may not work for unmanned aircraft."[838] More generally, what criteria should govern prioritization among conflicting flight plans, e.g., *first-come, first-served*,[839] or should the performance capabilities of the aircraft determine priority, e.g., *best equipped, best served*? Or, BVLOS over

VLOS operations? Should a lottery control access, or should the USS be allowed to negotiate over volumes of airspace? Should access depend on the relative need of each aircraft to fly a specific route? For example, "[t]here may be multiple ways to get from point 1 to point 2 (e.g., for package delivery), whereas someone doing house inspection truly has to be there."[840] Alternatively, should access be assigned randomly except for public safety and national security operations? Should multiple negotiations be permitted, and can more than one negotiation be undertaken simultaneously? And should system abusers, such as those submitting unbounded (time or airspace volume) intent notifications or engaging in intractable negotiations be penalized by deprioritizing or denying their flight intent requests?[841]

Design principles should "[e]nsure that availability and access to airspace take into account the needs of all stakeholders."[842] NASA asserts that a prioritization scheme must "be deterministically calculable by each USS given the same operation data."[843] Furthermore, it urges the development of "metrics to establish fairness of Operational Volume definitions to self-regulate the community to encourage UAS Operator 'good behavior' (i.e., anti-gaming)."[844] Otherwise, "the FAA may be required to manage the demand for access."[845] To ensure fairness in deconfliction, flight planning should take into account the opaque and subtle latency variances among networks—not unlike network impact on Wall Street trading. It has been urged that "this game of drones is ultimately a network remote ID and DSS networking game to be resolved by lowest latency for which tier 1 owners of fiber may ultimately battle to control the UTM ecosystem."[846]

"[E]quity of airspace access for UTM operations is fostered through operation orchestration/[o]perator negotiation …. "[847] Various technologies for advancing fair access are under consideration including automated negotiation and artificial intelligence (AI).[848] Over time, perhaps sufficient data will be collected to inform an automated mediation or "conflict resolution protocol" system to best achieve fairness.[849] As a practical matter, in the near-term, prioritization will be "decided the hard way by trial and error. In the beginning, prioritization will not be very efficient."[850] As observed by Reinaldo Negron, Head of UTM, Wing, "There is still more learning to be done especially around fairness."[851]

4.5 SPECTRUM

> "Spectrum management is going to become very difficult—very complex."
>
> **Waseem Naqvi**
> *CTO, Raytheon Unmanned Systems*
> *Presentation at AUVSI Xponential (Oct. 8, 2020)*[852]

Radio frequency spectrum[853] is essential to UAS and UTM for command and control, navigation, DAA, payload (e.g., real time transmission of video or sensor data), vehicle-to-infrastructure (V2I), and future vehicle-to-vehicle (V2V) communications. The forecast demand for applications such as commercial BVLOS operations enabled by UTM exceeds the currently allocated aviation spectrum, including for UAS.

In the effort to marshal this scarce resource, there have been several discussions on how to best utilize non-aviation spectrum, particularly for sUAS. Such spectrum could be either licensed or license exempt. Commercially licensed spectrum is not necessarily aviation-specific and certain spectrum bands may be prohibited for aeronautical use. Unlicensed spectrum is not entitled to interference protection, can be congested, and is viewed as the least reliable. Unlicensed spectrum is specifically viewed as not suitable for BVLOS operations, operations over people, complex operating environments, and operations over mission critical assets.

The spectrum that will be suitable for any given UAS operation will depend on many factors including the concept of operations and the capabilities of the spectrum, the radio technology, including the propagation characteristics, transmitter power, operational altitude, receiver performance, and operational environment.[854] This could lead to a hybrid approach where different types of spectrum can be utilized, such as commercial LTE and commercial satellite.

Some initiatives propose combining commercial licensed spectrum, and unlicensed spectrum, through spectrum sharing strategies.[855] There is need to deploy systems to coordinate usage of the shared spectrum between different users to minimize interference. Regulations should clearly define the condition for using the shared spectrum, including transmission power levels, locations, times, exclusion zones, and guard bands, to minimize interference. There are few examples of spectrum sharing globally, and it will take considerable time before deployment occurs in actual infrastructures. Spectrum sharing strategies may require (or be optimized with) software-defined radios providing dynamic channelization, and allocation of spectrum resources for different operations.[856] "There are so many spectrum resources and they have to be woven together, and they should be smart, and use dynamic frequency allocation."[857]

The critical need for commercial spectrum resources beyond Aeronautical Mobile (Route) Service (AM(R)S) spectrum allocated to aviation for UAS has compelled industry collaboration to advance legislative and regulatory reform to permit use of non-aviation-specific spectrum. For example, the FAA Reauthorization Act of 2018, Section 374 "Spectrum":

> requires the FAA, NTIA, and the Federal Communications Commission to submit to Congress a report on whether UAS operations should be permitted to operate on spectrum designated for aviation use. The report shall also include recommendations of other spectrum frequencies (such as LTE) that may be appropriate for flying UAS.[858]

The mandated report was submitted to Congress on August 20, 2020, concluding in part:

> The 5030-5091 MHz band as well as certain flexible-use bands are potential options for supporting such UAS communications. The 5030-5091 MHz band appears to offer promise for intensive UAS use because it is unencumbered, but that band poses some technical and regulatory issues that require further review before UAS operations may be permitted.[859]

Other initiatives may affect the future of UTM spectrum allocation and management, such as rulemakings at the Federal Communications Commission (FCC), the FAA, the NTIA's Commerce Spectrum Management Advisory Committee (CSMAC),[860] and reports or petitions by other stakeholders such as the Aerospace Industries Association (AIA).[861] The standards-making community, including RTCA and 3GPP, also are addressing spectrum for various UAS operations. The work of the International Telecommunications Union (ITU) in the World Radio Conferences also has import.[862] Federal programs to make certain radio frequency spectrum available for non-federal use[863] may also benefit UTM in the longer term.

Although commercial licensed spectrum is already in use for UAS at lower altitudes, a nationwide network of protected spectrum with spectrum sharing has been envisioned for UAS operating at higher altitudes, in spite of the long-time horizon for implementation.[864] Notwithstanding, "alternative approaches that are already being used such as existing terrestrial LTE networks alongside satellite—are provided by existing networks and regular cellular and satellite spectrum."[865] 5G cellular service should increasingly provide further relief and new capabilities for UAS and AAM/UAM.[866] *See* Chapter 3, Section 3.2.3 *Cellular Industry*. Additionally, effective UTM governance may stimulate the development of cloud-based solutions responsive to UAM that embrace edge computing, the converged tower concept,[867] fiber optics, and beneficial features offered by "smart cities."[868]

4.6 STATE, REGIONAL, AND LOCAL

"If the community is not involved, they will reject it."

Jay Merkle
Exec. Dir., UAS Integration Office, FAA[869]

UTM infrastructure has been developed primarily by private enterprise, national governments (CAAs), and major inter- and non-governmental organizations, but as UTM nears deployment, state, regional, and local governmental entities become more and more important. The primary challenge (and focus) has been in first standing up a "version 1.0" of UTM with its technical interoperability requirements. However, since the airspace and ground infrastructure span many administrative and political jurisdictions, there are challenges incorporating their participation in UTM governance.[870]

While in many countries national aviation regulators have exclusive jurisdiction over airspace, the role of state, regional, and local entities in UTM development and governance transcends their basic public safety mandate, extending to transportation planning and oversight,[871] economic development[872] (including enterprise zones and Silicon Valley-like technology hubs), neighborhood preservation and improvement,[873] zoning,[874] communications infrastructure,[875] critical infrastructure, environmental impact,[876] and more.[877] State, regional, and local stakeholder interests must therefore be considered[878] or else UTM risks being developed without necessary consideration of stakeholder operating environments which could lead to failed deployment or disrupted operations.

Ceding any national authority requires careful analysis, and a full understanding of the extent to which it may hamper a seamless, interoperable, and efficient UTM. For example, the FAA Reauthorization Act of 2018 ordered a study of the roles of federal, state, local, and tribal governments in regulating low-altitude operations of UAS,[879] and the FAA's UAS Integration Pilot Programs (IPPs) focused on "balanc[ing] local and national interests related to drone integration" to produce actionable insights.[880]

There are tensions between overlapping jurisdictions in assuring both a fully integrated UTM and accommodating legitimate local control,[881] particularly regarding the reasonable time, place, and manner of low-altitude flights,[882] and responsive localized UTM services. Consider, for example, comments regarding the *Uniform Tort Law Relating to Drones Act*,[883] a proposed uniform state law highlighting the vexing and far-reaching challenges in harmonizing local vs. national governance:

> [the] draft Uniform Law ... threatens to do damage to the integrity and substantive justice achievable under property law, both in the area of drones and more generally. [It] substitutes wholesale a bespoke regime under which every intrusion no matter how close to the ground or severe in its effects is subject to an open-ended 13-factor test that will scare off potential plaintiffs and more generally threaten to destabilize the law of property in other areas ... this radical departure from existing law ... bears no resemblance to any existing law and performs the impressive feat of outdoing nuisance law in its vagueness.[884]

Recognition of the centrality of regional and local stakeholder engagement has been elevated within the FAA for UAM by advancing community-based rules (CBRs):

> the FAA will engage stakeholders to set high level principles and assumptions on performance with an understanding that the principles and assumptions will be implemented as required. Stakeholders will develop and adopt applicable CBRs related to collaboration, information sharing, operational protocols, and equipment performance to support the automated nature of the envisioned cooperative environment. This concept will mature to encompass increasingly complex operations in heavily populated environments and more heavily utilized airspace.[885]

In sum, "local authorities—cities and regions—should be involved from the beginning to manage societal concerns and ensure integration in the urban and regional environment. This should not lead to fragmentation."[886] "There is never too much community engagement."[887]

4.7 UTM ADMINISTRATION

4.7.1 Intro—UTM Administration

Striking the right balance between governmental oversight and industry self-governance and administration is a key challenge to UTM's success. There is need to "identify the new roles and responsibilities in the UTM ecosystem,"[888] and encourage and engage expertise within industry and government.[889]

4.7.2 Industry Consortia

UTM industry consortia can provide a collaboration framework to advance selected governance-related issues and functions such as those associated with standardization, safety promotion, and certain airspace prioritization fairness and usage.[890] Consortia are diverse in terms of mission, formality, organizational structure; private vs. public nature, sectorial concentration, and temporality; and their international, regional, national or local focus.[891] UTM consortia can be stood up by a single industry or governmental entity, organized by multiple entities, or could represent a team of companies collaborating on a request for information (RFI) or contract bid.[892] Consortia can be quite dynamic—created *ad hoc*, or reconstituted in response to market and regulatory conditions. There is not necessarily a bright line demarcating various types of consortia "[nor an] agreed definition of the term 'industry consortium'."[893]

The leading UTM industry consortium is the Global UTM Association (GUTMA - *see* Chapter 3, Section 3.3.3), self-described as "a non-profit consortium of worldwide Unmanned Aircraft System Traffic Management (UTM) stakeholders [whose] purpose is to foster the safe, secure and efficient integration of drones in national airspace systems [and with a mission] to support and accelerate the transparent implementation of globally interoperable UTM systems."[894] It further seeks "to act as an advisory body, totally independent and equally focused on the key stakeholders of the industry [and] liaise with civil aviation authorities (CAAs) to develop interoperable and harmonized regulations .... "[895] GUTMA has forged collaborative initiatives that themselves can be characterized as consortia, such as with the cellular communications industry. *See* Chapter 3, Section 3.2.3.3 *ACJA*. Additionally, the UTM's industry promotion of open source technology through the InterUSS Platform and Linux Foundation present noteworthy examples of consortia advancing UTM (*see* Chapter 2, Section 2.4.5.3).

As a model illustrating how consortia could contribute to UTM governance, consider an example from another highly regulated industry—the banking industry. Nacha (previously the National Automated Clearing House Association) governs the automated clearing house (ACH) network, and marshals necessary operating rules, technical standards, and administration.[896] As a consortium Nacha demonstrates many attributes necessary for effective UTM governance: trust, expertise, capacity to help facilitate high-volume operations, development and administration of standards, promotion, dispute resolution, operation within an intensively regulated environment, and the influence and gravitas to help forge effective regulation. According to Jane Larimer, President and CEO of Nacha,

> Self-regulation is a powerful tool for the private sector. Even in highly regulated industries, industry-led rules and standards bodies (industry consortia) allow the private sector, those with the most intimate understanding of the specifics of the industry being governed (including the systems and technology used, and the role and responsibilities of each type of participant in the system) to develop fit-for-purpose rules/standards in a fluid and responsive manner. While good regulation creates a broad framework of accountability, industry-set rules and standards fill any regulatory gaps and operationalize accountabilities with more specific details and understanding of the particulars of the system. This layering of standards, rules and

> regulations ensures system participants have certainty about their responsibilities and potential liabilities to other participants in the system. This creates an environment of trust, where participants agree on the rules of the game and where there is certainty around outcomes.[897]

Soliciting expertise from mature entities such as Nacha (or individuals with relevant expertise) for UTM consortia is consistent with seeking out best practices and lessons learned at the leading edge of such regulated industries.[898]

UTM consortia have played key roles in diverse governmental initiatives to advance UTM research and pilot projects, including with ANSPs, CAAs, JARUS, NASA, and SESAR.[899] For example, *see* the collaborative NASA initiatives described in Chapter 1, Section 1.5 *Research and Development*, and correspondingly by SESAR in Chapter 5, Section 5.3 *U-space*. Additionally, various legacy aerospace associations have supported consortia.[900] Standards development organizations may act as consortia, yet have also been described as "immediate precursors to consortia."[901] As the preeminent intergovernmental organization addressing UTM, ICAO facilitates and participates in consortia.[902] There is a growing list of organizations engaging in consortia to address UTM governance, including, for example, traditional humanitarian non-governmental organizations (NGOs).[903]

4.7.3 Business Models and Competition

Economic models for UTM affect governance, and raise important questions: How will UTM's financial burden be shared?[904] Will fees generated from UTM services sustain the UTM industry? Will USS be permitted to monetize their operational safety data or be required to make such data freely available, including to manned aircraft pilots?[905] What roles or services should or must legacy ANSPs provide or avoid?[906] How will UTM governance accommodate business models advanced by service providers?

In those nations permitting delegation of certain UTM operations, private profit-making models are being deployed.[907] "Other industries provide a glimpse into the potential models available, such as fee-based or software as a service (SaaS) models, where operators may receive basic services free of charge, but require higher investment for more advanced needs, scaled operations or higher service levels."[908] MNO business models generally reflect their highly regulated environments. They are frequently tailored to specific vertical markets and customer performance requirements and could affect UTM service provider viability and governance.[909]

Current and proposed UTM business models vary widely, each responsive to differing needs of the states, suppliers, operators and other participants[910]. Many industries are already heavily invested in one of these business models and may not easily transition to another. While operator revenue is a prime consideration, some business models may demand industry consolidation. One view holds that "[t]he UAM industry is likely to be consolidated due to the autonomous nature of its vehicles, which need to be coordinated by a centralized platform. As a result, the elimination of unnecessary competitors will lead to sufficient pricing power for the remaining operator(s) to ensure reasonable investment returns."[911]

It is unlikely that large companies such as an Amazon or Alphabet would either require or would accept a third-party system controlling their drones—and their data. Alternatively, a small medical delivery company with a niche market would seek a third-party provider because the alternative would be cost effective. Some companies build their product/service catalog to meet needs across the spectrum of business models, and governments seem receptive to this. For example, the FAA allows companies to operate their own system within LAANC for internal flight operations while simultaneously approving LAANC providers as third-party providers.[912]

In the aviation sector, safety permeates business model considerations, requiring close governmental coordination. Other UTM business model factors affecting governance include interoperability, fairness, and nondiscrimination,[913] Viable, well-established operating rules understood by the stakeholders are important, and for AAM/UAM, business models must contemplate stakeholder development of Community Based Rules (CBRs).[914] Additionally, antitrust law plays a material role in shaping future business models in the future.[915] Historically, ICAO recognized "the absence of competitive market forces calls for prudent governance with respect to the provider in order to avoid abuse of monopoly power."[916] The extent to which ICAO and state regulators engage competition issues should reflect future UTM market structure developments.

4.7.4 Liability

The envisioned UTM ecosystem is complex with diverse participants and relationships—including operators, service providers, customers, governments, NGOs, and communities. Establishing the legal rights and responsibilities of UTM participants and other affected persons, and determining liability have been urged by ICAO and others.[917] The body of potentially applicable law is extensive, transcending basic aviation law,[918] and includes rapidly evolving autonomous systems, data rights and privacy, and competition law.[919]

Absent law to the contrary, UTM service providers and UAS operators can be expected to apportion liability among themselves contractually. Service providers will generally seek to disclaim and limit liability; governments tend to assert sovereign immunity as a defense to claims[920] while seeking to hold service providers and operators accountable; and certain classes of stakeholders such as consumers and passengers may garner enhanced protections. As a practical matter, the most protected class will be the non-participant—particularly those on the ground.[921] Non-participants may have various common law or statutory claims against UTM service providers causing them injury or property damage, such as negligence, invasion of privacy, trespass, nuisance, trespass to chattels, and conversion.

From a regulatory perspective, the responsibilities between UAS operators and USS are in play and potentially complex.[922] One industry leader suggests, "I think it's going to be quite a complex answer [and] not a topic that we want to shy away from."[923] Another industry leader suggested that "putting the onus on the USS is the right direction;"[924] while another asserts, "no one party will own 100%" of the responsibility."[925] Yet another industry leader asserts, "[a]s we move towards

USSs, at some point some liability might be on the USS in service level agreements that assign responsibility for certain functions like strategic deconfliction. So, USSs will take on responsibility that does not exist in LAANC where the ultimate responsibility is on the pilot."[926] Additionally, NASA asserts, "[i]n a future operational system, operators are ultimately responsible for the safety of their operations and knowledge of the involved systems," and the FAA UTM ConOps holds operators "ultimately responsible for maintaining separation from other aircraft, airspace, weather, terrain, and hazards, and avoiding unsafe conditions throughout an operation."[927]

Richard Parker, Founder and CEO of Altitude Angel described the UTM service provider responsibility situation as follows.

> The general principle the PIC is the responsible party is something we hold very dear to our hearts. We're very clear, if we are going to be involved somehow in the management or execution of a flight, then the parts we are responsible for have to be carried out safely and securely.
>
> The first thing we have to do is explain the capabilities and the limitations of the services being used. So, for example, if we're building a deconfliction engine it should be clear what it can do and what it can't do. And it should also be clear what are the expected outcomes of the deconfliction engine. We are frequently asked, "can you guys use artificial intelligence in your deconfliction engine? We could, but do we really want an emergent system with a nondeterministic series of outcomes?
>
> We have to back that with a legal framework which allows and tracks UAs properly. When an instruction is given, and how that instruction is carried out, who is that party responsible for carrying that out? Our view of deconfliction is to supplement onboard systems and onboard autopilot flight control systems effectively, a little bit like a co-pilot in many respects. So, for example, rather than taking direct command and control of the flight, we're able to 'nudge' the flight with regular information about changes it might need to make to its flight path in order to avoid an upcoming conflict. So that approach is minimally invasive, it doesn't require the UTM to actually have an interface aboard the vehicle; it doesn't require the UTM to actually speak a specific proprietary drone control protocol, but it does allow the UTM to standardize a set of messages that are common and ubiquitous to everyone. So there are lots of different angles here which need to be explored."[928]

The U-space regulation provides for minimum requirements for UAS operators and U-space service providers, that are "proportionate to the nature and risk of the operations"[929] and requires common information service (CIS) and U-space service provider demonstration that they "use systems and equipment that guarantee the quality, latency and protection [and have agreements] specifying the allocation of *liability* between them."[930]

In fashioning a UTM liability regime, the capabilities of each party, current rules, agreements,[931] and practices between CAAs and service providers deserve consideration. The FAA's MOA for LAANC provides an early example informing future UTM liability regimes, but addresses neither the extensive and necessary performance requirements that still need resolution, nor limitations of contract between participating service providers and other stakeholders:

ARTICLE 2. SCOPE …

2.2 Goals and Objectives to be accomplished.

The parties are bound by a duty of good faith and best effort in achieving the goals of this Agreement ….

ARTICLE 14. LIMITATION OF LIABILITY

The parties agree that the FAA assumes no liability under this Agreement for any losses arising out of any action or inaction by _____________, its employees, or _____________'s Representatives. The parties agree that _____________ assumes no liability under this Agreement for any losses arising out of any action or inaction by the FAA or its agents, officers, employees, or representatives, provided, however, that this limitation of liability does not apply to or limit the liability provided for in Articles 15 and 18 [Indemnity and Protection of Information, respectively] of this Agreement. _____________ agrees to reimburse the FAA for any damage to or destruction of FAA property caused by _____________ or _____________'s Representatives arising out of activities under this Agreement to the extent permitted by law.

Claims for damages against the FAA of any nature whatsoever pursued under this Agreement must be limited to direct damages only up to the aggregate amount of the funding obligated under this Agreement at the time the dispute arises. In no event must the FAA be liable for claims for consequential, punitive, special, or incidental damages; lost profits; or other indirect damages.[932]

Other instructive datapoints include the SWISS Remote Identification (SRID) *Master Agreement* addressing service provider liability and responsibilities;[933] and the FAA Data Disclaimer of the UAS Data Delivery Service (UDDS) which includes the FAA's UAS Facility Maps used in LAANC and UTM trials:

> The Federal Aviation Administration (FAA) ("We") will not be held liable for any improper or incorrect use of the UAS Data Delivery Service (UDDS), and we assume no responsibility for anyone's use of the data from UDDS ….
>
> To the maximum extent permitted by applicable law, in no event will we be liable for any direct, indirect, punitive, special, incidental, indirect, or consequential damages whatsoever (including, but not limited to, damages for: loss of use, data or profits arising out of or in any way connected with the use or performance of UDDS; loss of confidential or other information; business interruption; personal injury; damage to reputation; death; loss of privacy; failure to meet any duty (including of good faith or of reasonable care); negligence; and any other loss whatsoever) arising out of or in any way related to the use of or inability to use UDDS or the provision or failure to provide UDDS or otherwise under or in connection with any provision of this agreement, whether based on contract, tort, negligence, strict liability, or otherwise even if we have been advised of the possibility of such damages.[934]

Terms for various aeronautical information services with the FAA, such as with Jeppesen,[935] and for certain ADS-B services are instructive and may contribute to a sustainable scheme for UTM liability apportionment.[936]

Within the data provider industry, there is a lengthy history of service agreements covering governance and liability considerations developed over the history of manned aviation. While some new sources and types of data are anticipated in the UTM framework, most data sources and streams will overlap those used in the current ATM and operator control paradigm. Therefore, the liability restrictions and concerns are likely to be similar.[937] Nonetheless, if, for example, the envisioned International Aviation Trust Framework is deployed, its liability constructs and liability limitations may affect UTM service providers.[938]

Because data providers are generally aggregators of source data generated by other entities over which they have little or no control—and portions are often even generated by the end-user of the dataset for proprietary purposes—liability is generally limited to modification of data after it has been received from the source, or to the segment of data the providers produce themselves. The validity and reliability of the data is ensured by, among other measures, adherence to industry standards.[939] Compliance with these standards is confirmed or designated via the issuance of Letters of Acceptance, government certifications, or other qualification regime. Current USS-operator agreement and ANSP liability schemes should also be considered.[940]

Finally, UTM liability will, in part, be informed by the robust body of law and practice addressing ATC.[941] This embodiment includes both contract and tort—domestically as well as internationally—and involves CAA delegation to ANSPs, and possible ATC privatization.[942] It is clear that many challenges remain, including that "[t]he standard of care to apply to [ATC] is bound to change constantly, following developments in aeronautical engineering and ATC technology."[943] Indeed, sorting out UTM liability will be challenging, and will go hand-in-hand with resolving general UTM operational rules.[944]

4.7.5 INSURANCE

"The insurance industry provides another model of private governance."

Daniel W. Woods, DPhil
University of Oxford[945]

ICAO recognizes a gap regarding "liability and insurance implications for [USS] in relation to UAS operators."[946] The insurance industry seeks to reduce the likelihood and size of losses suffered by its insured. It could do so by requiring compliance with risk management measures, by sharing information about risk factors, or by offering discounts for adopting risk mitigation measures. Policymakers across the EU and USA have discussed the potential for insurers to play this role, including in improving cybersecurity.[947] Insurance can affect cybersecurity requirements, broader risk mitigations, and the economics and governance of UTM. The formulation of UTM insurance regulation is in play.[948]

An early insurance datapoint is the FAA's MOA with LAANC platforms that requires USS to arrange for insurance against all liability to third parties as follows:

ARTICLE 13. INSURANCE

_________________________ must arrange by insurance for reasonable protection of itself from and against all liability to third parties arising out of, or related to, its performance of this Agreement.[949]

One influential CAA provides that "[t]he parties shall have an adequate insurance coverage, for their activities under this MoC and any damage resulting thereof, at their own expenses."[950] Similarly, the EU U-space regulation requires a U-space service provider to demonstrate that it "has the required liability and insurance cover appropriate to the risk of the service(s) provided."[951] Service provider agreements' insurance provisions, and early UTM insurance products for operators may also inform future UTM insurance practices and requirements.[952]

Additionally, recognized security frameworks (*see* Chapter 2, Section 2.4.7 *Cybersecurity*) may provide a model for the insurance industry to improve underwriting consistency, establish actuarial data needed to reduce risk, and ultimately reduce claims.[953] These mechanisms represent policy-measures that should broaden adoption of UTM insurance, thereby increasing the scope of firms influenced by insurers. In essence, meeting certain security requirements might represent a minimum standard to be eligible for coverage. Satisfying additional security requirements could be used to justify increased coverage, reduced premiums, reduced deductibles, and fewer carve outs/exceptions, etc. In this regard, to the extent cyber liability insurance is required, the insurance industry could influence governance.

Insurance industry market dynamics demand consideration as well. For example, "[i]nsurers face incentives to put mechanisms of control in place to reduce cyber losses" that may influence the trustworthiness of UTM and its governance.[954] Nonetheless, traditional insurance compensation via brokerage commission tends to "direct applicants towards insurers with the least stringent application process ... creating a race-to-the-bottom in risk assessment."[955] To the extent these dynamics affect UTM insurance, the need for heightened oversight is likely indicated. Developing efficacious insurance will benefit from formal UTM-insurance industry engagement.

4.7.6 Data Rights and Privacy Management

Since a complete understanding of the traffic situation (i.e., the airspace representation) is required to operate a DSS, and diverse, extensive information is required for the UTM, consideration of their data exchanges and associated data rights and privacy management is imperative. The unconstrained monitoring, processing, collection, transfer, and storage of certain communications could result in the improper disclosure of information,[956] in the reverse engineering of routes, could disadvantage some competitors while helping others optimize their respective routes, provide a lens on the competition's customers and operations, and could have implications for national security. Such unconstrained activity may also violate the privacy rights of individuals or create privacy liabilities.

In response, the UTM system should be designed to mitigate some of these risks by avoiding the collection or other processing of personally identifiable information (PII) and storing only the minimum required data.[957] The DSS already takes this into account. For example, the identity of UAS and its operator, and certain location and mission data for constraints is created without sensitive/private information shared between USS. For flight operations, traditional PII is *not* stored in the DSS or passed in the USS-to-USS interfaces. Moreover, for RID, since the display provider view area is limited in size, and the "breadcrumb" trails are limited to 60 seconds, the system is intended to limit full (origin to destination) mission visibility while still providing real time awareness of a particular airspace.[958] For example, business to household relationships cannot easily be inferred, and other airspace usage notices can be created without disclosure (to the public) of who is occupying the area.

> The complete details of the entities and any associated Personally Identifiable information (PII) are not stored in the DSS but instead are retained by the owning USS; only limited information such as the type of entity, its location (in terms of what cells of the airspace model it intersects), the current opaque version number (OVN) of the entity (OVNs are updated whenever the entity is modified), and how to contact the owner of the entity are stored in the DSS.[959]

Since the DSS service knows the complete airspace representation (for its region), one could expect DSS to become a material evidentiary source[960] of operational/transactions data. However, this representation is at a "certain coarseness corresponding to the resolution of the grid cells that compromise the airspace representation," and the DSS must discard precise extents of operational volumes once mapped into the airspace representation.[961] Technical and legal requirements for DSS and USS data persistence and retention also deserve attention.[962] (*See* Chapter 2, Section 2.4.9 *UTM Data Exchange*). Because some of such data may be considered confidential or proprietary, the DSS should at least meet the same privacy and security requirements as a USS.[963] The following data retention schema in the ASTM Remote ID specification is instructive.

> (3) Discard data: all UAS Remote ID data acquired by a Net-RID Display Provider from other Net-RID Service Providers must be discarded within a specified time period after the data is received .... Disposing of data when no longer needed (in combination with limiting data access to only what is required for display areas being viewed by a Net-RID Display Provider's associated Display Applications) helps protect privacy and sensitive data of operators and, if applicable, their customers.[964]

For nominal (non-contingency) operations, a USS would disclose only those grid cells with operations that conflict with those of other USS. The sharing of position reports and other sensitive data would be limited to non-conforming operations or demand by law enforcement, a regulator, or a litigant for accident investigation or other authorized purposes.[965] Beyond contractual and legal restrictions on the handling of confidential or proprietary information, DSS, USS, other UTM service entities, including registry[966] may have legal or contractual obligations regarding the handling of PII. They must comply with applicable data privacy laws. A complete description of data privacy requirements is beyond the scope of this chapter. Rather, only some key issues and requirements are introduced. For example, the FAA Reauthorization Act of 2018 includes the following UAS privacy policy that, in part, is relevant to UTM:

> It is the policy of the United States that the operation of any unmanned aircraft or unmanned aircraft system shall be carried out in a manner that respects and protects personal privacy consistent with the United States Constitution and Federal, State, and local law.[967]

For Europe, an EASA draft opinion states, "U-space service providers shall comply with applicable privacy and data protection regulations and shall only store or process information about the aircraft operations when necessary for safety, occurrences investigation and operational purposes."[968] LAANC providers must conform to a specified FAA privacy statement[969] and must develop a data protection plan and allow the FAA to audit it.[970] More generally, UTM service providers must be aware of and comply with applicable privacy and security laws of general application.[971] Some privacy laws define "personal information" very broadly.[972]

Applicable law within the United States is most likely state law, although as of the date of this publication, various proposals are before Congress for federal data protection legislation.[973] The most significant state privacy law now in existence is the California Consumer Privacy Act of 2018 (CCPA).[974] CCPA became effective on January 1, 2020 and the Attorney General of California is now able to enforce CCPA against violators. CCPA grants certain privacy rights to California residents with significant data protection implications to covered UTM service providers.[975] CCPA defines "personal data" broadly to include any data relating to an identified or identifiable natural person and provides a long list of examples that include geolocation data relating to an individual, which UTM services may collect and process.[976]

Foreign laws may also apply. In particular, USS should determine if they must comply with the European Union's General Data Protection Regulation (GDPR). The GDPR applies to US-based and other non-European companies that have operations in the European Economic Area (EEA—EU countries plus Iceland, Liechtenstein, and Norway). It also applies to companies that market their services in EEA countries. Finally, it applies to companies that monitor the behavior of any individuals that are currently located in EEA countries.[977] As with the CCPA, GDPR defines "personal data" broadly to include any data relating to an identified or identifiable natural person. Therefore, GDPR potentially protects name and geolocation information collected or processed by UTM services. To the extent GDPR applies to a UTM service provider collecting personal data of individuals located in EEA countries, it must comply with the various privacy and data security requirements of GDPR and applicable national laws in EEA countries.

UTM service providers should also be aware that GDPR restricts the export of personal data from EEA countries to other countries. Even remote access by a user in the U.S. to personal data stored on a server or cloud service in an EEA country will likely be considered an international transfer of that personal data. If the business imports personal data from a business collecting personal data in the EEA (called a "data controller") for purposes of further processing, the exporting business must make sure that the importing business provides adequate protection to the personal data. If the importing business is in a country whose laws the European Commission has found provides adequate protection of personal data, such as Canada, Israel, Japan, or Switzerland, no further documentation is necessary. Laws applying to the importing business already provide protections.

Other countries, however, do not have laws that provide adequate protection for personal data. If a DSS located in one of these other countries wants to import personal data, it will require another mechanism. This limitation underlies why PII is restricted in DSS. Businesses importing personal data to the U.S. could self-certify compliance with the European Union-U.S., and Switzerland-U.S. Privacy Shield Frameworks that permits the import of personal data from EEA countries or Switzerland.[978] The European Commission had deemed the U.S. to have adequate protection of personal data if and only if the importing business has self-certified compliance with the Privacy Shield Framework. Nonetheless, the European Court of Justice invalidated this mechanism in 2020.[979] Therefore, even though the Commerce Department will continue to maintain the Privacy Shield program for now, from a GDPR compliance perspective, businesses can no longer rely on Privacy Shield for the importation of personal data from EEA countries. However, the Switzerland-U.S. Privacy Shield Framework remains unaffected by the Court of Justice decision.

Another mechanism to protect personal data imported to a country whose laws do not provide adequate protection are the EC *Standard Contractual Clauses* (SCC). Businesses frequently incorporate the SCCs into data processing addenda to their service agreements to protect personal data.[980] As a practical matter, most inter-company transfers from EEA countries to the U.S. will require SCC as part of the data processing agreement covering the importation of personal data from EEA countries. Binding corporate rules for multinational corporations provide another approach for intra-enterprise transfers.

In sum, after July 2020, a business located in the U.S. seeking to import personal data from a business located in an EEA country needs one of two mechanisms. It must either use Standard Contractual Clauses or obtain approval of binding corporate rules. Should EU-U.S. negotiation resolve a post Privacy Shield Framework option, an additional mechanism could result. Even with one of these mechanisms in place, the importing business must still commit in an agreement to protect personal data imported from the EEA to standards that satisfy GDPR. Most frequently, the importing business agrees to an exporting business's data processing addendum or agreement (DPA) that embodies the importer's commitments. For transfers of EEA personal data to a U.S. data processor, DPAs must satisfy Article 28 of GDPR.

Finally, consider that compliance is only part of the picture with privacy and security. Relevant UTM participants should have processes in place to review and enter into DPAs with controllers of personal data and downstream vendors and should understand requirements arising from contracts in addition to requirements under public law. They should understand potential data protection liabilities and take steps to manage their risk. They should have procedures to investigate and respond to privacy and security breaches. Additionally they should have governance mechanisms in place, principally privacy and security policies and procedures used in a privacy or security program, to manage the entire lifecycle of personal data from collection through processing and ending in archiving or destruction.[981] Finally data privacy requirements must be considered in USS qualification regimes (*see* Chapter 4, Section 4.2, *USS Qualification*).

5 Selected Initiatives and Implementations

5.1 GENERAL—SELECTED INITIATIVES AND IMPLEMENTATIONS

Early UTM initiatives have sought to acquire data, to develop, test, and refine concept of operations (ConOps) and architectures, to perfect technologies, forge strategic alliances, validate business models, and secure regulatory approvals. Each component of a UTM deployment has required examination, including how they interact and maintain the safety of a nation's airspace system. The following initiatives and implementations represent a few of the early significant, influential, and innovative efforts to advance the state of UTM.

5.2 LOW ALTITUDE AUTHORIZATION AND NOTIFICATION CAPABILITY

"We consider LAANC to be UTM."

Steve Bradford
Chief Scientist for Architecture and NextGen Development, FAA[982]

5.2.1 General—LAANC

The Low Altitude Authorization and Notification Capability (LAANC), is a system designed to integrate sUAS into a national airspace system by automating the application and approval process to fly sUAS in controlled airspace below 400′ AGL proximate to airports. It has been described as "an initial UTM capability,"[983] a step toward the first operational UTM implementation,[984] and a test of "public-private partnership."[985] "LAANC handles authorization requests under 14 CFR § 107.41 (Part 107) and 49 U.S.C. § 44809 (Section 44809 [recreational operations])."[986] Such operations can be undertaken without further coordination with the controlling facility.[987] However, LAANC "does not help USS-USS coordination as one of the parts of UTM."[988]

LAANC is deployed at approximately 400 air traffic facilities[989] covering more than 700 airports,[990] and diverse airport facilities.[991] More than 500,000 LAANC authorization requests have been processed by the FAA.[992] Foreign LAANC deployments are also underway. Following its May 2019 expansion of LAANC capabilities to recreational (hobbyist) UAS operators, such operations now account for more than 25% of all authorizations, and represent "over a third of total LAANC authorizations granted [having] occurred in just the first half of 2020."[993]

The deployment of LAANC has been an important learning experience that is informing future UTM developments.[994] It is widely cited as an innovative first-to-market implementation.

5.2.2 LAANC Concept of Operations

The following three subsections introduce LAANC UAS Facility Maps, LAANC USS, and associated safety challenges.

5.2.2.1 UAS Facility Maps

LAANC overlays the very low-level airspace of its operation areas into rectangular "segments" or "grid cells" initially of 1 min. latitude and 1 min. longitude length, and designated maximum permissible altitudes (0–400 feet AGL, in 50 feet increments). These segments are divisions within *UAS Facility Maps* (UASFMs) representing the areas where the FAA has pre-determined that it may be safe for sUAS to operate with authorization.[995] The UASFMs are available on the internet and within commercial apps.[996] Future versions should provide greater grid cell granularity. Figure 5.1 presents the UASFM area for the San Francisco Bay region, and a corresponding LAANC app presentation of that area.

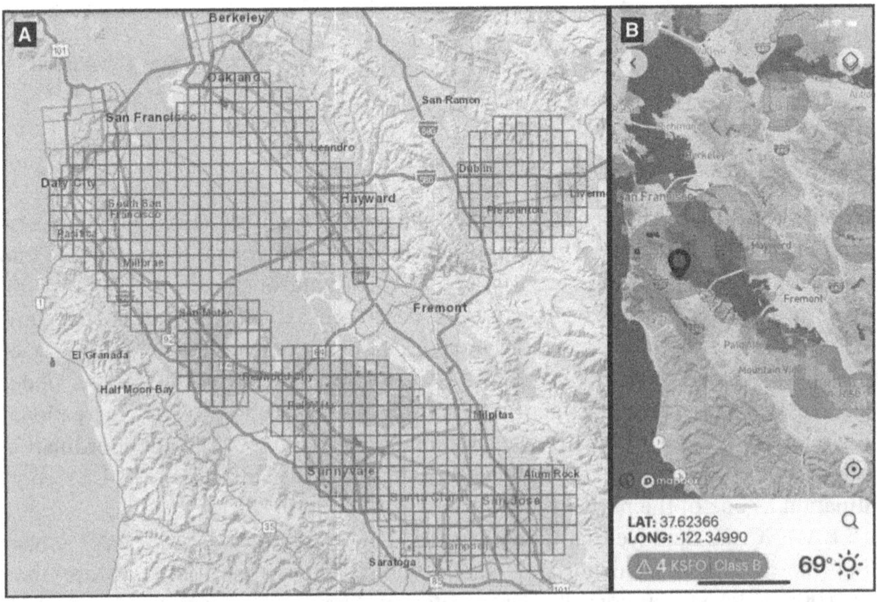

FIGURE 5.1 San Francisco Bay region UASFM [A] and [B]. (Courtesy of FAA; Reprinted with permission of Kittyhawk.io.)

Figure 5.2 of the Moffett Federal Airfield (KNUQ/NASA Ames Research Center's LAANC-enabled airspace) enlarges that portion of Figure 5.1, and depicts the UASFM segment maximum altitudes. The abutting Palo Alto Airport (KPAO) UASFM segments to the northwest, and the San Jose International Airport (KSJC) UASFM segments to the southeast, are also represented.

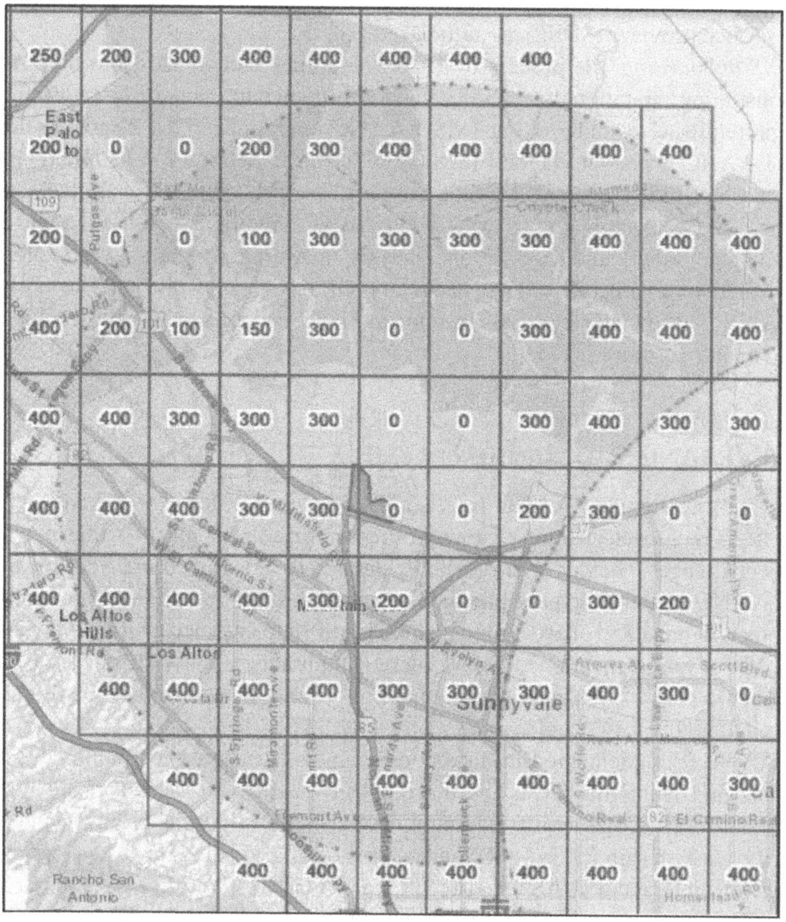

FIGURE 5.2 Moffett Airfield UASFM. (Courtesy of FAA.)

Remote pilots seeking authorization to operate within one or more UASFM segment(s) submit a request via a LAANC UAS service supplier (discussed below). ATC may deny, terminate, qualify, or ignore authorizations for safety reasons. If a remote pilot seeks a higher altitude than designated in the UASFM, a waiver request is required.

The altitudes displayed in Figure 5.2 represent the maximum altitudes (AGL) for authorizations that may be approved automatically. If a commercial remote pilot seeks a higher altitude than designated in the UASFM, the pilot must submit a "further coordination" request to the system via manual process for ATC approval.[997] About 20 percent of Part 107 authorizations have required further coordination.[998]

5.2.2.2 LAANC UAS Service Suppliers

UAS service suppliers (USS) were introduced in Chapter 2, Section 2.2.2, as third parties with authority to provide one or more UTM-related services. A USS serves as the interface between the remote pilot/operator and the FAA for communications required to obtain, revise or terminate authorization for LAANC-enabled operations. Other LAANC-related services may be provided by a USS or supplemental data service provider (SDSP), as authorized.

USS onboarding, the process of a USS obtaining acceptance from the FAA by demonstrating operational capability via conformance to onboarding requirements, is a prerequisite to undertaking USS LAANC operations.[999] USS interconnection with LAANC is not yet via the FIMS (*see* Chapter 2, Section 2.4.3 *FIMS*)—rather, it is interfaced via the FAA's Mission Support Network, the administrative network for the FAA ATO.[1000] Doing so expedited LAANC operations because it could avoid considerable safety and integration analysis measured in years' worth of effort that would have otherwise been necessary on a NAS network.[1001] The provision of LAANC has been viewed as an interim USS service, pending the evolution of LAANC into a future UTM.

5.2.3 LAANC Safety Challenges

There is officially no "LAANC" or "UTM" airspace.[1002] However, LAANC creates a de facto recharacterization of low altitude airspace near airports that may raise safety challenges whether or not formally defined or recognized. As a practical matter, LAANC-enabled airspace precludes manned aircraft operations[1003] yet sectional and other aeronautical charts fail to include UASFM to enhance manned pilot awareness. Moreover, as mentioned below, there is no practical LAANC operational tool available to ATCT personnel and manned pilots and, of course, sUAS are particularly challenging to see and avoid.

LAANC is implemented in airport environments, where most manned aircraft take-off and land; the density of aircraft traffic and pilot workload are highest; manned aircraft are generally closest to obstacles (including the ground); and pilots have the least margin for error, the narrowest stall-margin, and the fewest options for maneuvering to avoid sUAS.[1004] Aircraft, particularly rotorcraft, may approach and depart via diverse vectors often unanticipated by other pilots,[1005] and rotorcraft generally operate at very low altitudes. Also, pilots on an approach may *duck under* to avoid weather or traffic.[1006] All of these factors may place aircraft in closer proximity to LAANC-enabled airspace than anticipated.[1007]

The following Figure 5.3 presents a screenshot from the FAA's LAANC administrative tool depicting the San Francisco Bay Area. The tool is not intended for the operational separation of traffic and therefore is not an aide to operational controllers.[1008] Rather, each UASFM segment includes a number encircled in its lower-right corner representing the total number of sUAS currently authorized for simultaneous operation within that segment, including UAS operations authorized via waiver.

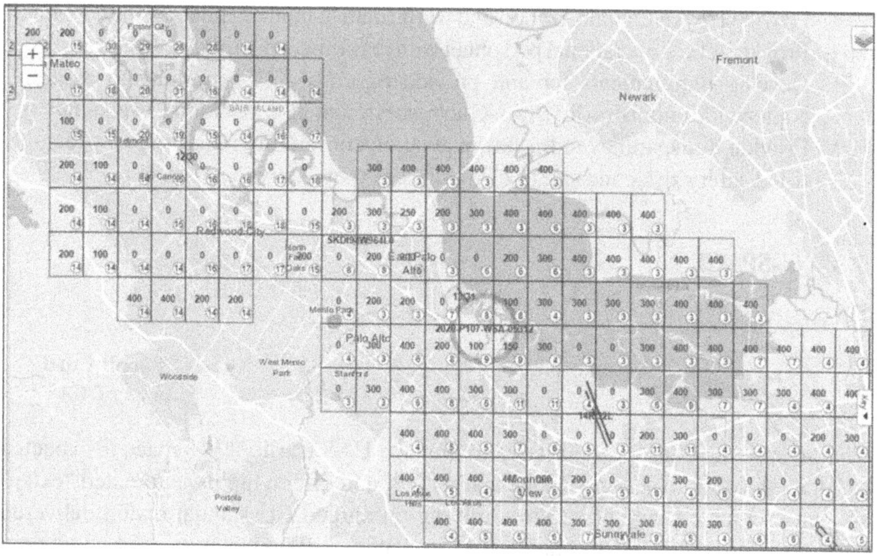

FIGURE 5.3 FAA LAANC administrative tool. (Courtesy of FAA.)

5.2.4 LAANC Safety Mitigations

Until UAS remote ID and ubiquitous DAA are deployed, near-term safety mitigations responsive to LAANC hazards deserve particular consideration. Currently UAS traffic information is directed almost exclusively to the remote pilot, often leaving the manned aircraft pilot situationally unaware. The most effective solutions provide a layered safety approach and engage/support *both* manned and unmanned aircraft operations.[1009] Various proposals enable display of nearby active UASFM segments in the cockpit.[1010] Relevant UASFM operation data would be transmitted for graphical presentation on cockpit avionics, electronic flight bags (EFB), tablets, and other displays via diverse networks. More generally, the LAANC "program office is busy looking at internal and external changes for the delivery of the service."[1011]

5.2.5 Lessons

There are many lessons associated with LAANC, but how many lessons are learned remains to be seen. Despite LAANC's noteworthy "first to market" achievement, and its heavily lauded promotion, it remains in beta testing. Some of the lessons that *should* be learned from LAANC include:

- Engage the manned pilot community vigorously and early as essential stakeholders
- Resolve gaps (and material dissent) in safety risk management documents before going live, and within designated time periods
- Include LAANC-enabled airspace in relevant aeronautical charts

- Provide LAANC authorization and traffic data to manned pilots in the cockpit
- Provide ATCT operational personnel with real-time, effective LAANC tools[1012]
- Develop requirements for and provide rigorous ATCT training to assure Nationwide uniform safe LAANC airspace designation, operation, oversight[1013]
- Provide transparency to the entire aviation community regarding new identified safety risks and safety challenges as they become apparent.

5.3 U-SPACE

"U-space is open now."

Benoit Curdy
Swiss FOCA[1014]

Characterized as Europe's initiative to manage UAS traffic,[1015] U-space is expected to "open the door to a UAS service market,"[1016] and offer a highly automated "extensive and scalable range of services relying on agreed EU standards and delivered by service providers"[1017] to enable data exchange, communication, and situational awareness for UAS operations. It is described as a combination of "a digital system, provision of services and procedures (which are linked to the type of operation and may evolve over time) and airspace …."[1018] U-space has also been called "a super set of UTM" to the extent it is not restricted to VLL.[1019] Unlike UTM in the USA, U-space has airspace designated specifically as U-space with associated flight rules (*see* Figure 4.1—U-space VLL volume types).

U-space includes four mandatory services: network identification, geo-awareness, UAS flight authorization, and traffic information,[1020] and two supporting services: weather information and conformance monitoring.[1021] Furthermore, U-space provides both a technical platform and "support[s] the EU aviation strategy and regulatory framework on drones."[1022]

Advanced by the European Commission, EASA, SESAR JU, EUROCONTROL, and European Member States, the U-space program is structured in the following four incremental steps:

U1—U-space <u>foundation/basic services</u> provide e-registration, e-identification, and pre-tactical geo-awareness. Its operations are substantially VLOS, and provides operators with information such as airspace restrictions.

U2—U-space <u>initial services</u> support the management of drone operations and may include flight planning, flight approval, tracking, airspace dynamic information, and procedural interfaces with ATC to support limited operations in controlled airspace. Limited constrained BVLOS operations are supported.

U3—U-space <u>advanced services</u> support more complex operations in dense (urban) areas and may include capacity management and assistance for conflict detection. The availability of automated "detect and avoid" (DAA) functionalities, and more reliable means of communication should increase operations in all environments.

U4—U-space <u>full services</u>, particularly services offering integrated interfaces with manned aviation, support the full operational capability of U-space and will rely on extensive automation, connectivity, and digitalization for both UAS and the U-space system.[1023]

The timeline for U-space services implementation has been envisioned in the U-space roadmap, presented in Figure 5.4.[1024]

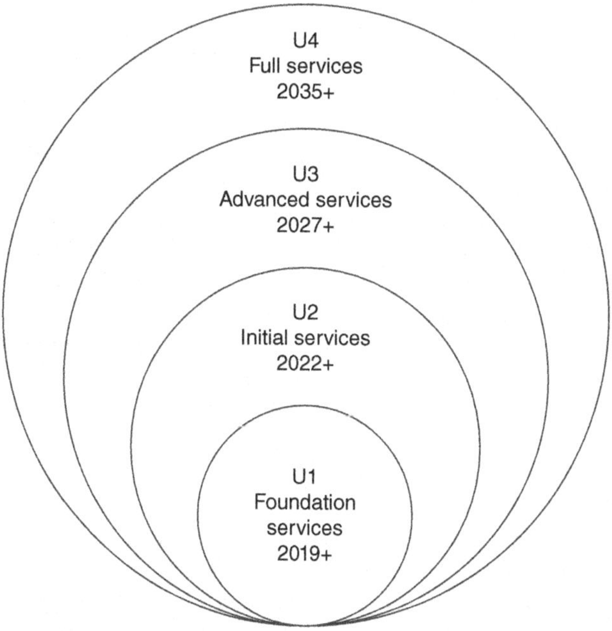

FIGURE 5.4 U-space roadmap. (Reprinted with permission of SESAR JU.)

U-space has various architectural visions,[1025] including a service-oriented architecture, embracing federated service providers with differing degrees of centralized management of certain functions.[1026]

The SESAR U-space research program comprises Exploratory Research (ER) projects and Very Large-scale Demonstrations (VLD).[1027] Some of these U-space initiatives contributing to the above four-phased plan included the following.[1028] Importantly, the list provides a temporal snapshot of noteworthy U-space project activity, demonstrating its broad reach.

- **AIRPASS (ER)**—*Aircraft Integrated RPAS Avionics Safety Suite*—researching DAA for cooperative and noncooperative traffic, autopilot and other systems for U-space.[1029]
- **CLASS (ER)**—*CLear Air Situation for uaS*—tackling U-space tracking and surveillance service of U-space and merging cooperative and noncooperative aircraft data for deconfliction and constraint adherence "to build the core functions of an Unmanned Traffic Management System (UTMS)."[1030]
- **CORUS (ER)**—*Concept of Operation for EuRopean UTM Systems*—published a European ConOps for U-space with a focus on sUAS operations in

VLL airspace proximate to airports and controlled airspace, and transition between controlled and non-controlled airspace.[1031] This project is transversal to the program (architecting, master planning, and standardization); the ConOps took input from and guided the other projects.

- **DACUS (ER)**—*Demand and Capacity Optimisation in U-Space*—developing a service-oriented demand and capacity balancing (DCB) process for drone traffic management responding to an operational and technical need in European drone operations of SESAR U-space services functions for traffic management to produce timely, efficient, and safe decisions.[1032]
- **DIODE (VLD)**—*D-Flight Internet of Drones Environment*—demonstrating the capability to safely manage multiple drones and the full-set of U-space services, to accomplish multiple tasks and missions, fully integrated with manned aviation, and coordinated by the Italian ANSP, ENAV.[1033]
- **DOMUS (VLD)**—*Demonstration of Multiple U-space Suppliers*—aims to illustrate all core U2 services, and demonstrate specific U3 services, such as tactical deconfliction and ATM collaboration via three U-space service providers interacting with an ecosystem manager, and several drone operators flying drones from different manufacturers.[1034]
- **DREAMS (ER)**—*DRone European AIM Study*—evaluating aeronautical information management solutions for UAS to identify possible data service providers and required facilities.[1035]
- **DroC2om (ER)**—*Drone Critical Communications*—aims to design a hybrid architecture integrating cellular and satellite networks to ensure reliable and safe operations for U-space services.[1036]
- **EURODRONE (VLD)**—*A European UTM Testbed for U-space*—aims to connect various stakeholders (operators, regulators, law enforcement agencies, product developers) and different systems in a unified environment to test U-space functionalities up to U3.[1037]
- **GEOSAFE (VLD)**—*Geofencing for Safe Autonomous Flight in Europe*—enables new geofencing solutions enforcing U-space regulation via flight test campaign.[1038]
- **GOF U-SPACE (VLD)**—*Gulf of Finland (GOF) SESAR U-space very large demonstrator*—implements common shared situational awareness seeking to increase transparency of all users, it enables cooperation in both uncontrolled and controlled airspace between unmanned and general aviation aircraft safely sharing airspace, sharing data and information between three USP platforms, operators, and ATM coherently.[1039]
- **ICARUS (ER)**—*Integrated Common Altitude Reference system for U-Space*—project proposes an innovative solution to the challenge of the Common Altitude Reference inside VLL airspaces with the definition of a new U-space service and its validation in a real operational environment. In manned aviation, the methods of determining the altitude of an aircraft are based on pressure altitude difference measurements (e.g., QFE, QNH and FL) referred to a common datum.[1040]
- **IMPETUS (ER)**—*Information Management Portal to Enable the integraTion of Unmanned Systems*—explores development of a cloud-based

server-less environment that responds to multiple users with diverse business models, including ATM integration.[1041]

- **LABRINTH (ER)**—an initiative to "produce a centralised planning system capable of communicating with all drones active in a given airspace volume, and by identifying their directional layout to avoid collisions by calculating alternative routes where necessary."[1042]
- **PercEvite (ER)**—*Sense and avoid technology for small drones*—develops sensor, communication, and a processing suite to enhance automation for ground and airborne DAA of cooperative and noncooperative obstacles.[1043]
- **PODIUM (VLD)**—*Proving Operations of Drones with Initial UTM*—four complementary large-scale demonstrations with over 185 VLOS and BVLOS VLL flights in both rural and urban areas, near airports, uncontrolled and controlled airspace, and in mixed environments with manned aviation to validate service levels U1, U2, and some U3 services.[1044]
- **SAFEDRONE (VLD)**—demonstrating how to integrate general aviation, state aviation, optionally piloted aircraft and drones into non-segregated airspace in a multi-aircraft and manned flight environment. The demonstration addresses U1 and U2 services, and a limited version of U3 advanced services including automated DAA technologies.[1045]
- **SAFIR (VLD)**—*Safe And Flexible Integration of initial U-space services in a Real environment*—multiple studies benefitting from interoperation between U-space service providers.[1046]
- **SECOPS (ER)**—*SECurity concept for drone OPerationS*—mitigating security risks of UAS; includes C-UAS methods and technologies.[1047]
- **TERRA (ER)**—*Technological European Research for RPAS in ATM*—addresses U-space performance requirements and identifies technologies (existing and new) that could meet these requirements, including for UAS-manned aviation interactions.[1048]
- **USIS (VLD)**—*U-space Initial Services*—validating the initial UTM core services, "especially targets regulations and coordination with air traffic control units."[1049]
- **VUTURA (VLD)**—*Validation of U-space by Tests in Urban and Rural Areas*—to validate the use of shared airspace between manned and unmanned aircraft, and multiple service providers servicing a specific airspace with ATC involvement.[1050]

Many of the projects listed above have now closed, while some have recently started, others awarded, and will be disclosed in 2021.[1051] The SESAR program will release a report presenting the consolidated results of the U-space program to date. New U-space and certain supporting research projects transcend SESAR JU, advance an "integrated AMT" and will progress contemporaneously.[1052] A Strategic Research and Innovation Agenda—*Digital European Sky* includes roadmaps supporting further U-Space research.[1053]

U-space Regulation—Rulemaking is under active development with a key opinion published in March 2020[1054] and a regulation issued in early 2021 (*see below*). Precursors to this U-space regulation included, but are not limited to, an EASA proposed regulatory framework for small drones (2015), a prototype basic regulation

(2016), and amendments (2017).[1055] An opinion proposing UAS regulation for "open" and "specific" categories (2018) "contain[ed] requirements for … three elements required to put in place this U-Space system, namely registration, geofencing and electronic identification."[1056] Subsequently, the "Drones Amsterdam Declaration" (November 2018) recounted European initiatives to advance U-space, and *inter alia*,

> [i]nvited the European Commission and EASA, with the support of the SESAR Joint Undertaking and EUROCONTROL, and in close cooperation with Member States, to develop, as a matter of urgency, an institutional, regulatory and architectural framework for a competitive U-space services market [and urged to provide priority to support] to Member States in the implementation of the European drone regulations.[1057]

Following public comment on the opinion, the European Commission issued *delegated*[1058] and *implementing*[1059] UAS regulations (in March and May 2019, respectively) presenting UAS operations as a foundation for U-space. Both are characterized by EASA as "the cornerstones of the new European UAS regulation [defining] the overarching subdivision of the UAS operations into three categories—Open, Specific, and Certified—and the thresholds between these categories."[1060] The delegated regulation addressed, in part, "remote identification add-ons" for UAS, and the implementing regulation (consistent with the earlier 2018 opinion, above), provided:

> While the 'U-space' system including the infrastructure, services and procedures to guarantee safe UAS operations and supporting their integration into the aviation system is in development, this Regulation should already include requirements for the implementation of three foundations of the U-space system, namely registration, geo-awareness and remote identification, which will need to be further completed.[1061]

An EASA U-space Regulatory Framework Workshop (May 2019) further addressed the need for and content of a separate U-Space regulation.[1062] Involving both regulator and industry stakeholders, this consultation sought consensus on diverse issues affecting a U-space regulation. Among its general conclusions:

- There is a need for a separate U-space regulation to reflect the innovative character and the paradigm shift, distinct from, yet building on other aviation safety regulations; e.g., ATM present regulatory framework;
- There is a regulatory need to define:
  - flight rules and airspace where U-space services will apply; and
  - roles and responsibilities of the actors, what applies and who is affected.[1063]

Another EASA workshop was held in October 2019 to discuss the U-space regulatory framework draft Opinion issued by EASA earlier that month. An EASA Opinion, No 01/2020, *High-level regulatory framework for the U-space* ("Opinion")[1064] was issued, leading to a U-space regulation planned for early 2021 release.[1065] The Opinion is "a first regulatory step to allow immediate implementation of the U-space after the entry into force of the Regulation and to let the unmanned aircraft systems and U-space technologies evolve."[1066] It includes an annex presenting the draft regulation[1067] to provide "the minimum necessary rules."[1068]

The *Commission Implementing Regulation on a regulatory framework for the U-space* ("U-space regulation"), is the centerpiece of the regulatory package governing U-space, and becomes law in January 2023. Its structure is depicted in Figure 5.5.[1069]

Chapter I	**Principles and general requirements** Article 1 / Subject matter and scope Article 2 / Definitions
Chapter II	**U-Space Airspace and Common Information Services** Article 3 / U-space airspace Article 4 / Dynamic airspace reconfiguration Article 5 / Common information services
Chapter III	**General Requirements for UAS Operators and U-space Service Providers** Article 6 / UAS operators Article 7 / U-space service providers
Chapter IV	**U-space services** Article 8 / Network identification service Article 9 / Geo-awareness service Article 10 / UAS flight authorization service Article 11 / Traffic information service Article 12 / Weather information service Article 13 / Conformance monitoring service
Chapter V	**Certification of U-space Service Providers and Single Common Information Service Provider** Article 14 / Application for a certificate Article 15 / Conditions for obtaining a certificate Article 16 / Validity of the certificate
Chapter VI	**General and Final Provisions** Article 17 / Capabilities of the competent authorities Article 18 / Tasks of the competent authorities

FIGURE 5.5 Structure of the U-space regulation.

Selected provisions of the U-space regulation are considered within the relevant sections of this book. A few highlights of the U-space regulation include:

- **Article 1 Subject matter and scope**
  - Applies in UAS geographical zones established as U-space airspace to UAS operators, U-space providers, and common information services (CIS) providers.

- **Article 3 U-space airspace**
  - All U-space supported by risk assessment and operations subject to at least the following mandatory services: Network identification service [**Article 8**], Geo-awareness [**Article 9**], UAS flight authorization service [**Article 10**], Traffic information service [**Article 11**] (including remaining conspicuous to proximate air traffic; and complying with U-space airspace requirements in Implementing Regulation (EU) 2019/947).
- **Article 4 Dynamic airspace reconfiguration**
  - Where U-space is designated in controlled airspace, ensure manned aircraft provided ATC services and UAS remain segregated via dynamic reconfiguration.
- **Article 5 Common information service**
  - Member states to make diverse U-space airspace information available including airspace dimensions, list/details of relevant U-space providers, UAS geographical zones, static and dynamic airspace restrictions, and terms. CIS providers ensure such information complies with data quality, latency and protection requirements. A single CIS provider may be designated. Nondiscriminatory access.
- **Article 6 UAS operators**
  - Comply with capabilities and performance requirements; conditions and constraints.
- **Article 7 U-space service providers**
  - Provide applicable services; arrange with ATC for adequate coordination and data exchange; adhere to a common open communications protocol; reporting.
- **Article 13 Conformance monitoring service**
  - Upon deviation detection shall alert UAS operator, proximate: UAS operators, U-space service providers offering services, and relevant ATC and authorities.
- **Article 15 Conditions for obtaining a certificate**
  - UAS Service Provider shall demonstrate ability to provide services in a safe, secure, efficient, continuous and sustainable manner, consistent with the intended UAS operations and established performance level; guarantee services; maintain appropriate net capital, management system and security management system; report occurrences, adhere to record retention periods; arrange to cover liabilities, have agreements allocating liability if engaged with other service providers; develop contingency plan; have emergency management and communication plans to assist UAS operator emergency.
- **Article 18 Tasks of the competent authorities**
  - Establish, maintain, and make available a registration system for U-space service and single CIS providers; determine traffic data attributes and availability; determine services to be provided; ensure compliant data exchanges. Establish participant coordination mechanisms; certification and continuous risk-based oversight programme; information security management system; audits.

The regulation's adoption is a "significant milestone" and pioneering, yet "not the end-state of U-space [and is] very, very high-level."[1070] It is understandable that any program as ambitious and complex as U-space, together with the impact of COVID-19, can be expected to face deployment challenges and delays that should be factored into a realistic implementation timeline including completion of responsive Acceptable Means of Compliance / Guidance Material (AMC/GM).[1071]

5.4 SELECTED COUNTRY-SPECIFIC DEVELOPMENTS

> "Now it's transitioning from innovation
> and experimentation to operationalization."
>
> **Amit Ganjoo**
> *Founder and CEO, ANRA Technologies*[1072]

The following Table 5.1 presents selected early developments, many of which are demonstrations, trials, or initial deployments. Recognizing the breathtaking pace of UTM developments, the table serves primarily for historical purposes.[1073]

TABLE 5.1
Selected Early Developments

Australia	• Wing Aviation, LLC provides food delivery using UTM services in Canberra.[1074] • Uber selects Melbourne as the first international city test site.[1075] • Zephyr stands-up high-altitude pseudo-satellite (HAPS) aircraft.[1076]
Belgium	• Unifly, the Belgium-based UTM technology and platform provider, supports diverse UTM demonstrations.[1077]
Canada	• ATM and UTM integration addressed through a multi-stakeholder team (including government, industry, and academia) for all Canadian airspace.[1078] • Early trials leverage industry and research partnerships from 2020 through 2022. • NAV CANADA-Unifly contract for civil air navigation system UTM integration.[1079]
China	• The Civil UAS Aviation Operation Management System (UOMS),[1080] China's UTM, undergoes trials, including in Shenzhen and Jiangxi. UOMS is integrated with the CAA of China (CAAC) General Aviation Flight Service (GAFS). • A Nanjing Technical University UTM concept initiated.[1081] • EHang makes first passenger-carrying autonomous flight.[1082]
Finland	• SESAR U-space demonstration in the Gulf of Finland (GOF) implements central FIMS-routed data between operators, UTM platforms, and ATC, including GA.[1083] • Open industry standard proposed "making U-space immediately relevant for low-level aircraft."[1084] • Wing initiates food delivery in Helsinki, its first European rollout.[1085]
France	• The French civil aviation authority (DGAC), air navigation services department (DSNA)[1086] publishes a request for information (RFI)[1087] to support a U-space implementation and finalizes relevant RFPs. • France participation in the PODIUM, GEOSAFE, and ONERA projects.[1088]

(Continued)

TABLE 5.1 (Continued)
Selected Early Developments

Germany	• UTM deployed by Droniq exploiting LTE module integrated SIM card as a "hook-on device" and sensing FLARM and ADS-B.[1089]
India	• To establish a national UTM platform, the *Digital Sky Platform*, India's Directorate General of Civil Aviation and Ministry of Civil Aviation tasked Happiest Minds and ANRA.[1090]
Italy	• Established in Feb. 2019, d-Flight S.p.A. the Italian Public Private Partnership constituted by ENAV S.p.A. and Leonardo S.p.A., Telespazio S.p.A. develop / deploy U-space services. Via d-Flight, ENAV aims to synergize traditional air traffic with UAS requirements to facilitate diverse new, services.[1091]
Japan	• In 2017 METI published its roadmap to achieve industry goals of urban BVLOS package delivery and other UAS applications.[1092] • NASA and the Japan Aerospace Exploration Agency (JAXA) have collaborated on UTM research related to Japan's D-NET disaster response management project.[1093] • Through NEDO, a government affiliated R&D management organization, the Ministry of Economy, Trade and Industry (METI) funded a three-year UTM research project to advance UTM in Japan, later extended for follow-on research for two years.[1094] • Japan's Civil Aviation Bureau (JCAB) is considering regulation needed to allow more advanced BVLOS operations, including BVLOS, with changes forthcoming. No concrete timeline is set for deployment of UTM services.[1095]
Korea	• The Korea Agency for Infrastructure Technology Advancement (KAIA), in coordination with Korean Telecom is developing a UTM system.[1096]
Latvia	• First BVLOS cross-border UAS flight (Latvia to Estonia) completed via dual SIM cards.[1097]
Malawi	• First drone logistics company to operate fleet of aircraft from outside country of operation (Australia) via Swoop Aero platform.[1098]
Netherlands	• Altitude Angel awarded nationwide UTM contract by NL's ANSP LVNL[1099]
New Zealand	• Early testbed for UTM experimentation and integration.[1100] • Canterbury and Queenstown Lakes region UTM trials by Airways and AirMap commenced.[1101]
Norway	• Planned deployment of a national UTM by Altitude Angel and Frequentis.[1102]
Poland	• Polish Air Navigation Services Agency (PANSA) launches PansaUTM.[1103]
Singapore	• Singapore's Air Traffic Management Research Institute (ATMRI), CAA of Singapore (CAAS) collaborated with Nova Systems[1104] and OneSky on a UTM pilot to support package delivery.[1105] • Future Flight Consortium[1106] selected by CAAS and Ministry of Transport to develop a Connected Urban Airspace Management System.[1107]
Sweden	• In preparation for a Swedish national UTM, LFV, the Swedish national ANSP released a tender for study of UTM implementation.[1108]
Switzerland	• Swiss ANSP skyguide,[1109] FOCA, and industry stakeholders introduced Europe's first national deployment of U-space;[1110] and FIMS.[1111] • Matternet[1112] Swiss healthcare operations (largely blood & pathology) develops general BVLOS logistics platform, partners with Boeing, collaborates with UPS, and, with Swiss Post, seeks Part 135 certification for deliveries, including for US healthcare.[1113] • Swiss U-space ConOps issued.[1114]

(Continued)

TABLE 5.1 (Continued)
Selected Early Developments

Tanzania	• A medical drone project operated by DHL delivery and drone specialist Wingcopter services a remote island with marginal transportation infrastructure, demonstrating UAS value for international development.[1115] • The 2018 Lake Victoria Challenge, hosted by the Tanzanian CAA, World Bank, UNICEF, World Food Program, and World Economic Forum provided regulators opportunity to observe UTM systems capability to enhance awareness and in preparation for a tender for service in 2019.[1116]
United Kingdom	• UK CAA NATS Airspace User Portal (AUP), developed within *Operation Zenith* in 2018, provides enhanced airport safeguarding and automated approvals to fly in controlled airspace.[1117] • Altitude Angel's GuardianUTM O/S supports the UK's traffic management.[1118] • NATS launches a virtual testing space, known as an "Innovation Sandbox" to facilitate UTM testing.[1119] • Connected Places Catapult—Groundwork for a UTM system to allow commercial drone usage alongside traditional manned aircraft.[1120]
United States	• The US is recognized as having originated the UTM concept, supports extensive UTM research and testing, and hosts diverse implementations, as addressed throughout this book.[1121]

6 The Future of UTM

"It's inspiring to see how far the industry has advanced in such a short period of time."

Stephen P. Creamer
Director, Air Navigation Bureau, ICAO[1122]

"Given the pace of innovation, the solutions we propose today will be obsolete in 5 years."

Reinaldo Negron
Head of UTM, Wing[1123]

6.1 GENERAL—THE FUTURE OF UTM

While we may not be able to see over the horizon, we have some idea where the road leads. There are signposts, concrete test results, significant investments, and innovative implementations that help chart the way. The momentum to stand-up UTM is rapidly accelerating, globally. Near-daily announcements of new technical, industry, and regulatory developments lend confidence to forecasts of UTM's transition from "if" to "when," from "test" to "going live," and from legacy ATM systems into integrated traffic management platforms. Industry and government are committing to aggressive UTM timelines, whether to implement rules or to achieve technical thresholds. Additionally, the momentum for enhanced BVLOS approvals,[1124] including issuance of UAS air carrier certification, is compelling.[1125] As urged by Amit Ganjoo, Founder and CEO, ANRA Technologies, "The future is nearer than you think,"[1126] and by Steve Bradford, Chief Scientist—Architecture & NextGen Development, FAA, "the future is actually here."[1127]

Nonetheless, the path to UTM remains difficult, and new aviation entrants who are not steeped in aviation safety culture may not grasp the rigorous aeronautical certification and operational requirements—*and their propriety*. Furthermore, there are co-dependencies, such as between infrastructure and applications, requiring iterative *apps-infrastructure cycles* that take time.[1128] The technology integration process is challenging. As observed by Dallas Brooks of Wing, "[s]uccess in integration comes in bites, not in meals .... Those who 15 years ago were trying to solve all our problems at once are still trying."[1129] It seems that UTM will evolve incrementally, rather than emerging all at once, whole.

6.2 TECHNOLOGY INNOVATION

"Emerging technologies are going to be the answer."

Mark Wuennenberg
ICAO[1130]

"What's next, for me? It's about the digitization process."

Michael Gadd
Altitude Angel[1131]

As a fundamental shift in how traffic is managed, UTM will continue to be an area of hot technology innovation for some time. This subsection highlights selected areas where innovation is imperative and inevitable—including communications infrastructure and UAS/UTM autonomy—and considers the promising capabilities of blockchain.

6.2.1 Communications Infrastructure

Ubiquitous, reliable, low-latency communications, both terrestrial and space-based, are needed to support envisioned UAS operations—particularly BVLOS operations—that will rely on UTM services.[1132] Mobile network operators are transforming their current wireless infrastructure with new radio spectrum based on 4G and 5G standards, with new architectures and configurations, and new equipment to better support airborne applications. There are initiatives to accelerate IP implementation for traffic management,[1133] to tightly integrate UTM services with the mobile network to enhance reliability and reduce latency. (*See* Chapter 3, Section 3.2.3 *Cellular Industry*). Autonomous operations may affect communication requirements within a given airspace. In space-based infrastructure, a vast and developing network of low Earth orbit, low-cost satellites will increasingly offer low-latency communications for primary and backup applications, including where terrestrial networks are unavailable such as in certain nonurban areas and high altitudes.[1134] Reliance on secondary networks such as satellite or High-Altitude Pseudo Satellites (HAPS) may also cover gaps and service unavailability, offer effective UTM mitigations, and serve as a viable alternative for some UAS communications.[1135] Finally, extended RF spectrum will contribute to the future of UTM (*see* Chapter 4, Section 4.5 *Spectrum*).[1136]

6.2.2 Advanced Automation and Autonomy

"We don't have a book that says *Autonomy for Dummies*."

Parimal Kopardekar, PhD
Dir. NARI, NASA Ames[1137]

"[Control station] sticks are the arch enemy."

Chris Anderson
CEO, 3DR[1138]

UTM's capabilities, efficiencies, and safety will be difficult to achieve with traditional systems that require human-in-the-loop intervention. Nonetheless, UTM dynamics may push certain functions to the edge, such as conformance monitoring, further accelerating the need for autonomy. "If you look across work done around the world, automation/manage by exception is the only way to manage complexity and scale."[1139] According to former NASA Administrator Jim Bridenstine, "in the future it's going to be safest to fly in uncrewed aircraft than crewed aircraft ... I know it sounds crazy, but it is absolutely true."[1140]

Although implementing increasing levels of automation and autonomy is inevitable,[1141] understanding and responding effectively to these issues have proven challenging.[1142] "Most of the difficulty and touch points [are in] the transition from manned [operations] to automation."[1143] This transition (from manned to automated, including to autonomous) is critical but vexing.[1144] Dr. Koparkedar has cautioned, "unless automation can solve all contingencies and address off-nominal conditions, there will always be a role for humans."[1145]

To assist, various schemas have been developed in the form of "levels of automation."[1146] Yet, "it became clear that the most effective path forward for the [aviation] industry was not a simple adaptation of a level-based approach but rather a new framework in which various risk and safety benefits could be considered in a more nuanced and operationally relevant context."[1147] Such a framework assesses functions or tasks in terms of (a) risks and safety benefits, (b) roles of automation, and (c) complexity, and then undertakes a functional decomposition[1148] of the work to be performed by the combined human-autonomy team. Analysis and research indicate that these functions can be allocated across multiple agents (humans, computers, etc.) to ensure safe interactions between the various agents. The question to ask is: What is the best allocation of function and teamwork "between humans and autonomy that minimizes risk as the flight and airspace conditions change."[1149] As autonomy technology improves, these roles and allocations can be evolved in a manner that maintains safety and performance objectives.[1150]

> Both UA and UTM are progressing towards automated interactions with very well-defined semantics and roles. Increasing aircraft autonomy can reduce the number of interactions required between a single aircraft and its UTM handler. There are some limitations to reliance on UTM communications for collision avoidance which becomes increasingly prevalent as the number of aircraft scale up. For example, package delivery to vertiports and within urban cities is limited by the communication uncertainty to a UTM network. For safety, the interactions can become complex and must occur at fast time-scales. Although the UTM's function does not have to have increased autonomy, the physical location of the function of coordinating multiple spatially close aircraft has to be addressed in a robust manner. With limited batteries, the endurance margins at mission end-points are slim. The function of now coordinating multiple aircraft in a safe manner at fast time-scales becomes acute.[1151]

Introducing decision-making algorithms and machine-learning (ML) models into UTM services must be done with care and under risk-based certification guidelines. It has been urged that early candidates for automation should be in strategic, less time-critical areas:

- The small UAS industry could start putting a separate UTM structure in place that will eventually be interoperable with ATM.

- AI/ML could be used to study trajectory management data from existing manned aircraft operations, and could then be used to identify flight paths that are not following the "norm" for a particular route, procedure, etc. so that air traffic controllers and pilots start using AI/ML in an advisory manner first.
- In the same way, AI/ML could be used to deconflict proposed UAS trajectories in a dense air traffic environment, as it is clear that human controllers will not be able to manage a large number of targets. However, it would need to start as an advisory system in conjunction with human controllers and grow its capabilities as the number of targets increases.[1152]

The need for responsive AI research is well-recognized.[1153] Additionally, formal design-time verification of complex algorithms may not be possible in some cases. One responsive industry standard, ASTM F3269 addresses run-time assurance to facilitate use of complex algorithms by safely bounding complex algorithm behavior, thereby advancing certification or approval.[1154] Simulation and modelling are also imperative.[1155]

> We are going to be asking autonomy to do a lot … across every application that we have … in that regard, I think understanding the performance of autonomous systems and using a variety of ways to understand the bounds on their outputs of using simulation and modeling to find edge cases and prove to ourself that autonomy works and also using structured demonstrations and tests to understand the end-to-end integration of autonomy with the rest of the aircraft system, that's all vitally important.[1156]

With regard to the inevitable changes to flight operations precipitating from AI, Steve Bradford, Chief Scientist for Architecture and NextGen Development, FAA, remarked, "Maybe after 2050 we'll have autonomy on the control side—but it's not going to happen before 2050."[1157] Among other constraints is need for extensive data sets to train the algorithms—but such data will take time to accumulate and mine. Finally, ethical principles affecting autonomous systems in UTM need formal consideration, including to assure against "blindly automating tasks," and to inform certification regimes.[1158]

6.2.3 Blockchain in UTM

> "When you combine machine learning with the data blockchain can provide on UAS registration, accountability, and tracking, an entire world becomes available for drone safety analysis, decision making, and even regulation."
>
> **Regina Houston**
> *U.S. DoT*[1159]

UTM has security and governance requirements that remain incomplete (*see* Chapter 2, Section 2.4.7 *Cybersecurity*, and Chapter 4 *Governance*). Blockchain technology may potentially provide pathways to a globally coherent, cyber-secure governance structure for UTM. UTM faces a problem in the form of a coordination challenge, better understood as a social contract problem in political philosophy. UTM engages the global community trying to solve a technology problem that invokes a Common Pool Resource (CPR)—in airspace.[1160]

Blockchain is an electronic distributed ledger (database) with each ledger entry (block) cryptographically hashed to the prior record, making successful tampering infeasible. In some respects, it reflects a technological extension of more decentralized aviation governance, and it has been the subject of considerable ATM research.[1161]

Bitcoin[1162] was the first blockchain and computer network to exploit *mechanism design principles*, a subset of game theory, to steer participant behavior in directions that "align the incentives of the individual such that they act in the common interest of the network."[1163] The result permits foregoing a centralized "trusted" third party governing a CPR. The Bitcoin network's underlying economic policies enforce its rules.[1164] Bitcoin has become the world's first decentralized autonomous organization (DAO).[1165]

The Bitcoin blockchain lives in constant operation, out *there*. It *pays* people[1166] to keep it alive; grows and matures in alignment with stakeholder interests, yet with autonomy;[1167] and is infeasible to manipulate, interrupt, or stop. Since Bitcoin's deployment, blockchain governance experimentation has experienced a sort of Cambrian explosion—from "liquid democracies" to stake-based weighted voting and futures markets (Futarchy).[1168] Decentralization as a CPR governance framework is becoming mainstream through the Bitcoin Network,[1169] but not without controversy.[1170]

In some limited respects, aviation governance too is experimenting with decentralization.[1171] Decentralization is not binary but should instead be viewed as a continuum as it changes over time. For example, COCESNA (Central American Region),[1172] the Agency for Aerial Navigation Safety in Africa and Madagascar (ASECNA),[1173] and EASA/EUROCONTROL[1174] exemplify nation states embracing the strengths of collaborative regional governance. Also, the ANSP NAV CANADA highlights a decentralized, nonprofit user co-op governed by a 15-member board that includes diverse industry participants. Moreover, UTM industry consortia are growing with expectation of a governance role, and open source UTM technology collaboration may further contribute to decentralization.[1175] And yet, these examples are skewed toward the centralized end of the decentralization continuum.

Network access could be opt-in and permissionless,[1176] just like the $1 trillion Bitcoin Network. "If such a leaderless network can safely and securely be opt-in and permissionless, why can't an aviation network do the same? It very likely *could*."[1177] Perhaps only open, transparent, verifiable blockchain architecture can deliver this powerful value proposition for a CPR, and with unmatched system security.[1178]

The blockchain is a technically slow, cumbersome database, if viewed in isolation, yet higher performance is not required[1179] for it to deliver the source of truth of airspace "state"—of who is occupying how much airspace, and when. This information is essential for airspace authorizations, network planning, contingency management, rule setting and compliance, and many other airspace access functions that contribute to a national airspace system. The blockchain would include the consensus rules that are set via the decentralized governance framework.[1180] The blockchain is not a silver bullet[1181] yet may offer unique, viable, future capabilities for UTM governance, compliance, incentives, and security.[1182]

6.3 INTERNATIONAL COORDINATION

"It's an international race to get there together."

Jay Merkle
Exec. Dir., UAS Integration Office, FAA[1183]

UTM coordination efforts were presented earlier, including for standards development (Chapter 3, *Standards-Making for UTM*), and governance (Chapter 4, *UTM Governance*). Diverse inter-governmental and non-governmental organizations contributing to such coordination were considered, including ICAO and GUTMA. Alignment of international aviation standardization as formalized between ICAO and various standards development organizations (SDOs),[1184] and regional coordinating initiatives, including between Europe and the United States,[1185] were also introduced. European and US government aviation leadership have collectively urged international harmonization, emphasizing the role of international standards, with validation by both authorities, and underscoring that "the main principles [of UTM] are the same … just implementation is different."[1186]

The urgency for greater international coordination at many levels is well-recognized and accelerating, including the use of bilateral agreements.[1187] Of note is a Declaration of Intent between the FAA and SWISS CAA, Federal Office of Civil Aviation as "global leaders in aviation safety [that] intend to cooperate in advancing domestic and international safety standards, and their harmonization for Unmanned Aircraft Systems/Remotely Piloted Aircraft Systems (UAS/RPAS)" including working together on "(UTM) concept validation," "system engineering documentation," "safety documentation," and "UAS regulation."[1188] A bilateral agreement between Europe and China created a Certification Oversight Board to, among other things, "minimize the differences in the regulatory systems, standards and certification processes of the Parties" including "interface requirements" that should facilitate UTM.[1189]

"Aviation is inherently an international business … and we're heading towards coalescing."[1190] We can expect further engagement and coordination initiatives among CAAs, inter- and non-governmental organizations, the private sector, and other stakeholders as this international UTM race progresses. In fact, "UTM is playing a harmonizing role across all of the government regulators."[1191]

6.4 REGULATION

"The regulation is for tomorrow."

Maria Algar Ruiz, Drone Prog. Mgr.
EASA[1192]

While early deployment of UTM systems allows UAS operators to conduct limited operations today, CAAs in collaboration with other stakeholders are developing regulations that will enable large scale routine, safe operations, with considerable focus on BVLOS operations. Nonetheless, aviation regulatory reform is historically deliberative and industry concern over regulatory delay has spurred support for accelerating the rulemaking process.[1193] Bill Stanton, Sr. Advisor—Policy, FAA stated,

> Rulemaking takes quite a bit of time because we have to be very cognizant [that] what we write today may become obsolete tomorrow [especially considering] the speed with which your organizations [and technology] are developing … .[1194]

UTM's possible regulatory scope is expansive, touching: rules of the air, airspace, operations, performance, risk assessment, security, spectrum, privacy, environmental, and the extent of mandatory participation by unmanned and manned aircraft.[1195] Regulations addressing architecture (*see* Chapter 2, Section 2.4.2 *Centralized vs. Federated Architecture*) will have many implications, for example, apportioning stakeholder responsibility and liability.[1196]

Lorenzo Murzilli, formerly of Swiss FOCA characterized needed regulatory initiative as follows:

> This work requires leadership, requires mastery, and requires courage … the success of any radical or disruptive innovation in a regulated field such as aviation depends on the success of the effort of those regulators and industries that go out there to change the regulatory framework in a way that allows disruption to eventually materialize.[1197]

UTM rulemaking is informed by research results at NASA in collaboration with the FAA and industry, the SESAR research program in Europe, and other regional and national initiatives. In the United States, UTM-promoting provisions within the FAA Reauthorization Act of 2018 mandate an implementation plan[1198] and the Remote ID Rule deserve attention[1199] —the latter viewed as "the lynchpin for future regulation [and a] major step."[1200] Effective UTM rulemaking will need to consider AAM/UAM and upper E technical requirements.[1201]

In Europe, UAS[1202] and U-space regulation UAS[1203] offer a multi-state collaborative vision influential well beyond European borders.[1204] Additionally, the contributions, cited in earlier chapters, of ICAO, and NGOs, and the standards development community greatly factor into UTM regulation. As Bill Stanton proclaimed, "we want to get to a stage that is not yet described"—in which airspace is shared by legacy and new entrants alike.[1205]

6.5 ADVANCED AIR MOBILITY/URBAN AIR MOBILITY

> "We're getting a little closer to being Jetson's World."
>
> **Joseph Rios, PhD**
> *ATM-X Project Chief Eng'r, NASA Ames*[1206]

> "It's more than just hype."
>
> **Jay Merkle**
> *Exec. Dir., UAS Integration Office, FAA*[1207]

Urban Air Mobility (UAM) represents the infrastructure, aircraft,[1208] and decision support services[1209] for the integrated transport of people and cargo in urban environments and beyond. This capability should become particularly important as projected worldwide urban demographics overtax current terrestrial transportation infrastructure. UAM should support scheduled[1210] and "on-demand air transportation within core urban areas and residential suburban destinations outside city centers using new, electric-powered, vertical takeoff and landing (eVTOL) aircraft."[1211] Eric Allison, Head of Product at Joby Aviation posits four essential UAM components: multimodal aerial ride sharing, electric aircraft, an automation platform, and connected "skyports."[1212]

UAM is considered a subset of advanced air mobility (AAM).[1213] AAM provides local, regional, and intraregional service in addition to the urban service of UAM, with intra- and inter-city transport and an expected 50–300 mile maximum range, respectively. NASA is a key AAM pioneer whose extensive AAM vision is depicted in Figure 6.1.[1214]

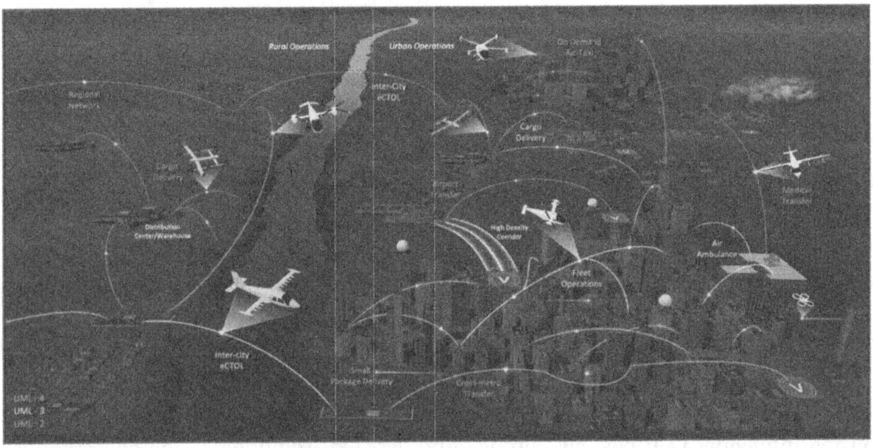

FIGURE 6.1 NASA AAM Mission. (Courtesy of NASA.)

"UAM leverages a common, shared, technical environment, similar to UTM,"[1215] and extends operations beyond typical sUAS capabilities to include higher altitudes, controlled airspace, and the carriage of people and cargo. UAM is expected to interconnect intensively with airports, increase traffic density, and may become "[t]he highest risk area for VLL [very low level airspace]."[1216]

Because "gaps exist [in UTM's] suitability for UAM flights,"[1217] UAM may require an "advanced stage of UTM."[1218] The National Academies of Science has recommended that "NASA, in coordination with the [FAA], should perform research to extend unmanned aircraft system traffic management concepts to accommodate emerging advanced aerial mobility traffic in all classes of airspace."[1219] Responsively, NASA, the FAA and industry have developed an initial ConOps for UAM representing a "shared vision" for this new mode of transportation.[1220] NASA's Joseph Rios posits that "[w]hat we're learning [in UTM] is translating to UAM."[1221]

Tom Prevot underscores that "[w]e have to interact with ATC and air traffic management. I don't think we're going to try to integrate much directly into the current system because the current system is already very complex. It's more of us listening to the various SWIM feeds. The UTM environment is an important piece."[1222]

UAM may present new challenges in contingency operations, including need for digital voice and data communications from pilot to pilot, pilot to autonomous operator, and autonomous operation pilot to ATC, and accommodation of yet-to-be ascertained human factors requirements for both pilot and controller. As urged by Dr. Prevot, for UAM operations (within UTM), "design for the unexpected" and provide graceful degradation.[1223]

To meet the requirements of UAM, UTM may need to provide robust ground-based radar surveillance around vertiports, unique airspace designation above most sUAS and below most manned aircraft operations in all classes of airspace, flow control and prioritization, plus use of high-density routes or corridors (*see* Chapter 4, Section 4.4.3 *Corridors*), and—to satisfy future business model economics—a high degree of automated and autonomous operation.[1224] In addition, UTM will require robust low altitude weather gathering and reporting systems in urban areas (*see* Chapter 2, Section 2.2.3 *Supplemental Data Service Providers*).

UAM also has unique aeronautical design requirements that address noise abatement, alternative landing sites, and battery management. "It's like being in the early 1900s and imagining what an airplane should be when you don't quite know yet. Kind of a blank canvas for creativity and innovation; kind of unique to aerospace; coming before the optimization and design consolidation phase."[1225] One such unknown is how many aircraft designs must be accommodated—each with varying performance characteristics. Unlike aircraft certified under Part 27, many UAM aircraft may be certificated under Part 23, which does not require demonstration of controllability in all wind azimuths (as do helicopters). Their cruise speeds, and landing transition profiles affect airspace geometry and vertiport landing areas.

As the mission complexity and risk level increases from sUAS in UTM to larger eVTOLs in a NAS, similar services or capabilities may be needed, but with additional design assurance considerations. Design assurance levels are well understood in manned aviation with an accepted process for setting them. Such processes remains in development for UAM services.[1226] This has implications for software systems, which may be web-based or highly automated, that currently lack verification feasibility at high design assurance levels.[1227] Additionally, more simulation is required to resolve use of infrastructure and airspace,[1228] high tempo operations (as many as 1,000 lift-offs or landings per hour across a large metro area, often in higher risk airspace), operations above population centers (reduced separation compared to manned traffic separation management), and to help promote a viable regulatory scheme.[1229] As with current aircraft conducting commercial air transport operations, eVTOL aircraft must be safely capable of aborting an approach or departure in any off nominal situations throughout the entire landing or takeoff profile.

Although UAM is in an early stage of development, it "is coming quicker than most people realize."[1230] UAM business models are undergoing experimentation using conventional aircraft and existing on-demand flight services as newly-designed UAM aircraft and infrastructure are developed, routes and vertiports are laid out, and user apps are advanced.[1231] "There is also a path to commence [UAM piloted] operations . . . under VFR constructs without need for revising current rules."[1232] Announcements by UAM manufacturers suggest a quick path to initial certification with initial deployment as early in 2023 or shortly thereafter. A National Academies of Science finding regarding UAM air taxi services is of import:

> **Finding**: Urban air taxi service for the general public, due to its requirements for vehicle performance, safety, sophisticated operations, infrastructure, operating costs, and system scale and tempo, is one of the most demanding applications of advanced aerial mobility. However, it is an attractive application once the system capabilities are in place.[1233]

Because ground-based infrastructure such as vertiports and airports are integral components of UAM and require infrastructure development,[1234] UAM community collaboration with *smart city* initiatives is important to accommodate its communications, power, environmental and logistics requirements.[1235] The criticality of local stakeholder engagement cannot be overstated.[1236] UAM is also a multimodal transportation component. "[M]ultimodality is complex—it requires deep understanding of how cities move,"[1237] and it must be reflected in innovative UTM design, business models, regulation, and implementation.[1238] Effective and economically feasible business models demand reproduceable, standardized infrastructure and practices. Ultimately, the public, both UAM user and non-user alike, must trust, accept and be able to depend on UAM.[1239]

6.6 STRATOSPHERIC OPERATIONS

> "Very rapidly, this is gonna move to something that looks remarkably like UTM."
>
> **Andy Tailby**
>
> *Zephyr Future Approvals Lead, Airbus*[1240]

The demand for access to altitudes in "High E" (high altitude Class E, typically above FL600) airspace, above current commercial aircraft trajectories, continues to grow[1241] such as for persistent internet communications platforms where no other such infrastructure is available. Correspondingly, technological advances can increasingly support operations at these altitudes.[1242] Such operations may include, for example, high-altitude loitering UAS, balloons, airships, supersonic, and suborbital flights—some inflight for a year or more—with increasing traffic density causing need for enhanced deconfliction.[1243] The lack of traditional or comprehensive ATM support,[1244] variability in airspace structure among nation states (e.g., some set the floor at FL450; others above FL500), the entry of new unmanned aircraft operations at those altitudes, and their need to transition all airspaces demand responsive action.[1245] "We will do a disservice to the aviation system if we don't have the same kind of thinking and the same kind of DNA from lower to upper [airspace] regardless of what we call it … it has to have the same DNA and the same kind of principles."[1246]

NASA has depicted notional high-altitude traffic management in Figure 6.2.[1247] It shows cooperative separation above FL600, transitional traffic management down to mid FL500 altitudes (both cooperative separation and ATC services provided), and below those altitudes, use of conventional ATC services.

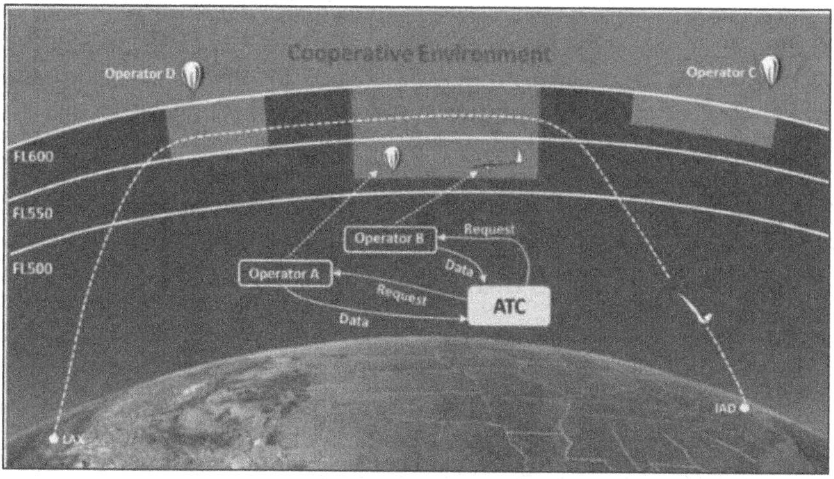

FIGURE 6.2 High altitude traffic management. (Courtesy of NASA.)

"It becomes very apparent that most of the principles of UTM and hopefully some of the tools that UTM will exploit are equally applicable to operations in the stratosphere …. "[1248] The harmonization and integration of VLL and high altitude operations via UTM is increasingly the subject of investigation.[1249] The type of intent shared between vehicles, contingency operations, and real-time risk assessment, among other issues, require development. Steve Bradford, Chief Scientist for Architecture and NextGen Development, FAA asserts, "We already have an analogy for what we want to do at this altitude."[1250] One approach would provide for a variable "deconfliction flight level" set by CAAs where "once the vehicles are at altitude they can separate themselves."[1251] In April 2020, the FAA released an initial ConOps for *Upper Class E Traffic Management (ETM)* that proposes an approach to such operations, and where possible, leverages "UTM foundational principles, architecture, and concept elements."[1252] In fact, according to Dr. Kopardekar, research extending UTM beyond high-altitude operations to "UTM-like" space traffic management (STM) is underway, "enabling beyond possible!"[1253]

6.7 ATM-UTM INTEGRATION

"The UTM will become the ATM."

Nancy Graham
Graham Aero. space Int'l[1254]

"In the end, there will be only one traffic system."

Mark Wuennenberg
Technical Officer, ICAO[1255]

UTM and ATM infrastructures will find increasing convergence and integration as automation increases, equipment becomes increasingly IP- and cloud-based, requirements focus increasingly on supporting integration, and the collaboration accelerates between ATM and UTM experts accelerates. "[H]ow will UTM affect ATM? The simple answer is *in every way possible*."[1256] Additionally, "as the UATM [urban air traffic management] system evolves, it may eventually integrate all UAS operations"[1257] and become "the inspiration and basis for all air traffic management [introducing] the concept of 'Universal Traffic Management' [that is,] UTM that encompasses the new capabilities of UAS traffic management and the future evolution of ATM."[1258] As urged by Peter F. Dumont, Pres. and CTO, ATCA, "[W]e need something more than evolutionary ATM improvements to safely manage the increased airspace demand."[1259] "In the long term, there will be no difference between ATM and UTM …"[1260] and (aspirationally) with "zero inefficiencies due to ATM by 2040" and "hyper-connectivity between all stakeholders."[1261] "A UTM inspired ATM will be developed,"[1262] sometimes characterized as ATM 2.0. Such integration has been described as "one of the major challenges in aviation in the first half of the 21st century,"[1263] presenting "technical, regulatory, and cultural conflicts."[1264] "We don't have a defined pathway for integrating and operating yet, [and] there is obviously more work to be done."[1265] Nonetheless, the path is being laid—aggressively—and there exists a pressing need for ATM evolution.[1266]

7 Conclusions

"Oh, my goodness, it's all coming to fruition."

Jay Merkle
Exec. Dir., UAS Integration Office, FAA[1267]

Chapter 1 introduces UTM. Since its inception as a NASA research project, UTM has represented a major step integrating unmanned aircraft operations into the NAS, as well as a turning point in aviation history. UTM enables the automation needed to handle the new services and predicted volume of unmanned flights. Nevertheless, challenges to safety and efficiency may arise from volume (leading to traffic saturation), small vehicle size, challenging DAA and SWaP, and overall interaction with the ATC system, taxing ATC-UTM coordination. Experts from diverse fields have committed to forging an aviation future that is agile and resilient, with innovative and viable new services. The UTM concept has progressed through extensive collaborative research and testing with the private and public sectors.

Chapter 2 describes UTM ConOps and identifies global variations of the architecture. The chapter describes UTM participants including UAS operators, USS, SDSPs, ANSPs, CAAs, and other participants and stakeholders; it also describes data and data exchanges necessary for safe and secure UAS operations. The scope and characterization of possible UTM services remain in play, while a developing consensus embraces a core set of services and performance metrics. Such services may include, in part, flight planning, strategic deconfliction, conformance monitoring, constraints, geo-awareness, and remote identification.

Essential UTM features include interoperability, scalability, automation, and, of course, safety. Driven by national requirements, either centralized or federated traffic management architecture can be implemented. The chapter gives focus to the Flight Information Management System (FIMS), discovery mechanisms, and supporting elements of a notional UTM architecture. The architecture must be responsive to applicable legal requirements, including data rights, privacy management, and competition. Although not part of UTM, DAA capability is critical for UTM's successful operation.

Chapter 2 also introduces UTM's approach to, and exigent need for, an actionable, integrated cybersecurity response. UTM harbors a heightened attack surface from the developing federated infrastructure, analogous to those of Internet and cloud services. In addition to UAS/UTM cybersecurity standards development and awareness, initiatives such as the International Aviation Trust Framework seek to advance cyber resilience and security services. Remote ID has been prioritized to support necessary accountability, safety, and security.

Chapter 3 surveys standards-making for UTM. Since UTM is infeasible without standards, and because governments anticipate using standards as a means of compliance, many leading/relevant SDOs and their relevant output are presented, including ASTM, EUROCAE, ICAO, ISO, JARUS, and RTCA. Expert support of UTM standards activities, including by government, industry, and supporting organizations such as ANSI has been significant.

Chapter 4 explores UTM governance—approaches to the assessment, approval, management, and oversight of UTM components and ecosystem. Conformity assessment, its basis, and continuing compliance and safety assurance are considered. Furthermore, the trend toward quantitative data-driven risk assessment, influenced by the shift from prescriptive to performance-based regulation, is described. Potential UTM qualification systems examined include, in part, self-certification, third-party audits, service level agreements, and traditional or updated airworthiness certification. The limited delegation of oversight authority is addressed.

Rules of the air must (1) respond to UTM/ATM boundaries and advancing surveillance technologies, (2) resolve when manned aircraft will be required or permitted to participate in UTM, and (3) contemplate simultaneous, integrated operations of ATM and UTM, as well as incremental autonomous operations. Traditional IFR and VFR rules may be unsuitable for UTM, prompting proposals for new airspace volumes and constructs. Moreover, fairness in airspace access presents novel challenges requiring industry consensus and technical solutions.

Other governance challenges include assuring suitable RF spectrum allocation to support expanding communication requirements. Additionally, there is broadening recognition of the role of state, regional, and local governments as UTM stakeholders, the need to secure necessary infrastructure (including via smart cities initiatives) and address community issues such as noise, privacy, and safety policed at the local level. Industry consortia, and non-governmental organizations factor into governance. Business models and competition law affect governance particularly because some of the world's largest companies are at the helm. Apportionment of liability, and development of effective insurance regimes must also be addressed.

Chapter 5 presents selected early example UTM initiatives and implementations that have influenced, serve as pathfinders, and advance UTM. While limited in its scope, the FAA's Low Altitude Authorization and Notification Capability (LAANC) informs development of certain technical and procedural UTM constructs. Europe's U-space has demonstrated important initiatives including significant technical and regulatory platforms for UAS traffic management.

Chapter 6 considers the future of UTM, touching upon important potential barriers to its success, including challenges to ATM/UTM integration, international coordination, and regulatory reform. A few specific areas highlight where technology innovation is imperative, such as for communications infrastructure and autonomy—describing cellular and satellite developments, and pathways for increasing and managing automation and autonomy. Other promising technologies such as blockchain are considered.

UTM will eventually support more than very low level altitude (VLL) airspace sUAS operations. This will include air taxis, stratospheric (high "E"—>60,000′ MSL), and even space traffic management (STM) operations. AAM, including UAM, the infrastructure and aircraft for short-haul urban and regional air transportation is introduced. UAM components include multimodal aerial ride sharing, electric aircraft, an automation platform, and connected vertiports. Lessions learned from UTM will inform and enable AAM.

Nearly every dimension of aviation infrastructure is in the crosshairs of UTM and fair game for scrutiny, imagination, innovation, and reformulation. These changes can truly be viewed as disruptive and even revolutionary.[1268] The skies will be forever changed. Indeed, it is a fascinating and exhilarating voyage with yet-to-be-fully appreciated consequences for the skies and humanity, but the enterprise is certain to fail without assurance of the security and safety of flight.

Notes

1. Presentation at the FAA UAS Symposium (July 9, 2020), https://www.faa.gov/uas/resources/events_calendar/archive/. *See* Reinaldo Negron, Presentation at ICAO DRONE ENABLE Symposium 2021 (April 14, 2021) ("We have capability to implement UTM today."). *Cf.* Richard Parker, Founder and CEO, Altitude Angel, *Standards: A little less conversation, a little more action please!* (webinar) (July 21, 2020) ("UTM is not a product—it is really about the gradual introduction of automation into the airspace…").
2. *See* ICAO, *Unmanned Aircraft Systems Traffic Management (UTM)—A Common Framework with Core Principles for Global Harmonization,* ed. 3 (Sept. 2020), p. 8, https://www.icao.int/safety/UA/Pages/UTM-Guidance.aspx ["ICAO UTM Framework Ed. 3"] (UTM is undertaken through "the collaborative integration of humans, information, technology, facilities and services, supported by air, ground and/or space-based communications, navigation and surveillance."); ASTM, F_ [WK63418], *New Specification for UAS Traffic Management (UTM) UAS Service Supplier (USS) Interoperability*, https://www.astm.org/DATABASE.CART/WORKITEMS/WK63418.htm ["ASTM UTM Spec."]. *See infra* Chap. 2, Sect. 2.1 *ConOps—General* (citing leading UTM Concept of Operations (ConOps) documents, each with UTM definitional content); FAA NextGen, *Concept of Operations, Unmanned Aircraft Systems (UAS) Traffic Management (UTM),* v2.0 (Mar. 2, 2020), https://www.faa.gov/uas/research_development/traffic_management/media/UTM_ConOps_v2.pdf ["FAA ConOps UTM v2.0"], Exec. Summary (UTM "a community-based traffic management system"). GUTMA, *UAS Traffic Management Architecture* (April 2017), https://www.gutma.org/docs/Global_UTM_Architecture_V1.pdf ["GUTMA Architecture 2017"] (defining UTM as "a system of stakeholders and technical systems collaborating in certain interactions, and according to certain regulations, to maintain safe separation of unmanned aircraft, between themselves and from ATM users, at very low level, and to provide an efficient and orderly flow of traffic."); Paul Campbell, Aerospace Eng'r, FAA, Remarks at the ASTM F38 DAA WG (May 4, 2020) ("UTM is nothing more than drone-to-drone deconfliction."); Amit Ganjoo, Founder and CEO, ANRA Technologies, Remarks, *Road to Higher Autonomy*, ICAO DRONE ENABLE Webinar Series (Mar. 5, 2021) (UTM: "The glue that holds everything together.").
3. Email from Parimal Kopardekar, Ph.D., Dir., NARI, NASA (Sept. 13, 2019) (on file with author). *See generally An Introduction to Unmanned Aircraft Systems* (Douglas Marshall et al. eds., CRC 3rd ed. 2021) (providing a comprehensive introductory resource on UAS).
4. Email from Parimal Kopardekar, *id. See* EUROCONTROL & EASA, *UAS ATM Integration, Operational Concept*, Ed. 1.0 (Nov. 27, 2018), Sect. 4.1.3, https://www.eurocontrol.int/sites/default/files/publication/files/uas-atm-integration-operational-concept-v1.0-release%2020181128.pdf ["EUROCONTROL UAS ATM Integration"] ("Some [UTM] services are similar to ATM Services. Such services must have a high-level of coordination with ATM. Therefore, for these services, UTM is part of ATM.").
5. FAA, NPRM, *Remote Identification of Unmanned Aircraft Systems*, 48 Fed. Reg. 72438-72524 (Dec. 31, 2019) (to be codified at 14 C.F.R. pts. 1, 47, 48, 89, 91, and 107), Sect. IV.B. Unmanned Aircraft Systems Traffic Management (UTM), https://www.federalregister.gov/documents/2019/12/31/2019-28100/remote-identification-of-unmanned-aircraft-systems ["RID NPRM 2019"]. *See* Email from Peter Sachs, FAA (formerly, Security and Risk Architect, Airbus) (Oct. 16, 2019) (on file with author) ("Legacy ATM is a set of services provided by a single company or agency for each country. Operators don't get to choose which ANSP they use. UTM is envisioned

to be decentralized and distributed, with many companies providing competing services to meet the needs (and price points) of the range of operators. Some capabilities may be 'centralized', especially in Europe, where there is a tendency to have single instances of some ground truth services, e.g., surveillance information and airspace constraints.").

6. Email from Parimal Kopardekar, *supra*. *See generally* Yigang Liu, et al., *SOA-Based Aeronautical Service Integration, INTECH* (2011), https://cdn.intechopen.com/pdfs/20427/InTech-Soa_based_aeronautical_service_integration.pdf; Reinaldo Negron, Head of UTM, Wing, Presentation at the FAA UAS Symposium (June 9, 2020) ("If you look across work done around the world, automation, manage by exception, is the only way to manage complexity and scale.").
7. *See infra* Chap. 6, Sect. 6.7 *ATM-UTM Integration*. *Cf.* Mark Wuennenberg, Technical Officer, ICAO, Presentation at DRONE ENABLE/3, Montreal, CA (Nov. 23, 2019) (Asserting that some implementers are looking for a stand-alone UTM system; others as part of an existing ATM system. "In the end, it could be part of the ATM or standing alone."); Gov't of India, Ministry of Civil Aviation, *National Unmanned Aircraft System (UAS) Traffic Management Policy, Architecture, Concept of Operations and Deployment Plan for enabling UTM ecosystem in India, Discussion Draft,* Ver.1.0 (Nov. 30, 2020), p. 7, https://www.civilaviation.gov.in/sites/default/files/National-UTM-Policy-Discussion-Draft-30-Nov-2020-updated.pdf ["India UTM Policy"] (UTM "envisioned as a cooperatively driven and collaborative extension" of ATM); *id.* Sect. 9.1 UTM-ATM Integration ("ATM and UTM systems should communicate with each other at the systems level.").
8. *See infra* Chap. 4, Sect. 4.1 *General—UTM Governance* (addressing delegable and nondelegable duties).
9. *See infra* Chap. 6, Sect. 6.5 *Advanced Air Mobility / Urban Air Mobility)*; Chap. 6. Sect. 6.6 *Stratospheric Operations.*
10. Remarks at the FAA Drone Advisory Committee meeting (June 19, 2020).
11. Email from Parimal Kopardekar, Ph.D., *supra.*
12. *See, e.g.,* FAA, *FAA Aerospace Forecast Fiscal Years 2020–2040* (2020), p. 41, et seq., https://www.faa.gov/data_research/aviation/aerospace_forecasts/media/FY2020-40_FAA_Aerospace_Forecast.pdf ["FAA Aerospace Forecast"]. The expected orders of magnitude of new UAS operations (including swarms and formations of them) would saturate existing surveillance systems and airspace separation capabilities; were neither contemplated nor facilitated in the ATC Concept of Operations (ConOps); and make manual traffic management of UAS unworkable. *See infra* note 22 (presenting Automatic Dependent Surveillance-Broadcast (ADS-B) Out limitations). *See generally* [FAA ConOps UTM v2.0]; FAA, *NextGen Implementation Plan 2018–19*, https://www.faa.gov/nextgen/media/NextGen_Implementation_Plan-2018-19.pdf; SESAR Joint Undertaking, *European ATM Master Plan, Digitalising Europe's Aviation Infrastructure* (2020), https://www.atmmasterplan.eu/exec/overview ["SESAR ATM Master Plan 2020"].
13. Joseph L. Rios, ATM-X Project Chief Eng'r, et al., *UTM UAS Service Supplier Development, Sprint 2 Toward Technical Capability Level 4*, NASA/TM-2018-220050 (Dec. 2018), p. 4, https://utm.arc.nasa.gov/docs/2018-UTM_UAS_TCL4_Sprint2_Report_v2.pdf ["Rios, TLC4 Sprint 2"].
14. *See* Tim Beam, CEO, Fortem, *quoted in*, Nick Zazulia, *Fortem and Unifly Plan to Solve Unmanned Airspace*, Rotor & Wing (April 30, 2019), https://preview.tinyurl.com/Tim-Beam ("[A] world where every bit of land is an airport and every business and home is its own airline.").
15. Parimal Kopardekar, Ph.D., et al., NASA Ames, *Unmanned Traffic Management (UTM) Concept of Operations*, 16th AIAA Avi. Tech., Integration, and Ops. Conf., Wash., D.C. (June 2016), https://utm.arc.nasa.gov/docs/Kopardekar_2016-3292_ATIO.pdf ["Kopardekar 2016"].
16. *See, e.g.,* 14 C.F.R. Sect. 91.113, *Right of Way Rules: Except water operations. See also* Chap. 4, Sect. 4.3 *UAS Flight Rules.*

17. Ryan J. Wallace, et al., *Cleared to land: Pilot visual detection of small unmanned aircraft during final approach*, INT'L J OF AVIATION, AERONAUTICS, AND AEROSPACE, 6(5) (2019), https://commons.erau.edu/ijaaa/vol6/iss5/12/. *See* Michael S. Baum, et al., Aviators Code Initiative, *UAS Pilots Code,* Annotated Ver. (2018), note 60, http://www.secureav.com/UASPC-annotated-v1.0.pdf (addressing see and avoid) ["UAS Pilots Code"].
18. [FAA ConOps UTM v2.0], Sect. 2.7 Airspace Management ("layered approach to safety"); *id.* Sect. 2.7.1 ("UTM has multiple layers of separation assurance"); FAA, *Unmanned Aircraft Systems (UAS) Traffic Management (UTM) Pilot Program Phase Two, Concept of Use V1.0* (Dec. 2019), Sect. 4.5 ("UTM has multiple layers of separation assurance to ensure the safe conduct of operations, from strategic management to contingency management to real-time collision avoidance capabilities.") ["FAA UPP Phase 2 ConUse"]. *See* [ICAO UTM Framework Ed. 3], Appn. G, Deconfliction and Separation Management, p. 38 (referencing *Global Air Traffic Management Operational Concept* (ICAO Doc 9854)—stating "that conflict management will consist of three layers: strategic deconfliction, separation provision (tactical deconfliction) and collision avoidance.").
19. Robert Roth, Head of Software Eng'g, Uber ATG (formerly, Eng'g Dir., Prime Air at Amazon), Remarks, Palo Alto, Cal. (Aug. 29, 2019) ("One of the largest benefits of UTM as a part of the overall safety system is that it is the one place in the system that we can see intent—UTM can see into the future because we're filing flight plans and therefore we can optimize the system to avoid unintended conflict."). *See* [FAA UPP Phase 2 ConUse], Sect. 4.3.5 ("Separation is achieved through shared intent, shared awareness, strategic deconfliction, aircraft tracking and conformance monitoring, and the establishment of procedural rules of the road.").
20. *See infra* Chap. 2, Sect. 2.4.6 *Detect and Avoid (DAA)*; UTM may, in the future also provide services akin to tactical deconfliction in the form of dynamic re-planning services; Chap. 2, Sect. 2.38 *Tactical Deconfliction*; Email from Peter Sachs, FAA (formerly, Security and Risk Architect, Airbus) (Oct. 16, 2019) (on file with author) ("We think of 'tactical' as similar to what controllers do today, like vectoring and speed control, not last-minute evasive maneuvers such as commanded by TCAS, ACAS or DAA.").
21. *See* FAA, *Modernization of U.S. Airspace*, https://www.faa.gov/nextgen/?locationId=17.
22. For example, a core NextGen aircraft requirement, ADS-B Out, is prohibited for use by most sUAS due to concern regarding system saturation and multipath. Jay Merkle, Exec. Dir., FAA UAS Integration Office, Remarks at the Drone Advisory Committee, Crystal City, Va. (June 6, 2019) ("Initially there were concerns that ADS-B Out would saturate the downlink but that is only the tip of the iceberg. With the number of drones forecasted, multipath in urban [areas] would be received by the FAA SBS [Surveillance Broadcast Service] and would have to be filtered out [causing] 'a very complex problem'... Multipath (when ADS-B bounces off of multiple surfaces and creates ghosts) [is] very difficult for controllers to tell real from ghost [targets and] puts our system at risk."). *See* FAA, *Remote Identification of Unmanned Aircraft,* Final rule (issued Dec. 28, 2020), 86 Fed. Reg. 4390 (Jan. 15, 2021), https://www.faa.gov/news/media/attachments/RemoteID_Final_Rule.pdf, https://www.federalregister.gov/d/2020-28948 ["Remote ID Rule"], Sect. 89.125 *Automatic Dependent Surveillance-Broadcast (ADS-B) Out prohibition* ("Automatic Dependent Surveillance-Broadcast (ADS-B) cannot be used to comply with the remote identification requirements of this part.").
23. *See infra* Chap. 5, Sect. 5.2 *Low Altitude Authorization and Notification Capability.*
24. Frost and Sullivan, *Global Commercial UAS Market Outlook, 2020* (April 8, 2020), https://ww2.frost.com/news/press-releases/commercial-drone-market-to-hit-2-44-million-units-by-2023-says-frost-sullivan/; [FAA Aerospace Forecast], p. 41, et seq. (describing "healthy growth" and "enormous promise" while acknowledging "operational challenges including safe integration into the NAS). *See* [Remote RID Rule], Preamble, XX.B Comments on Benefits and Cost Savings ("The FAA recognizes its fleet forecast for recreational unmanned aircraft is most likely underestimated...").

Cf. Teal Group, *World Civil Unmanned Aerial Systems Market Profile & Forecast (July 2019),* https://www.ainonline.com/aviation-news/business-aviation/2019-07-18/teal-report-sees-civil-uas-market-tripling-size-2028 (civil UAS market tripling by 2028).

25. Jay Merkle, Exec. Dir, UAS Integration Office, FAA, Presentation at the Transportation Research Board (TRB) annual meeting, Orlando, Fl. (Jan. 13, 2020), https://www.c-span.org/video/?468065-1/automation-technology-transportation. 14 C.F.R. Part 107 certifications are granted to certificated remote pilots flying drones less than 55 pounds after they pass a knowledge test and register their drone with the FAA.
26. Carl Burleson, FAA Acting Deputy Admin'r, Presentation at AUVSI Xponential, Chicago, Ill. (May 2, 2019), *quoted in, Dronelife*, https://dronelife.com/2019/05/01/faa-acting-deputy-administrator-carl-burleson-our-job-is-to-find-a-way-forward/; [FAA ConOps UTM v 2.0], Sect. 1.1 Need for UTM (volume of sUAS could be "greater, to that of present-day manned air traffic"); [ICAO UTM Framework Ed. 3], Foreword (civil UAS "operations are expected to surpass the number of manned aircraft operations in the near future.").
27. FAA, *LAANC Drone Program Expansion Continues* (Dec. 2, 2019), https://www.faa.gov/news/updates/?newsId=94750; [FAA Aerospace Forecast], p. 53-56.
28. *See, e.g.*, SESAR, *European ATM Master Plan: Roadmap for the safe integration of drones into all classes of airspace* (2017), http://www.sesarju.eu/sites/default/files/documents/reports/European%20ATM%20Master%20Plan%20Drone%20roadmap.pdf ["SESAR ATM Master Plan 2017"]; SESAR, *Strategic Research and Innovation Agenda, Digital European Sky* (Sept. 2020), Fig. 7, 2050 Outlook of Future Increase Use of the Airspace by Manned and Unmanned Operations, https://www.sesarju.eu/sites/default/files/documents/reports/SRIA%20Final.pdf ["SESAR SRIA"]. *See infra* Chap. 3, Sect. 3.3.5 *SESAR JU.*
29. Frost & Sullivan, *Analysis of Urban Air Mobility and the Evolving Air Taxi Landscap*e, 2019 (Oct. 25, 2019), https://store.frost.com/analysis-of-urban-air-mobility-and-the-evolving-air-taxi-landscape-2019.html.
30. *See* Porsche Consulting, *The Future of Vertical Mobility, Sizing the market for passenger, inspection, and goods services until 2035* (2019), p. 3, https://www.porsche-consulting.com/fileadmin/docs/04_Medien/Publikationen/TT1371_The_Future_of_Vertical_Mobility/The_Future_of_Vertical_Mobility_A_Porsche_Consulting_study__C_2018.pdf.
31. *NEXA UAM study sets 20-year market value of $318 billion*, Air Traffic Management.net (Aug. 20, 2019), https://airtrafficmanagement.keypublishing.com/2019/08/20/nexa-uam-study-sets-20-year-market-value-of-318-billion-across-74-cities/?dm_i=4JU%2C6G8B6%2CPM6JE7%2CPKRIH%2C1; Volocopter, *The roadmap to scalable urban air mobility, White paper 2.0* (Mar. 2021), https://press.volocopter.com/images/pdf/Volocopter-WhitePaper-2-0.pdf ["Volocopter Roadmap"], respectively.
32. *See, e.g.*, The National Academies of Sciences, and Engineering, Medicine, Committee on Enhancing Air Mobility, *Advancing Aerial Mobility, A National Blueprint,* The Nat'l Academies Press (2020), https://doi.org/10.17226/25646 ["Nat'l Academies AAM"] ("There are mismatches between the exuberance of entrepreneurs and early investors and the realities of implementation...").
33. Dir., NARI, NASA. U.S. patent application by Dr. Kopardekar entitled, *Unmanned Aerial Systems Traffic Management,* submitted late 2013, U.S. Patent No. US20160275801A1 (Issued June 25, 1029), *available at* https://patents.google.com/patent/US10332405B2/enhttps://patents.google.com/patent/US10332405B2/en. *See* NASA Technology Transfer Program, *Unmanned Aerial Systems (UAS) Traffic Management,* https://technology.nasa.gov/patent/TOP2-237; NASA, *How to License NASA Technology,* https://technology.nasa.gov/license.
34. FAA, *Roadmap for Integration of Civil Unmanned Aircraft Systems (UAS) in the National Airspace System (NAS),* First Edition (2013), https://www.faa.gov/uas/resources/policy_library/media/uas_roadmap_2013.pdf. *See* FAA, *Integration of Unmanned Aircraft Systems into the National Airspace System, Concept of*

Operations, v2.0 (Sept. 28, 2012), *available at* http://www.sarahnilsson.org/app/download/965094433/FAA-UAS-Conops-Version-2-0-1.pdf (adopting, e.g., "TFM"—traffic flow management, and "TMIs"—traffic management initiatives, but not "UTM").

35. NASA Aeronautics Research Mission Directorate: Team Seedling Solicitation, *Enabling Low Altitude Civilian Applications of Unmanned Aerial Systems by Unmanned Aerial System Traffic Management (UTM),* Submitted by Parimal Kopardekar, Ph.D., Principal Investigator (2013) (copy on file with author).
36. *See* CBS This Morning, *Amazon CEO Unveils Drone Delivery Concept* (Dec. 2, 2013), *available at* https://www.youtube.com/watch?v=-qOBm3Iwlzo; *but see* Kelsey Atherton, *Amazon's Delivery Plans aren't Realistic*, POPULAR SCIENCE (Dec. 4, 2013), *available at* https://www.businessinsider.com/amazons-delivery-drones-arent-realistic-2013-12. The Amazon proposal followed shortly after the proposal by Dr. Kopardekar. Parimal Kopardekar, Ph.D., Dir., NARI, NASA, Remarks, in Moffett Field, Cal. (Aug. 26, 2019).
37. Parimal H. Kopardekar, Ph.D., Dir., NARI, NASA, *Unmanned Aerial System (UAS) Traffic Management (UTM): Enabling Low-Altitude Airspace and UAS Operations*, NASA, NASA/TM-2014-21899 (April 2014), https://ntrs.nasa.gov/archive/nasa/casi.ntrs.nasa.gov/20140013436.pdf ["Kopardekar 2014"]. NASA UTM core operating principles were proposed during a Feb. 24, 2014 UTM stakeholder meeting precipitating groundwork for project initial prototyping. The first UTM flight tests commenced May 2015.
38. NASA Ames, *Unmanned Traffic Management (UTM): Enabling Low Altitude Civilian UAS Operations* (Feb. 12-14, 2014). While the concept was favorably received, concern was expressed that the term UTM "would confuse people." John Scull Walker, Sr. Partner, The Padina Group & Chair, ISO/TC 20/SC 16, Remarks (May 16, 2019).
39. Parimal H. Kopardekar, Ph.D., Dir., NARI, NASA, Remarks, in Moffett Field, Cal. (Aug. 26, 2019) ("Until then, the FAA and NASA were unsure. After the convention, they went back and said, 'I don't think we can stop this thing anymore.' When they saw this mass of people…they changed their mindset; momentum picked up." This was the *NASA Unmanned Aerial Systems (UAS) Traffic Management Convention,* Moffett Field, Cal. (July 28–30, 2015), https://utm.arc.nasa.gov/utm2015.shtml.
40. ICAO, *Unmanned Aircraft Systems Traffic Management (UTM) – A Common Framework with Core Principles for Global Harmonization*, Ed. 2 (Nov. 2019), p. 5, https://www.icao.int/safety/UA/Documents/UTM-Framework%20Edition%202.pdf.
41. The Japan Aerospace Exploration Agency (JAXA) and the Civil Aviation Administration of China (CAAC), and other programs also deserve attention. *See infra* Table 5.1–*Selected Early Developments.*
42. NASA, *UAS Integration in the NAS Project*, https://www.hq.nasa.gov/office/aero/iasp/uas/index.htm, and Small Business Innovation Research / Small Business Technology Transfer, NASA, *UAS Integration in the NAS* (Oct. 31, 2013), https://sbir.nasa.gov/content/uas-integration-nas.

    While neither a sUAS nor UTM project, other initiatives may have contributed to an early foundation. For example, *Access 5 Project* - "[A] national project sponsored by NASA and Industry with participation by the FAA and DoD to introduce high altitude long endurance (HALE) unmanned aircraft systems (UAS) for routine flights in the National Airspace System (NAS)." Underway from 2004 through 2006, it "sought to lay the groundwork for the future introduction of other classes of UAS, such as those that operate below Flight Level 180." Access 5 (archival website), https://web.archive.org/web/20060627203947/http://www.access5.aero/site_content/index.html.

    "Free Flight" initiatives in the 1990s and subsequent autonomous flight rules (AFR) initiatives also deserve attention. *See, e.g.,* David J. Wing, et al., Langley Research Center, *Autonomous Flight Rules A Concept for Self-Separation in U.S. Domestic Airspace*,

NASA/TP–2011-217174 (Nov. 2011), https://ntrs.nasa.gov/archive/nasa/casi.ntrs.nasa.gov/20110023668.pdf ["Wing AFR"]; Steve Bradford, Chief Scientist for Architecture and NextGen Development, FAA, Remarks at the NASA, AAM Airspace WG (webinar) (Aug. 4, 2020) (recounting that for free flight, "planes were going to take off and land completely on their own – but we learned you need structure because the ATC is providing services to others... otherwise you become a burden on the system.").

43. DoD UAs returning from war zones such as Afghanistan and Iraq needed a place to fly to train pilots and maintain proficiency. Most restricted military airspace was already in use or could not handle the scale of operations required for this new wave of aircraft.
44. Such research into SAA capabilities escalated in 2007, with the DoD assigning the Army the lead for Ground Based Sense and Avoid (GBSAA) technologies and the Air Force the lead of Airborne Sense and Avoid (ABSAA) technologies. In 2016, the Army, in conjunction with the FAA, approved the Army GBSAA system for routine BVLOS UAS operations in the US NAS within the operational bounds of the system. *See Remotely piloted drones no longer accompanied by piloted planes flying into Hancock Field thanks to new radar system,* Nextar Broadcasting, Inc. (Sept. 10, 2019), https://www.localsyr.com/news/local-news/remotely-piloted-drones-no-longer-accompanied-by-piloted-planes-flying-into-hancock-field-thanks-to-new-radar-system/.
45. *See, e.g.,* NASA, *UAS Traffic Management (UTM) Project* (Sept. 4, 2018), https://www.nasa.gov/aeroresearch/programs/aosp/utm-project-description/; Ronald D. Johnson, UTM Project Mgr., NASA, *Unmanned Aircraft Traffic Management (UTM) Project* [slides] (April 10, 2018), https://ntrs.nasa.gov/archive/nasa/casi.ntrs.nasa.gov/20180002542.pdf; *FAA Extension, Safety, and Security Act of 2016*, Pub. L. No. 114-190, 130 Stat. 615 (2016), codified at 49 U.S.C. 40101 note, https://www.congress.gov/114/plaws/publ190/PLAW-114publ190.pdf ["FESSA"], Sect. 2208 Unmanned Aircraft Systems Traffic Management (directing the FAA to coordinate UTM research with NASA).
46. [Kopardekar 2016], p. 11.
47. [Kopardekar 2016], pp. 6–7. *See* NASA Ames, *What is Unmanned Aircraft Systems Traffic Management?* (May 3, 2019), https://www.nasa.gov/ames/utm. *Cf.* Transport Canada, White Paper, *Drone Talks: Planning for Success* (May 29–30, 2019), Workshop #1: Airspace and RTM System, https://www.dropbox.com/s/uiozw2pwr8v4nku/Canada-RTM.doc?dl=0 (guiding principles: fair access to airspace; encourage innovation & economic development; accommodate diversity of traffic, traffic density and autonomy; prioritize cyber-security/system resilience; build public policy/social acceptance and data privacy/protection mitigation; take a multi-disciplinary/collaborate approach).
48. *See* Email from Jeffrey R. Homola, UTM Integration & Testing Lead, NASA Ames (Dec. 20, 2019) (on file with author) ("This ConOps was heavily influenced by the interactions between NASA and the FAA through the RTT."). *See infra* Chap. 1, Sect. 1.5.2 *Research Transition Team*; Chap. 2, Sect. 2.1 *ConOps—General.*
49. Arwa S. Aweiss, et al., NASA Ames Research Center, *Unmanned Aircraft Systems (UAS) Traffic Management (UTM) National Campaign II* (2018), p. 3, https://ntrs.nasa.gov/archive/nasa/casi.ntrs.nasa.gov/20180000682.pdf (as a function of population density, amount of people and property on the ground, number of manned aircraft in close proximity to the sUAS operations, and density of the UAS operations) ["Aweiss 2018"].
50. NASA, *UTM*, https://utm.arc.nasa.gov/index.shtml.
51. For example, validity of the RESTful API approach to data sharing; including for airspace reservations and notifications completion.
52. Testing included all six UAS Test Sites flying multiple UAS simulations simultaneously.
53. Joseph Rios, Ph.D., ATM-X Project Chief Eng'r, NASA Ames, Remarks at ASTM Committee F38 meeting, Seattle, WA (April 17, 2019).
54. *See* NASA, *Technical Capability Level 2.0. Software Version Description*, Ver. 1.0 (Feb. 2017), https://ntrs.nasa.gov/archive/nasa/casi.ntrs.nasa.gov/20170001941.pdf; *infra* Chap. 2, Sect. 2.4.5 *Discovery.*

55. [Aweiss 2018], p. 2. All six test sites participated. *See* Chap. 1, Sect. 1.5.1.2 *Test Sites.*
56. *See, e.g.,* FAA & NASA, Unmanned Aircraft System Traffic Management (UTM) Research Transition Team (RTT), Concept Working Group, *Concept & Use Cases Package #2 Addendum: Technical Capability Level 3*, Ver. 1.0, Doc. No. 20180007223 (July 2018), Sects. 1.2.4.2 & 2.2.4.2, *Shared Information Across Actors*, https://ntrs.nasa.gov/search.jsp?R=20180007223 ["TCL 3"]. All six test sites participated. *See also* Arwa Aweiss, et al., NASA Ames, *Flight Demonstration of Unmanned Aircraft System (UAS) Traffic Management (UTM) at Technical Capability Level 3* (IEEE DASC, Sept. 8-12, 2019), https://utm.arc.nasa.gov/docs/2019_Aweiss_DASC_2019.pdf, and https://ntrs.nasa.gov/archive/nasa/casi.ntrs.nasa.gov/20190030743.pdf ["Aweiss 2019"].
57. Joseph L. Rios, Ph.D., et al., ATM-X Project Chief Eng'r, NASA, *UTM UAS Service Supplier Development, Sprint 2 Toward Technical Capability Level 4*, NASA/TM-2018-220050 (Dec. 2018), https://utm.arc.nasa.gov/docs/2018-UTM_UAS_TCL4_Sprint2_Report_v2.pdf ["Rios TCL4 Sprint 2"]; Joseph Rios, et al., *Flight Demonstration of Unmanned Aircraft System (UAS) Traffic Management (UTM) at Technical Capability Level 4*, AIAA Aviation Forum (June 25–19, 2020), https://utm.arc.nasa.gov/docs/2020-Rios-Aviation2020-TCL4.pdf ["Rios TCL4"]; FAA & NASA, Unmanned Aircraft System Traffic Management (UTM) Research Transition Team (RTT), Concept Working Group *Concept & Use Cases Package #3: Technical Capability Level 4,* Ver. 1.0 (March 2019) ["FAA & NASA UTM RTT WG #3"]; NASA, Joseph L. Rios, et al., *UAS Service Supplier Network Performance, Results and Analysis from Flight Testing Multiple USS Providers in NASA's TCL4 Demonstration*, NAS/TM-2020-220462 (Jan. 2020), https://utm.arc.nasa.gov/docs/2020-Rios_TM_220462-USS-Net-Perf.pdf ["Rios 2020"]. *See* John Bowden, *NASA begins testing system to manage drone traffic in cities*, The Hill (May 24, 2019), https://thehill.com/policy/technology/445393-nasa-launches-tests-of-system-to-manage-drone-traffic-in-cities; *ANRA UTM powers NASA TCL4 campaigns in Nevada and Texas,* sUAS News (May 15, 2019), https://www.suasnews.com/2019/05/anra-utm-powers-nasa-tcl4-campaigns-in-nevada-and-texas/?mc_cid=415259365a&mc_eid=a9df228121.
58. Parimal Kopardekar, Ph.D., Dir., NARI, NASA, *Unmanned Aerial System Traffic Management System*, Talks at Google (May 3, 2016), https://www.youtube.com/watch?v=iDjZzyswlMY.
59. *See infra* note 72 (Advanced Air Mobility National Campaign); Chap. 2, Sect. 2.4.9 *UTM Data Exchange* (re NASA's Digital Information Platform (DIP) initiative); Chap. 6, Sect. 6.5 *Advanced Air Mobility / Urban Air Mobility. See also* FAA, *BEYOND*, https://www.faa.gov/uas/programs_partnerships/beyond/ (program following the IPP).
60. Initiated per the *FAA Modernization and Reform Act of 2012*, Pub. L. No. 112-95, Sect. 332(c), 126 Stat. 11 (Feb. 12, 2012), https://www.congress.gov/bill/112th-congress/house-bill/658/text (The test sites include: U. of Alaska, State of Nevada, NY's Griffiss Int'l Airport, N. Dakota Dep't of Commerce, Texas A&M, and Virginia Polytechnic Institute and State U. (Virginia Tech)). *See* FAA, *Fact Sheet—FAA UAS Test Site Program* (Dec. 30, 2013), https://www.faa.gov/news/fact_sheets/news_story.cfm?newsId=15575. New Mexico State University was subsequently designated in 2016.
61. The test sites are not formally a UTM component yet are making diverse contributions to its advancement. *See* FAA, *UAS Test Site Program*, https://www.faa.gov/uas/programs_partnerships/test_sites/.
62. The states of North Dakota, Virginia, and Nevada. *See infra* Chap. 1, Sect. 1.5.4 *UTM Pilot Program (UPP)* (introducing the UPP); Sect. 1.5.3 *Relevant Follow-on Efforts.*
63. Email from Nicholas Flom, Exec. Dir., Northern Plains UAS Test Site (June 10, 2020) (on file with author).
64. ASSURE, http://www.assureuas.org/.
65. FAA Air Transportation Centers of Excellence, https://www.faa.gov/about/office_org/headquarters_offices/ang/grants/coe/. The US General Accountability Office (GAO) has completed an assessment of the Test Sites program. *See* GAO, *Unmanned Aircraft Systems, FAA Could Better Leverage Test Site Program to Advance Drone Integration*,

Rpt. to the Ranking Member, Comm. on Transport. and Infrastructure, House of Representatives, GAO-20-97 (Jan. 9, 2020), https://www.gao.gov/assets/710/703726.pdf (recommending, in part, development of plan to advance UAS integration, and that the FAA "should publicly share more information on how the test site program informs integration…" p. 40. Need for a data analysis plan for Mission Logging System (MLS) data was also underscored.). The GAO report also analyzes the types and extent of testing undertaken at the Test Sites.

66. FAA Reauthorization Act of 2018, Pub. L. No. 115-254 (Oct. 5, 2018), *available at* https://www.congress.gov/115/bills/hr302/BILLS-115hr302enr.pdf ["FAARA"], Sect. 44803(g).
67. FAA & NASA, *UAS Traffic Management (UTM), Research Transition Team (RTT) Plan*, V1.0, Sect. 1.4 (Jan. 31, 2017), https://www.faa.gov/uas/research_development/traffic_management/media/FAA_NASA_UAS_Traffic_Management_Research_Plan.pdf ["RTT 2017"] (The general RTT objectives "are to: (1) provide a structured forum for researchers and implementers to constructively work together on a continuing basis; (2) ensure that planned research results will be fully utilized, and will be sufficient to enable implementation of NextGen air navigation services concepts; and (3) provide a forum for the inclusion of all of the NASA and FAA stakeholders who would be involved in the planning, conducting, receiving, and utilizing of the research conducted by the RTTs." *id.* p. 6).
68. [RTT 2017], pp. 8–9 (also noting that research was segmented into four subgroups: concepts and use cases, data exchange and information architecture, communications and navigation, and sense and avoid).
69. [RTT 2017], p. 5.
70. *See, e.g.,* the NextGen SENSR program, *infra* note 868. *See infra* Chap. 1, Sect. 1.5.6 Private Industry.
71. *See infra* Chap. 6, Sect. 6.6 *Stratospheric Operations.*
72. NASA, Aeronautics Research Mission Directorate, https://www.nasa.gov/uamgc. The Advanced Air Mobility National Campaign (formerly entitled the "Grand Challenge") series will "bring together aircraft manufacturers and airspace service providers to identify maturity levels [UAM Maturity levels ("UML*")] for vehicle performance, safety assurance, airspace interoperability, etc., and to develop and demonstrate integrated solutions for civil use." NASA, Aeronautics Research, *Advanced Air Mobility National Campaign, Overview* (March 12, 2020), https://www.nasa.gov/aeroresearch/aam/description; NASA, *Advanced Air Mobility (AAM), Urban Air mobility (UAM) and Grand Challenge*, AIAA (July 24, 2020), https://ntrs.nasa.gov/archive/nasa/casi.ntrs.nasa.gov/20190026695.pdf. *See* [Nat'l Academies, AAM], Sects. 3-10 - 3-11 (in part, describing the Grand Challenge Goals, and scenarios relevant to UTM); NASA, *Grand Challenge Developmental Testing Partners* (March 3, 2020), https://www.nasa.gov/aeroresearch/grand-challenge-developmental-testing; NASA, *UAM Vision Concept of Operations (ConOps), UAM Maturity Level (UML) 4*, Doc. ID 20205011091, Deloitte (Dec. 2, 2020), https://ntrs.nasa.gov/citations/20205011091 ["NASA UAM Vision ConOps"], Sect. 1.1 Background: UAM Maturity Levels.
73. *See* NASA, https://www.nasa.gov/aeroresearch/programs/aosp/atm-x (will include demonstration of an open architecture approach, integration of air traffic technologies, system-wide data use, advances in human-machine teaming, and increasingly autonomous decision-making); William N. Chan, Proj. Mgr., et al., *Overview of NASA's Air Traffic Management—eXploration (ATM-X) Project* (2018), https://ntrs.nasa.gov/archive/nasa/casi.ntrs.nasa.gov/20180005224.pdf; David Murakami, Research Eng'r, et al., NASA Ames, *Space Traffic Management with a NASA UAS Traffic Management (UTM) Inspired Architecture*, 10.22514 (June 2019), http://sreejanag.com/Documents/murakami_scitech_final.pdf.
74. FAA, *UTM Pilot Program* (webpage), https://www.faa.gov/uas/research_development/traffic_management/utm_pilot_program/. *See* FAA & NASA, *UTM Pilot Program Brochure*, https://www.faa.gov/uas/research_development/traffic_management/

utm_pilot_program/media/UTM_Pilot_Program_Smart_Sheet.pdf; Steve Bradford, FAA, and Parimal Kopardekar, NASA, *Unmanned Aircraft Systems (UAS) Traffic Management (UTM Pilot Program (UPP), UPP Summary Report* (Oct. 2019); FAA, *Unmanned Traffic Management System Demonstrations* (Sept. 4, 2019), https://www.youtube.com/watch?v=zpc4aoJKefA.

75. FAA, *Statement of Objectives for Test Site selection to participate in UTM Pilot Program Demo Program*, Rev#3 (July 20, 2018), https://www.faa.gov/uas/research_development/traffic_management/utm_pilot_program/media/UTM_Pilot_Program_Smart_Sheet.pdf ["UPP Objectives"].
76. [FESSA], Sect. 2208. Unmanned Aircraft Systems Traffic Management.
77. [FESSA], *id.* (demonstrating, *inter alia*, "the initial integrated UTM ecosystem"). *See* Jeffrey R. Homola, UTM Integration & Testing Lead, NASA Ames, Remarks, in Moffett Field (Oct. 30, 2019) ("I almost consider UPP the next phase of testing that built upon NASA's experience with TCLs and moved us one step closer to implementation.").
78. *See* [UPP Objectives], p. 3. For example, one UPP initiative was the UTM Pilot Demonstration (Demo) Program awarded to the Northern Plains Test Site. This program required the UAS Test Site to be "capable of supporting the FAA and participants in exercising the following capabilities:
    Capability 1: UAS Operator to USS Communication
    Capability 2: USS to USS Communication
    Capability 3: Enterprise Services via Application Programming Interface (API)
    Capability 4: UAS Volume Reservations (UVR) Viewing
    Capability 5: Shared Information
    Capability 6: UVR Service via API [formerly entitled Dynamic Restrictions Service via API]"
79. Email from Nicholas Flom, Exec. Dir., Northern Plains UAS Test Site (Aug. 22, 2019) (on file with author) (also stating, "[t]hrough the UPP, the FAA has now seen how the communication protocols work between the USSs and FIMS for the capabilities tested, but [it] was not a specific springboard to operationalize all UAS users.").

    UAS Integration Pilot Program (IPP)—Established in 2017, the IPP heightened awareness of the requirements and effective relationship of local, regional, and state government with the national requirements for UAS and UTM—intending to advance public acceptance and integration. It also helped shape policy. FAA, *UAS Integration Pilot Program*, https://www.faa.gov/uas/programs_partnerships/integration_pilot_program/. *See* Jay Merkle, Exec. Dir., FAA, UAS Integration Office, Presentation at the Transportation Research Board (TRB) Annual Meeting, Orlando, Fl. (Jan. 13, 2020), https://www.c-span.org/video/?468065-1/automation-technology-transportation ("The biggest…lesson learned for future unmanned aircraft and Urban Air Mobility…is community engagement and community acceptance."). *See infra* Chap. 4, Sect. 4.6 *State, Regional, and Local*; Jay Merkle, Dir., UAS Integration Office, FAA, UAS Symposium, *Innovation: Shaping the Future* (Aug. 19, 2020) ("While the IPP might be over, the journey is not done.…There will be something after the IPP."). On October 26, 2020 the FAA commenced an IPP follow-on entitled BEYOND to address BVLOS with emphasis on infrastructure inspection, public operations and small package delivery; "operating under established rules rather than waivers", data collection, and streamlined approvals. FAA, *BEYOND, Program Overview*, https://www.faa.gov/uas/programs_partnerships/beyond/).
80. Joseph Rios, Ph.D., UTM Project Chief Eng'r, NASA Ames, Presentation at the FAA/NASA UTM Pilot Program Phase Two Industry Workshop, Moffett Field, CA (Dec. 9, 2019) (describing FAA vision to include: remote ID services, public safety operations, UVR services, USS transmission of flight information to FAA due to off-nominal UTM event, and support message security best practices or authorization, authentication, and message signing); FAA & NASA, *Unmanned Aircraft Systems (UAS) Traffic Management (UTM) Pilot Program (UPP) Phase 2, Industry Workshop* (Dec. 2019),

https://www.faa.gov/uas/research_development/traffic_management/utm_pilot_program/media/UPP2_Industry_Workshop_Briefing.pdf ["UPP Phase 2"]; FAA & NASA, *Unmanned Aircraft Systems (UAS) Traffic Management (UTM), UTM Pilot Program (UPP) Phase Two (2) Progress Report, V. 1.0* (March 2021), https://www.faa.gov/uas/research_development/traffic_management/utm_pilot_program/media/UTM_Pilot_Program_Phase_2_Progress_Report.pdf ["UPP Phase 2 Progress Report"] (UPP 2 results will "serve as a basis for policy considerations, standards development, and the implementation of a UTM system." Sect. 1.1.3; and explaining that UPP2 "incorporated a layered message security approach using digital certificates and message signatures" Sect. 4.1). *See generally* [FAA UPP Phase 2 ConUse] (also defining "high-density" airspace "as 12 operation volumes in a 0.2 square mile area." Sect. 4.3); FAA, UPP Pilot Program, *How does the UAS Traffic Management Pilot Program (UPP) support drone integration?* (Dec. 18, 2020), https://www.youtube.com/watch?v=PShrpiz7BTw&feature=youtu.be.

81. *See* FAA, *FAA Continues Drone Integration Initiatives*, https://www.faa.gov/news/updates/?newsId=95371 (also noting Phase 1 completion Aug. 2019, and that work included the generation of UVRs); Steve Bradford, *Welcome to the Future: The Next State of UAS Traffic Management*, FAA (Dec. 9, 2020), https://www.suasnews.com/2020/12/welcome-to-the-future-the-next-stage-of-uas-traffic-management/.
82. Rios, *supra*.
83. Including, for example, China, Japan, and Singapore. *See infra* Chap. 5, Sect. 5.4 *Selected Country-Specific Implementations.*
84. SESAR, *About, Discover SESAR*, https://www.sesarju.eu/discover-sesar. SESAR JU is further considered *infra* Chap. 3, Sect. 3.3.5 *SESAR JU.*
85. SESAR, *U-space*, https://www.sesarju.eu/U-space.
86. *See, e.g.,* Mark Blanks, Head of Flight Ops., Wing, formerly Dir., Mid-Atlantic Aviation Partnership at VA Tech, Remarks at the ATCA Annual Conference (Dec. 9, 2020) ("We've moved from a government-mandated architecture from NASA to an industry-standard based architecture.").
87. Email from Parimal Kopardekar, Ph.D. Dir., NARI, NASA (Sept. 14, 2019) (on file with author) (underscoring the "2014/2015 UTM RFI Industry partnerships—Over 70+ Space Act agreements with FAA"), and NASA, UTM Partners, Current List of UTM Partners (last updated April 10, 2017), https://www.nasa.gov/aeroresearch/programs/aosp/saso/partners/.
88. Many of the key participating companies represent some of the largest technology-driven companies.
89. Email from Parimal Kopardekar, Ph.D., Dir., NARI, NASA Ames (Sept. 14, 2019) (on file with author).
90. Email from Jeffrey Homola, UTM Integration and Testing Lead, NASA Ames (Dec. 23, 2019) (on file with author).
91. *See, e.g.,* [SESAR ATM Master Plan 2020]; [ASTM UTM Spec.], Specification Roadmap appendix.
92. *See infra* Chap. 5, Sect. 5.2 *Low Altitude Authorization and Notification Capability.*
93. *See infra* Chap. 5, Sect. 5.4 *Selected Country-Specific Developments.*
94. *See, e.g.*, EASA, Opinion No 01/2018 *Unmanned aircraft system (UAS) operations in the 'open' and 'specific' categories* (2018), https://www.easa.europa.eu/sites/default/files/dfu/Opinion%20No%2001-2018.pdf ["EASA Opinion 01/2018"] (requiring increasing categorical constraints and mitigations for open, specific, and certified categories).
95. *See* [Nat'l Academies AAM], Sects. 3-2 – 3.5 (Finding, in part, that "[t]he commercial cargo market is ready to adopt an operationalized advanced aerial mobility capability.").
96. David D. Murakami, et al., NASA Ames Research Center, *Space Traffic Management with a NASA UAS Traffic Management (UTM) Inspired Architecture*, 10.22514 (June 2019), *available at* https://www.researchgate.net/publication/330206407_Space_Traffic_Management_with_a_NASA_UAS_Traffic_Management_UTM_Inspired_Architecture.

97. *See infra* Chap. 4, Sect. 4.2.5 *Continuing Compliance and Safety Assurance;* [SESAR ATM Master Plan 2020], Sect. 4.2.4 ("safe, efficient and secure"); Parimal Kopardekar, et al., *Unmanned Aircraft System Traffic Management (UTM) Concept of Operations*, 6th AIAA Aviation Technology, Integration, and Operations Conference (June 2016), p. 2, https://www.nasa.gov/utm (urging "an organized approach to enabling these operations to balance efficiency and safety").
98. *See* ["ICAO UTM Framework Ed. 3"], pp. 7–8. *Cf.* [ASTM UTM Spec.] (safety and compliance).
99. *See* [Airbus UTM Architecture], Sect. 1.3 ("Compatibility means that multiple providers sharing the same airspace can coexist, without causing negative consequences for users.")
100. *See* Airbus & Boeing, *A New Digital Era of Aviation: The Path Forward for Airspace and Traffic Management* (Sept. 2020), p. 3, https://www.airbusutm.com/a-new-digital-era (parsing UTM interoperability into four types: between service providers, between different vehicle and operation types, between countries, and with existing ATM systems); GUTMA, *Designing UTM for Global Success* (Nov. 2020), p. 9, https://gutma.org/designing-utm-for-global-success/ (parsing interoperability between service providers, regions and countries, different technologies and systems, and with existing ATM systems).
101. *See, e.g.,* Capt. Andreas Meyer, Avi. Cybersecurity Officer, ICAO, *Integrated Risk Management: A Holistic Approach to Managing Aviation Risk,* Uniting Aviation (Feb. 4, 2019), https://www.unitingaviation.com/strategic-objective/safety/integrated-risk-management/ (providing context of the safety-security relationship).
102. *See, e.g.*, Michael S. Baum, et al., Aviators Code Initiative, *Improving Cockpit Awareness of Unmanned Aircraft Systems Near Airports* (March 8, 2019), http://www.secureav.com/UAS-Awareness-Listings-Page.html ["Baum 2019"] (proposing interim mitigations to improve manned pilot situation awareness); Steve Weidner, UAS Rep., Nat'l Air Traffic Controllers Ass'n, Presentation at the FAA UAS Symposium, Balt., Md. (June 4, 2019) ("It does become a hazard to manned aircraft if they are not getting UTM information.... It would be good if manned aircraft participated in the UTM system. UTM offers a lot of opportunities for manned aviators."); Mildred Troegeler, Dir., Global Airspace Integration, Boeing Co., Presentation at ICAO, DRONE ENABLE/3, Montreal, CA (Nov. 12, 2019) ("We need a more holistic view of airspace."). *Cf.* Reinaldo Negron, Head of UTM, Wing, Presentation at the FAA UAS Symposium (July 9, 2020) (recognizing "value of having the intent of manned aircraft shared over time.").
103. *See infra* Chap. 4, Sect. 4.5 *Spectrum.*
104. (formerly Dir., Airspace Systems, Uber), Presentation at the Uber Elevate Summit 2019, Wash., D.C. (June 11, 2019).
105. **ConOps:** Instructive ConOps (listed alphabetically) that may inform UTM development include, but are not limited to:
    - Embraer^X, et al., *Flight Plan 2030, An Air Traffic Management Concept for Urban Air Mobility* (May 2019), https://daflwcl3bnxyt.cloudfront.net/m/4e5924f5de45fd3a/original/embraerx-whitepaper-flightplan2030.pdf ["Embraer^X 2019"];
    - Embraer^X and Airservices, *Urban Air Traffic Management, Concept of Operations*, Ver. 1 (Dec. 2020), https://embraerx.embraer.com/global/en/uatm ["Embraer UATM ConOps"];
    - FAA NextGen, *Concept of Operations for Urban Air Mobility (UAM)*, v1.0 (June 26, 2020), https://nari.arc.nasa.gov/sites/default/files/attachments/UAM_ConOps_v1.0.pdf ["FAA ConOps UAM v1.0"];
    - FAA NextGen, *Concept of Operations, Unmanned Aircraft Systems (UAS) Traffic Management (UTM),* v2.0 (Mar. 2, 2020), https://www.faa.gov/uas/research_development/traffic_management/media/UTM_ConOps_v2.pdf ["FAA ConOps UTM v2.0"];
    - FAA NextGen, *Concept of Operations, Upper Class E Traffic Management*, v1.0 (April, 22, 2020), https://nari.arc.nasa.gov/sites/default/files/attachments/ETM_ConOps_V1.0.pdf ["FAA High E ConOps v1.0"];

- ICAO, *Remotely Piloted Aircraft System (RPAS) Concept of Operations (CONOPS) for International IFR Operations* (2018), https://www.icao.int/safety/UA/Documents/RPAS%20CONOPS.pdf;
- India, Gov't of, Ministry of Civil Aviation, *National Unmanned Aircraft System (UAS) Traffic Management Policy, Architecture, Concept of Operations and Deployment Plan for enabling UTM ecosystem in India, Discussion Draft,* Ver.1.0 (Nov. 30, 2020), https://www.civilaviation.gov.in/sites/default/files/National-UTM-Policy-Discussion-Draft-30-Nov-2020-updated.pdf ["India UTM Policy"];
- P.R.C., State Post Bureau, Specification for Express Delivery Service by Unmanned Aircraft, YZ/T 0172 --2020, National Postal Standardization Technical Committee (SAC/TC462) (Dec. 2020), http://www.spb.gov.cn/zc/ghjbz_1/201508/W020201204542195544172.pdf E;
- Parimal Kopardekar, Ph.D., Dir., NARI, NASA, et al., NASA Ames Research Center, *Unmanned Traffic Management (UTM) Concept of Operations*, 16th Avi. Tech., Integration, and Ops. Conf. (June 13-17, 2016), https://www.nasa.gov/utm;
- SESAR JU, CORUS, *U-space Concept of Operation,* Vol. 1 Enhanced Overview, and Vol. 2 ConOps (Oct. 25, 2019), https://www.sesarju.eu/projects/corus, and https://www.sesarju.eu/node/3411 ["CORUS ConOps"]; and
- SWISS FOCA, *Swiss U-Space, ConOps,* Ver. 1.1 (2020), https://www.bazl.admin.ch/bazl/en/home/good-to-know/drones-and-aircraft-models/u-space.html ["Swiss U-Space ConOps v.1.1"].

106. *See, e.g.,* Steve Bradford, Chief Scientist for Architecture and NextGen Development, FAA, Remarks at the FAA UAS Symposium (July 8, 2020) (addressing future FAA ConOps: "Version 3.0 will go further into remote ID and security; more on BVLOS; intent for strategic deconfliction."); [ASTM UTM Spec.], Specification Roadmap appendix.
107. The participating entities are described by their role rather than by their respective interfaces; and are not necessarily listed in their order of importance, or applicability to a specific UTM. A single entity may play more than one of these roles in practice (i.e., operator and USS and even SDSP may all be a single commercial entity in some deployments, but may be three distinct entities in others, etc.). A UA could be considered a participant or merged with an operator.
108. *See* ["Comm Implementing Reg. (EU) 2021/664 - U-space"], Art. 7 Obligations for operators of manned aircraft operating in U-space airspace (the term *operator* referring to operators of manned aircraft operating in U-space airspace).
109. [FAA ConOps UTM v2.0], Sect. 2.2.1 UAS Operators.
110. This may include both performance and airspace authorizations. [FAA ConOps UTM v2.0], Sects. 2.4.1 & 2.4.3. *See* Email from Craig Bloch-Hansen, Prog. Mgr. - UAS Design Standards, Transport Canada (Dec. 19, 2019) (on file with author) (underscoring that users of UTM airspace remain responsible for aviating, navigating, and communicating).
111. *See, e.g.*, [FAA ConOps UTM v2.0], Sect. 2.4.3 Airspace Authorization; [Comm Implementing Reg. (EU) 2021/664 - *U-space*], Art. 6 UAS operators, ¶¶ 5-8 (operator shall not start operation til flight authorization granted and activated; shall comply with the flight authorization else request a new authorization); Art. 10 UAS flight authorisation service. *Cf.* Embraer^{X} and Airservices, *Urban Air Traffic Management, Concept of Operations*, Ver. 1 (Dec. 2020), https://embraerx.embraer.com/global/en/uatm ["Embraer UATM ConOps"], Annex B.3 Flight Planning and Authorisation Service, p. 63 ("A flight authorization includes the clearance for a UAM flight, the flight plan and a reservation for vertiport use."). *See infra* Chap. 2, Sect. 2.3.12 *Simplified Phases of Flight and Operational States.*
112. *Editorial Note*: In this publication, the singular form ("USS") is adopted for both singular and plural usage of UAS Service Supplier(s). Such usage is consistent with certain FAA and other influential documents. *See, e.g.,* [RID NPRM 2019], "UAS Service Suppliers (USS)" appears in the table of contents as the title of Section XIV-A, which states in part, "[The] FAA is partnering with third parties referred to as UAS

Service Suppliers (USS);" GAO, *Unmanned Aircraft Systems, FAA Could Strengthen Its Implementation of a Drone Traffic Management System by Improving, Communication and Measuring Performance*, GAO-21-165 (Jan. 2021), n. 28, https://www.gao.gov/assets/720/712037.pdf ["GAO UAS"]. *Contra* [FAA ConOps UTM v2.0].

113. Joseph L. Rios, et al., *UAS Service Supplier Specification, Baseline requirements for providing USS services within the UAS Traffic Management System,* NASA/TM-2019-220376 (Oct. 2019), Sect. 3, Terminology, https://utm.arc.nasa.gov/docs/2019_Rios-TM-220376.pdf ["NASA USS Spec."].
114. Although there is no universal or immutable class, bundle, or grouping of USS minimum services or roles, there may be dependencies that cause regulators to evoke such requirements. *See* Joseph Rios, Ph.D., Remarks at the ASTM UTM WG meeting (March 17, 2020) (observing, "Things are kind of bundled."). To assure flexibility and efficacy, industry should articulate to regulators the safety case of each service distinctly. *Cf.* Chap. 5, Sect. 5.3 *U-space* (introducing the U-space regulation's four mandatory services).
115. [Rios TCL4 Sprint 2], p. 5.
116. *Id.*
117. *See infra* [ASTM Remote ID], Sect. 4.5.3 ("For Network Remote ID, two USS roles are identified: Network Remote ID (Net-RID) Service Providers and Net-RID Display Providers"); Chap. 2, Sect. 2.4.8 *UAS Remote ID*; [Remote ID Rule] (not invoking network RID).
118. [FAA ConOps UTM v2.0], Sect. 2.3.2.1 USS Service Supplier/USS.
119. *See, e.g.,* [Comm Implementing Reg. (EU) 2021/664 - *U-space*], Art. 3 Designation of U-space airspace, ¶ 2 (designating as mandatory, Art. 12, Flight authorization service). *Cf.* [Swiss U-Space ConOps v1.1], Sect. 3.5.6 Airspace Authorization Service.
120. [FAA ConOps UTM v2.0], Sect. 2.3.2.1 USS Service Supplier/USS. Note that "operator" includes its computer platform since operations at scale will require largely automated systems.
121. [FAA ConOps UTM v2.0], Sect. 2.3.2.1 UAS Service Supplier/USS. *See* [RID NPRM 2019], Sect. 89.135 *Record retention* (requiring any "Remote ID USS to retain any remote identification message elements…for 6 months from the date" received or first possessed); [Comm Implementing Reg. (EU) 2021/664 - *U-space*], Art. 15, Conditions for obtaining a certificate, ¶ 1.(g) (U-space service providers shall "retain for a period of at least 30 days recorded operational information and data or longer, where the recordings are pertinent to accident and incident investigations until it is evident that they will no longer be required."). Record retention may raise privacy issues. *See infra* Chap. 4, Sects. 4.2.5 *Continuing Compliance and Safety Assurance* (addressing UTM data logging), and 4.7.6 *Data Rights and Privacy Management.*
122. *See, e.g.,* EASA, *Draft acceptable means of compliance (AMC) and Guidance Material (GM) to Opinion No 01/2020 on a high-level regulatory framework for the U-space, Issue 1 (XX Month 2020),* https://www.easa.europa.eu/sites/default/files/dfu/Draft%20AMC%20%26%20GM%20to%20the%20U-space%20Regulation%20—%20for%20info%20only.pdf ["EASA Draft AMC & GM"] ("[USSP] capability of providing at least the four mandatory U-space services (network identification, geo-awareness, traffic information and UAS flight authorization.… GM1 to Art. 8 U-space service providers."). *Cf.* Robert Roth, Head of Software Eng'g, Uber ATG (formerly, Eng'g Dir., Prime Air at Amazon), Remarks at the ASTM F38 UTM WG meeting, Palo Alto, Cal. (Aug. 30, 2019) (proposing minimum USS requirements: "The floor of what is to be a USS: you have to be DSS aware and provide at least one UTM service."). Note that USS are sometime characterized by their specific function or role, such as those initiating new or modified operational intents or constraints, or providing conformance monitoring.
123. [Comm Implementing Reg. (EU) 2021/664 - *U-Space*], Art. 3 U-space airspace, ¶ 2(a-d). In the UK, a USS is known as a UTM Service Provider (UTMSP).

124. FAA NextGen, *Concept of Operations for Urban Air Mobility (UAM)*, v1.0 (June 26, 2020), Sect. 5, Notional Architecture, https://nari.arc.nasa.gov/sites/default/files/attachments/UAM_ConOps_v1.0.pdf ["FAA ConOps UAM v1.0"]. *See infra* Chap. 6, Sect. 6.5 *Advanced Air Mobility / Urban Air Mobility.*
125. *See generally* Marcus Johnson, Dep. Project Mgr. for UTM Project, NASA, Presentation at the Community Integration Working Group: *Supplemental Data Service Providers (SDSP)*, NASA (Dec. 3, 2020) (webinar), https://www.youtube.com/watch?v=uyDeTKHaeeI&feature=youtu.be; Edward "Ted" Lester, fmr. Chief Technologist, AiRXOS, Presentation at the Community Integration Working Group: *Supplemental Data Service Providers (SDSP), id.* (asserting the FAA is an SDSP providing increasingly important and diverse aeronautical information relevant to UTM).
126. [FAA ConOps UTM v2.0], Sect. 2.3.2.3 UAS Supplemental Data Service Providers. Many of these services neither yet have a single, globally recognized name nor uniform parameters. SPSPs may consolidate certain overlapping or redundant services and brand them distinctly. Consequently, certain boundaries between USS and SDSP functions are imprecise.
127. *See* [Comm Implementing Reg. (EU) 20211/664 - U-space], Art. 12 Weather information service; [CORUS ConOps], Annex K: U—space Architecture, Sect. 6 Service Architecture, ("hyperlocal" weather information); WXXM, *Weather Information Exchange Models*, http://wxxm.aero/ (includes WXCM-WXXM-WXXS); ANSI, *Standardization Roadmap for Unmanned Aircraft Systems*, Ver. 2.0 (June 2020), Sect. 7.5, Weather (listing weather standards), https://share.ansi.org/Shared%20Documents/Standards%20Activities/UASSC/ANSI_UASSC_Roadmap_V2_June_2020.pdf ["ANSI Roadmap"]; [NASA UAM Vision ConOps], Sect. 4.3.6 Urban weather. *See generally* S.E. Campbell, et al., *Preliminary Weather Information Gap Analysis for UAS Operations*, Project Rpt. ATC-437, Rev. 1, Lincoln Laboratory (Oct. 2017), https://www.ll.mit.edu/sites/default/files/publication/doc/2018-05/Campbell_2017_ATC-437.pdf.

    ASTM F38.02 has established a Weather Supplemental Data Service Provider WG to examine weather data performance standards for UTM operations and is developing ASTM, F_ [WK73142], *New Specification for Weather Supplemental Data Service Provider (SDSP) Performance*, https://www.astm.org/DATABASE.CART/WORKITEMS/WK73142.htm. The WK73142 WG chair has urged, "Aviation has depended on government weather data and systems for services. The weather innovation and infrastructure investment required to scale a reliable and predictable UAS industry must incentivize the private sector, local, and state governments to fill the recognized weather information gaps, with agility and speed" Email from Don Berchoff, CEO, TruWeather Solutions (Feb. 7, 2021) (on file with author). *See* Tom Prevot, Ph.D., Air Taxi Product Lead, Joby Aviation, formerly, Dir., Airspace Systems, Uber, Remarks, in Palo Alto, Cal. (Aug. 9, 2019) ("We need much better sensors and models than what we have today to understand weather micro-effects—nobody has the solution today and that is why we are still doing research."); Shimon Elkabetz, CEO, ClimaCell, *quoted in,* Eric Niller, *What's in the Forecast: Private Weather Predictions,* Wired (Dec. 28, 2019), https://tinyurl.com/UTM-Weather ("We are at a tipping point where the technology of weather forecasting, which was dominated by government and still is, is going to change."); Steve Bradford, Chief Scientist for Architecture and NextGen Development, FAA, Remarks at the NASA, AAM Airspace WG (webinar) (Aug. 4, 2020) ("I'm 100% for crowdsourcing weather."); RTCA, DO-369, *Guidance for the Usage of Data Linked Forecast and Current Wind Information in Air Traffic Management (ATM) Operations* (July 13, 2017), https://my.rtca.org/NC__Product?id=a1B36000003GlOfEAK.
128. *See* ACJA, *Interface for Data Exchange between MNOs and the UTM Ecosystem, NetworkCoverage Service Definition*, v1.00 (Feb. 2021), https://gutma.org/acja/wp-content/uploads/sites/10/2021/02/ACJA-NetworkCoverage-Service-Definition-v1.00.pdf (A joint cooperation between GSMA and GUTMA) ["MNO-UTM Interface"]; Mark

Davis, Ph.D., Pres., Crossbar Inc. (formerly, VP, Intel), et al., *Aerial Cellular: Aerial Cellular: What can Cellular do for UAVS with and without changes to present standards and regulations*, Presented at AUVSI Xponential, in Chicago, Ill. (May 2, 2019), p. 13 ["Davis 2019"] (proposing that "[t]he cellular network could be represented as a specific type of SDSP"); Terry Martin, Ph.D., Applied Research Lead, Nova Systems, Presentation at GUTMA Annual Conference, Portland, Or. (June 19, 2019) (Because cellular networks are highly dynamic systems, there is need for near-real-time network coverage maps to notify UTM providers and users of network availability status, particularly as a function of flight planning. NOTAMs are needed to inform of cell outages or problems as part of flight planning and risk assessment, perhaps providing a joint forecast of multiple networks.). *See infra* Chap. 3, Sect. 3.2.3 *Cellular Industry.*

129. *See* Email from Christopher T. Kucera, Head of Strategic Partnerships, OneSky Systems (Dec. 24, 2019) (on file with author) ("Common terrain and common use of terrain is essential if we are going to remove enough uncertainty in tactical deconfliction.... A common understanding of things like terrain, obstacles, pressure altitude, etc. could be controlled by a FIMS [Flight Information Management System]. Our altitude service could be used as an example. It could be viewed as an SDSP, but I would prefer them to be labeled as 'authoritative' or 'centralized' SDSP."). *See also* OneSky, *Supplemental Data Service Provider (SDSP),* https://onesky.xyz/sdsps. Additionally, SDSPs could offer enhanced *Simultaneous Location And Mapping* (SLAM) cartographic services—a basic requirement for autonomous systems.
130. *See, e.g.,* ASTM F_ [WK69690], *New Specification for Surveillance UTM Supplemental Data Service Provider (SDSP) Performance,* https://www.astm.org/DATABASE.CART/WORKITEMS/WK69690.htm ["ASTM Surveillance SDSP"] (defining minimum performance standards for SDSP services to USS/USP; and providing aircraft track information to DAA systems enabling BLVOS operations). Surveillance SDSPs may serve as one nonexclusive input for situational awareness, strategic or tactical deconfliction, and UAS monitoring. *Cf.* Email from James Licata, Bus. Dev. Mgr., Hidden Level, Inc. (Jan. 22, 2020) (on file with author) ("Low altitude airspace monitoring/surveillance can be a transitional tool for UTM, echoing the NextGen process. The benefit becomes a backup and validation tool that supports the cooperative airspace operations (Remote ID), just as MLAT [multilateration] will do beyond 2025 for commercial aviation and ADS-B."). *See infra* Chap. 2, Sect. 2.4.5 *Detect and Avoid* (addressing surveillance SDSPs).
131. *See, e.g.,* [Comm Implementing Reg. (EU) 2021/664 - U-space], Ch. IV, Art. 11 Traffic information service, ¶ 3 ("...service shall provide information about the position of other known air traffic..."). Other candidate surveillance-related SDSP services may include confirming/validating sensor input performance, sensor data fusion, and enabling zero-conflict airspace (ZCA).
132. SDSPs are not anticipated to plan flights—it would take substantial additional data to do so—but there may be SDSPs that provides "de-confliction services."
133. *See, e.g.,* [CORUS ConOps], *Annex K: U-space Architecture*, Sect. 6 Service Architecture, p. 45.
134. [CORUS ConOps], Annex K: *U-Space Architecture*, Sect. 7.1 Stakeholders and Roles (introducing "Insurance Data Service Provider"); [Swiss U-Space ConOps v.1.1], Sect. 3.5.17 Liability Insurance Service. *See infra* Chap. 4, Sect. 4.7.5 *Insurance.*
135. *See, e.g.,* Peter Sachs, UTM Implementation Prog. Mgr., FAA (formerly, Safety and Risk Architect, Airbus), *A Quantitative Framework for UAV Risk Assessment*, Vol. 1, Rpt. TR-008 (Sept. 13, 2018), https://storage.googleapis.com/blueprint/TR-008_Open_Risk_Framework_v1.0.pdf; Ersin Ancel, Ph.D., et al., Aerospace Eng'r, NASA, *In-Time Non-Participant Casualty Risk Assessment to Support Onboard Decision Making for Autonomous Unmanned Aircraft* (June 14, 2019), https://arc.aiaa.org/doi/abs/10.2514/6.2019-3053; Ersin Ancel, *id.*, *Ground Risk Assessment Service Provider (GRASP) Development Effort as a Supplemental Data Service Provider (SDSP) for Urban Unmanned Aircraft System (UAS) Operations,* DASC (Sept. 1, 2019),

https://ieeexplore.ieee.org/document/9081659 ("a pre-flight planning tool that allows comparison of alternative flight trajectories and flight dates/times," Sect. V.). *See infra* Chap. 4, Sect. 4.2.4 *Continuing Compliance and Safety Assurance.*

136. *See generally* ICAO, *Air Navigation Services Providers (ANSPs) Governance and Performance*, Working Paper, CEANS-WP/36, presented by CANSO (Aug. 28, 2008), 1.4, https://www.icao.int/Meetings/ceans/Documents/Ceans_Wp_036_en.pdf; EU, *Commission Implementing Regulation (EU) 2017/373 of 1 March 2017, laying down common requirements for providers of air traffic management/air navigation services and other air traffic management network functions and their oversight*… (March 8, 2017), https://eur-lex.europa.eu/legal-content/EN/TXT/PDF/?uri=CELEX%3A32017R0373&from=EN.
137. Email from Anthony Rushton, Ph.D., UTM Technical Lead, NATS (March 3, 2020) (on file with author) ("ANSPs are pleased to be amongst the first to implement wide-spread UTM systems (such as LAANC and the Airspace User Portal); and have been working hard to innovate new air traffic interfaces for U-Space Volumes."). *See, e.g.*, Chap. 2, Sects. 2.4.3 *Flight Information Management System (FIMS)*, and 2.4.4. *System-Wide Information Management.*
138. https://www.nats.aero/. *See* NATS, *Our Ownership*, https://www.nats.aero/about-us/what-we-do/our-ownership-2014/ (NATS not completely privatized – it is a public-private partnership).
139. https://www.skyguide.ch/en/ (serving as U-space's implementation executive).
140. http://www.navcanada.ca/.
141. Email from Craig Bloch-Hansen, Prog. Mgr. - UAS Design Standards, Transport Canada (Dec. 19, 2019) (on file with author).
142. CANSO, *ANSP Considerations for Unmanned Aircraft Systems (UAS) Operations*, v 1.1 (2016), p. 16, https://canso.org/publication/ansp-considerations-for-unmanned-aircraft-systems-uas-operations/. *See* [Comm Implementing Reg. (EU) 2021/664 - U-space], Art. 5 U-space service providers, 5.(a) ("exchange any information that is relevant for the safe provision of U-space services amongst themselves").
143. *See infra* Chap. 4, Sect. 4.7.3 *Business Models and Competition* (re: monetization of operational safety data).
144. Connected Places Catapult, *Enabling UTM in the UK* (May 2020), Sect. 5.1 General Duty, https://s3-eu-west-1.amazonaws.com/media.cp.catapult/wp-content/uploads/2020/05/22110912/01296_Open-Access-UTM-Report-V4.pdf ["Connected Places Catapult"] (recognizing "a general duty to protect the public and consumers of aviation services").
145. One challenge will be any cross-coordination required between UTM and ATM when operations demand a dynamic shift in airspace use, such as around a medevac operation or flight into private airport or heliport situated within a volume of UTM airspace, etc.
146. Email from Craig Bloch-Hansen, Prog. Mgr. - UAS Design Standards, Transport Canada (Dec. 19, 2019) (on file with author). *See* [ICAO UTM Framework Ed. 3].
147. *See, e.g.*, [India UTM Policy], Sect. 4.1.8 General Public, p. 11 (recognizing individual right to data on need-to-know basis and to protect privacy).
148. *See* Chap. 4, Sect. 4.6 *State, Regional, and Local.*
149. [FAA ConOps UTM v2.0], Sect. 2.3.2.2 USS Network ("the amalgamation of USSs connected to each other").
150. *See, e.g.*, [CORUS ConOps], Enhanced Overview, p. 20 ("VFR Pilot Operating in VLL airspace which is also used by drones"); [Baum 2019]. *Cf.* [FAA ConOps UTM v2.0], Sect. 2.4.1.3 Manned Aircraft Operators (listing manned aircraft operators optionally as passive or active participants); [Comm Implementing Reg. (EU) 2021/664 - U-space], Art. 11 Traffic information service, ¶ 2 ("shall include information about manned aircraft and UAS traffic shared by other U-space service providers …").
151. [Rios TCL4 Sprint 2], p. 4. *See supra* note 105 (listing instructive ConOps presenting varying UTM services). **Figure 5.5** *Structure of the Implementing Regulation*, Ch. IV-U-space services, Arts. 10-16 (anticipated U-space services).

152. Source: Transport Canada, *RPAS Traffic Management (RTM) Services Trials, CALL FOR PROPOSALS, Phase 1 – Round 1* (May 11, 2020), Sect. 3.2, p. 3, https://www.tc.gc.ca/en/services/aviation/drone-safety/drone-innovation-collaboration/remotely-piloted-aircraft-systems-rpas-traffic-management-services-testing-call-proposals.html ["Transport Canada RTM Services"]. Its four-phased implementation schema provides a prioritization and feasibility analysis perspective within its broader scope. *See generally* [FAA ConOps UTM v2.0], App. D – UTM Services; Joseph Rios, Ph.D., UTM Project Chief Eng'r, *UAS Traffic Management (UTM) Project Strategic Deconfliction: System Requirements Final Report,* UTM-SD.05 (July 2018), https://www.researchgate.net/publication/332107751_UAS_Traffic_Management_UTM_Project_Strategic_Deconfliction_System_Requirements_Final_Report/download, and (deck) https://utm.arc.nasa.gov/docs/2018-UTM-Strategic-Deconfliction-Final-Report.pdf ["Rios - Strategic Deconfliction"] (characterized as a tactical separation function).
153. Presentation at ICAO, DRONE ENABLE/3, Montreal, CA (Nov. 12, 2019).
154. [ICAO UTM Framework Ed. 3], p. 11, List of Services (citing Registration Service). *See id.* Appn. A, Registration, Identification and Tracking.
155. *See, e.g.,* [CORUS ConOps], Sect. 5.1.1.1 Registration service; [Connected Places Catapult], Sect. 1.4.1 Stakeholder Registration (includes registration of associated service providers). The information contained in the registry differs by jurisdiction. In Europe, the operator should be registered and not the aircraft.
156. *See* [Remote ID Rule], Preamble, XV. Registration (The rule "tie[s] the remote identification requirements to the registration of unmanned aircraft [and] serve[s] the dual purposes of both identifying aircraft and promoting accountability and the safe and efficient use of the airspace…"). *See generally* 49 U.S.C. Sects. 44101-44106 & 44110-111 (aircraft registration requirements).
157. European Commission, *Commission Implementing Regulation (EU) 2019/947 of 24 May 2019 on the rules and procedures for the operation of unmanned aircraft* (O.J.) (L 152/1) (July 11, 2019), https://eur-lex.europa.eu/eli/reg_impl/2019/947/oj ["Comm Implementing Reg. (EU) 2019/947 - UA"], 26.
158. *See, e.g.,* FAADroneZone/Registration, https://faadronezone.faa.gov/#/ (addressing authority for registration: "Pursuant to 49 U.S.C. § 44807, § 44809, and 14 C.F.R. § 107.13, persons operating small unmanned aircraft under 14 C.F.R. part 107 must comply with the registration requirements. Additionally, persons operating in accordance with the statutory exception for limited recreational operations of unmanned aircraft under 49 U.S.C. § 44809 must comply with the registration requirements pursuant to subsection (a)(8) of that section."); ICAO, *ICAO Model UAS Regulations, Part 101 and Part 102* (June 23, 2020), https://www.icao.int/safety/UA/Documents/Final%20Model%20UAS%20Regulations3%20-%20Parts%20101%20and%20102.pdf, Sect. 101.5(a) Unmanned Aircraft Registration and Certificate of Registration ("Every person lawfully entitled to the possession of a UA who will operate a UA in [specify country] shall register that UA and hold a valid certificate of registration for that aircraft…"); European Parliament *Regulation (EU) 2018/1139 of the European Parliament and of the Council (O.J.) (L 212/8)* (July 4, 2018), *Art. 48* Repository of information, https://eur-lex.europa.eu/legal-content/EN/TXT/?uri=CELEX:32018R1139 (requiring establishment and management of electronic repository). *See also* [FAARA], Sect. 112(b)(7), (authorization expenditures for digitization of the Civil Aviation Registry); *id.* Sect. 546 (FAA Civil Aviation Registry Upgrade).
159. [Comm Implementing Reg. (EU) 2019/947 - UA], Art. 14 *Registration of UAS operators and certified UAS,* ¶ 1. *See* [Comm Implementing Reg. (EU) 2021/664 U-space], Art. 18 *Tasks of the competent authorities* (d) ("establish, maintain and make publicly available a registration system for certified U-space service providers"), and Art. 5 *U-space airspace,* (5.) (U-space service providers authenticating identities of UAS operators via access to registration database… and (b) any other national registration system).
160. [Comm Implementing Reg. (EU) 2019/947 - UA], ¶ 17.

161. UAS Advisory Group, Aircraft Registry Network and the UAS impact, Presentation at ICAO, DRONE ENABLE/2, Chengdu, P.R.C. (Sept. 13-14, 2018), https://unitingaviation.com/news/safety/skylive-drone-enable-streaming-icaos-third-unmanned-aircraft-systems-symposium/ (describing a UAS registry database and services that would support states that did not already have or want to build their own). *See* ICAO, *Regulations and Procedures for the International Registry*, Doc 9864 (8th ed. 2019), https://www.icao.int/publications/Documents/9864_8ed.pdf (presenting certain relevant registry paradigms); *infra* Chap. 2, Sect. 2.4.7.2 *International Aviation Trust Framework* (dependent on and supporting registry).
162. GUTMA, *Interoperable Drone Registry Demo 2019*, Presentation at the GUTMA Annual Conference, Portland, Or. (June 18, 2019). *See, e.g.*, Dr. Hrishikesh Ballal, *Aircraft Registry (2019)*, GitHub repository, https://github.com/openskies-sh/aircraftregistry/blob/master/registry/models.py#L68 (presenting diverse "class Authorization" choices); *Registration Landscape White Paper* (Nov. 2019) (updated Jan. 6, 2020), GitHub, https://github.com/openskies-sh/aircraftregistry/blob/master/documents/registration-white-paper.md; *Drone Registry Brokerage* (May 27, 2019), GitHub, https://github.com/openskies-sh/aircraftregistry-broker/blob/master/documents/registration-brokerage-specification.md; *Identity and Authentication in the registry*, GitHub, https://github.com/openskies-sh/aircraftregistry/blob/master/documents/registration-identity-authentication.md; *Comprehensive Registry Testing*, GitHub, https://github.com/openskies-sh/aircraftregistry/blob/master/documents/comprehensive-registry-testing.md; https://github.com/openskies-sh/flight_passport (developing an open source UTM/U-space specific OAuth service to issue tokens and aviation-specific roles supporting a registry).
163. *See, e.g.*, ICAO, *Convention on International Civil Aviation*, Doc 7300/9 (9th Ed. 2006), https://www.icao.int/publications/Documents/7300_9ed.pdf ["Chicago Convention"], Arts. 17-21, https://www.icao.int/publications/Documents/7300_cons.pdf (Chicago Convention - re cross-border operations and exchange of registration documentation); [Comm Implementing Reg. (EU) 2019/947 - UA], *Art. 13* Cross-border operations or operations outside the state of registration.
164. *See, e.g.*, [SESAR ATM Master Plan 2020], Annex 1: U-space services. *Cf.* DLR, *DLR Blueprint, Concept for Urban Airspace Integration DLR* (Dec. 2017), Sect. 2.2, https://www.dlr.de/fl/Portaldata/14/Resources/dokumente/veroeffentlichungen/Concept_for_Urban_Airspace_Integration.pdf (mission planning: "a. UAS operator or pilot defines a flight plan…") ["DLR Blueprint"]. *See supra* Chap. 2, Sect. 2.2.3 *Supplemental Data Service Providers (SDSPs)*.
165. [ICAO UTM Framework Ed. 3], p. 11. The USS may also perform this function while the flight is active—provided the operations remain "strategic" in nature. *Cf.* Email from Jeffrey Homola, UTM Integration and Testing Lead, NASA Ames (Dec. 20, 2019) (on file with author) ("At least in our approach, that is more of an operator function facilitated by information from the USS. It may just be a matter of interpretation.").
166. [FAA ConOps UTM v2.0], Sect. 2.4.4 Operation Planning. *See* Benoit Curdy, Digital Transformation, Swiss FOCA, Presentation at ICAO, DRONE ENABLE/3, Montreal, CA (Nov. 13, 2019) ("I believe that we are converging toward 'operation planning' in order precisely to avoid confusion with manned aviation. A reference to 'traditional' flight planning is good though as it is similar. Operation planning allows to account for the limited battery life of certain drones but to still consider one mission with several battery swaps as one unit (instead of several flights)."). *See* Email from Mike Glasgow, Technical Standards and UTM System Architect, Wing, and Co-chair, ASTM F38 UTM WG (Aug. 24, 2020) (on file with author) (urging that most operators and USS do not want "routes assigned to them, but rather [they prefer] retaining the ability to determine the most desirable and conflict free route on their own.").
167. *See* Email from Mike Glasgow, Technical Standards and UTM System Architect, Wing, and Co-chair, ASTM F38 UTM WG (Aug. 23, 2020) (on file with author). ASTM adopts this meaning of strategic coordination, but it is not universal. *See* Kathy Hunt, *Flight Plan*, ASTM Standardization News (July/Aug. 2020), https://www.stand-

ardizationnews.com/standardizationnews/july_august_2020/MobilePagedArticle.action?articleId=1598423#articleId1598423 (also characterizing *strategic coordination* and airspace constraints as "cornerstone capabilities"); [ICAO UTM Framework Ed. 3], p. 11 (re strategic deconfliction service – providing for the "arrangement, negotiation and prioritization of intended operational volumes"); [Rios - Strategic Deconfliction]. As noted elsewhere, UTM constructs, terminology, standards, and implementations are in a state of development and vary considerably.

168. In certain deployments of UTM, VLOS operational intents may conflict with BVLOS operational intents. This issue is ultimately determined by regulators. *Cf.* [Comm Implementing Reg. (EU) 2021/664 - U-space], Art. 10 UAS flight authorisation service, ¶ 6 (requiring "proper arrangements to resolve conflicting UAS flight authorisation requests" – providing no exception for VLOS).

Some experts argue that strategic deconfliction ends prior to takeoff; others assert it ends after takeoff and remains available provided no immediate conflict is evident for a recognized time period (e.g., at least X seconds before conflict— notionally, e.g., more than 20 seconds before conflict). "Conflict" should be defined to indicate whether it means a finite time or distance before Remain Well Clear is violated, before impending threat of collision, or something else. Optimally an operator should be able to change the operation plan either before flight or in flight, which may involve prioritization and/or negotiation among USS on behalf of UAS operators. An operator would need to account for the time required to complete these mechanisms to ensure adherence with (i.e., remaining inside) approved volumes. If in-flight replanning is prohibited, then the UTM arguably becomes a pure reservation system with heightened prioritization and fairness challenges, particularly as traffic volumes climb. Indeed, once airspace is "reserved" (via a set of 4D volumes), UAS operators who come later—even seconds later— may need to de-conflict around the reserved airspace. This may accommodate two or three flights, but not a crowded area. *See* Maxim Egorov, Research Scientist, et al., Airbus, *Encounter Aware Flight Planning in the Unmanned Airspace*, ICNS, Herndon, Va. (April 2019), https://storage.googleapis.com/blueprint/icns2019.pdf (underscoring "an encounter with more vehicles in it is a complex multi-agent problem and is less likely to be resolved by a planning algorithm."). High traffic volumes may drive operators to plan their flights well in advance to ensure advantage. There is also need to address in-flight priority operations (i.e., ambulance helicopters or manned aviation). *See* Chap. 4, Sect. 4.4.4 *Fair Access*.

169. [Rios - Strategic Deconfliction] (Additionally, strategic deconfliction, *inter alia*, must: "have the 4-D non-intersection of operation plans as its primary objective" [UTM-SD.10], "be transparent to operators" [UTM-SD.15], "be supported by all USSs" [UTM-SD.20], "be mandated by the airspace regulator"[ ], "be deterministic" [UTM-SD.30], be "efficiently calculable" by USS [UTM-SD.35], and be "calculable by each USS" [UTM-SD.37]).
170. *See* Chap. 2, Sect. 2.3.7 *Constraint Services.*
171. Email from Jeffrey Homola, UTM Integration and Testing Lead, NASA Ames (Dec. 20, 2019) (on file with author) (stating operations "should also be free of conflicts with airspace structures (e.g., TFR), airspace class boundaries, and other areas such as national parks."). *See* Chap. 2, Sect. 2.3.7 *Constraint Services.*
172. Andrew R. Lacher, Sr. Principal, Aerospace Research and Autonomous Systems, Noblis, formerly, Sr. Mgr. for Autonomous Systems Integration, Boeing NeXt, Remarks at the ASTM F38 WG meeting, in Brussels, Belg. (Feb. 24, 2020) ("Strategic deconfliction provides a level of safety by reducing [tactical deconfliction] burden, but doesn't eliminate it.").
173. Trajectory management may be strategic and limited to predeparture actions or extend to in-flight services in which case it demands awareness of aircraft position at critical points along its flight path. Research and testing of various trajectory optimization approaches are underway. *See, e.g.*, ICAO, *The Role of FF-ICE*, https://www.icao.int/airnavigation/FFICE/Pages/Role-Of-FFICE.aspx. *See generally* ICAO, *Manual*

on Flight and Flow—Information for a Collaborative Environment (FF-ICE), Doc 9965, AN/483, 1st ed. (2012), Sect. 3.15, https://www.icao.int/Meetings/anconf12/Documents/9965_cons_en.pdf (addressing trajectory versus volume-based airspace operations).

174. Email from Peter Sachs, UTM Implementation Prog. Mgr., FAA (formerly, Security and Risk Architect, Airbus) (Oct. 16, 2019) (on file with author). *See* ICAO, *Global Air Traffic Management Operational Concept*, 1st ed., Doc 9854 (2005), Sect. 2.7.10, https://www.icao.int/Meetings/anconf12/Document%20Archive/9854_cons_en%5B1%5D.pdf ("Strategic conflict management is the first layer of conflict management and is achieved through the airspace organization and management, demand and capacity balancing and traffic synchronization components.").
175. Email from Mike Glasgow, Technical Standards and UTM System Architect, Wing, and Co-chair, ASTM F38 UTM WG (Mar. 11, 2021) (on file with author).
176. Presentation at the Technical Workshop on U-space and ATM Aspects - Stream 2A - EASA High Level Conference on Drones 2019, Scaling Drone operations, Amsterdam, NL (Dec. 10, 2019), *available at* https://tinyurl.com/Amsterdam-U-spaceServ.
177. [ICAO UTM Framework Ed. 3], p. 12. *See* [NASA USS Spec.], Sect. 9.2 Conformance Monitoring ("Each Conformance Volume is contained (four-dimensionally) within an Operation Volume. It is the set of Operation Volumes that is included as part of the Operation data supplied to other UTM components via the appropriate APIs.").
178. The operational intent volume (OIV) or operational intent (OI) is the 4D volume shared (communicated) with other UTM participants. *See infra* Chap. 4, Sect. 4.4.2 *UTM Volumes* (in part, defining notional *conformance and operation volumes*).
179. [Comm Implementing Reg. (EU) 2021/664 - U-space], Art. 13 Conformance monitoring service, 2. *Compare*, the ASTM UTM standard may not include alerts based on the anticipation that operational intents will be violated – instead, only actual violations. "The problem becomes significantly more complicated if alerts are to be provided on an anticipatory basis. It remains to be seen if risk analysis forces us in this direction." Email from Mike Glasgow, Technical Standards and UTM System Architect, Wing, and Co-chair, ASTM F38 UTM WG (Aug. 23, 2020) (on file with author). *See* Andrew R. Lacher, Sr. Principal, Aerospace Research and Autonomous Systems, Noblis, formerly Sr. Mgr. for Autonomous Systems Integration, Boeing NeXt, Remarks at the ASTM F38 meeting, Raleigh, N.C. (Nov. 9, 2019) ("Conformance monitoring by USS for other users, I would think, will enhance safety. Where you do have connectivity, it seems like conformance should be monitored...From a layered safety approach, if you don't have conformance monitoring, effectiveness could be diminished.").
180. [NASA USS Spec.], Sect. 9.2 Conformance Monitoring (also, "[UTM-USS-046] *A USS MUST be aware within 1 second that an operation under its management is out of conformance.*").
181. *Id.*
182. Email from Peter Sachs, UTM Implementation Prog. Mgr., FAA (formerly, Security and Risk Architect, Airbus) (Oct. 17, 2019) (on file with author).
183. Email from Mike Glasgow, Technical Standards and UTM System Architect, Wing, and Co-chair, ASTM F38 UTM WG (Aug. 23, 2020) (on file with author). *See infra* Chap. 2, Sect. 2.4.6 *Detect and Avoid. See also* [Embraer UATM ConOps], Sect. 2.1.1 (expecting eVTOLs capable of "self-conformance monitoring").
184. Edward "Ted" Lester, formerly Chief Technologist, AiRXOS, Remarks at the ASTM F38 UTM WG meeting, in Brussels, Belg. (Feb. 24, 2020). Presumably this information could inform both safety oversight and determination of "bad actor" status. *See infra* Chap. 2, Sect. 2.4.5.2 *Discovery and Synchronization Service (DSS)* (re: Bad Actor List).
185. *See* Chap. 4, Sect. 4.2.5 *Continuing Compliance and Safety Assurance.*
186. Joseph L. Rios, et al., *Strategic Deconfliction Performance, Results and Analysis from the NASA UTM Technical Capability Level 4 Demonstration* (Aug. 2020),

p. 16, https://utm.arc.nasa.gov/docs/2020-Rios-NASA-TM-20205006337.pdf ["Rios Strategic Deconfliction Performance"]; *see, e.g.,* [FAA ConOps v2.0], Sect. 2.4.5 Constraint Information and Advisories; [FAA & NASA UTM RTT WG #3], Sect. 1.1.5.2 (describing the UVR process). *Cf.* [Comm Implementing Reg. 2021/664 U-space], Art. 4 Dynamic airspace reconfiguration.

187. [ASTM UTM Spec.]. Constraints planned over active flights should require notification to the operator. *See* [Indian UTM Policy], Sect. 5.4.2.5 Constraint Management.
188. Controversy regarding commercial constraints has a history in legacy aviation. *See, e.g.,* Matt Pierce, *No-fly zones over Disney parks face new scrutiny*, L. A. Times (Nov. 10, 2014), https://www.latimes.com/nation/la-na-disney-airspace-20141110-story.html. *See generally* FAA, *National Security UAS Flight Restrictions*, https://udds-faa.opendata.arcgis.com/search (explaining and graphically depicting such restrictions).
189. Email (March 9, 2020) (on file by author).
190. Airbus, *Blueprint For The Sky* (2018), p. 20, https://storage.googleapis.com/blueprint/Airbus_UTM_Blueprint.pdf ("is effective at avoiding near-term threats") ["Airbus Blueprint"]. *See* [ICAO UTM Framework Ed. 3], List of Services, pp. 11-12 ("Conflict management and separation service including "[t]actical separation with manned aircraft service": a service that provides real-time information about manned aircraft so that UA remain well clear of manned aircraft."). Nonetheless, tactical deconfliction is not universally recognized as a UTM service.
191. *See, e.g.,* [FAA ConOps UTM v2.0], Sect. 2.3.2.1 UAS Service Supplier/USS ("USS services . . . support tactical deconfliction . . ."). *See also* Altitude Angel, *Altitude Angel Launches World First UTM Conflict Resolution Service* (July 18, 2019), https://www.altitudeangel.com/news/posts/2019/july/altitude-angel-launches-world-first-utm-conflict-resolution-service/ ("Tactical CRS [conflict resolution service] will provide information to drone pilots or the drone itself to ensure separation is maintained during the in-flight phase.").
192. [SESAR ATM Master Plan 2020], p. 67 (for U3). *Cf.* [Comm Implementing Reg. 2021/664 U-space] (tactical deconfliction not included as an explicit service).
193. Email from Andrew Carter, Pres. and Co-founder, ResilienX, Inc. (March 3, 2020) (on file with author) (further stating, "UTM can provide any services deemed useful.... The UTM system has more info than can be provided by on-craft sensors (generally) and can make better avoidance recommendations."); Email from Mark Wuennenberg, Technical Officer, ICAO (March 5, 2020) (on file with author) ("Given the various models for UTM, there is the potential that the UTM system could support tactical deconfliction. Whether that be UTM managed tactical deconfliction or facilitating vehicle to vehicle comms to support tactical deconfliction."); Marcus Johnson, Remarks at the UTM Project Technical Interchange Meeting (TIM) (Feb. 23, 2021) (explaining that strategic deconfliction does not alone address all collision hazards and "there has to be a *continuum* of conflict management that goes across all systems—USS to aircraft." Emphasis added).
194. Andrew Carter, *id.* Further noting that latency is not relevant to airborne DAA, and if ground-based DAA is implemented, there may be viable implementation options in a distributed manner.
195. *See* Chap. 2, Sect. 2.4.6 *Detect and Avoid.*
196. The terms "dynamic rerouting" and "dynamic deconfliction" need to be standardized. The term "dynamic" indicates that the mission is in progress and needs updating. "Enroute rerouting" or "in-flight rerouting" could be reasonable substitutes.
197. For example, a medevac helicopter could require reroute of all UAS in an airspace segment, or a police emergency could require that all proximate drones land.
198. For example, challenges may arise when a routing overlaps existing routes or reservations and negotiation between airspace participants is needed. Additionally, although a strategic deconfliction system may exist, other aircraft not flying "prescribed routes" may create unacceptable risk, e.g., an ad-hoc operation such as a VLOS photography

mission where the separation would be reduced between strategically deconflicted aircraft and aircraft relying solely on tactical deconfliction.

199. Email from Maxim Egorov, Research Scientist, Airbus (Dec. 23, 2019) (on file with author). *See generally* Sheng Li, & Maxim Egorov, et al., *Optimizing Collision Avoidance in Dense Airspace Using Deep Reinforcement Learning*, Thirteenth USA/Europe Air Traffic Management Research and Development Seminar (ATM2019), *available at* https://ieeexplore.ieee.org/document/6669888 ["Sheng Li DRL 2019"].
200. Formerly, Dir. Airspace Systems, Uber, Remarks, in Palo Alto, Cal. (Aug. 9, 2019).
201. ASTM Int'l Committee F38, F3411-19, *Standard Specification for Remote ID and Tracking*, Sect. 1.1, https://www.astm.org/Standards/F3411.htm ["ASTM Remote ID"]. *See* [RID NPRM 2019], p. 72471 ("Remote identification information, when correlated with UAS registry information, would inform law enforcement officers about two essential factors: Who registered the UAS, and where the person manipulating the flight controls of a UAS is currently located. This is particularly relevant to a law enforcement officer's decision on whether use of force would be appropriate."); [Remote ID Rule], Preamble; [Comm Implementing Reg. (EU) 2021/664 - U-space], Art. 8 Network identification service (service description and message content).
202. Carl Burleson, Acting Deputy Admin'r, FAA, Remarks at AUVSI Xponential, Chicago, Ill. (May 1, 2019). *See* FAA ATO, *Remote Identification (Remote ID) of Unmanned Aircraft System, Concept of Use (ConUse): FAA Data Exchanges with UAS Service Suppliers (USS),* v. 1.0 (Jan. 17, 2020) (addressing data exchange programs and systems to support public safety, including civil enforcement use of RID; establishing the concept of a baseline stream of data; and mentioning RID use by DoD, DoI, DoJ, and DHS).
203. Jay Merkle, *Merkle Leading FAA's UAS Integration Effort,* Business Avi. Insider (July/Aug. 2019), https://nbaa.org/news/business-aviation-insider/.
204. Authorized parties are typically law enforcement and national security personnel. Tracking is also presented as a U-space U2 initial service. [SESAR ATM Master Plan 2017], Annex 1: U-space services.
205. The RID service per se will not provide collision avoidance. Rather, aircraft equipped to surveil the RID position tracking RF signal may independently use it for situational awareness and to enhance a collision avoidance system. *See* [Remote ID Rule], Preamble, VII.D. (U.S. Nat'l Transport. Safety Bd. noting broadcast RID "may support aircraft-to-aircraft collision avoidance capability").
206. *See, e.g.,* [Comm Implementing Reg. (EU) 2021/664 - U-space], Art. 8 Network identification Service, 2 (listing required network ID message content).
207. The RID service per se will not provide collision avoidance. Rather, aircraft equipped to surveil the RID position tracking RF signal may independently use it for situational awareness and to enhance a collision avoidance system.
208. [ICAO UTM Framework Ed. 3], p. 12.
209. Jon Standley, Business Development Lead, Aviation Systems / L3Harris Technologies (Aug. 20, 2020) (on file with author).
210. Prompt and effective notification of local public safety operations such as Medivac and firefighting require near-instantaneous capability to clear and segregate theatre-of-operations airspace.
211. *See* Jaewoo Jung, et al., NASA Ames Research Center, *Initial Approach to Collect Small Unmanned Aircraft System Off-nominal Operational Situations Data* (June 25-29, 2018), https://utm.arc.nasa.gov/docs/2018-Jung-Aviation2018-Jun.pdf (providing a structured list of off-nominal events); [NASA USS Spec.], Sect. 9.3 Contingency Management ("[UTM-USS-056] A USS MUST provide at least one Contingency Plan per Operation Volume within an Operation plan as defined per the [USS REQ-API]"); [NASA USS Spec.], Sect. 11.1 (addressing in-flight emergencies and corresponding USS obligations). *See also* [NASA UAM Vision ConOps], Appn. D: Use Cases (describing contingency scenarios, including passenger in distress, weather restricts landing, non-cooperative aircraft, UAM aerodrome closure, and off-nominal scenarios).

212. Jaewoo Jung, et al., NASA Ames Research Center, *Automated Management of Small Unmanned Aircraft System Communications and Navigation Contingency* (Jan. 2020), https://utm.arc.nasa.gov/docs/2020-Jung_SciTech_2020-2195.pdf ["Jung 2020"] (also recommending agreement on two performance parameter values: minimum data transfer rate and maximum round trip latency, Sect. IV.A).
213. [FAA ConOps UTM v2.0], Sect. 2.7.1.3 Contingency Management. Depending on applicable rules, the actions taken upon lost link may vary as a function of UA location.
214. *See* [FAA & NASA UTM RTT WG #3], Sect. 1.1.5.3 (and envisioning V2V broadcast capability to broadcast alerts to nearby vehicles to enhance awareness).
215. [FAA & NASA UTM RTT WG #3], Sect. 4.2.4. When a UA's position goes outside of the operational volume, one approach is to enlarge its operational intent volume (original volume plus where the UA may in the next X seconds); and change the state of operational intent to nonconformance state). *See* [NASA USS Spec.], Sect. 13 Accounting and Auditing (referencing reporting requirements for off-nominal conditions, including for contingency/rogue, unplanned return or landing, and unplanned loiter).
A UA could be in its operational volume yet nonconforming because its position certainty is too low (e.g., due to navigational performance degradation from lost satellite), thus transgressing the operational volume of an adjacent operational volume, causing a contingency. In early UTM deployments, it is anticipated that once a flight is designated as in contingency, its operational volume is no longer shared, and it cannot return to normal status; it can only be ended. "For situations where corrections cannot be made, operators are responsible for notifying affected airspace users as soon as practical and executing a predictable response." [FAA UPP Phase 2 ConUse], Sect. 4.3.1. Modifications to an operational intent when one is in nonconforming or contingent state may not require deconfliction and proving to the DSS that all other flight operations were considered by providing the opaque version numbers (OVNs).
216. *See* [FAA ConOps UTM v2.0], Sect. 2.7.1.3 Contingency Management. *See* 14 C.F.R. Sect. 91.3 Responsibilities and authority of the pilot in command (b) (*directed to pilots*: "In an in-flight emergency requiring immediate action, the pilot in command may deviate from any rule of this part to the extent required to meet that emergency."); Andrew R. Lacher, Sr. Principal, Aerospace Research and Autonomous Systems, Noblis (formerly, Sr. Mgr. for Autonomous Systems Integration, Boeing NeXt), Remarks at the ASTM F38 UTM WG meeting, Palo Alto, Cal. (Jan. 22-23, 2020) (asserting, "[o]ne contingency vehicle and the whole thing could break.").
217. Email from Mark Wuennenberg, Technical Officer, ICAO (Sept. 23, 2019) (on file with author); 14 C.F.R. Sect. 91.139 Emergency air traffic rules (a) & (b) (*directed to the FAA*: to utilize NOTAMS and "issue an immediate effective air traffic rule or regulation in response to that emergency condition"). UTM rules should provide for priority and notification.
218. The description of states is adapted from developing standard and rules, envisioning a federated UTM architecture. The phase, or state of an operational intent determines data exchange and other responsibilities associated with managing the flight. Operational intent states may include, for example: accepted, activated, nonconforming, contingency, and ended. Other ways of implementing UTM could rely on other state flows.
219. UTM implementations may adopt differing approaches to the meaning of "ended," such as that the UA has landed, or that the UA has permanently departed its operational intent volumes—the latter being analogous to in-flight cancellation of IFR.
220. *See, e.g.,* [Comm Implementing Reg. (EU) 2021/664 - U-space], *Art.* 9 Geo-awareness service; NASA, UTM Project Technical Interchange Meeting (TIM), Remarks by Marcus Johnson, Research Aerospace Eng'r, NASA Ames (Feb. 23, 2021) (questioning where geofencing fits into the UTM conflict management mode – "it's a complicated question"). *See also* Unmanned Aircraft Safety *Team, Safety Enhancement No. 1, Airspace Awareness and Geofencing, Final Report, Out-of-the-Box Protection of High-Risk Airport Locations* (May 1, 2020), http://unmannedaircraftsafetyteam.org/

safety-enhancement-no-1-%E2%80%A8airspace-awareness-and-geofencing/ (suggesting "authorized UTM/LAANC systems [could] automatically unlock 'out-of-the-box' geofences"); Open Geospatial Consortium (OGC), https://www.ogc.org (providing diverse tools and open geospatial standards relevant to geo-fencing).

221. *See, e.g.,* [Embraer UATM ConOps], Sect. 4.4 Flow Management Service (to "maximize the capacity of vertiport FATO [final approach takeoff areas], Flow Management will be required to manage arrival and departure times and slots."). *See infra* Chap. 6, Sect. 6.5 *Advanced Air Mobility / Urban Air Mobility.*
222. Source: NASA, [Kopardekar 2016] (presented here as updated in 2020).
223. Joseph Rios, Ph.D., ATM-X Project Chief Eng'r, NASA Ames Research Center, Presentation at the FAA/NASA UTM Pilot Program Phase Two Industry Workshop, Moffett Field, CA (Dec. 9, 2019).
224. [CORUS ConOps], Sect. 5.2.1 *Architectural Principles* (describing a model with broad applicability; with a service-oriented approach, modular, safety focused, open, standard-based, interoperable, technology agnostic, based on an evolutionary development-incremental approach, automated, allowing variants, deployment agnostic, and securely designed). *See generally* [GUTMA Architecture 2017].
225. Presentation at the FAA UAS Symposium (July 9, 2020).
226. Email from Hrishikesh Ballal, Ph.D., Founder and CEO, Openskies Aerial Technology Ltd (Aug. 29, 2019) (on file with author).
227. Nonetheless, certain responsibilities are retained by CAAs. *See* Chap. 4, Sect. 4.1 *General - UTM Governance* (addressing delegation and identifying immutable governmental roles); [CORUS Intermediate ConOps], Sect. 7.5.1 *Federated* (government-retained services characterized as "state-mandated core").
228. Email from Craig Bloch-Hansen, Prog. Mgr. - UAS Design Standards, Transport Canada (Dec. 19, 2019) (on file with author).
229. Email from Hrishikesh Ballal, Ph.D., Founder & CEO, Openskies Aerial Technology Ltd (Aug. 29, 2019) (on file with author). *See* Moxie Marlinspike, *Reflections: The ecosystem is moving*, Blog, Signal.org (May 10, 2016), https://signal.org/blog/the-ecosystem-is-moving/ (thoughtful considerations of centralized versus federated IT ecosystems); The White House, *National Strategy for Aviation Security of the United States of America* (Dec. 2018), App. B, p. 17, https://www.aviationtoday.com/2019/10/08/homeland-security-dod-transportation-officials-focus-aviation-cyber-security/ ["Nat'l Strategy for Avi. Security 2018"] (considering the interconnectivity of aviation ecosystem: "The term 'Aviation Ecosystem' refines the term 'Aviation Domain' and is intended to include all aspects of Airports, Airlines, Aircraft, Airlift, Actors, and Aviation Management. This term is a more holistic, robust description of the reality of modern aviation and more fully captures the global scope and complexity of the industry and the economic impact it generates.").
230. Email from Jeffrey Homola, UTM Integration and Testing Lead, NASA Ames (Dec. 20, 2019) (on file with author) ("One thing about tradeoffs is that, of course, the more centralized a system becomes the more brittle it becomes. Something to be said about graceful degradation."). *See* Email from Maxim Egerov, Research Scientist, Airbus (Sept. 15, 2020) (on file with author) ("My personal take on the biggest shortcoming, is that ATM systems today consistently lag behind technologically in nearly every regard – it's very difficult to deploy an upgraded ATM system into the current environments.").
231. Murzilli Consulting, fmr. Leader, Innovation and Digitalization Unit at Federal Office of Civil Aviation Switzerland - Swiss U-Space Program Manager, Presentation at the FAA UAS Symposium (July 8, 2020).
232. *Cf.* Gur Kimchi, formerly VP, Amazon, Presentation at ICAO, DRONE ENABLE/3, Montreal, CA (Nov. 13, 2019) (noting that the current centralized ANSP model is "incredibly robust but expensive and complex to scale").
233. Assistant Dir. - Airspace and Future Technology - Aviation and Airports, Dept. of Infrastructure, Transport, Reg'l Dev. and Comm., Presentation at the World ATM Congress Virtual Panel: UTM & ATM: Integrate Now or Integrate Later? (Oct. 13, 2020).

See Michele Merkle, Dir. of ATO Operations Planning and Integration, FAA, Presentation at the FAA UAS Symposium (July 8, 2020) ("We view [UTM] utmost as a great opportunity for industry to manage some of those operations."). *But see* Email from Andrew Carter, Pres. and Co-founder, ResilienX, Inc. (Dec. 12, 2019) (on file with author) ("The market-driven approach often results in many solutions looking for a problem, rather than agreed-upon requirements.").

234. Email from Craig Bloch-Hansen, Prog. Mgr. - UAS Design Standards, Transport Canada (Dec. 19, 2019) (on file with author). Other challenges of federated systems have been voiced: "While it sounds great in theory, it is very difficult to implement in practice given the complexity of stakeholders and technology in the ecosystem. How will the intercommunication system perform at different loads and conditions? How are the specific technical implementations of data and APIs between the stakeholders harmonized? Are existing standards good enough or might new ones need development?" Email from Hrishi Ballal, Ph.D., Founder and CEO, Openskies Aerial Technology Ltd (Aug. 29, 2019) (on file with author).
235. Kaitlynn M. Whitney, et al., *The Root Cause of Failure in Complex IT Projects: Complexity Itself,* Procedia Computer Science, Elsevier, V.20 (2013), pp. 325–330, https://doi.org/10.1016/j.procs.2013.09.280.
236. Email from Reinaldo Negron, Head of UTM, Wing (Dec. 28, 2019) (on file with author).
237. For example, *Grid Architecture*, "the application of system architecture, network theory, and control theory to the electric power grid…introduce[s] rigor into the specification or development of grid structures." Jeffrey D. Taft, *Grid Architecture*, IEEE POWER AND ENERGY MAG. (Sept./Oct. 2019), pp. 19, 23, https://www.nxtbook.com/nxtbooks/pes/powerenergy_091019/index.php#/20 (also claiming to offer "the means to future-proof technological investments and reduce integration costs," p. 28). *See* Mary Prandini, et al., *Toward Air Traffic Complexity Assessment in New Generation Air Traffic Management Systems*, IEEE Trans. on Intelligent Transport. Sys., vol. 12:3 (Sept. 11, 2011), https://ieeexplore.ieee.org/abstract/document/5723748 (describing complexity analysis undertaken for the largely centralized ATM, and "an autonomous aircraft framework envis[ioning] new tasks where assessing complexity may be valuable and requires a whole new perspective in the definition of suitable complexity metrics." Extensions of such analyses may inform federated or hybrid centralized/federated UTM systems).
238. *See* Email from Craig Bloch-Hansen, *supra* (Sept. 4, 2020) (on file with author) ("Standards in effect reduce the complexity of the system by clearly defining the 'diversity' expectations with each actor in the system, and laying out the minimum interconnection requirements."). *See generally* Chap. 3 *Standards-Making for UTM.*
239. *See, e.g.,* [ICAO UTM Framework Ed. 3], Appn. B Communications Systems.
240. *See, e.g.,* [ASTM UTM Spec.]; Github, https://github.com/nasa/utm-apis (providing a collection of APIs for NASA's UTM project as OpenAPI documents); *infra* Chap. 2, Sect. 2.4.5.2 *Discovery and Synchronization Service* (further addressing APIs); Chap. 3, *Standards-Making for UTM.*
241. Maxim Egorov, Research Scientist, et al., Airbus, *Encounter Aware Flight Planning in the Unmanned Airspace*, ICNS Conference, Herndon, Va. (April 2019), p. 7, *available at* https://storage.googleapis.com/blueprint/icns2019.pdf. *See infra* Chap. 4, Sect. 4.4.4 *Fair Access.*
242. Maxim Egorov, *id.*
243. Email from Mike Glasgow, Technical Standards and UTM System Architect, Wing, and Co-chair, ASTM F38 UTM WG (Aug. 23, 2020) (on file with author). (further stating that "if the structure/rules are established where necessary, one can certainly envision that each of the participants can apply the structure/rules"). *See, e.g.,* Chap. 4, Sect. 4.4.3 *Corridors.*
244. *See, e.g.,* [ASTM UTM Spec.] (reflecting the distillation of extensive R&D, NASA trials, commercial implementation, and standards).
245. Email from Hrishikesh Ballal, Ph.D., Founder and CEO, Openskies Aerial Technology

Ltd (Aug. 29, 2019) (on file with author) (asserting, "a truly federated, decentralized system would be difficult. Therefore, using a system like FIMS, some components would make a 'hybrid' architecture.").

246. Jay Merkle, Exec. Dir., UAS Integration Office, FAA, Presentation at *GUTMA, High Level Webinar Session - Ask the Experts* (June 9, 2020), https://gutma.org/ask-the-experts/ ("FAA has on-demand access to UTM operational info. We do have ability to reach into the ecosystem." (via the FIMS)).
247. *See generally* ["Comm Implementing Reg. (EU) 2021/664 - U-space"], Art. 5, Common information service. The CIF provides expanded support to host common data. EASA, Draft Opinion, *High-level regulatory framework for the U-space*, RMT.0230 (Oct. 2019), Art. 5, *available at* https://rpas-regulations.com/wp-content/uploads/2019/10/EASA_Draft-Opinion-on-U-space.pdf ["EASA Reg. Framework U-space"] ("established to communicate static and dynamic information" *id.* Art. 5.1.); [Swiss U-Space ConOps v1.1] Sect. 3.4.1 CIF ("an overarching function regrouping all centralized services and capabilities"). *See* Email from Christopher T. Kucera, Head of Strategic Partnerships, OneSky Systems (Dec. 24, 2019) (on file with author) ("The CIF is the best place to implement common data considering that ANSPs would want industry to come to consensus on something like a terrain database for consistency. So, the CIF expansion of FIMS is interesting to me.... [E]ven in a Federated architecture where you have a central FIMS, the FIMS would regulate common services for very important things like altitude conversion."); EASA, Opinion No 01/2020, *High-level regulatory framework for the U-space*, RMT.0230 (March 2020), https://www.easa.europa.eu/document-library/opinions/opinion-012020 ["EASA Opinion 01/2020 - *U-space*"], Sect. 2.4 ("The CIS is at the heart of the U-space system.").
248. [FAA UPP Phase 2 ConUse], Sect. 4.2.2.4.
249. [FAA ConOps UTM v2.0], Sect. 2.3.2.4 Flight Information Management System/FIMS.
250. [Aweiss 2018], p. 3. *Cf.* EASA, A-NPA 2015-10, *Introduction of a regulatory framework for the operation of drones* (2015), https://www.easa.europa.eu/sites/default/files/dfu/A-NPA%202015-10.pdf (Art. 3, Sect. 1., "[FIMS] means a service enabling the exchange of data and information necessary to facilitate operations in the U-space airspace."); [FAA & NASA UTM RTT WG #3], Sect. 1.1.6 (FIMS data exchange includes airspace constraint data and "archived UTM data").
251. [Aweiss 2019], p. 2, *citing* [FAA ConOps UTM v1.0]. *See* [NASA USS Spec.], Sect. 6.1 API Requirements ("[UTM-USS-004] A USS MUST communicate with FIMS per the [FIMSUSS-API], and the [USSREQ-API]."). *See also* NASA FIMS-API (GitHub), https://github.com/nasa/utm-apis/tree/v4-draft/fims-api; [Swiss U-Space ConOps v1.1], Sect. 6.1.2 *Legal Considerations and Open Platform* ("The FIMS shall ... serve all approved USPs under reasonable and non-discriminatory terms.").
252. Email from Hrishikesh Ballal, Ph.D., Founder and CEO, Openskies Aerial Technology Ltd (Aug. 8, 2019) (on file with author). *See* Christopher K. Kucera, Head of Strategic Partnerships, OneSky Systems (on file with author) ("The FIMS/USS line is on a spectrum based on how much control the CAA or ANSP will desire. ASTM standards are relevant wherever the line is placed as they control the interactions with other participants. A public entity can use ASTM standards to interact with the UTM network, and could be given extra capabilities based on their user roles and authorization to help manage the airspace.").
253. Joseph Rios, Ph.D., UTM Project Chief Eng'r, NASA, Remarks at the ASTM F38 UTM WG Telecon (March 18, 2020). *Cf.* [EASA Opinion 01/2020 - U-space], Sect. 2.2 and n. 6 ("there is need to have an appropriate common information service (CIS) that will enable the exchange of essential information between the U-space [participants and] is 'the single point of truth'"). FIMS should also support security tokens.
254. Tom Prevot, Ph.D., Air Taxi Product Lead, Joby Aviation, formerly Dir., Airspace Systems, Uber, Remarks, in Palo Alto, Cal. (Aug. 9, 2019). *See* Steve Bradford, Chief Scientist for Architecture and NextGen Development, FAA, Presentation at ICAO, DRONE ENABLE/3, Montreal, CA (April 9, 2019) (underscoring that the "FAA shows a remarkably small footprint.").

255. Email from Parimal Kopardekar, Ph.D., Dir., NARI, NASA (Sept. 13, 2019) (on file with author) ("UTM construct is agnostic. The original UTM idea offered many alternatives and maintained flexibility for each country to make that choice.").
256. Amit Ganjoo, Founder and CEO, ANRA Technologies, *quoted in, A Deep Dive into UTM and the Flight Information Management System for Drones,* DRONELIFE (Aug. 22, 2019), https://dronelife.com/2019/08/22/a-deep-dive-into-utm-and-the-flight-information-management-system-for-drones-long-form/ (also stating, "[i]nstead, we recommend a FIMS solution that adapts to a UTM architecture that must consider the CAA's regulatory and business model"). Some UTM implementations have the FIMS only providing airspace info to USS which do all the administrative decision-making. This is the FIMS-as-gateway-to-ATM-only model.
257. *See* Email from Andrew Carter, Pres. and Co-founder, ResilienX, Inc. (Dec. 12, 2019) (on file with author) ("What I see…is that FIMS becomes a trusted data provider (SDSP) on which there cannot be competition (i.e., only a single source exists). Also, there has been a trend in the U.S. to move towards a more centralized system and in Europe to move to a less centralized system. Each group started off on the very edge of the opposite sides of the spectrum and seems to be harmonizing somewhere in the middle.").
258. *Id. See* Email from Christopher T. Kucera, Head of Strategic Partnerships, OneSky Systems (Dec. 24, 2019) (on file with author) ("I think FIMS is centralized, but I also believe that many public entities could participate in FIMS. Example, law enforcement could publish a UVR (drone restriction) possibly without permission of the aviation authority? Does a Public USS offer this UVR interaction to the FIMS?").
259. Source: Confederation Swiss, FOCA, *Swiss U-Space* (2019), https://www.bazl.admin.ch/dam/bazl/en/dokumente/Gut_zu_wissen/Drohnen_und_Flugmodelle/Swiss_U-space_Implementation.pdf.download.pdf/Swiss%20U-Space%20Implementation.pdf.
260. Lorenzo Murzilli, Founder and CEO, Murzilli Consulting, formerly Leader, Innovation and Digitization Unit, Swiss Federal Office of Civil Aviation, Interview, in Chicago, Ill. (May 1, 2019) ("We [U-space in Switzerland] are investigating which option works best. [It currently] makes sense to do it in the FIMS because it is the source of trust. Yes, we are now starting to work with InterUSS—discussion proponents include full FIMS vs. FIM-less—and early results are very promising. There is a will to push for a decentralized infrastructure, and InterUSS will be key for such a solution to work. The Swiss architecture includes both FIMS and InterUSS for maximum flexibility.").
261. Source: [Transport Canada RTM Services], p. 26.
262. Source: Australian Government, Airservices Australia, *FIMS Request for Information (RFI),* ASA RFI 394578597 (Aug. 19, 2020), Sect. 1.8, Fig. 1, https://engage.airservicesaustralia.com/50159/widgets/263980/documents/177901 ("a gateway for data exchange between the UTM participants and the Air Traffic Management (ATM) system, through which the ANSP fulfils its legislative requirements, while making relevant airspace information available to airspace users.… [t]o provide applicable and necessary centralised UTM services.… ").
263. [India UTM Policy], Fig. 5.1, and pp. 13-16.
264. For example, it is uncertain to what extent certain UTM functions (at least noncritical functions) should instead be interfaced via System-Wide Information Management (SWIM). *See infra* Chap. 2, Sect. 2.4.4 *System-Wide Information Management.* The types of vehicles supported by UTM may also inform the scope/type of data communicated via FIMS, e.g., whether voice data supporting "mixed traffic" communications capability will pass through the FIMS. *See infra* Chap. 6, Sect. 6.2.1 *Communications Infrastructure* (note addressing VoIP).
265. *See* Joseph Rios, Ph.D., UTM Project Chief Eng'r, NASA Ames, Presentation at the FAA & NASA UTM Pilot Program Phase Two Industry Workshop, Moffett Field, Cal. (Dec. 9, 2019), https://www.faa.gov/uas/research_development/traffic_management/utm_pilot_program/media/UPP2_Industry_Workshop_Briefing.pdf.

266. Source: FAA, *UTM Ecosystem*, https://www.faa.gov/uas/research_development/traffic_management/utm_pilot_program/media/UTM_Architecture_lrg.jpg. *Cf. supra* Figure 2.3 – *UTM notional architecture.*
267. Email from Mike Glasgow, Technical Standards and UTM System Architect, Wing, and Co-chair, ASTM F38 UTM WG (Aug. 23, 2020) (on file with author) (For example, "[w]hether route planning or tactical deconfliction is centralized or distributed has huge implementation impacts on USS, and having to accommodate either (depending on the country) is a very heavy lift and complexity adder, likely being a factor in whether or not a USS/operator elects to participate in a country.").
268. Email from Maxim Egorov, Research Scientist, Airbus (Sept. 20, 2020) (on file with author).
269. Presentation at AUVSI Xposition (Oct. 7, 2020) (further stating that the philosophy of the UAM network is "SWIM philosophy—IP and identity-based").
270. *See generally* ICAO, *Manual on System-Wide Information Management (SWIM) Concept,* Doc 10039 AN/511 (20xx) (Adv. ed., unedited), https://www.icao.int/airnavigation/IMP/Documents/SWIM%20Concept%20V2%20Draft%20with%20DISCLAIMER.pdf ["ICAO SWIM Concept"]; EUROCONTROL, *System-wide information management,* https://www.eurocontrol.int/concept/system-wide-information-management; FAA, *System Wide Information Management (SWIM),* https://www.faa.gov/air_traffic/technology/swim/.
271. [ICAO SWIM Concept], Sect. 2.7.1 SWIM Principles.
272. *See* CANSO, *A first successful step for SWIM governance* (Aug. 13, 2020), https://www.canso.org/first-successful-step-swim-governance (in part, describing the DSNA-led SWIM governance framework and that "…SWIM Governance provides the backbone for true ATM digitization."); FAA, *SWIM Governance*, https://www.faa.gov/air_traffic/technology/swim/governance/; FAA, *System Wide Information Management (SWIM), Governance Policies*, Ver. 3.1 (Feb. 6, 2020), https://www.faa.gov/air_traffic/technology/swim/governance/standards/media/SWIM%20Governance%20Policies%20v3.1_20200206_Final.pdf ["FAA SWIM Governance"]; *infra* Chap. 4 *Governance.*
273. Email from Jon Standley, Business Development Lead, Aviation Systems / L3Harris Technologies (Sept. 25, 2020) (on file with author).
274. [ICAO SWIM Concept], p. 3-15. SWIM architecture supports both a Request/Reply and/or Publish/Subscribe distribution frameworks providing users easy access to discover and exchange data among participating systems.
275. [ICAO UTM Framework Ed. 3], Appn. E, Essential Information Exchange Between UTM And ATM Systems.
276. *See, e.g.,* Frequentis, *ANSPs UAS integration service platform* (2020) (MosaiX SWIM platform interconnecting UTM and ANSPs); Frequentis, *MosaiX SWIM Use Case,* https://www.frequentis.com/sites/default/files/support/2020-05/Frequentis_CRO_use-case_MosaiX_SWIM.pdf. *See also* Chap. 2, Sect. 2.4.3 *Flight Information Management System.*
277. Robert Roth, Head of Software Eng'g, Uber ATG, formerly, Eng'g Dir., Prime Air at Amazon, Remarks, in Palo Alto, Cal. (Aug. 28, 2019). *See* John Fort, CEO, Frequentis, *UTM/ATM Integration-Challenges/Lessons Learned from U-Space GOF project*, Presentation at the Air Transp. Info. Ex. Conf. - ATIEC 2019, Tysons Corner, Va. (Sept. 24 2019), https://www.faa.gov/air_traffic/flight_info/aeronav/atiec/media/Presentations/Day%202%20AM%20003%20John%20Fort%20ATM%20UTM%20Integration.pdf (concluding that "UTM/ATM integration is today one of the best use case[s] for SWIM").
278. ICAO, *System-Wide Information Management,* AN_Conf/13-WP/4 (April 2, 2018), https://www.icao.int/Meetings/anconf13/Documents/WP/wp_004_en.pdf (noting, *inter alia*, SWIM will "enable trajectory-based operations [including] UTM." 2.2 & 2.4). *See generally* ICAO, *SWIM as a Foundation for UTM/ATM Integration*, presented by Frequentis, SWIM TF/3–IP/11, Agenda Item 3(h), The Third Meeting of System Wide Information Management Task Force (SWIM TF/3), Bangkok (May 7-10, 2019), https://www.icao.int/APAC/Meetings/2019SWIMTF3/IP11_Frequentis%20AI3h%20-%20Task%201-8_SWIM%20Foundation%20UTM%20ATM%20Integration%20Rev1.pdf#search=Search%2E%2E%2Eutm; ICAO, *Concept of Operations for a global resilient aviation network* (2018), p. 34, https://

www4.icao.int/ganpportal/trustframework (addressing SWIM and service-oriented architecture-SOA); Liu, Yigang, et al., *SOA-Based Aeronautical Service Integration, INTECH* (2011), DOI: 10.5772/29307, https://cdn.intechopen.com/pdfs/20427/InTech-Soa_based_aeronautical_service_integration.pdf (presenting SOA and SWIM).

279. [SESAR U-space Blueprint], p. 7.
280. *See* [ICAO UTM Framework Ed. 3], Appn. E Essential Information Exchange Between UTM and ATM Systems (presenting an extensive list of elements of information exchange); Nouri Ghazavi, Sys. Eng'r, FAA NextGen Office, *quoted in,* Woodrow Bellamy III, *FAA Researching Use of Airplane Connectivity for SWIM,* Avionics Int'l (June 18, 2019), https://www.aviationtoday.com/2019/06/18/faa-researching-use-airplane-connectivity-swim/ (The FAA "is also looking at taking SWIM a step further, by leveraging high speed IP datalinks for more advanced airborne and on-the-ground uses by pilots for things like filing flight plans, negotiating trajectories and finalizing clearances for takeoff. [T]o increase the collaborative decision making and incorporate pilots and air crews input . . . ").
281. Security requirements and limitations of the FAA's intranet and administration have contributed to this pace.
282. Melissa Matthews, SWIM Program Manager (A), Communications, Information and Network Programs, AJM-316, FAA, Presentation on SWIM Cloud Distribution Service, 2019 Air Transp. Info. Ex. Conf., McLean, Va. (Sept. 23, 2019), https://www.faa.gov/air_traffic/flight_info/aeronav/atiec/media/Presentations/Day%201%20PM%20009%20Melissa%20Matthews%20SCDS.pdf.
283. A lack of familiarity of SWIM within the broader UTM community has sparked concerns regarding suitability of SWIM data latency, agility, and its broader role in UTM; correspondingly, the legacy community has raised concerns about the IT-centric UTM development community, and new entrant fitness to develop aviation-grade infrastructure.
284. Remarks at the ASTM F38 meeting, Raleigh, N.C. (Nov. 6, 2019).
285. [Kopardekar 2016], p. 9. A *consistent view of the airspace* can only happen after this conflict is discovered and the two (or more) USS instances talk directly to each other to confirm (or disconfirm) the overlap. The prevention of overlapping operational volumes ensures strategic separation. *See infra* Chap. 4, Sect. 4.4.2 *UTM Volumes.*
286. *See, e.g.,* [FAA ConOps UTM 2.0], App. 3 ("Exchanges between identified parties require that USSs have discovery to FIMS . . . "). *Cf.* [EASA Reg. Framework U-space], Art. 5, Common information function ("The common information function shall offer a *discovery function* for interfaces and capabilities of the common information service.") (emphasis added).
287. [ASTM Remote ID], Sect. 3.1.5. *See infra* Chap. 2, Sect. 2.4.8 *UAS Remote ID.*
288. Remarks at the ASTM F38 UTM WG meeting, Palo Alto, Cal. (Jan. 22, 2020).
289. These include, for example:
    - ASTM, F3411-19, *Standard Specification for Remote ID and Tracking*, https://www.astm.org/Standards/F3411.htm ["ASTM Remote ID"] (implementing a Discovery and Synchronization Service (DSS)). *See infra* Chap. 2, Sect. 2.4.5.2 *Discovery and Synchronization Service (DSS)*;
    - ASTM, F_ [WK63418], *Standard Specification for UAS Service Supplier (USS) Interoperability*, https://www.astm.org/DATABASE.CART/WORKITEMS/WK63418.htm ["ASTM UTM Spec."];
    - NASA Ames UTM model deploying a peer-to-peer data exchange among USSs—the *Local USS Network* ("LUN"). *See, e.g.,* [Rios TCL4 Sprint 2], Sect. 2.1, p. 6 (also accommodating certain distributed approaches);
    - Swiss U-space's discovery approach invoking the FIMS as a central traffic ledger. *See* Swiss Confederation, *Swiss U-Space, ConOps* (March 29, 2019); Joseph Rios, Ph.D., UTM Project Chief Eng'r, NASA, Presentation at ASTM Committee F38 meeting, Seattle, WA (April 17, 2019) (remarking that there is "need for a central ledger *somewhere*"); and
    - *InterUSS Platform* invoking a distributed, discovery open source architecture and communications platform. *See* Linux Foundation, *InterUSS Project*, http://interussplatform.org; and Wing, *InterUSS Platform™, Overview, Governance*

Requirements, Design and Implementation, v.1.0.1 (Aug. 28, 2018), https://github.com/wing-aviation/InterUSS-Platform.

290. For some implementations, UTM discovery methods could be fashioned as session-based keys to anonymize the USS-to-USS integration while also maintaining an idempotent ID in the UTM system for security. *See infra* Chap. 2, Sect. 2.4.5.3 *Inter-USS Platform*. Also, the ASTM UTM WG (and the FAA Remote ID NPRM) includes the notion of sharing a (dynamic) flight "Session ID" in lieu of a specific unique/registered aircraft ID to allow a degree of public privacy, while still allowing authorized officials to resolve a Session ID to an aircraft ID, and thus the responsible party.
291. *See, e.g.*, [NASA USS Spec.], Sect. 12.2 Position Reports (*"ROGUE or NONCONFORMING states,"* and requirements for access to position updates regarding Part 107 operations).
292. Email from Mike Glasgow, Technical Standards and UTM System Architect, Wing, and Co-chair, ASTM F38 UTM WG (Aug. 23, 2020) (on file with author) ("No PII in the traditional sense is shared between USSs per the standard. Operational intents themselves can be considered PII because they represent routes connecting, for example, businesses to home addresses, even though the name, phone number, SS#, actual address is not part of the data."). ASTM F38 advocates that the DSS store a minimum generalized amount of information, even though very specific information was shared. The requirement here is that the DSS always identify a potential overlap of two or more volumes, even if the volumes don't actually overlap. This model ensures that a USS will never miss an overlap, but may find no overlap when specific details are shared between two USS.
293. *See* Robert Roth, Head of Software Eng'g, Uber ATG formerly, Eng'g Dir., Prime Air at Amazon, Presentation at the ASTM F38 meeting, Seattle, WA (April 17, 2019) ("A DSS doesn't do deconfliction—it's simply match-making" among USS. It "provides the awareness of which USS must collaborate to achieve deconfliction.").
294. *See supra* Chap. 2, Sect. 2.3.7 *Constraint Services.*
295. The DSS discovery architecture requires synchronization so that any two (or more) DSS instances will always be using the same data. Thus, you can trust that you will get the same answer from any DSS—as long as you ask both at the same time.
296. This is a well-known problem in any multi-agent computing environment and arises from well-established game theory. *See, e.g.,* Douglas A. Hass, ImageStream Internet Solutions, Inc., *The Never-Was-Neutral Net and Why Informed End Users Can End the Net Neutrality Debates*, BEPRESS LEGAL SERIES (2008), http://www.kentlaw.edu/faculty/rwarner/classes/ecommerce/materials/privacy/materials/network_neutrality/hess_neverwasneutralnet.pdf (describing a "broadband duopoly," multi-tiered service, and corresponding network performance implications). *See* Email from Greg Deeds, CEO, Technology Exploration Group (Aug. 10, 2019) (on file with author) ("In a 'race condition', in the time between having checked and planned a route, another USS might see the route and re-plan their respective route, recursively." Also characterizing the preferred dynamic as *UTM Neutrality*). The network is the point of concentration and thus affects prioritization. Without network neutrality, the UTM players with control over networks may prioritize traffic of their company, customers, and partners; Robert Roth, Head of Software Eng'g, Uber ATG formerly, Eng'g Dir., Prime Air at Amazon, Remarks at the ASTM F38 UTM WG meeting, Palo Alto, Cal. (Aug. 28, 2019) ("There are some potential race opportunities. The whole goal was to bound that. It is possible to change the airspace so frequently that you can't write your operation. There are some mechanisms that can reduce it; we've not yet found a perfect solution."). As a practical matter, even the current DSS approach does not eliminate the possibility of a race condition—we can't simply "stop" the airspace from changing, and data propagation will always take time. Nonetheless, the DSS prevents changes without the knowledge that the data one USS holds is now *stale*. *See infra* Chap. 4, Sect. 4.4.4 *Fair Access*.
297. Email from Mike Glasgow, Technical Standards and UTM System Architect, Wing, and Co-chair, ASTM F38 UTM WG (Aug. 23, 2020) (on file with author) (Further stating, "Originally we considered an approach to the DSS that generalized the interoperability paradigm and tried to support different services with the same code – but we consciously elected to move away from that paradigm. Had we not, then a specification for the DSS

would have made sense – now it does not. Unless perhaps we write a standard to just describe the paradigm, which has not been discussed."). Note, however, there is a baseline of interoperability to which any DSS could add additional value-added services.

298. Remarks at the ASTM F38 UTM WG meeting, Brussels, Belg. (Feb. 24, 2020).
299. Email from Reinaldo Negron, Head of UTM, Wing (Dec. 28, 2019) (on file with author). DSS emerged from the ASTM F38 Remote ID and UTM standardization efforts.
300. *See* Edward "Ted" Lester, formerly Chief Technologist, AiRXOS, Remarks at the ASTM F38 UTM WG meeting, in Brussels, Belg. (Feb. 25, 2020) ("We decided DSS should be really dumb."). The DSS can be viewed as a lightweight wrapper around a synchronization server to enable secure message exchange, such as for sharing operational intent, certain constraints, RID, and other data between USS (and others) to ensure a consistent view of the data.
301. Source: Derived from [ASTM Remote ID].
302. Mike Glasgow, Technical Standards and UTM System Architect, Wing, and Co-chair, ASTM F38 UTM WG, Remarks, in Palo Alto, Cal. (Aug. 29, 2019); Email from Mike Glasgow (Aug. 23, 2020) (on file with author); Mike Glasgow, *Shared Airspace Representation Concept, Discovery and Synchronization Service* (DSS), Presentation at the ASTM F38 UTM WG (April 25, 2019); [Rios TCL4 Sprint 2], Sect. 2.2, p. 10; [ASTM Remote ID]; [ASTM UTM Spec.].
303. The DSS was designed to be used for information that is dynamic, like operations or constraints that are temporary in nature; not information that is long-lived and static, like terrain data or fixed obstacles such as power-lines and towers. The latter can simply be published in a geo-referenced form and ingested by USS.
304. The OVN is specific to an operation (or constraint)—not to a grid cell. A key DSS concept is that the details of the airspace representation (i.e., cells) are completely hidden from the user. This was a deliberate change from earlier versions of the InterUSS Platform. (*See infra* Chap. 2, Sect. 2.4.5.3 *InterUSS Platform*); Email from Mike Glasgow, Technical Standards and UTM System Architect, Wing (Sept. 4, 2020) (on file with author) ("Implementations that formed dependencies on the airspace representation created a real problem" regarding ability to invoke necessary changes as needed).
305. *See* [ASTM UTM Spec.], App. (presenting APIs). *See also* Swaggerhub, *InterUSS Platform - Data Node Stateless API*, https://app.swaggerhub.com/apis/InterUSS_Platform/data_node_api/tcl4.2.0 (The API for USS-to-USS communications is characterized as "a NASA UTM Project delta on the original. In the future, this can be a branch or a fork from a publicly-available repo of InterUSS."). Such APIs include requisite commands, for example, "PUT" for the USS to create or update a constraint, operation or subscription to a geographical area; "GET" to retrieve constraints and operations; and "DELETE" to expunge a subscriber area. The API to the DSS does not expose details of the airspace representation used by the DSS. APIs are also addressed *infra* Chap. 2, Sect. 2.4.9 *UTM Data Exchange*; [ICAO UTM Framework Ed. 3], Appn. F, UTM Service Providers (USP) Organizational Construct and Approval Process, p. 34 ("The case for interoperability however goes beyond APIs to include a common communication language and requirements on the core information to be shared within the UTM ecosystem.").
306. Email from Hrishikesh Ballal, Ph.D., Founder and CEO, Openskies Aerial Technology Ltd (Sept. 24, 2019) (on file with author) ("Every USS will have an API—that much is clear, and DSS standardizes a subset of these APIs. DSS is the first software enabling this type of intercommunication between USS and other parties; maybe it will be the only one. From an industry point of view, I think that the healthy thing would be to have multiple implementations/flavors of the DSS or DSS type of software (maybe two or three).").
307. An API is specified for the DSS, the InterUSS—the leading implementation (*see* Chap. 3, Sect. 3.4.5.3 *InterUSS Platform*). The InterUSS project exploits the robust synchronization capability of the CockroachDB to achieve data consistency among nodes, ensuring all successful writes to the system cannot be repudiated, and that a USS will never be informed that a "write" didn't actually propagate completely and is now invalid (critical for DSS success).

308. *See, e.g.,* Cockroach Labs, *Start a Local Cluster (Insecure)*, https://www.cockroachlabs.com/docs/stable/start-a-local-cluster.html#step-6-scale-the-cluster (adopting "cluster" rather than mesh).
309. Robert Roth, Head of Software Eng'g, Uber ATG, formerly, Eng'g Dir., Prime Air at Amazon, Presentation at the ASTM F38 meeting, Seattle, WA (April 17, 2019).
310. It is anticipated that 8-10 DSS nodes in the United States will be hosted across various industry members. *See* Reinaldo Negron, Head of UTM, Wing, Remarks at the ASTM F38 WG meeting, in Brussels, Belg. (Feb. 24, 2020) ("You'll have European-wide DSS").
311. Industry associations have played well-recognized trusted third-party roles in many industries. Their paradigms and roles may inform DSS administration. *See infra* Chap. 4, Sect. 4.7.2 *Industry Consortia*; Chap. 2, Sect. 2.4.7 *Cybersecurity.*
312. Email from Reinaldo Negron, Head of UTM, Wing (Dec. 28, 2019) (on file with author); Reinaldo Negron, *id.* Remarks at the ASTM F38 UTM WG meeting, in Brussels, Belg. (Feb. 24, 2020) ("I think there is self-interest in providing a DSS…we want to make sure there is high availability."). *Cf.* Email from Andrew Carter, Pres. and Co-founder, ResilienX, Inc. (Dec. 30, 2019) (on file with author) ("I believe that the current [USS implementation of DSS] is more of a stop gap than anything to get operations happening (which needs to happen or everyone is going to go broke!)….I would expect every USS to try to run a DSS node, just so they have the data and the understanding about what is going on. The reason I think some may want it to be independent is so that competitive DSSs don't get their data. I see this being an ANSP function in Europe."); Amit Ganjoo, Founder and CEO, ANRA Technologies, Remarks at the ASTM F38 UTM WG meeting, in Brussels, Belg. (Feb. 24, 2020) ("From a tech view, it doesn't matter who hosts this - it is a business decision, market by market."); Email from Mike Glasgow, Technical Standards and UTM System Architect, Wing (Aug. 23, 2020) (on file with author) ("From a purely technical perspective of how things work, it is true that it does not matter who hosts the DSS instances – but from a business perspective, it matters very much. Industry-hosted versus any government entity-hosted will affect costs to the industry (if government hosts, they will attempt to do cost recovery through user fees, which could result in untenable business cases – e.g., Italy wanting to charge 30 Euros a flight) AND if government hosts and puts a government acquisition process in front of it, they will not be able to respond quickly or cost effectively to evolving standards. This really is a significant issue.").
313. *See* Mike Glasgow, Technical Standards and UTM System Architect, Wing, Remarks at the ASTM F38 UTM WG meeting, Palo Alto, Cal. (Aug. 28, 2019) ("We need to test DSS separately. I don't think we can avoid having performance requirements on DSS."). *See generally* Gov. UK, *Guidance, Setting API service levels* (Jan. 16, 2019), https://www.gov.uk/guidance/setting-api-service-levels.
314. *See* Benjamin Pelletier, Simulation and Data Developer, Stds. Mgr., Wing, *Conceptual Documentation for the DSS* (#182), GitHub, https://github.com/interuss/dss/commit/4777f976cd90f931c27a38f688f6a3f43771d3e1 ("From a client perspective, interacting with any DSS instance within the DSS Region is equivalent to interacting with any other DSS instance within the DSS Region.").
315. In the future we will likely need to load balance the distribution of DSS so it will be important to have a better mechanism than having each party merely picking an arbitrary instance to use.
316. *See* Matthew Schwegler, Prod. Mgr., Uber Elevate Cloud Serv., Remarks at the ASTM F38 UTM WG meeting, Palo Alto, Cal. (Jan. 22, 2020) ("Planning but not using is bad behavior."); Tom Prevot, Ph.D., Air Taxi Product Lead, Joby Aviation, formerly Dir., Airspace Systems, Uber, ASTM F38 UTM WG meeting, Palo Alto, Cal. (Jan. 22, 2020) (like "Uber mode – you cancel or pay a fee"). Other candidate conduct for a bad actor list may include repeated instances of contingency that cause airspace shutdown.
317. Email from Mike Glasgow, Technical Standards and UTM System Architect, Wing, and Co-chair, ASTM F38 UTM WG (Aug. 23, 2020) (on file with author).

318. It is preferable for CAAs to resolve acceptable conduct transparently within a recognized process rather than on an ad hoc basis. Inclusion of a stakeholder association, to include operators, USS, and others to advance such a process deserves consideration. *Cf.* Email from Andrew Carter, Pres. and Co-founder, ResilienX, Inc. (Dec. 12, 2019) (on file with author) ("Architecturally, the concerns of the DSS do not include system administration, which is functionally where a *Bad Actor* List would fall. Maintenance of this list should not fall to the DSS, although enforcement of the list... may be within the DSS purview."). A Bad Actor List may also be implemented for other "fairness" reasons. *See infra* Chap. 4, Sect. 4.4.4 *Fair Access.*
319. Email from Reinaldo Negron, Head of UTM, Wing (Dec. 28, 2019) (on file with author).
320. Email from Mike Glasgow, Technical Standards and UTM System Architect, Wing (May 21, 2019) (on file with author) ("I think we [referring to ASTM Committee F38] are improving it (in some ways significantly), but the core concepts are from the originators... the intent always being that it become a community resource."). Primary modifications from the InterUSS Platform concept were hiding details of the airspace representation from clients, and adding subscription mechanisms.
321. *See InterUSS Project,* http://interussplatform.org. The Linux Foundation is the preeminent open source collaboration association; Linux Foundation, *The Linux Foundation to Host Open Source Project for Drone Aviation Interoperability* (Sept. 18, 2019), https://www.prnewswire.com/news-releases/the-linux-foundation-to-host-open-source-project-for-drone-aviation-interoperability-300920905.html; Tom Prevot, Ph.D., Air Taxi Product Lead, Joby Aviation, formerly Dir., Airspace Systems, Uber, Remarks, in Palo Alto, Cal. (Aug. 10, 2019) (asserting Linux Foundation efficacy). The Linux Foundation has a history of providing UAS support via Dronecode, https://www.dronecode.org/ (and its Dronecode Airspace Working Group). *See also* Chap. 4, Sect. 4.7.2 *InterUSS* and Linux Foundation roles as consortia; Google Open Source, *InterUSS-Platform,* https://opensource.google.com/projects/interuss-platform. The open source version was announced in March 2019. The InterUSS was originally based on the open source Apache Zookeeper project. Apache Zookeeper, *General Information,* https://cwiki.apache.org/confluence/display/ZOOKEEPER/Index (through "a shared hierarchical name space of data registers [znotes]" supporting highly distributed systems); Apache Zookeeper, *Project Description,* https://cwiki.apache.org/confluence/display/ZOOKEEPER/ProjectDescription. The ASTM F38 UTM WG subsequently determined that global UTM demands a more robust, scalable database (avoiding the 100-byte limitation of Zookeeper) and chose the open source CockroachDB database for further reference implementation consideration.
Open Source—*See* Josh Ziering, Kittyhawk, *quoted in,* Sally French, *Exclusive: Remote ID for drones is possible, and Google's Project Wing, AirMap and Kittyhawk just proved it,* The Drone Girl News (Jan. 14, 2019), https://thedronegirl.com/2019/01/14/remote-id-drones/ (asserting that open source software "is a critical tool to creating interoperability and enabling the transparency that prevents bad actors from exploiting power consolidation when a single entity is in charge of a piece of software."); Global UTM Association, Drone Alliance Europe, Small UAV Coalition & UAV DACH, *Joint Statement on U-space* (Amsterdam, Dec. 4, 2019), https://secureservercdn.net/198.71.233.179/4z7.785.myftpupload.com/wp-content/uploads/2019/12/Joint-statement-on-U-Space-041219.pdf ("Given the variety of European stakeholders, the availability of open source solutions and application program interfaces (API) are needed to provide the necessary flexibility and interoperability between member State U-space services.").
322. Wing, *InterUSS Platform™, Overview, Governance Requirements, Design and Implementation,* v.1.0.1 (Aug. 28, 2018), https://github.com/wing-aviation/InterUSS-Platform ["InterUSS Platform"], pp. 1, 3 (InterUSS posits *two key tenets*: "facilitating communication amongst actively operating USSs [and] UAS operations are not stored or processed on the InterUSS Platform"). The InterUSS describes a physical architecture engaging three types of entities: *government,* an *InterUSS Technical Steering Committee,* and *USS participants*—and considers their respective roles. The *InterUSS*

Data Node contains "API, logic and underlying data structures that hold and access the information about how and when to communicate with other USS participants." *Id.* pp. 10-11. The *InterUSS Authentication Node* supports associated security/credentials (including link discovery and synchronization, enabling on-demand querying).

323. The InterUSS's Gridded USS Discovery (GUD) was accommodated in NASA TCL trials, and influenced industry standards. NASA modifications to the InterUSS included: a key representing "pre-operation" flight data stored in the grid—including a Globally-Unique Flight Identifier (GUFI) to expedite queries from other USS such that only changed data requires retrieval or writing; and "writing to multiple grids as a single transaction." [Rios TCL4 Sprint 2], Sect. 2.2, p. 11. It also implemented standard OAuth for authentication nodes. *See* IEFT, RFC 6749, *The OAuth 2.0 Authorization Framework* (Oct. 2012), https://tools.ietf.org/html/rfc6749; *infra* Chap. 2, Sect. 2.4.7 *Cybersecurity*.
324. Email from Robert Roth, Head of Software Eng'g, Uber ATG, formerly, Eng'g Dir., Prime Air at Amazon (Sept. 17, 2019) (on file with author) [NASA USS Spec.], Sect. 4 USS Overview (The LUN represents "[t]he set of other USSs with which a USS are required to communicate.").
325. Remarks at AUVSI Xponential, Chicago, Ill. (May 2, 2019).
326. Presentation at the FAA UAS Symposium, Balt., Md. (June 5, 2019). Jay Merkle, Exec. Dir., UAS Integration Office, FAA, Presentation at the FAA UAS Symposium, Balt., Md. (June 5, 2019) ("Surveilling uncooperatives is essential. You will never solve uncooperatives. You can never assume you have a fully cooperative site. You must build with the fundamental assumption that uncooperatives will exist."). *See* Jay Merkle, Exec. Dir., UAS Integration Office, FAA, Remarks at *GUTMA, High Level Webinar Session - Ask the Experts* (June 9, 2020), https://gutma.org/ask-the-experts/ ("The problem with non-cooperate aircraft is by far the most difficult safety risk to mitigate that we have encountered so far."); [SESAR ATM Master Plan 2017], App. 1; EUROCAE, *Operational Services and Environment Definition for Detect & Avoid in Very Low Level Operations*, Draft ED-269 for Open Consultation (June 2019), https://eurocae.net/news/posts/2019/june/eurocae-open-consultation-ed-267/ ["EUROCAE OSED-DAA"], Sect. 3.2.1; [ASTM Surveillance SDSP]. *But see* [Comm Implementing Reg. (EU) 2021/664 - U-space], Art. 11, Traffic information service, ¶¶ 1, 2 (requiring "information on any other conspicuous air traffic [including] manned aircraft..."]; *id.*, recitals, ¶ 22 (denoting necessary rules for "unmanned aircraft to safety operate alongside manned aircraft in U-space airspace"); EC, *Commission Implementing Regulation (EU) 2021/666 of 22 April 2021 amending Regulation (EU) No 923/2012 as regards requirements for manned aviation operating in U-space airspace*, https://eur-lex.europa.eu/legal-content/EN/TXT/?uri=CELEX%3A32021R0666 ["Comm Implementing Reg. (EU) 2021/666 - U-space"], Art. 1(2)(c) ("Manned aircraft operating in airspace designated by the competent authority as a U-space airspace, and not provided with an air traffic control service by the ANSP, shall continuously make themselves electronically conspicuous to the U-space service providers.").
327. *See, e.g.,* 14 C.F.R. Sect. 91.111 (Operating near other aircraft. "(a) No person may operate an aircraft so close to another aircraft as to create a collision hazard."); 14 C.F.R. Sect. 91.113 (Right-of-way rules: Except water operations. "(b) ***General.*** When weather conditions permit, regardless of whether an operation is conducted under instrument flight rules or visual flight rules, vigilance shall be maintained by each person operating an aircraft so as to see and avoid other aircraft...").
328. Generally, "noncooperatives" are aircraft without an operating electronic means of identification, or that fail to operate such equipment due to malfunction or pilot action. While an aircraft with a Mode C transponder, but no ADS-B Out might still be considered "cooperative," as a practical matter it would not be detected by widely available and economical ADS-B receivers. *See* [India UTM Policy], p. 5 ("UTM Ecosystem [providing] a similar means of cooperative traffic management for UAS and other participating manned aircraft in uncontrolled airspace.").

329. *See* Email from Mike Glasgow, Technical Standards and UTM System Architect, Wing, and Co-chair, ASTM F38 UTM WG (Aug. 23, 2020) (on file with author) ("Airborne DAA is definitely not part of UTM, but it is desirable that airborne DAA is aware of UTM activities such as strategic deconfliction to avoid unnecessary maneuvers. It can be argued that ground-based DAA…is part of UTM"); Email from Andrew Carter, Pres. and Co-founder, ResilienX, Inc. (Dec. 12, 2019) (on file with author) ("A challenge in UTM is drawing the box or boundary for what UTM is. Is the UA a part of the UTM system? What if the equipage feeds the UTM system? Does the UA become a SDSP? I only mention it here because a UA may have a DAA system and may rely on nothing from the UTM system. That UA may feed its surveillance back into the UTM ecosystem so that everyone has better situational awareness. In the future, performance authorizations will need to look at the craft, equipage, USS, environment and operator to determine if an operation meets an adequate safety level. This is challenging when you can put mitigations to the same risk into multiple of those categories, DAA being an easy example.").
330. *See* FAA, *Integration Approach for Class III DAA in the UTM Environment*, JHU/APL: AOS-19-1663, ACAS_RPS_20_002_V1R0, Ver. 0, Rev. 0 (Dec. 31, 2019), Traffic-Alert & Collision Avoidance System (TCAS) Program Office (PO) (copy on file with author).
331. Andy Thurling, CTO, NUAIR Alliance, *The Drone Market*, ENTERPRISE TECH. REV. (Aug. 1, 2019), p. 30, https://www.enterprisetechnologyreview.com/magazines/August2019/Display_Tech/#page=29 — referencing 14 C.F.R. Sects. 91.111 and 91.113.
332. Paul Campbell, Aerospace Eng'r, FAA, Remarks at the ASTM F38 meeting in Raleigh, N.C. (Nov. 5, 2019) (on file with author).
333. [FAA ConOps UTM v2.0], Sect. 2.7.1.2. (Further stating: "Low altitude manned aircraft operating in both uncontrolled and controlled airspace have access to, and are encouraged to utilize UTM Operation Planning services to de-conflict their aerial work; low-altitude manned aircraft pilots share some responsibility with BVLOS UAS Operators for maintaining separation from each other (though they do not share responsibility for separation from VLOS UAS Operators.").
334. *See, e.g.*, UK CAA, CAP 1861, *Beyond Visual Line of Sight in Non-Segregated Airspace,* v.2 (Oct. 8, 2020), https://publicapps.caa.co.uk/modalapplication.aspx?appid=11&mode=detail&id=9294 ["UK CAA CAP 1861"], p. 7 (identifying "a matrix of potential solutions with 4 types of technology": ground-based infrastructure, electronic identification and conspicuity, on-board detect and avoid equipment, and unmanned traffic management); Joseph Rios, Ph.D., ATM-X Project Chief Eng'r, NASA Ames, Interview by NASA in Silicon Valley Live - *Air Taxis and the Future of Flight,* Moffett Field, Cal. (Dec. 19, 2019), https://www.youtube.com/watch?v=y8rRFFUGYpA&fbclid=IwAR1rABJecaOoCJvWT3WlYU-bm4QAutkzFY5oCkpWHU68zEXVMTzYlXo4SeA ("We look at three buckets: the aircraft, the airman or women, and the airspace. Some of the things that are done in airspace now for the current traffic may be better handed by one of those other buckets. Some of the things that a pilot takes care of now may be better handled by the airframe itself. So understanding where these things should be handled in this new way of travel is an important area of research as well.").
335. Interview with Terrence Martin, *in* Dawn M.K. Zoldi, *Unmanned Traffic Management (UTM): An Industry Champion Forging the Future in Singapore, Australia and Beyond (Part 1)*, INTERDRONE (June 29, 2020), https://interdrone.com/news/industry/unmanned-traffic-management-utm-an-industry-champion-forging-the-future-in-singapore-australia-and-beyond-part-1/. *Cf.* Experimental Aircraft Ass'n, *Comment on Amazon Prime Air Petition for Exemption, Regulatory Docket No. FAA-2019-0573* (Aug. 28, 2019), https://www.regulations.gov/document?D=FAA-2019-0573-0049 (proposing that "Amazon must show that their sense and avoid technology performs to, at minimum, a 10^{-9} failure rate for both collaborative and non-collaborative aircraft. This is the equivalent standard for critical systems on manned aircraft."). Performance requirements for validating ADS-B use for sUAS DAA (without secondary radar or other validation) is also under consideration.

336. [UK CAA, CAP 1861], p. 2.
337. *See infra* Chap. 2, Sect. 2.4.8 *UAS Remote ID.*
338. Email from Walter Bender, Johns Hopkins U. Applied Physics Lab (Sept. 15, 2020) (on file with author).
339. ASTM Int'l, F3442 / F3442M – 20, https://www.astm.org/Standards/F3442.htm ["ASTM F3442 M - 20 DAA Performance"]. An associated test method standard is in development: ASTM, F_ [WK62669], *New Test Method for Detect and Avoid,* https://www.astm.org/DATABASE.CART/WORKITEMS/WK62669.htm ("for DAA systems and sensors applicable to smaller UAS BVLOS operations for the protection of manned aircraft in lower altitude airspace"). A future version of F3442 may also address "correlated encounters – UTM tie in."
340. *See, e.g.,* EUROCAE, *Operational Services and Environment Definition for Detect & Avoid in Very Low Level Operations*, Draft ED-269 for Open Consultation (June 2019), https://eurocae.net/news/posts/2019/june/eurocae-open-consultation-ed-267/; *infra* Chap. 3, Sect. 3.2.5 *EUROCAE. But see* Email from Andrew Thurling, CTO, NUAIR Alliance (Aug. 26, 2019) (on file with author) ("Notwithstanding, the OSED treatment of DAA in VLL is viewed as having too broad a scope to be useful."). RTCA has also been influential but has asserted it is not developing DAA UTM standards. RTCA has developed approved DAA standards for larger UAS (e.g., the SC-228 DAA MOPS – DO-365). *See infra* Chap. 3, Sect. 3.2.13 *RTCA.*
341. **ACAS sXu—**

> sXu is intended to provide a DAA capability for sUAS operating BVLOS. sXu is designed to be flexible in adapting to airspace beyond the existing Part 107 restrictions; it is complementary to the UTM concept, but can also support operations outside of UTM if allowed. This would include operations in all airspace where sUAS operations are allowed without ATC tactical separation services and a 1090 MHz transponder is not required. sXu is intended for sUAS that are not operating under IFR and are not equipped with a 1090 MHz transponder / ADS-B Out or 978 MHz ADS-B Out. Consequently, the sUAS is not seen by the collision avoidance systems on manned and larger unmanned aircraft, nor does the sUAS receive ATC services. In this environment, sXu provides the primary tactical mitigation of collision risk with manned aircraft and larger UAS as well as against other sUAS (either equipped with sXu or another sUAS DAA system or unequipped). sXu uses cooperative and non-cooperative surveillance that may be onboard ownship, provided by a ground-based link, or some combination of onboard and ground-based. Any surveillance source may be used that meets minimum performance requirements (that support minimum system safety performance requirements) and these surveillance sources are correlated to provide a single input per intruder to the alerting logic. Additionally, the logic, as well as surveillance sources, can either be located entirely on ownship, entirely on the ground (with avoidance commands uplinked to the vehicle) or split between ownship and the ground.
>
> Since the sUAS is not receiving ATC services, only one level of alerting is provided, with two levels of alerting parameters for airborne intruders – a larger alerting volume for manned aircraft and larger unmanned aircraft and a smaller volume for other sUAS. The most likely case for sUAS DAA is automated response to avoidance commands, but the sXu design does not preclude pilot-in-the-loop response. There is considerable uncertainty about the structure of the future airspace environment and sXu is designed to be flexible and adaptable to a broad set of operational concepts. sXu is intended to be compatible with UTM and the alerting volumes are adaptable based on evolving standards.

Walter Bender, Johns Hopkins U. Applied Physics Lab., *Airborne Collision Avoidance System Xu for Smaller UAS (ACAS sXu), Position White Paper,* ACAS_RPS_20_001_

V2R0, Ver. 2, Rev. 0, FAA TCAS Program Office (Feb. 26, 2020). *See infra* Chap. 3, Sect. 3.2.13 *RTCA*.

> The goal of the ACAS X program is to provide total airspace protection. In order for the breadth of smaller UAS to be served, particularly those with limited on-board sensing and compute capability, ACAS sXu has embraced the UTM ecosystem as a place where data may be shared, aggregated and used for the purpose of detect and avoid. The more vehicles that are equipped and the more vehicles that can be reliably surveilled, the more compelling of a safety case can be made which will unlock routine BVLOS operations. While these operations will transform our low altitude airspace, systems like ACAS sXu will ensure the transformation occurs safety and efficiently.

Email from Josh Silbermann, Asymmetric Operations, Critical Infrastructure Protection Group, Johns Hopkins U. Applied Physics Lab. (March 9, 2020) (on file with author).

UTM runs the "separation stack" – from flight planning to strategic deconfliction to tactical deconfliction to DAA. ACAS sXu can help with the "last ditch" tactical piece of the puzzle and also inform how the others (e.g., flight planning, strategic deconfliction, and dynamic rerouting) are developed. sXu is also capable of supporting non-UTM operations, including non-participating VLOS recreational drones equipped with sXu. Recognizing the sUAS ADS-B Out prohibition, non-IFR flight, BVLOS operations, and aircraft without ADS-B Out-equipped transponders, off-frequency ADS-B Out could be a good V2V link option. "ACAS sXu facilitates use of a broad set of surveillance technologies and initial guidance on surveillance performance requirements to meet the presently available safety targets." Josh Silbermann, *id.*

342. *See* [ASTM Surveillance SDSP] (defining minimum performance standards for SDSP services to USS/USP; and providing aircraft track information to DAA systems enabling BLVOS operations). Surveillance SDSPs may serve as one nonexclusive input for situational awareness, strategic or tactical deconfliction, and UAS monitoring. *Cf.* Email from James Licata, Bus. Dev. Mgr., Hidden Level, Inc. (Jan. 22, 2020) (on file with author) ("Low altitude airspace monitoring/surveillance can be a transitional tool for UTM, echoing the NextGen process. The benefit becomes a backup and validation tool that supports the cooperative airspace operations (Remote ID), just as MLAT [multilateration] will do beyond 2025 for commercial aviation and ADS-B."). *See supra* Chap. 2, Sect. 2.2.3 *Supplemental Data Service Suppliers (SDSPs)* (re surveillance SDSPs). [ASTM F3442M – 20 DAA Performance], ASTM [WK62669] *DAA Test Methods*, together with [ASTM Surveillance SDSP] provide an interconnected set of standards intended to enable these operations in a UTM context, but with flexibility for many possible architectural implementations.
343. Edward "Ted" Lester, formerly Chief Technologist, AiRXOS, *quoted in,* Brian Garrett-Glaser, *Bringing Non-Cooperative Drone Traffic into UTM Solutions*, Aviation Today (Aug. 11, 2020), https://www.aviationtoday.com/2020/08/11/bringing-non-cooperative-drone-traffic-into-utm-solutions/. *See generally* FAA, DOJ, FCC, & DHS, *Advisory on the Application of Federal Laws to the Acquisition and Use of Technology to Detect and Mitigate Unmanned Aircraft Systems*, 9.95.300-UAS (Aug. 2020), https://www.justice.gov/file/1304841/download (addressing legal constraints on certain surveillance activities).
344. Joseph Rios, Ph.D., ATM-X Project Chief Eng'r, NASA, Remarks at the ASTM F38 UTM WG teleconference (Aug. 4, 2019); *Id.* Remarks at the NASA-UTM Project Technical Interchange Meeting (TIM) (Feb. 23, 2021) (Cybersecurity: "we tried to bake it in the best we could.").
345. *See* Steve Bradford, Chief Scientist for Architecture and NextGen Development, FAA, Presentation at ICAO, DRONE ENABLE/3, Montreal, CA (Nov. 12, 2019) (underscoring that "everything is going IP" and that FAA communication providers are decommissioning non-IP ATO private circuits); Robert Segers, Information Security Architect, FAA, Presentation at ICAO, DRONE ENABLE/3, Montreal, CA (Nov. 12, 2019)

(asserting that a "man in the middle attach is the biggest nightmarish threat of using ubiquitous IP addresses"). Additionally, the intensive use of commercial cellular, particularly developing 5G technology, may introduce other cyber risks. NIS Cooperation Group, *EU coordinated risk assessment of the cybersecurity of 5G networks Report* (Oct. 9, 2019), Sect. 1.6, https://ec.europa.eu/newsroom/dae/document.cfm?-doc_id=62132 ("These new features [e.g., software defined networks, network slicing, and mobile edge computing] will bring numerous new security challenges."); Bruce Schneier, *China Isn't the Only Problem With 5G,* FOREIGN POLICY (Jan. 10, 2020), https://foreignpolicy.com/2020/01/10/5g-china-backdoor-security-problems-united-states-surveillance/; Andy Thurling, CTO, NUAIR Alliance, *Assessing Risk with SORA and Andy Thurling of NUAIR*, INTERDRONE PODCAST, Episode 24 (May 26, 2019), https://soundcloud.com/michael-pehel-690682997/epsiode-24-andy-thurling ("From a cyber perspective, the UAS industry is like the canary in a coal mine."); *infra* Chap. 3, Sect. 3.2.3 *Cellular Industry.*

346. Email from Peter Sachs, UTM Implementation Prog. Mgr., FAA (formerly, Safety and Risk Architect, Airbus) (Oct. 16, 2019) (on file with author).

347. "Inconceivable threats"—Dan Diessner, Sr. Research Scientist, Center for Aerospace Resilience, ERAU, Remarks at ICAO DRONE ENABLE Symposium 2021 (April 21, 2021) (asserting need for attention at both the component and system of systems levels, and that 90% of functionality is now software driven); "CIA"—*See* ICAO, *Annex 17 Security, to the Convention on International Civil Aviation* (10th Ed. 2017), Sect. 9.9.1, https://www.icao.int/Security/SFP/Pages/Annex17.aspx (emphasis added):

> *4.9.1 Recommendation.— Each Contracting State should, in accordance with the risk assessment carried out by its relevant national authorities, ensure that appropriate measures are developed in order to protect the* ***confidentiality, integrity and availability*** *of critical information and communications technology systems and data used for civil aviation purposes from interference that may jeopardize the safety of civil aviation.*

See also [CORUS ConOps], Sect. 4.6 Cyber security of U-space (addressing CIA plus security awareness and enforcement).

348. *See, e.g.,* SESAR, CORUS, *Intermediate ConOps Annex E: List of Threats and Events*, Ed. 01.00.00 (Mar. 19, 2019), Sect. 3, U-Space threats/events, https://tinyurl.com/U-space-Threats-Events (they include: technical failure, datalink loss, loss of data, processing error, total loss, latency, and dependency); Wee Keong Ng, Nanyang Technological Univ., et al., *A Study of Cyber Security Threats to Traffic Management of Unmanned Aircraft Systems* (Jan. 2007), *available at* https://www.researchgate.net/publication/313476896_A_Study_of_Cyber_Security_Threats_to_Traffic_Management_of_Unmanned_Aircraft_Systems (presenting a threat model with the properties: intent, scope, target, and manageability, pp. 32-23; determining that "cyber threats to UTMS is very high," p. 61; and offering an "actionable checklist" for cyber safety, pp. 21-76); Krishna Sampigethaya, Parimal Kopardekar, et al., *Cyber security of unmanned aircraft system traffic management (UTM),* ICNS (2018), https://ieeexplore.ieee.org/document/8384832 ["Sampigethaya, Kopardekar 2018"] (asserting "cyber security considerations are largely missing" and presenting the following "unique threats to UTM"):

T1. intentionally alter perceptions of space and time in the UTM to cause disruption. For example:
- T1.1. Manipulation of sUAS & other aircraft perception of its own position & spatial orientation.
- T1.2. Manipulation of sUAS aircraft's and UTMS perception of the presence of physical objects in airspace, such as bogus sUAS aircraft nodes.
- T1.3. Manipulation of sUAS aircraft's perception of its spatial environment and restrictions, e.g., geo-fences, virtual corridors, meteorology, and terrain.

T1.4. Manipulation of the perceptions common to two or more entities.
T1.5. Manipulation of an entity's perception of time.
T2. Threats from attacks that misuse some UTM resources for self-gain:
T2.1. Manipulation of rules-of-the-sky, e.g., geofence data, to fly a rogue aircraft.
T2.2. Misuse of a sUAS aircraft's data to invade privacy or steal proprietary data.
T3. Threats from attacks that evade liability for caused accidents.
T4. Threats from attacks that deny or degrade UTM accessibility and operations.

349. NASA, Joseph L. Rios, et al., *UAS Service Supplier Framework for Authentication and Authorization*, NASA/TM–2019–220364 (Sept. 2019), Sect. 4 Threat Modeling and Testing, https://utm.arc.nasa.gov/docs/2019-UTM_Framework-NASA-TM220364.pdf ["UFAA 2019"].
350. [Sampigethaya, Kopardekar 2018], p. 20 (characterizing such attacks as "cyber-physical").
351. *See* DoT, Office of the Inspector General, *FAA Has Made Progress but Additional Actions Remain To Implement Congressionally Mandated Cyber Initiatives* (Mar. 20, 2019), https://www.oig.dot.gov/sites/default/files/FAA%20Cybersecurity%20Program%20Final%20Report%5E03.20.19.pdf ("FAA's ATC system is becoming more interconnected as the agency introduces a range of new communication, navigation, and surveillance capabilities [and the] *sophistication of cyber threats continues to increase and evolve.*") (emphasis added); [SESAR ATM Master Plan 2017], Sect. 3.3 (citing "new vulnerabilities to cyber-attacks"). *See also* Jonathan Daniels, CEO, Praxis Aerospace Concepts Int'l, and Chair, ASTM F38.03, Remarks (Sept. 27, 2019) ("If we allow a federated approach, we have giant security holes. The value security will have cannot be understated."); Robert Segers, Information Security Architect, FAA, Presentation at the GUTMA Annual Conference, Portland, Or. (June 20, 2019) (querying how, in this distributed infrastructure, to assure that "information of the drone will be unmodified by man-in-the-middle attacks—so that an end-user (such as law enforcement) does not rely on [nonauthentic] networked communications").
352. [Rios 2020], p. 30 (re: UTM-MOP 11).
353. *See* NIST, SP 800-204, *Security Strategies for Microservices-based Application Systems* (Aug. 2019), https://csrc.nist.gov/publications/detail/sp/800-204/final. Microservices "is an architectural style that structures an application as a collection of services that are highly maintainable and testable, loose, coupled, independently deployable and organized around business capabilities [enabling] an organization to evolve its technology stack." Microservices.io, www.Microservices.io. *See, e.g.,* SESAR JU, *Solving big challenges in small packets: SESAR project makes progress on U-space information management* (May 27, 2019), https://www.sesarju.eu/news/solving-big-challenges-small-packets-sesar-project-makes-progress-u-space-information (in part, describing SESAR microservices); Reinaldo Negron, Head of UTM, Wing, Presentation at ICAO, DRONE ENABLE/3, Montreal, CA (Nov. 12, 2019) (highlighting "*fossilized APIs* where ATC can't improve").

Harmonized and increasingly unified standards, and adoption of a microservices approach to UTM software, should bolster flexibility, and in many jurisdictions, facilitate new participation and competitive services by multiple providers. SESAR JU, *U-Space Blueprint* (2017), p. 20, https://www.sesarju.eu/sites/default/files/documents/reports/U-space%20Blueprint%20brochure%20final.PDF ["SESAR U-Space Blueprint"]. *Cf.* Drone Alliance Europe, *U-space Whitepaper* (July 2019), p. 6, http://dronealliance.eu/wp-content/uploads/2019/07/Drone-Alliance-Europe-U-Space-Whitepaper-2-July-2019.pdf ["Drone Alliance Europe, U-Space Whitepaper 2019"] (Service providers "should operate in a similar fashion to telecom carriers, where carriers seamlessly provide overlapping service and consumer access between carriers.").

354. The U.S. Patriot Act of 2001 (H.R. 3162), Pub. L. No. 107-56 (codified at 42 U.S.C. Sect. 5195c(e)) (Oct. 26, 2001), *available at* https://www.law.cornell.edu/uscode/text/42/5195c (defining critical infrastructure as "systems and assets, whether physical or virtual, so vital to the United States that the incapacity or destruction of such systems

and assets would have a debilitating impact on security, national economic security, national public health or safety, or any combination of those matters."). The following documents may support (nonexclusively) the inclusion of UTM as critical infrastructure: [Nat'l Strategy for Avi. Security 2018] (linking the aviation ecosystem and critical infrastructure, including "aviation management; stating that such aviation management "operational services range from air traffic management, including Air Traffic Control, used to prevent collisions and expedite the orderly flow of air traffic, to enabling networks of communications, navigation, surveillance, and automation critical infrastructure;" and asserting that "The United States must ensure the safety and security of the Aviation Ecosystem."). *See* US Dept. of Homeland Security, Cybersecurity and Infrastructure Security Agency (CISA), *Transportation Systems Sector, Sector Overview,* https://www.dhs.gov/cisa/transportation-systems-sector ("The Transportation Systems Sector [includes:] **Aviation** [including] aircraft, *air traffic control systems*, and about 19,700 airports, heliports, and landing strips... In addition, the aviation mode includes... a *wide-variety of support services*, such as... *navigation aids....*") (emphasis added); CISA, *National Critical Functions Set*, https://www.cisa.gov/national-critical-functions-set (includes, e.g., "Provide Information Technology Products and Services," "Transport Cargo and Passengers by Air," and "Provide and Maintain Infrastructure"); 18 U.S.C. Ch. 113B Terrorism, Sect. 2339D(C), *available at* https://www.law.cornell.edu/uscode/text/18/2339D ("(3) the term 'critical infrastructure' means systems and assets vital to national defense, national security, economic security, public health or safety including both regional and national infrastructure. Critical infrastructure may be publicly or privately owned; examples of critical infrastructure include... telecommunications networks... and transportation systems and services"); Presidential Policy Directive/PPD–21, *Directive on Critical Infrastructure Security and Resilience* (Feb. 12, 2013), https://www.hsdl.org/?view&did=731087 (addressing "distributed networks, varied organizational structures and operating models"); Email from Parimal Kopardekar, Ph.D., Dir., NARI, NASA (Sept. 3, 2019) (on file with author) (asserting that "UTM is also useful for DoD and DHS to identify legitimate operations...."). More than 85 percent of critical infrastructure is held by the private sector. Telephone Interview with Tim Bennet, PMP, Air Domain Awareness, Science & Technology Directorate, DHS (Oct. 11, 2019).

355. Email from Parimal Kopardekar, Ph.D., Dir., NARI, NASA (Sept. 3, 2019) (on file with author).
356. Gur Kimchi, formerly VP, Amazon, Presentation at ICAO, DRONE ENABLE/3, Montreal, CA (Nov. 13, 2019) (addressing system requirements).
357. Transport Layer Security (TLS) provides the underlying security for most UTM communications. *See, e.g.,* [EASA Reg. Framework U-space], Art. 6. 4(b), "all data exchanges via APIs are completed over transport layer security (TLS)-secured connections." *See generally* IETF, RFC 8446, *The Transport Layer Security* (TLS) Protocol, *Version 1.3* (Aug. 2018), https://tools.ietf.org/html/rfc8446; [India UTM Policy], Sect. 7.1 (requiring TLS v1.3 or similar latest standards for information exchange between systems); NIST, Special Publication (SP) 800-52 Rev. 2, *Guidelines for the Selection, Configuration, and Use of Transport Layer Security (TLS) Implementations*, Kerry A. McKay and David A. Cooper, Computer Security Div. Info. Tech. Lab. (Aug. 2019), https://nvlpubs.nist.gov/nistpubs/SpecialPublications/NIST.SP.800-52r2.pdf. *See also* Joseph Rios, Ph.D., ATM-X Project Chief Eng'r, NASA, Remarks at ASTM F38 UTM WG meeting, Palo Alto, Cal. (Aug. 29, 2019) (asserting that for the open standard for access delegation, OAuth, "there are about 83,000 design decisions" required); [UFAA 2019], App. 1 (presenting open issues, many of which concern TLS); Email from Hrishikesh Ballal, CEO, Openskies Aerial Technology Ltd (March 14 & 17, 2020) (describing developing an open source OAuth alternative, *Flight Passport*, https://github.com/openskies-sh/flight_passport reflecting research that identified "limitations of the current offerings" including lack of fit for UTM, need for alternative to "big Cloud" providers, suitable pricing/licensing, lack of certain features, and complex installation).

358. Amit Ganjoo, Founder and CEO, ANRA Technologies, and Co-chair, ASTM F38 UTM WG, Remarks (Aug. 11, 2019) (underscoring that "UTM security is definitely a concern, however existing security frameworks, NIST guidelines and best practices can be used to address these just like any other connected IoT network"). *See infra* Chap. 3, Sect. 3.2.4 *CTA* (describing IoT security standards for sUAS); S. 734, IoT Cybersecurity Act of 2019, 116th Cong. (2019-2020), https://www.congress.gov/bill/116th-congress/senate-bill/734/text.
359. Edward "Ted" Lester, formerly Chief Technologist, AiRXOS, Remarks at the ASTM F38 UTM WG meeting, in Palo Alto, Cal. (Aug. 29, 2019). *See* Aerospace Industries Association of America, Inc. (AIA), Civil Aviation Cybersecurity Subcommittee, *AIA Civil Aviation Cybersecurity Industry Assessment & Recommendations, Report to the AIA Civil Aviation Council* (Aug. 2019), Sect. 3 Ensuring a Product Cybersecurity Culture, https://www.aia-aerospace.org/wp-content/uploads/2019/10/AIA-Civil-Aviation-Cybersecurity-Recommendations-Report-2019-Final-1.pdf; GAO, *Unmanned Aircraft Systems, FAA's Compliance and Enforcement Approach for Drones Could benefit from Improved Communication and Data,* GAO-20-29 (Oct. 2019), https://www.gao.gov/assets/710/702137.pdf (noting continuing FAA enforcement challenges).
360. Email from Phillip Hallam-Baker, Ph.D., Web Pioneer, security consultant, and expert witness (Oct. 8, 2019) (on file with author).
361. Robert Roth, Head of Software Eng'g, Uber ATG, formerly, Eng'g Dir., Prime Air at Amazon, Remarks at the ASTM F38 UTM WG meeting, in Palo Alto, Cal. (Aug. 29, 2019). *See infra* Chap. 4, Sect. 4.7.2 *Industry Consortia.*
362. *See* ICAO, *Assembly Resolution A40-10: Addressing Cybersecurity in Civil Aviation* (Oct. 2019), https://www.icao.int/Meetings/a40/Documents/Resolutions/a40_res_prov_en.pdf; ICAO, *Addressing Cybersecurity in Civil Aviation*, Working Paper presented to the Council of ICAO, Agenda Item 16: Aviation Security – Policy, A39-WP/17, EX/5 (May 30, 2016), https://www.icao.int/Meetings/a39/Documents/WP/wp_017_en.pdf (in part, instructing the Secretary General to: "a) Assist and facilitate States and industry in taking these actions; and b) Ensure that cybersecurity matters are fully considered and coordinated across all relevant disciplines within ICAO.").
363. ICAO, *Thirteenth Air Navigation Conference (AN-CONF/13) Outcome*, Agenda Item 2: Global Development in Aviation, MIDANPIRG/17 and RASG-MID/7 Meeting (Cairo, Egypt, 15–18 April 2019), WP/4 (April 4, 2019), recommendation 5.2/1 – Very low altitude operations, C-WP/14837, App. A-35 (h), https://www.icao.int/MID/MIDANPIRG/Documents/MID17%20and%20RASG7/WP4.pdf (emphasis added).
364. [UFAA 2019].
365. [FESSA], Sect. 2111 Aviation Cybersecurity. (a)(1). *See generally* 49 U.S.C. Sect. 40103 Sovereignty and use of airspace.
366. [FESSA], *id. See* [FESSA], Sect. 2208 Unmanned Aircraft Systems Traffic Management.
367. US DoT, Office of Inspector General, *FAA Has Made Progress but Additional Actions Remain to Implement Congressionally Mandated Cyber Initiatives*, Report No. AV2019021 (March 20, 2019), p. 3, https://www.oig.dot.gov/sites/default/files/FAA%20Cybersecurity%20Program%20Final%20Report%5E03.20.19.pdf (concluding that "implementation delays of cybersecurity tools inhibit FAA's ability to keep its cybersecurity up-to-date and make it difficult for the Agency to fully identify and mitigate vulnerabilities." *id.* p. 9; that the DoT's "cybersecurity program remains ineffective; and that its Office of the Chief Information Officer (OCIO) must establish effective internal controls especially continuous management oversight." *id.* p. 19). *See* DoT, Office of the Inspector Gen., *FAA Lacks Sufficient Security Controls and Contingency Planning for Its DroneZone System*, FAA, IT2020027 (April 15, 2020), https://www.oig.dot.gov/sites/default/files/FAA%20DroneZone%20Security%20Controls%20Final%20Report.pdf?utm_medium=email&utm_source=govdelivery; DoT, Office of the Inspector, Matthew E. Hampton, Asst. Inspector Gen. for Avi. Audits, *Memorandum*, Subject: Information: Audit Announcement | Unmanned Aircraft Systems Traffic Management (March 21, 2021), https://www.oig.dot.gov/sites/default/files/FAA%20UTM%20

Audit%20Announcement_03.29.2021.pdf (audit objectives: "FAA's (1) progress with UTM development and implementation, including results of its UTM pilot program; and (2) collaboration with other Government agencies regarding UTM").

See also The White House, *National Strategy for Aviation Security of the United States of America* (Dec. 2018), https://www.aviationtoday.com/2019/10/08/homeland-security-dod-transportation-officials-focus-aviation-cyber-security/ (recognizing that "[t]he Aviation Ecosystem demands extremely high standards of security be implemented in an efficient manner; requiring a "holistic and adaptive approach to securing the Aviation Ecosystem that prioritizes enhanced domain awareness, collection of anticipatory information, augmentation and sustainment of layered security measures, improved system resilience, and effective engagement with government and private-sector partners") ["Nat'l Strategy for Avi Security 2018"]; Joseph Rios, Ph.D., ATM-X Project Chief Eng'r, NASA Ames, Interview by NASA, Silicon Valley Live - *Air Taxis and the Future of Flight,* in Moffett Field, Cal. (Dec. 19, 2019), https://tinyurl.com/Rios-SVL ("You want to be secure as possible at *all layers.*") (emphasis added).

368. *See infra* Chap. 2, Sects. 2.4.8 *UAS Remote ID*, and 2.4.7.2 *International Aviation Trust Framework.*
369. Dr. Kopardekar underscored, "After Part 107 everyone thought BVLOS was next. They had a roadmap. But we [industry and government] got sidetracked by RID. We didn't postulate the need for security. That was a good lesson that the need for security has to be taken into account. [Nonetheless,] you cannot anticipate everything. Instead of anticipatory planning, we need adaptive planning to respond effectively to change based on uncertainties as they come alone. Therefore, set up smaller goals and *crawl, walk, run.* Part 107 took a lot of years. We could have done things that are lower total risk and progressed specific use cases more quickly rather than waiting and giving everyone 107." Parimal Kopardekar, Ph.D., Dir., NARI, NASA, Remarks, in Moffett Field, Cal. (Aug. 26, 2019).
370. *See* ISO, *ISO/IEC 27000:2018, Information technology — Security techniques — Information security management systems — Overview and vocabulary,* https://www.iso.org/standard/73906.html; ISO, *ISO/IEC 27032:2012, Information technology — Security techniques — Guidelines for cybersecurity,* https://www.iso.org/standard/44375.html; ISO, *ISO/IEC 27001:2013 Information technology — Security techniques — Information security management systems — Requirements,* https://www.iso.org/standard/54534.html; ISO, *The ISO 27001 Certification Process,* http://www.27000.org/ismsprocess.htm. *See infra* Chap. 3, Sect. 3.2.8 *ISO.*
371. The ASTM UTM specification also permits use of equivalent standards. Its use of ISO/IEC 27001 was informed, in part, by the Virginia Tech Mid-Atlantic Aviation Partnership (MAAP), *MAAP UPP2 Final Report Attachment A, Security Considerations for Operationalization of UTM Architecture* (Jan. 12, 2021) (copy on file with author).

[Comm Implementing Reg. (EU) 2021/664 - U-space], Art. 15.1.f, requires USSPs to implement and maintain a security management system per Commission Implementing Regulation (EU) 2017/373 (Mar. 1, 2019), Annex III, Subpart D, ATM/ANS.OR.D.010, *available at EASA, Easy Access Rules for ATM-ANS (Regulation (EU) 2017/373,* p. 201, https://www.easa.europa.eu/sites/default/files/dfu/Easy_Access_Rules_for_ATM-ANS.pdf (covering both physical and cyber risks). Correspondingly, EUROCAE, ED-205 (presented below, this subsection, specifies use of ISO/IEC 27001). *See* Email from Benoit Curdy, Digital Transformation, SWISS FOCA (Mar. 1, 2021) (on file with author) ("The idea is to simply stick with the same standard for the certification as USSP and in the services themselves.").

ISO/IEC 27001 implementations include, for example, Indra Avitech, *Indra Avitech strengthens its Information Security Management System through successful global certifications* (Oct. 20, 2020), https://tinyurl.com/Indra-ISO27001; *New ISO-27001 certification expands DroneDeploy's data security commitment,* Comm. Drone Prof'l

(May 9, 2019), https://www.commercialdroneprofessional.com/new-iso-27001-certification-expands-dronedeploys-data-security-commitment/ (a mission planner—not a USS, yet its use of ISO/IEC 27001 demonstrates this standard's potential contribution to the UTM ecosystem).

372. *See* https://www.iso.org/iso-9001-quality-management.html (certifiable criteria for quality management).
373. The RMF publications are technology neutral, yet "can be applied to secure the IS platform for UTM, especially when leveraging overlay development and tailoring guidance to develop a more specific set of controls that are most appropriate for this platform." Email from Eduardo K. Takamura, NIST (Sept. 3, 2019) (on file with author).
374. NIST, FIPS Publication 200, *Minimum Security Requirements for Federal Information and Information Systems* (March 2006), https://csrc.nist.gov/csrc/media/publications/fips/200/final/documents/fips-200-final-march.pdf.
375. NIST, FIPS PUB 199 (March 1, 2004), https://www.nist.gov/publications/federal-information-processing-standard-fips-199-standards-security. *See* [NASA USS Spec.], Sect. 5 Authority to Operate (NASA assumes airspace regulator classification of such data will "[l]ikely follow federal guidelines and leverage the initial step of the Risk Management Framework" of FIPS PUB 199.).
376. NIST, *Security and Privacy Controls for Information Systems and Organizations,* SP 800-53, Rev. 5 (DRAFT) (Aug. 2017), Sect. 1.2, https://csrc.nist.gov/CSRC/media//Publications/sp/800-53/rev-5/draft/documents/sp800-53r5-draft.pdf (Note, current version is Rev. 4) (April 2013), https://nvlpubs.nist.gov/nistpubs/SpecialPublications/NIST.SP.800-53r4.pdf ["NIST SP 800-53 Rev. 5"]. *See* Edward "Ted" Lester, formerly Chief Technologist, AiRXOS, Remarks at the ASTM F38 UTM WG meeting, in Palo Alto, Cal. (Aug. 29, 2019) (describing implementation of NIST SP 800-53 as "a freak'in nightmare – overburdensome for some less important services.").

    See also the E-Government Act of 2002, Pub. L. No. 107-374 (Title III), 44 U.S.C. 101 note, https://www.govinfo.gov/content/pkg/PLAW-107publ347/pdf/PLAW-107publ347.pdf (in part, establishing a "comprehensive framework for ensuring the effectiveness of information security controls over information resources that support Federal operations and assets" Sect. 3.5.4.1); the Federal Information Security Management Act (FISMA) of 2014, Pub. L. No. 113-283 (Dec. 18, 2014), https://csrc.nist.gov/Topics/Laws-and-Regulations/laws/FISMA; and for historical purposes, FAA, *Order 1370.112*, Subj: FAA Application Security Policy (Oct. 5, 2010), https://www.faa.gov/documentLibrary/media/Order/1370.112.pdf (in part, requiring implementation of "policies, standards, requirements, and procedures that are consistent with NIST" and "perform[ance] of periodic testing and evaluation of the effectiveness of information security policies, procedures, practices, and security controls with a frequency depending on risk, but no less than annually."). FAA Order 1370.112 superseded by Order 1370.121 - FAA Information Security and Privacy Program & Policy (Dec. 23, 2016), https://www.faa.gov/regulations_policies/orders_notices/index.cfm/go/document.information/documentID/1030708 (access restricted to *FAA network only*).
377. [NIST SP 800-53 Rev. 5], Sect. 1.2.
378. FAA, *Memorandum of Agreement for Low Altitude Authorization and Notification Capability (LAANC) Between FEDERAL AVIATION ADMINISTRATION (FAA) And______*, v2.3 (April, 2020), https://www.faa.gov/uas/programs_partnerships/data_exchange/laanc_for_industry/media/Memorandum_of_Agreement.pdf ["LAANC MOA"], Sect. 2.3.10, providing that:

    > [USS] agrees to interact with LAANC AP API in accordance with information security requirements defined by National Institute of Standards and Technology (NIST) FIPS Publication 200, Minimum Security Requirements for Federal Information and Information Systems and in conjunction with the selection and implementation of the appropriate security controls and assurance requirements from

NIST Special Publication 800-53, Revision 4, Recommended Security Controls for Federal Information Systems, including any subsequent revisions, updates, amendments, or changes thereto. [USS] shall secure systems and components interacting with LAANC AP API at a FIPS 199 Moderate Impact Level. Alternatively, [USS] may utilize an equivalent or higher security level, including the ISO 27000 series of standards, so long as [USS] provides the FAA with evidence of [USS]'s use of that equivalent or higher security standard.

379. NIST, *Framework for Improving Critical Infrastructure Cybersecurity*, Ver. 1.1 (April 16, 2018), https://nvlpubs.nist.gov/nistpubs/CSWP/NIST.CSWP.04162018.pdf.
380. Vol. 1, Ch. 2, p. 9 (Nov. 2016, updated Mar. 21, 2018), https://csrc.nist.gov/publications/detail/sp/800-160/vol-1/final.
381. SP 800-160, Vol. 2 (Nov. 2019), https://csrc.nist.gov/publications/detail/sp/800-160/vol-2/final. *See* NIST, Media Release, *NIST Releases SP 800-160 Vol. 2: Developing Cyber Resilient Systems – A Systems Security Engineering Approach* (Nov. 27, 2019) (in part, responding to "Advanced adversaries, collectively referred to as the Advanced Persistent Threat (APT), have the capability to breach our critical systems, establish an often undetected presence within those systems, and inflict immediate and long-term damage on the economic and national security interests of the Nation.").
382. https://www.nist.gov/itl/tig/projects/special-publication-800-63 ["SP 800-63-3"].
383. [NASA USS Spec.], Sect. 16. USS Checkout Process (further stating, "These assurance levels are obtained by considering the "Maximum Potential Impacts for Each Assurance Level [and recognize that] the 'inconvenience, distress, or damage to standing or reputation' and 'harm to agency programs or public interest' could be argued to be 'High' for each Assurance Level."); [NASA USS Spec.], Sect. 5 Authority to Operate (presenting requirement: "[UTM-USS-001] A USS MUST meet the requirements of the authorizing entity for authority to operate a USS.").
384. *See, e.g.,* NIST, Special Publication 800-37, Rev. 2, *Risk Management Framework for Information Systems and Organizations* (Dec. 2018), Sect. 2.8 Supply Chain Risk Management, https://nvlpubs.nist.gov/nistpubs/SpecialPublications/NIST.SP.800-37r2.pdf ("[FISMA] and [OMB A-130] require external providers handling federal information or operating systems on behalf of the federal government to meet the same security and privacy requirements as federal agencies."); NIST, NISTIR 7628 Rev 1, *Guidelines for Smart Grid Cybersecurity* (Sept. 2014), https://csrc.nist.gov/publications/detail/nistir/7628/rev-1/final.
385. *See* Email from Robert Moskowitz, HTT Consulting (Dec. 13, 2020) (on file with author) ("ATM security should be easy with such big, powered platforms (other than significant legacy platforms holding it back). UTM at least has a 'clean slate' but highly constrained platforms.").
386. (March 2019), https://eshop.eurocae.net/eurocae-documents-and-reports/ed-205/ (*see also* above mention of ED-205 for U-space, this subsection). *See* Aerospace Industries Ass'n, *Civil Aviation Cybersecurity Industry Assessment and Recommendations, August 2019, Report to the AIA Civil Aviation Council, Civil Aviation Regulatory & Safety Committee*, Sect. 5.7 Air Traffic Management (ATM), https://www.aia-aerospace.org/wp-content/uploads/2019/10/AIA-Civil-Aviation-Cybersecurity-Recommendations-Report-2019-Final-1.pdf (highlighting EUROCAE WG-72 Subgroup ED-205's end-to-end cybersecurity approach, asserting lack of US counterpart, and mentioning ISO/IEC 27001).
387. *Available at* https://standards.ieee.org/standard/1609_2-2016.html.
388. *See, e.g.,* Marc Kegelaers, formerly CEO, Unifly, Presentation at ICAO, DRONE ENABLE/3, Montreal, CA (Nov. 12, 2019) ("UTM is extremely complex."—and addressing its implications); [FAA ConOps UTM v2.0], p. xi, Exec. Summary.
389. Robert Segers, Information Security Architect, FAA, in Raleigh, N.C., Remarks (Nov. 6, 2019) (Further stating, "ISO 27000 alone means nothing without scoping because ISO 27000 alone is inward-focused. ISO 9000 has the same problem. It means nothing

in terms of interoperability. That is the fundamental difference of ISO vs a framework. Scoping of the security risks must be normalized because what is a risk to one organization may not be a risk to another.").

390. *See* Andrew Velasquez III, Managing Deputy Comm'r, CDA, Airports Blue Ribbon Panel, Presentation at AUVSI Xponential, Chicago, Ill. (May 2, 2019) ("It is a nontraditional, evolving threat. [We] need to understand the threat. You don't know what you don't know.").
391. Robert Segers, *supra* ("From a protocol perspective, using Internet standards is the best thing to do. Of course, you have to look at the protocols, but they are *battle tested.*").
392. Additionally, compare the use and benefits of systems such as EINSTEIN by DHS as a tool/system supporting defense-in-depth, and "situational awareness to use threat information detected in one agency to protect the rest of the government and to help the private sector protect itself." Might an analogous capability serve a federated UTM? *See* DHS CISA, EINSTEIN, https://www.dhs.gov/cisa/einstein.
393. Presentation at the World ATM Congress, Madrid, ES (March 12, 2019). *Cf.* Email from Robert Moskowitz, HTT Consulting (Dec. 13, 2020))(on file with author) (underscoring that "[y]our level of federation with people is based on your level of trust.").
394. Email from William Voss, Special Advisor to ICAO (Sept. 8, 2019) (on file with author). Saulo Da Silva, Chief, Global Interoperable Systems Section, ICAO, Remarks at ICAO DRONE ENABLE Symposium 2021 (April 21, 2021) (Identity and trust "[a] core function of ICAO since 1944"); *Cf.* Robert Moskowitz, *id.* ("Notwithstanding, there are unique fundamental root identity questions inherent in UA that are not present with large airframes.").
395. ICAO, *Thirteenth Air Navigation Conference (AN-CONF/13) Outcome*, Agenda Item 2: Global Development in Aviation, MIDANPIRG/17 and RASG-MID/7 Meeting, Cairo (April 15-18, 2019), WP/4 (April 4, 2019), C-WP/14837, Appn. A-39, https://www.icao.int/MID/MIDANPIRG/Documents/MID17%20and%20RASG7/WP4.pdf (also recommending at ¶ 5.4/1—Cyber resilience, "h) incorporate the trust framework into the Global Air Navigation Plan (Doc 9750) in an appropriate manner to highlight its urgent need, its importance and to improve its visibility").
396. *See* Global UTM Association, *ICAO Global Aviation Trust Framework* (Sept. 20, 2018), https://gutma.org/blog/2018/09/20/icao-global-aviation-trust-framework/; William Voss, Special Advisor to ICAO, Presentation at ICAO, DRONE ENABLE/2, Chengdu, P.R.C. (2018); ICAO, *Global Resilient Aviation Network Concept of* Operations, *For a secure and trusted exchange of information* (2018), https://www4.icao.int/ganpportal/trustframework; ICAO, *Working Outline for International Aviation Trust Framework (IATF),* Draft Ver. 2 (Oct. 2019) (on file with author) ["ICAO IATF 2019"]; *infra* Chap. 3, Sect. 3.2.6 *ICAO* (addressing GRAIN).
397. On March 14, 2019, the Secretariat presented an initial draft terms of references (ToRs) for a study group on trust framework (TFSG) to the Air Navigation Commission (ANC) requesting authorization for its establishment, in response to the recommendation 5.4/1 of the Thirteenth Air Navigation Conference (AN-Conf/13), as approved by the Council (C-DEC 216/5 refers).
398. ICAO, *Operational Use Cases to Validate The Trust Framework*, Agenda Item 5.6: FAA Operational Scenarios, Trust Framework Study Group (TFSG), First Meeting of the Working Group on Current and Future Operational Needs, TFSG-1-WP/19 (May 5, 2019).
399. *See generally* [ICAO IATF 2019]; ICAO, *X.509 Certificate Policy for the International Aviation Trust Framework (IATF) Certification Authority*, Draft Ver. 0.2 (Nov. 2019) (on file with author); TSCP, *Redacted X.509 CPS, v. 4 for the TSCP Bridge Certificate Authority (TBCA)* (Feb. 6, 2019), https://www.tscp.org/tscp-certification-practice-statement/; Email from Robert Moskowitz. HTT Consulting (April 25, 2021) (on file with author) ("ICAO anticipates adding DANE server (RFC 6698) and client (draft-huque-dane-client-cert) certificate support within the IATF for within the

ICAO gTLD. The current IATF Certificate Policy does not need revision to support this."). IEFT, RFC 6698, The DNS-Based Authentication of Named Entities (DANE) Transport Layer Security (TLS) Protocol: TLSA (July 29, 2020), https://datatracker.ietf.org/doc/rfc6698/; IETF, [draft-huque-dane-client-cert-05,] TLS Client Authentication via DANE TLSA records, S. Hoque, et al. (Oct. 30, 2020), https://datatracker.ietf.org/doc/draft-huque-dane-client-cert/.

See also GSA, *Federal Public Key Infrastructure Guides*, https://fpki.idmanagement.gov/; Michael S. Baum, *Federal Certification Authority Liability and Policy*, NIST, by MITRE Corp. under Contract #50SBN1C6732 (1992), *available at* https://tinyurl.com/PKI-Baum ["Baum FCA Liability"]; Warwick Ford and Michael S. Baum, *Secure Electronic Commerce* (Prentice Hall 2nd. ed., 2000), *available at* https://tinyurl.com/Secure-ECommerce (explaining PKI and associated infrastructure).

400. *See* ICAO and ICAN, *ICAO-ICANN MOU* (Feb. 5, 2019), *available at* https://www.icann.org/en/system/files/files/mou-icao-icann-05feb19-en.pdf; ICAO, *ICAO, ICANN, sign new agreement to guide progress on proposed aviation trust framework for digital environments* (Feb. 11, 2019), https://www.icao.int/Newsroom/Pages/ICAO-ICANN-sign-new-agreement-to-guide-progress-on-proposed-aviation-trust-framework-for-digital-environments.aspx.
401. William Voss, Special Advisor to ICAO, Presentation at ICAO, DRONE ENABLE/2, in Chengdu, P.R.C. (Sept. 13-14, 2018) (further stating, "ICAO is trying to find a 'middle ground' and a converging strategy—recognizing that interoperability requires global cooperation."). The Trust Framework has also been described as "building upon, but [] not limited to, the capabilities of a public key infrastructure (PKI) Internet protocol version 6 (IPv6) addressing, a domain name system (DNS), and information security management systems." [ICAO IATF 2019], Sect. 2.
402. .aero, https://information.aero/contact.
403. William Voss, Special Advisor to ICAO, Presentation at ICAO, DRONE ENABLE/3, Montreal, CA (Nov. 12, 2019).
404. *Available at* http://www.emsa.europa.eu/retro/Docs/marine_casualties/annex_13.pdf.
405. Done at Beijing 2020, https://www.icao.int/secretariat/legal/Docs/beijing_convention_multi.pdf (also attacks by civil aircraft carrying biological, chemical or nuclear materials).
406. Email from William Voss, *supra* (Aug. 10, 2020) (on file with author) (underscoring that an aircraft requires a robust digital identity linked to a national identity to take advantage of the full range of international laws).
407. Presentation at the GUTMA Annual Conference, Portland, Or. (June 18, 2019).
408. [RID NPRM 2019], Sect. IV.B.
409. Jay Merkle, Exec. Dir., UAS Integration Office, FAA, Remarks at the FAA UAS Symposium, Balt., Md. (June 4, 2019). *Cf.* [FAA ANPRM 2019], p. 3735 ("While UTM is focused on managing the safe and efficient operation of an increasing number of UAS operating in the NAS, especially beyond visual line of sight (BVLOS), there may be opportunities to mitigate public safety and national security risks at the same time.").
410. *See, e.g.,* [FESSA], Sect. 2209 (requiring the FAA to establish procedures for applicants to restrict UA operations "in close proximity to a fixed site facility" such as critical infrastructure); the Nat'l Defense Authorization Act for FY 2018, Sect. 1692, Protection of certain facilities and assets from unmanned aircraft, https://www.govtrack.us/congress/bills/115/hr2810; *id.* FY 2020, Sect. 1694. Extension of Authorization for Protection of certain facilities and assets from unmanned aircraft, https://www.congress.gov/bill/116th-congress/senate-bill/1790/text.
411. [ICAO UTM Framework Ed. 3], Appn. A, Registration, Identification and Tracking, p. 18 ("The ability to track UA with the UTM system was considered a critical service that has implications on system reliability, resilience and redundancy at the manufacturing and operational levels."). *See* FAA, NPRM, *Remote Identification of Unmanned Aircraft Systems*, 84 Fed. Reg. 72438-9 (Dec. 31, 2019), Sect. I.A, https://www.govinfo.gov/content/pkg/FR-2019-12-31/pdf/2019-28100.pdf.

412. [FAA ConOps UTM v2.0], Sect. 1.2 UTM Evolution. [Remote ID Rule], Preamble, Exec. Summary ("[T]he next incremental step towards enabling [BVLOS] operations.").
413. *See* Gur Kimchi, formerly VP, Amazon, Presentation at ICAO, DRONE ENABLE/3, Montreal, CA (Nov. 13, 2019) (urging that remote ID broadcast benefits may include V2V tactical deconfliction). *See supra* Chap. 2, Sect. 2.4.6 *Detect and Avoid*.
414. [FAA UPP Phase 2 ConUse], Sect. 4.6 ("The FAA uses UTM data to enforce its regulations and hold participants using UTM unlawfully responsible for their actions."). *See* Chap. 4, Sect. 4.7.4 *Liability* (further addressing responsibility).
415. *See* FAA, *UAS Identification and Tracking (UAS ID) Aviation Rulemaking Committee (ARC), ARC Recommendations, Final Report* (Sept. 30, 2017), https://www.faa.gov/regulations_policies/rulemaking/committees/documents/media/UAS%20ID%20ARC%20Final%20Report%20with%20Appendices.pdf (*E.g.,* "Cellular technology is being explored as part of a federated UTM-like service to provide situational awareness of all [manned and] unmanned aircraft.").
416. Jay Merkle, Exec. Dir., UAS Integration Office, FAA, Remarks at the FAA Drone Advisory Committee meeting, Crystal City, Va. (June 6, 2019). *See* [Remote ID Rule], Preamble, IX.D Incentives for Early Compliance (FAA commits to analysis re waivers, exemptions, and credits for RID use); FAA, Drone Advisory Committee, *eBook, DAC Member (Public) Information for the Oct. 17, 2019 DAC Meeting*, Wash., D.C. (Oct. 17, 2019), pp. 12-14, https://www.faa.gov/uas/programs_partnerships/drone_advisory_committee/media/eBook_10-17-2019_DAC_Meeting.pdf (providing "Voluntary Remote ID Equipage Incentive Recommendations" including but not limited to preferential contract and waiver application treatment, airspace access, and financial incentives); [FAARA], Sect. 377, Early implementation of certain UTM service.
417. *See, e.g.,* DJI, *Elevating Safety: Protecting The Skies In The Drone Era* (May 2019), Sect. 5.III, https://terra-1-g.djicdn.com/851d20f7b9f64838a34cd02351370894/Flysafe/190521_US-Letter_Policy-White-Paper_web.pdf ("Given the pace of government rule-making, implementation is likely to be years away").
418. European Commission (EC), *Commission Delegated Regulation (EU) 2019/945 of 12 March 2019 on unmanned aircraft systems and on third-country operators of unmanned aircraft systems*, Annex 1, Part 2, Sect. 12, published in the Official J. of the EU (June 11, 2019), https://eur-lex.europa.eu/legal-content/EN/TXT/PDF/?uri=CELEX:32019R0945&from=EN ["Comm Delegated Reg. (EU) 2019/945 - UAS"] (presenting requirements for a class C1 UAS), and Annex Part 6, *Requirements for a direct remote identification add-on. See* [Comm Implementing Reg. (EU) 2021/664 - U-space], Art. 8 Network identification service.
419. [Remote ID Rule], Preamble, IV.B. The Remote ID Rule will become a new 14 C.F.R. Part 89. *See generally* FAA, *Executive Summary Final Rule on Remote Identification of Unmanned Aircraft (Part 89)* (Dec. 28, 2020), https://tinyurl.com/FAA-RID-Summary.
420. *See* Chap. 2, Sect. 2.3.2 *Registration*.
421. [Remote ID Rule], Sects. 89.110 Operation of standard remote identification unmanned aircraft; 89.115 Alternative remote identification. (a) *Remote identification broadcast modules*, and (b) *Operations at FAA recognized identification areas*.
422. Note that the Remote ID Rule does not include the broadcast of flight intent (or operational intent), although the FAA "finds that such a requirement is appropriate to consider once UTM has been further developed and implemented. Flight intent is a foundational concept of UTM, and the FAA envisions such requirements may be a part of a future rulemaking to enable wide scale use of the UTM ecosystem." [Remote ID Rule], Preamble, VIII.A.
423. [ASTM Remote ID]. Developed by the ASTM F38 RID and Tracking WG.
424. A capability, "Cellular Vehicle-to-Everything" (C-V2X), is under development to allow local broadcasting using cellular radios. GSMA, *Cellular Vehicle-to-Everything (C-V2X), Enabling Intelligent Transport* (2018), https://www.gsma.com/iot/wp-content/uploads/2017/12/C-2VX-Enabling-Intelligent-Transport_2.pdf (employing both direct vehicle-to-vehicle and network communications; fully 5G compatible; and enabling autonomy). *See infra* Chap. 3, Sect. 3.2.3 *Cellular Industry*.

425. Email from Gabriel Cox, Principal Eng'r, Intel Corp., Chair, ASTM F38 Remote ID and Tracking Working Group (Aug. 6, 2019) (on file with author).
426. *See* [Remote ID Rule], Preamble, VII.A. Elimination of Network-based Remote Identification Requirement (Summarizing networked limitations: "Internet availability or connectivity issues; increased costs for UAS upgrades; Internet data plans; Remote ID USS subscriptions; and reduced air and ground risk when operating in remote areas with less air traffic and lower population density.").
427. [Remote ID Rule], Preamble, I.C.; *id.* Preamble VII. ("As the FAA builds the regulatory constructs that support increasingly advanced concepts, such as BVLOS and UTM, the United States Government will be prepared to solve safety and security issues related to those concepts based on more mature understandings.").
428. [Remote ID Rule], Preamble, XIII. Means of Compliance. *See* [Remote ID Rule], Preamble XXII. Regulatory Notice and analysis ("[T]he FAA believes it is practical for this industry consensus standard to be modified and submitted for acceptance as a means of compliance 6 months after the effective date of the final rule…."). The ASTM RID WG is drafting a RID means of compliance document presenting how to use [ASTM Remote ID] to comply with the Remote ID Rule.
429. Email from Gabriel Cox, Principal Eng'r, Intel and ASTM RID WG Chair (Jan. 5, 2020) (on file with author).
430. Source: [ASTM Remote ID].
431. [ASTM Remote ID], Sect. 5.4.5.12 ("provid[ing] flexibility to allow a multitude of signature formats that are not specified in this standard. The intended implementation is that an agreed upon signature format for each Auth Type required will be shared by both the signature encoding software and the verifier software."); *id.* Annex A1 – Broadcast Authentication Verifier Service (allowing "the signature specification to be defined by the verifier"). *Cf.* [Remote ID Rule], Preamble, VIII. B.11. Cybersecurity (upon removal of the network RID requirements "cybersecurity requirements for the broadcast functionality are no longer warranted."); FAA ATO, *Low Altitude Authorization and Notification Capability (LAANC), USS Performance Rules*, Ver. 4.1 (Apr. 17, 2020), https://www.faa.gov/uas/programs_partnerships/data_exchange/laanc_for_industry/media/LAANC_USS_Performance_Rules.pdf ["LAANC USS Performance Rules"], Sect. 3.1 ("USS must [] manage users using…*reasonably secure identification* mechanisms") (emphasis added).
432. [FAA ConOps UTM v2.0], Sect. 2.1 Overview.
433. *See supra* Chap. 2, Sect. 2.4.5.2 *Discovery and Synchronization Service (DSS)* (addressing APIs for the DSS); [Comm Implementing Reg. (EU) 2021/664 - U-space], Art. 18 Tasks of the competent authorities, ¶ 1.(d) (addressing data exchange). *See also* [RID NPRM 2019], Sect. IV.B (recognizing as a "critical element" establishing "a cooperative data exchange mechanism between the FAA and Remote ID USS…").
434. *See, e.g.,* [CORUS ConOps], Sect. 5.2.4 *Operational Processes and Information Exchanges*; [NASA USS Spec.], Sect. 9.2 (addressing "required data exchanges during operation.") [FAA UPP Phase 2 ConUse], Sect. 3.3 (presenting the following "technology gaps [for] which the FAA needs to help process data for UTM information exchange": technology standards (i.e., cloud systems, using REST API), protocol standards/data exchange formats (i.e., JSON over HTTP and GeoJSON) compared to XML, and endpoints/services standards (i.e., publishing/subscription services for messages); [Drone Alliance Europe, U-space Whitepaper 2019], p. 6 (urging "it will be imperative to establish a communications protocol and interfaces between USPs and ANSPs."); [Comm Implementing Reg. (EU) 2021/664 - U-space], Art. 7 and Annex V.
435. *See generally* [ASTM UTM Spec.]. The specification includes the definition, capabilities, and higher-level API interfaces for such services and provides a foundation to collect metrics. Performance standards common to certain data exchanges may include: maintaining persistent data connection, being in good standing/authorized holding required privileges, submitting compliant data for each request/response, and acting timely.

436. *See, e.g.,* Joseph Rios, Ph.D., ATM-X Project Chief Eng'r, NASA Ames Research Center, *Strategic Deconfliction: System Requirements*, Final Report (July 31, 2018), https://utm.arc.nasa.gov/docs/2018-UTM-Strategic-Deconfliction-Final-Report.pdf (such as facilitating requests, updating, and logging position and deconfliction).
437. *See* [ASTM UTM Spec.], Sect. 1. *See, e.g., infra* Chap. 3, Sect. 3.2 *Standards Development Organizations* (noting IEEE Standard(s) for IP network data exchange and access); IETF, *RFC 5322 Internet Message Format* (Oct. 2018), https://tools.ietf.org/html/rfc5322. *See infra* Chap. 3, Sect. 3.2.8 *IETF*.
438. Source: [CORUS ConOps], p. 82. Although the figure's architecture may appear centralized, it does not embrace a particular degree of centralization for implementations; actual data transfer endpoints are not necessarily represented in this figure; and the exchange of situational awareness data for manned aircraft is unidirectional, potentially leaving manned aircraft unaware of nearby threats.

    See [Rios TCL4], Sect. 5.A (noting that "data exchanged between system components of UTM are not always displayed to the user"). For comparison, in the ASTM UTM specification, "operation" has a very specific technical meaning akin to the colloquial term "flight"; and not all services represented in the figure would communicate with the flight directly. Moreover, the figure's central box would include USS and other elements of the strategic deconfliction infrastructure. *See infra* Chap. 5, Sect. 5.3 U-space (addressing DREAMS (ER)—DRone European *AIM Study*).
439. *See* [ASTM Surveillance SDSP]; [MNO-UTM Interface], Sect. 2.2. Scope ("The overall objective is to provide a minimum set of descriptions to standardize the way data between MNOs and the UTM ecosystem can be exchanged."); [India UTM Policy], p. 7 (describing a platform "built on a layered approach of information sharing and data exchange standards").
440. [Nat'l Academies AAM], Sect. 5-14 (emphasis added). *See* [FAA ConOps UAM v.1.0], Sect. 4.3.1.3 FAA-NAS Data Exchange (For UAM, "FAA data sources available via the FAA-industry data exchange include, but are not limited to, flight data, restrictions, charted routes, active Special Activity Airspaces (SAAs)"); Embraer UATM ConOps], Sect. 4.2 Information Exchange Service (objective "is to ensure shared situation awareness for all stakeholders").
441. Anthony Rushton, Ph.D., UTM Technical Lead, NATS, Remarks at *UAM Virtual, Integrating Urban Air Mobility into the Airspace*, Aviation Week (webinar) (Aug. 12, 2020) (seeking a "virtual black box in the cloud" to mitigate the anticipated data exchange "explosion of data beyond the ability to capture and use"). *See also* FAA, *Preparation of Interface Documentation*, FAA-STD-025f (Nov. 30, 2007), http://www.tc.faa.gov/its/worldpac/standards/faa-std-025f.pdf; Email from Jeffrey R. Homola, UTM Integration and Testing Lead, NASA Ames (Dec. 20, 2019) (on file with author) ("This is also still an active and open area of discussion and research.").
442. *See* ICAO TV, *Aeronautical Information Management for Unmanned Traffic Management (AIM for UTM)* webinar (Mar. 11, 2021), https://www.icao.tv/videos/ep03-aeronautical-information-management-for-unmanned-traffic-management (the preliminary guidance document is expected to include recommendations to ICAO); NASA, *Digital Information Platform* (April 2021), https://nari.arc.nasa.gov/atmx-dip (describing the Air Traffic Management-eXploration (ATM-X) project, Digital Information Platform (DIP) sub-project).
443. Remarks, in Chicago, Ill. (May 1, 2019).
444. Remarks at the *GUTMA, High Level Webinar Session - Ask the Experts* (June 9, 2020), https://gutma.org/ask-the-experts/.
445. *See* [Remote ID Rule], Preamble, I.A ("A means of compliance describes the methods by which the person complies with the performance-based requirements..."); FAA, *Accepted Means of Compliance: Airworthiness Standards: Normal Category Airplanes* (May 11, 2018), https://www.regulations.gov/document?D=FAA_FRDOC_0001-16680 (notice of availability of 63 MoC based on 30 ASTM F44 aircraft standards);

FAA, *Integration of Civil Unmanned Aircraft Systems (UAS) in the National Airspace System (NAS) Roadmap,* 2nd ed. (July 30, 2018), p. 27, Standards Development, https://www.faa.gov/uas/resources/policy_library/media/Second_Edition_Integration_of_Civil_UAS_NAS_Roadmap_July%202018.pdf ["FAA Roadmap"] ("For UAS to operate routinely in the NAS beyond what is currently allowed, they must conform to an agreed-upon set of minimum performance-based standards to ensure safety, efficiency, and reliability."); James D. Foltz, PE, Mgr., Strategic Policy Emerging Aircraft, James D. Foltz, PE, Mgr., Strategic Policy Emerging Aircraft, Aircraft Cert. Services, FAA, Presentation at the FAA UAS Symposium, Balt., Md. (June 4-5, 2019) (citing various ways for the FAA to recognize standards, including: TSO deliverables from RTCA, etc., Federal Register notices of availability, incorporation by reference, advisory circulars, standards widely used to type certify aircraft, "*and we plan to do the same for UAS as well*") (emphasis added); Underpinning this approach is the National Technology Transfer and Advancement Act of 1995, Pub. L. No. 104-113 (Mar. 7, 1996), 110 Stat. 782, Sect. 12(a) Use of Standards, 15 U.S.C. 3701 et seq., https://www.congress.gov/bill/104th-congress/house-bill/2196 (requiring "(3) to compare standards used in scientific investigations, engineering, manufacturing, commerce, industry, and educational institutions with the standards adopted or recognized by the Federal Government *and to coordinate the use by Federal agencies of private sector standards, emphasizing where possible the use of standards developed by private, consensus organizations*;") (emphasis added).

446. [SESAR ATM Master Plan 2017], Sect. 3.5 (also noting the possibility of adapting and using "existing standards from other relevant industries such as telecoms or autonomous land vehicles").
447. *See, e.g.,* ICAO, *Summary,* Third Meeting of the Asia/Pacific Unmanned Aircraft Systems Task Force, APUAS/TF/3 – WP/02 (March 7, 2019), 2.9, https://www.icao.int/APAC/Meetings/2019%20APUASTF3/WP02%20Related%20Meetings%20Outcomes.pdf#search=Search%2E%2E%2Eutm (referring to SARPS).
448. *See, e.g.,* Carl E. Burleson, Acting Deputy Admin'r, FAA, Remarks at AUVSI Xponential, Chicago, Ill. (May 1, 2019), https://www.faa.gov/news/speeches/news_story.cfm?newsId=23675 (The FAA desires "to figure out how to take full advantage of a performance based approach... [allowing] greater innovation.").
449. FAA, The Drone Advisory Committee (DAC), *Drone Access to Airspace, Final Report* (2017), https://www.faa.gov/uas/programs_partnerships/drone_advisory_committee/rtca_dac/media/dac_tg2_final_reccomendations_11-17_update.pdf ("Recommend[ing] the FAA establish, evaluate and implement performance-based navigation requirements for low altitude BVLOS operations [TO] promote integrated BVLOS airspace operations with shared intent, position data, and other information...").
450. Presentation at the FAA UAS Symposium (July 9, 2020). Andy Thurling, CTO, NUAIR Alliance, Remarks (Aug. 23, 2019) ("The biggest hole we have is to nail down acceptable levels of performance so we can kick something out the door and field it."). Performance-based regulation and standards are further addressed *infra* Chap. 4, Sect. 4.2.1 *General – USS Qualification*, and Chap. 4, Sect. 4.3.3 *Risk Assessment.*
451. ANSI, *Standardization Roadmap for Unmanned Aircraft Systems*, Ver. 2.0 (June 2020), https://share.ansi.org/Shared%20Documents/Standards%20Activities/UASSC/ANSI_UASSC_Roadmap_V2_June_2020.pdf ["ANSI Roadmap"].
452. James McCabe, Sr. Dir., Standards Facilitation, ANSI, Remarks at the FAA UAS Symposium, Balt., Md. (June 4, 2019).
453. James McCabe, *id.* (stating, "To prevent overlap that is why we've pulled together the UASSC.").
454. [ANSI Roadmap], Sect. 7.7. The ANSI Roadmap further identifies the need for EVLOS/BVLOS collaborative standards (GAP 03). Note that ANSI's wholly owned subsidiary, the ANSI National Accreditation Board (ANAB), is also recognized internationally and by many domestic agencies to assess the competence

of conformity assessment bodies of all types in accordance with ISO/IEC 17011 requirements. *See* ISO, ISO/IEC 17011:2017, *Conformity assessment - Requirements for accreditation bodies accrediting conformity assessment bodies,* https://www.iso.org/standard/67198.html. *See infra* Chap. 4, Sect. 4.2.4.2 *Third-Party Assessment.*

455. EUSCG, https://www.euscg.eu/. *Cf.* EC, European ATM Standards Coordination Group (EASCG), https://ec.europa.eu/transport/modes/air/single_european_sky/european-atm-standards-coordination-group_en.
456. [SESAR ATM Master Plan 2017], Sect. 3.5; EUSCG, *Rolling Development Plan*, v. 3 (June 3, 2019), https://www.eU.S.C.g.eu/media/1253/eU.S.C.g-066-version-30-rdp.pdf; Email with Sergiu Marzac, Technical Programme Mgr., EUSCG (Nov. 22, 2019) (on file with author). The extent of information sharing and coordination between the EUSCG and other standards groups is increasing. *See* EUSCG, RDP V5 (July 23, 2020), https://www.euscg.eu/media/1261/euscg-108_version-50-rdp_2020.pdf.
457. The Aerospace and Defence Industries Association of Europe—Standardization (ASD-STAN), https://asd-stan.org. *See* CEN, https://standards.cen.eu/.
458. Email from Benoit Curdy, Digital Transformation, Swiss FOCA (Nov. 12, 2019) (on file with author). *See* ETSI (European Telecommunications Standards Institute), https://www.etsi.org/about ("ETSI is a European Standards Organization (ESO). We are the recognized regional standards body dealing with telecommunications, broadcasting and other electronic communications networks and services. We have a special role in Europe. This includes supporting European regulations and legislation through the creation of Harmonised European Standards. Only standards developed by the three ESOs (CEN, CENELEC and ETSI) are recognized as European Standards (ENs).").
459. ASD-STAN, https://asd-stan.org/domain-d05-autonomous-flying/.
460. *Id.*
461. *See* ASD-STAN, *Direct Remote ID, Introduction to the European UAS Digital Remote ID Technical Standard* (2021), https://asd-stan.org/wp-content/uploads/ASD-STAN_DRI_Introduction_to_the_European_digital_RID_UAS_Standard.pdf (presenting the prEN 4709 series of RID standards); Email from Jan Broux, Product Mgr., Unifly (Dec. 4, 2020) (on file with author).
462. ASTM Int'l, News Release, *ASTM international Signs Memorandum of Understanding with ASD-STAN* (Oct. 27, 2020), https://newsroom.astm.org/astm-international-signs-memorandum-understanding-asd-stan; *Memorandum of Understanding Signed with ADS-STAN*, ASTM Standardization News (Jan./Feb. 2021), p. 47.
463. Remarks (Aug. 19, 2019).
464. ASTM, F38, *Subcommittees and Standards*, https://www.astm.org/COMMIT/SUBCOMMIT/F38.htm.
465. ASTM, *Committee F38 on Unmanned Aircraft Systems*, https://www.astm.org/COMMITTEE/F38.htm.
466. These include ASTM Committees: F37 on Light Sport Aircraft, F39 on Aircraft Systems, and F44 on General Aviation Aircraft.
467. Lorenzo Murzilli, Murzilli Consulting, formerly Founder and CEO, Leader, Innovation and Digitalization Unit, FOCA Switzerland - Swiss U-Space Program Manager, Presentation at the FAA UAS Symposium (July 8, 2020). *See* Philip M. Kenul, Chair, ASTM F38, Presentation at the FAA UAS Symposium, Balt., Md. (June 4, 2019) ("We're on our way getting off the block on expanded operations"—understood to encompass UTM).
468. ASTM, F_ [WK63418], https://www.astm.org/DATABASE.CART/WORKITEMS/WK63418.htm ["ASTM UTM Spec."].
469. ASTM F_[WK69690], https://www.astm.org/DATABASE.CART/WORKITEMS/WK69690.htm ["ASTM Surveillance SDSP"] (defines minimum performance standards for SDSP equipment and services to USS/USP; provides aircraft track information to DAA systems enabling BLVOS operations; and may support C-UAS systems). *See* ASTM F_[WK75981], *New Specification for Vertiport Automation Supplemental Data*

Services Provider (SDSP), https://www.astm.org/DATABASE.CART/WORKITEMS/WK75981.htm?A&utm_source=tracker&utm_campaign=20210223&utm_medium=email&utm_content=standards.

470. ASTM, F3411-19, https://www.astm.org/DATABASE.CART/WORKITEMS/WK65041.htm (applicable to "any low altitude geographical area generally under 400 feet AGL both urban and rural (in and beyond cellular infrastructure range), regardless of airspace class."). *See supra* Chap. 2, Sect. 2.4.8 *UAS Remote ID*; Reinaldo Negron, Presentation at the FAA UAS Symposium (June 8, 2020) ("The remote ID standard has laid a foundation of interoperation for UTM service and forced us to have the discussion of how to have more common standards.").
471. ASTM, F_[WK65042], https://www.astm.org/DATABASE.CART/WORKITEMS/WK65042.htm (Risk mitigation for flights over people is required for most envisioned UTM operations.).
472. ASTM, F3196-18, https://www.astm.org/Standards/F3196.htm; ASTM, F_WK75923, https://www.astm.org/DATABASE.CART/WORKITEMS/WK75923.htm?A&utm_source=tracker&utm_campaign=20210218&utm_medium=email&utm_content=-standards.
473. ASTM, F3422 / F3442-20, https://www.astm.org/DATABASE.CART/WORKITEMS/WK62668.htm ("[D]efine[s] minimum performance standards for DAA systems applicable to smaller UAS BLVOS operations for the protection of manned aircraft in lower altitude airspace."). *See supra* Chap. 2, Sect. 2.4.6 *Detect and Avoid.*
474. ASTM, F_[WK62669], https://www.astm.org/DATABASE.CART/WORKITEMS/WK62669.htm ("[D]efine[s] test methods for DAA systems and sensors applicable to smaller UAS BLVOS operations for the protection of manned aircraft in lower altitude airspace.").
475. ASTM, F_[WK73142], https://www.astm.org/DATABASE.CART/WORKITEMS/WK73142.htm.
476. For example, ASTM, Administrative Committee AC478—*BVLOS Strategy and Roadmapping for UAS*, www.astm.org. Email from Adam Morrison, Streamline Designs LLC and WG chair (Aug. 26, 2020) (on file with author) ("The working group has developed a technical report providing a 'modular and function-focused' strategy and roadmap for a standards-based path to system and operational approval for 'any conceivable' beyond line of sight (BLOS) mission. Functional aspects of a UTM system are further standardized as an outcome of this strategic effort."). *See* ASTM, F_ [WK69335], *New Guide for Framework for Using ASTM Standards for UAS*, https://www.astm.org/DATABASE.CART/WORKITEMS/WK69335.htm?A&utm_source=tracker&utm_campaign=20190801&utm_medium=email&utm_content=standards; Email from Jonathan Daniels, CEO, Praxis Aerospace Concepts Int'l, and Chair, ASTM F38.03 (July 20, 2020) (copy on file with author) ("The Framework serves as a tool to assist the ASTM community identify, understand, and marshal relevant F38 standards as a starting point for their practical use. For UTM, understanding UTM's discrete components and interactions as a function of the suite of applicable F38 standards cannot be overstated.").
477. Philip Kenul, Chair, ASTM F38, Presentation at the FAA UAS Symposium (July 9, 2020), and Email (Sept. 14, 2020) (on file with author).
478. Remarks at the FAA UAS Symposium (July 9, 2020) (further remarking, "we believe that cellular technology is essential to a scalable and economically viable operation.").
479. Email from Mark Davis, Ph.D., Pres., Crossbar Inc. (formerly VP, Intel) (Dec. 13, 2019) (on file with author).
480. Email from Mark Davis, Ph.D., *id.* (June 8, 2020) (on file with author) (Davis coined the term "converged tower" to describe moving various services to the edge on cellular towers—for "cellular networks as a fabric for deployment of aviation functions"); Mark Davis, *id.*, et al., *Aerial Cellular: What can Cellular do for UAVS with and without changes to present standards and regulations*, Presented at AUVSI Xponential, Chicago, Ill. (May 2, 2019), p. 1 ["Davis 2019"].

481. GSMA Intelligence, https://www.gsmaintelligence.com/.
482. GUTMA, *Meet Our Associate Members – Dr. Terrence Martin* (June 29, 2020), https://gutma.org/blog/2020/06/29/meet-our-members-dr-terrence-martin-nova-systems/ ["Terrence Martin Interview"].
483. Email from Barbara Pareglio, Exec. Dir. for Connectivity for Aviation and Drones, GSMA (Nov. 27, 2019) (on file with author). *See generally* GSMA, *THE 5G GUIDE, A Reference for Operators* (April 2019), https://www.gsma.com/wp-content/uploads/2019/04/The-5G-Guide_GSMA_2019_04_29_compressed.pdf ["5G GUIDE"].
484. Email from Barbara Pareglio, *id.* (underscoring that "5G will support a wide range of services ranging from low-power, sensor-driven smart parking to holographic conference calls"). *See generally*, JARUS, *White Paper: Use of Mobile Networks to Support UAS Operations*, JAR-DEL-WG5-D.05 (Jan. 21, 2021), http://jarus-rpas.org/.
485. Verizon, *The Eight Currencies*, https://www.verizonwireless.com/business/articles/business/5g-network-performance-attributes/.
486. *See* Ericcson, *Non-standalone and Standalone: two standards-based paths to 5G*, https://www.ericsson.com/en/blog/2019/7/standalone-and-non-standalone-5g-nr-two-5g-tracks.
487. Despite its many benefits, 5G faces unique challenges, such as in the vicinity of airports due to alleged interference with radio altimeters, safety critical systems in manned aircraft that could cause catastrophic failures. *See, e.g.*, FAA, *In the Matter of Expanding Flexible use of the 3.7 to 4.2 GHz Hand, Petition for Partial Reconsideration of the 3.7 GHz Band Report and Order,* GN Docket No. 18-122 (May 26, 2020), https://www.fcc.gov/document/fcc-expands-flexible-use-c-band-5g-0; Letter from AccuWeather, Inc., et al. to the Hon. Chuck Schumer, et al. (April 22, 2021), *available at* https://aea.net/pdf/LigadoOrderAnniversaryCoalitionLetterHill.pdf (Addressing the G5 interference risks resulting from the *Ligado Order*); the *Keep GPS Working Coalition*, https://www.keepgpsworking.com/; John Walker, Sr. Partner, The Padina Group and Chair, ISO/TC 20/SC 16, Remarks at the FAA UAS Symposium (July 9, 2020) (characterizing this issue, in part, as "hanging out there like a vulture on a limb").
488. *See, e.g.*, Ericsson, *Network Exposure and the case for connected drones*, https://www.ericsson.com/en/blog/2020/6/network-exposure-and-the-case-for-connected-drones (introducing network exposure, such as via exposure service-based architecture (SBA)).
489. *See, e.g.*, Robyn Mak, *Breakingviews – China's Huawei holds a 5G trump card*, REUTERS (July 31, 2020), https://www.reuters.com/article/us-huawei-tech-5g-security-breakingviews/breakingviews-chinas-huawei-holds-a-5g-trump-card-idUSKCN24S09Y.
490. Source: ITU and GSMA. *See* ITU-R, Recommendation ITU-R M.2083-0 (09/2015), *IMT Vision — Framework and overall objectives of the future development of IMT for 2020 and beyond*, https://www.itu.int/dms_pubrec/itu-r/rec/m/R-REC-M.2083-0-201509-I!!PDF-E.pdf ["ITU IMT 2020"]; [5G GUIDE], Sect. 1.2.1, p. 28, https://www.gsma.com/wp-content/uploads/2019/04/The-5G-Guide_GSMA_2019_04_29_compressed.pdf. *See generally* ITU, *ITU towards "IMT for 2020 and beyond,"* https://www.itu.int/en/ITU-R/study-groups/rsg5/rwp5d/imt-2020/Pages/default.aspx.
491. [Davis 2019], p. 24. *See* Fanni Lukácsy, Fmr. Sec'y Gen., GUTMA, *Connected Skies, Forum – GUTMA Annual Conference,* Portland, Or. (March 4, 2019), https://gutma.org/blog/2019/03/04/connected-skies-june-18-20-portland-oregon/. Other cellular-advancing technology associations are also expected to engage UTM. *See, e.g.*, the 5G Automotive Association (5GAA), https://5gaa.org/ ("a global, cross-industry organization of companies from the automotive, technology, and telecommunications industries (ICT), working together to develop end-to-end solutions for future mobility and transportation services."); CTIA, *Commercial Wireless Networks: The Essential Foundation of the Drone Industry* (Nov. 2019), https://www.ctia.org/news/white-paper-wireless-networks-are-key-to-unlocking-drones-potential.
492. *See infra* Chap. 3, Sect. 3.3.3 *GUTMA*; Email from Fanni Lukácsy, formerly Sec'y Gen., GUTMA (Aug. 2, 2019) (on file with author) ("GUTMA is in the perfect sit-

uation to pull aviation and telecommunication together to facilitate discussion and collaboration.").

493. 3GPP, *About 3GPP*, https://www.3gpp.org/about-3gpp. Additionally, the Alliance for Telecommunications Industry Solutions (ATIS), the North American organizational partner for 3GPP, has played a pioneering role in facilitating the cellular networks to support UAS and UTM. *See, e.g.*, ATIS, *Support for UAV Communications in 3GPP Cellular Standards*, ATIS-1-0000069 (Oct. 2018), https://access.atis.org/apps/group_public/download.php/42855/ATIS-I-0000069.pdf; ATIS, *Unmanned Aerial Vehicle (UAV) Utilization of Cellular Services, Enabling Scalable and Safe Operations,* ATIS-I-0000060 (2017), https://access.atis.org/apps/group_public/download.php/36134/ATIS-I-0000060.pdf. Note: the seven organizations are: ATIS, ARIB, CCSA, ETSI, TTA, TTC, and TSDSI.
494. *See* 3GPP, *Release 15* (April 26, 2019), https://www.3gpp.org/release-15; 3GPP, Technical Specification Group Radio Access Network, *Study on Enhanced LTE Support for Aerial Vehicles* (Rel. 15) 3GPP TR 36.777, v15.0.0 (Dec. 2017), www.3gpp.org/ftp/Specs/archive/36_series/36.777/36777-f00.zip. 5G should provide many features helpful to UTM such as reduced latency, dramatically higher peak and data volume rates (compared to 4G), and greater reliability. *See, e.g.*, Graham Trickery, Head of IoT, GSMA, Presentation at the GUTMA Annual Conference, Portland, Or. (June 20, 2019) (explaining, *inter alia*, a "network slicing concept" for service optimization amenable to UTM, including for enhanced monitoring and control support); 5G Slicing Association, *5G Network Slicing for Cross Industry Digitization: Position Paper* (2018), p. 20, https://cdn0.scrvt.com/fokus/b77ede33ea8d7dc8/1819ab58384e/5G-Network-Slicing-for-Cross-Industry-Digitization-Position-Paper–Digital.pdf (In addressing UTM, asserting "Network Slicing…currently developed in the context of 5G will undoubtedly enable better performance, safety and quality assurance to be realized for the new business activities of the future."); The 5G Slicing Association (5GSA), http://5gnsa.org/. *See generally* 3GPP, *UAS-UAV*, https://www.3gpp.org/uas-uav.
495. 3GPP, Technical Specification Group Services and System Aspects; *Study on Remote Identification of Unmanned Aerial Systems (UAS)*, Stage 1 (Rel. 16), 3GPP TR 22.8de V1.0.0 (May 2018); 3GPP, Technical Specification Group Services and System Aspects; *Unmanned Aerial System support in 3GPP,* Stage 1 (Rel. 16), 3GPP TS 22.125, v0.2.0 (Dec. 2018), http://www.3gpp.org/ftp//Specs/archive/22_series/22.125/22125-g20.zip.
496. 3GPP, *Release 17*, https://www.3gpp.org/release-17.
497. Email from Mark Davis, Ph.D., Pres. Crossbar, Inc. (formerly, VP, Intel) (Aug. 22, 2019) (on file with author).
498. Email from Barbara Pareglio, *supra*.
499. Email from Mark Davis, *supra*.
500. *See* [Davis 2019], p. 13.
501. Enhanced network knowledge and other competitive UTM capabilities are well-facilitated by mobile providers. *See, e.g.*, multi-access edge computing (MEC) that works at the edge of cellular networks for speed/efficiency of remote applications.
502. Nonetheless, the mobile industry may find the UTM space ripe for acquisitions rather than seeking to advance an independent UTM industry. *See, e.g.*, Larry Dignan, *Verizon acquires Skyward, aims to manage drone operations, connections*, ZDNet (Feb. 16, 2017), https://www.zdnet.com/article/verizon-acquires-skyward-aims-to-manage-drone-operations-connections/.
503. GSMA, *Using Mobile Networks to Coordinate Unmanned Aircraft Traffic* (2019), p. 2, https://www.gsma.com/iot/wp-content/uploads/2018/11/Mobile-Networks-enabling-UTM-v5NG.pdf ["GSMA 2019"].
504. *See generally* GSMA, www.gsma.com.
505. Email from Stephen Hayes, Dir. of N. Am. Standards, Ericsson, Inc. (Sept. 26, 2019) (on file with author).

506. GSMA, *GSMA IMEI Services,* https://www.gsma.com/services/gsma-imei/.
507. Email from Graham Trickey (Dec. 23, 2020) (on file with author). *See* GSMA, *Using Mobile Networks to Coordinate Unmanned Aircraft Traffic* (2019), https://www.gsma.com/iot/wp-content/uploads/2018/11/Mobile-Networks-enabling-UTM-v5NG.pdf; GSMA, *Accelerating the Commercial Drone Market using Cellular Technology* (Nov. 15, 2017), https://www.gsma.com/iot/wp-content/uploads/2017/11/Drones-Webinar-Slides-FINAL.pdf (asserting cellular connectivity is a "core enabler" for UTM, and correspondingly, key for efficient commercial BVLOS); The Drone Interest Group, www.gsma.com/drones.
508. [GSMA 2019], pp. 14–15, (asserting mobile services satisfy UAS connectivity needs, are scalable, highly reliable and available, cost-effective, easy to integrate, ready for immediate release, provide licensed spectrum, security, and avoid need for UTM-dedicated infrastructure).
509. *See infra* Chap. 3, Sect. 3.3.4 *GUTMA*.
510. Remarks of Eszter Kovacs, Acting Secretary General, *GUTMA – a time for new leadership and new aspirations,* Air Traffic Management (July 7, 2020), https://www.airinternational.com/air-traffic-management. *See* 3GPP, *Partners*, https://www.3gpp.org/about-3gpp/partners; GUTMA/GSMA, *Additional Annex 1 to Cooperation Agreement – Joint Activity #1* (Jan. 27, 2020) (formalizing the ACJA) (on file with author); https://gutma.org/acja/. *See also* GSMA, Internet of Things, *GSMA and GUTMA launch new joint initiative to align mobile and aviation industries* (June 10, 2020), https://www.gsma.com/iot/news/gsma-and-gutma-launch-new-joint-initiative-to-align-mobile-and-aviation-industries/.
511. *See* ACJA, *Technical Overview*, https://gutma.org/acja/technical-overview/, and ACJA, *Organization*, https://gutma.org/acja/organization/; ACJA, *LTE Aerial Profile* v1.00 (Nov. 2020), https://gutma.org/acja/wp-content/uploads/sites/10/2020/11/ACJA-WT3-LTE-Aerial-Profile_v1.00.pdf (identifying a minimum mandatory feature set of 3GPP specifications to guarantee interoperability over LTE); [MNO-UTM Interface].
512. (Sept. 2019), *available at* https://shop.cta.tech/collections/standards/products/small-unmanned-aerial-systems-serial-numbers (provides a 4-digit company ID plus additional 16 characters); www.cta.tech.
513. *See supra* Chap. 2, Sect. 2.4.8 *UAS Remote ID*; [ASTM Remote ID].
514. https://standards.cta.tech/apps/group_public/project/details.php?project_id=594.
515. https://standards.cta.tech/apps/group_public/project/details.php?project_id=625.
516. Email from Alexandra Blasgen, Mgr., Technology and Standards, Consumer Technology Assn. (Dec. 30, 2019) (on file with author).
517. *See* EUROCAE, *EUROCAE Technical Work Programme* (Public Ver. 2019), https://eurocae.net/media/1567/eurocae-public-twp-2019.pdf ["EUROCAE Programme 2019"], Sect. 4.7.2 UAS (describing its 6 relevant focus areas: 1. Detect and Avoid (DAA), 2. Command, Control and Communications, Spectrum and Security (C3&S), 3. **UAS Traffic Management (UTM),** 4. Design and Airworthiness (D&AW), 5. Enhanced RPAS Automation (ERA), and 6. Specific Operations Risk Assessment (SORA)) (emphasis added); [EUROCAE Programme 2019], Annex 1 (listing current activities/work product for these focus areas). It has also established a C-UAS working group (WG-115).
518. Email from Sergiu Marzac, Technical Programme Mgr., EUROCAE (Aug. 7, 2019) (on file with author) (also clarifying that "WG-73 and WG-93 were merged into WG-105 based on Council/TAC decision following the UAS workshop held in March 2016"). *See* Christian Schleifer-Heingärtner, Sec. Gen., EUROCAE, Presentation at the *Aviation Management for Unmanned Traffic Management (UTM)* AUVSI-webinar (April 1, 2021) (identifying three major standards "heavily focusing on UTM": ED-269 geo-fencing, ED-270 geo-caging, and ED-282 e-identification).
519. EUROCAE, WG-105 SG-13, *DAA for UAS operating in VLL* (June 2019), http://eurocae.net/news/posts/2019/june/eurocae-open-consultation-ed-267/. *See* Gilbert Amato, Sec'y Gen., EUROCAE, *EUROCAE WG-73 on Unmanned Aircraft Systems*, http://www.uasresearch.com/documents/yearbook/033_Contributing-Stakeholder_EUROCAE.pdf.

520. [EUROCAE Programme 2019], Sect. 4.7.2.1 (SORA work is generally undertaken in Focus Area 6). *See* EUROCAE, ED-269, *MOPS for UAS Geo-Fencing*, https://eshop.eurocae.net/eurocae-documents-and-reports/ed-269/.
521. www.icao.int.
522. *See* ICAO, *Remotely Piloted Aircraft Systems Panel (RPASP)*, https://www.icao.int/safety/UA/Pages/Remotely-Piloted-Aircraft-Systems-Panel-(RPASP).aspx (focused on international IFR operations).
523. ICAO first requested consultation on unmanned aerial vehicles in 2005, held an exploratory meeting in 2006 followed by an informal meeting in 2007 that concluded ICAO should address global interoperability and harmonization, among other actions. ICAO established a UAS Study Group (UASSG) in 2007, and its RPAS Panel in 2014.
524. [Chicago Convention]. *See id.*, Art. 8 ("Each contracting state undertakes to insure that the flight of such [pilotless] aircraft in regions open to civil aircraft shall be so controlled as to obviate danger to civil aircraft.")
525. ICAO, *Unmanned Aircraft Systems Advisory Group (UAS-AG)*, https://www.icao.int/safety/UA/Pages/Unmanned-Aircraft-Systems-Advisory-Group-(UAS-AG).aspx.
526. *Id.*
527. ICAO, *Unmanned Aircraft Systems Traffic Management (UTM) – A Common Framework with Core Principles for Global Harmonization* (April 3, 2019), https://www.icao.int/safety/UA/Pages/UTM-Guidance.aspx, and https://www.icao.int/safety/UA/Documents/UTM-Framework.en.alltext.pdf (However, the Framework neither addresses UTM–ATM transitions, UA design and certification standards, nor UTM in high-altitude airspace, each of which may be addressed in the future).
528. ICAO, *Unmanned Aircraft Systems Traffic Management (UTM) – A Common Framework with Core Principles for Global Harmonization,* Ed. 2 (Nov. 2019), https://www.icao.int/safety/UA/Documents/UTM-Framework%20Edition%202.pdf; [ICAO UTM Framework Ed. 3]. *See* ICAO, *ICAO UTM Framework* (Nov. 17, 2020), https://www.icao.int/NACC/Documents/Meetings/2020/UAS/UASWeb-P04EN.pdf#search=utm%20toolkit (An Edition 4 is in "Draft Stage" and expected to include: performance requirements, UTM system certification requirements, and UTM integration into aerodrome environments/activities.).
529. *See, e.g.,* ICAO DRONE ENABLE Symposium 2021, https://www.icao.int/Meetings/DRONEENABLE4/Pages/default.aspx; ICAO, DRONE ENABLE/3, in Montreal, CA (Nov. 12-14, 2019), https://www.icao.int/Meetings/DRONEENABLE3/Pages/default.aspx; *See also* ICAO, *Unmanned Aircraft System Traffic Management (UTM), Request for Information* (2020), https://www.icao.int/safety/UA/Documents/RFI%20-%20 2020.pdf.
530. Jay Merkle, Exec. Dir., UAS Integration Office, FAA, Keynote Presentation at the FAA UAS Symposium (July 8, 2020). *See* Dan Dalton, VP, Global Partnerships, Wisk, Remarks at ICAO DRONE ENABLE Symposium 2021 (April 21, 2021) (Urging ICAO's role as an "honest broker" to advance UTM).
531. *See* ICAO, *Thirteenth Air Navigation Conference (AN-CONF/13) Outcome*, Agenda Item 2: Global Development in Aviation, MIDANPIRG/17 and RASG-MID/7 Meeting (Cairo, Egypt, 15 - 18 April 2019), WP/4 (April 4, 2019), C-WP/14837App. A-36, https://www.icao.int/MID/MIDANPIRG/Documents/MID17%20and%20RASG7/WP4.pdf: to "incorporate a flexible framework for emerging air navigation concepts, such as… UAS traffic management (UTM)," Recommendation 1.2/1—Global technical level of the Sixth Edition of the Global Air Navigation Plan (Doc 9750, GANP f, p. A-5). Among the ICAO annexes affected by UTM, and under consideration for revision:
    - Annex 2 to the Convention on Int'l Civil Aviation – Rules of the Air, Q1 2018
    - Annex 6 to the Convention on Int'l Civil Aviation – Part IV – International Operations – RPAS, Q1 2020
    - Annex 8 to the Convention on Int'l Civil Aviation – Airworthiness of Aircraft, Q1 2018

- Annex 11 to the Convention on Int'l Civil Aviation – Air Traffic Services, Q1 2020
- Procedures for Air Navigation Services – Air Traffic Management (Doc 4444), Q1 2021
- Procedures for Air Navigation Services – Aircraft Operations – Vol I – Flight Procedures (Doc 8168), Q1 2021

532. *Cf.* ICAO, *ICAO Unmanned Aircraft System Advisory Group (UAS-AG) Update*, APUAS/TF/3-IP/02, 04, Bangkok (March 7, 2019), Sect. 2.12, https://www.icao.int/APAC/Meetings/2019%20APUASTF3/IP02%20ICAO%20Unmanned%20Aircraft%20System%20Advisory%20Group%20Update.pdf#search=UAS%20Advisory%20Group ("[p]rovisions to support RPAS are necessary in *all* ICAO Annexes") (emphasis added); ICAO, Resolutions, Assembly – 40th Session, Montreal (Sept. 24-Oct. 4, 2019), Provisional Ed. Oct. 2019, Assembly Resolution A40-7: New entrants, p. 27, https://www.icao.int/Meetings/a40/Documents/Resolutions/a40_res_prov_en.pdf ("1. *Directs* ICAO to review Standards and Recommended Practices (SARPs) relating to, inter alia, the rules of the air, air traffic services, certification, licencing, liability and the environment, for amendment or expansion as necessary, to facilitate the operation of New Entrants within a global, harmonized framework, taking into account regional frameworks and practices"); Leslie Cary, Chief, Remotely Piloted Aircraft Systems (RPAS) Section, Presentation at ICAO, DRONE ENABLE/3, Montreal, CA (Nov. 23, 2019) (announcing that ICAO, responsively, "will revise 18 of the 19 ICAO annexes").
533. *See, e.g.*, ICAO, *Proposal Regarding ICAO Governance Structure for Cybersecurity*, Presented by the International Coordinating Council of Aerospace Industries Associations (ICCAIA), A40-WP/2191 EX/82 (Aug. 5, 2019), https://www.icao.int/Meetings/a40/Documents/WP/wp_219_en.pdf (urging "need to establish an ICAO entity, governed by member States with the support from industry, which is not constrained by the existing ICAO organizational structure [and] responsible for a common ICAO strategy for cybersecurity"). [FAA ConOps UTM v2.0], Sect. 2.7.2.1 FAA ("The FAA leverages the GRAIN and the IATF policies to ensure the integrity and authenticity of the information received from all UTM stakeholders."). Additionally, cyber issues are being considered by the ICAO Working Group on Current and Future Air Navigation Systems. ICAO's cyber initiatives may also contribute to revisions to ICAO Annex 17 Security.
534. ICAO, 40th Triennial Assembly, https://www.icao.int/Meetings/a40/pages/default.aspx (ICAO's sovereign body). *See* William Voss, Special Advisor to ICAO, Remarks at the GUTMA Annual Conference, Portland, Or. (June 19, 2019) ("Doing so will put a marker down – a classic *UN get your foot in the door* strategy").
535. *See* ICAO, *Global Air Navigation Plan*, https://www.icao.int/airnavigation/Pages/GANP-Resources.aspx.
536. The RPAS Panel mandate addresses International IFR operations, possibly ruling out UTM activity depending on the mission at hand.
537. *See* ICAO, *ANC Technical Panels*, https://www.icao.int/about-icao/AirNavigationCommission/Pages/anc-technical-panels.aspx (for example, the Air Traffic Management Operations Panel, ATM Requirements & Performance Panel, and Information Management Panel).
538. IEEE, *IEEE Standards*, https://www.ieee.org/standards/index.html.
539. IEEE, P1939.1 - *Standard for a Framework for Structuring Low Altitude Airspace for Unmanned Aerial Vehicle (UAV) Operations*, https://standards.ieee.org/project/1939_1.html.
540. Email from Dr. Xiaohan Liao, Dep'y Dir. Gen. of the Institute of Geographic Sciences and Natural Resources Research, Dir. Gen. of the Research Center for UAV Applications and Regulation, Chinese Academy of Sciences, Chair of IEEE SA P1939.1 Working Group (July 24, 2019) (on file with author).
541. Email from Dr. Xiaohan Liao, *id.*
542. IEEE, P1920.1 - *Aerial Communications and Networking Standards*, https://standards.ieee.org/project/1920_1.html, https://development.standards.ieee.org/get-file/P1920.2.pdf?t=100264800003, and https://standards.ieee.org/project/1920_2.html

(referencing "5. The UAS Traffic Management (UTM) project led by NASA also emphasizes the need for V2V communications for drones."). *See* IEEE, 1609.2-2016 - *Standard for Wireless Access in Vehicular Environments—Security Services for Applications and Management Messages*, https://standards.ieee.org/standard/1609_2-2016.html.

543. IEEE, P2821 - *Guide for Unmanned Aerial Vehicle-based Patrol Inspection System for Transmission Lines*, https://standards.ieee.org/project/2821.html.
544. https://standards.ieee.org/standard/2450-2019.html, and http://ieee-ims.org/content/tc-41-traffic-enforcement-technologies/.
545. IETF, *About, Mission and principles*, https://www.ietf.org/about/mission/.
546. Stewart Card, et al., *UAS Remote ID, draft-card-tmrid-uas-00*, IETF (Nov. 4, 2019), https://tools.ietf.org/html/draft-card-tmrid-uas-00#page-9.
547. IETF, *Drone Remote ID Protocol (drip)* (Feb. 11, 2020), https://datatracker.ietf.org/group/drip/documents/ (presenting the WG charter).
548. *See, e.g.*, IEFT, *Trustworthy Multipurpose Remote ID*, https://datatracker.ietf.org/doc/charter-ietf-tmrid/; IETF, *Crowd Sourced Remote ID*, https://datatracker.ietf.org/doc/draft-moskowitz-drip-crowd-sourced-rid/.
549. ISO, ISO/TC 20/SC 16, *Unmanned aircraft systems*, https://www.iso.org/committee/5336224.html.
550. Dr. Masahide Okamoto, ISO, Presentation at the GUTMA Annual Conference, Portland, Or. (June 19, 2019). *See* Email from John Scull Walker, Sr. Partner, The Padina Group & Chair, ISO/TC 20/SC 16 (Aug. 25, 2019) (on file with author) (noting that "ISO/TC 20/SC 16/WG 4 (UTM) held its first formal meeting as a [WG] in Nov. 2018 [and] continues to develop plans with the engagement of the [UTM] industry; and, development of comprehensive safety and quality standards for UTM systems worldwide.").
551. Other related planned work includes ISO/PWI 23629-8, *UAS Traffic Management (UTM) — Part 8: Remote identification;* ISO/NP 23629-12.
552. ISO, https://www.iso.org/standard/70853.html.
553. ISO, https://www.iso.org/standard/76453.html?browse=tc.
554. ISO, https://www.iso.org/standard/78961.html.
555. ISO, https://www.iso.org/standard/76973.html?browse=tc.
556. ISO, https://www.iso.org/standard/78962.html.
557. ISO, https://www.iso.org/committee/45144.html.
558. ISO, https://www.iso.org/committee/45306.html, ISO/IEC 27007:2020 *Information security, cybersecurity and privacy protection – Guidelines for information security management systems auditing*, https://www.iso.org/standard/77802.html?browse=tc.
559. Email from John Scull Walker, Sr. Partner, The Padina Group & Chair, ISO/TC 20/SC 16 (Aug. 25, 2019) (on file with author).
560. ISO 21384-3:2019 *Unmanned aircraft systems* — Part 3: *Operational procedures*, https://www.iso.org/obp/ui/#iso:std:iso:21384:-3:ed-1:v1:en. Related draft ISO standards in development include ISO 21384-2, *Unmanned aircraft systems – Part 2: Product systems*, ISO 21384-4, *Unmanned aircraft systems – Part 4: Vocabulary*, and ISO 23665, *Unmanned aircraft systems – Training for personnel involved in UAS operations.*
561. *See generally* ITU, https://www.itu.int/en/ITU-R/conferences/wrc/Pages/default.aspx (most recently held in 2019. Among other issues, additional unique and shared aeronautical radio navigation spectrum allocation for UTM C2 communications is under consideration).
562. [ITU IMT 2020], Sect. 2.5. *See* Email from Craig Bloch-Hansen, Program Mgr. - UAS Design Standards, Transport Canada (June 8, 2019) ("Spectrum is interesting because while it is a finite 'natural resource' there is an 'equal distribution' of the resource amongst nations."). *See generally* Drone Alliance Europe, *Drones, UTM and Spectrum–A review* (June 8, 2016), http://dronealliance.eu/wp-content/uploads/2016/06/Spectrum-Allocation-White-Paper-Drone-Alliance-Europe-fin.pdf.

563. JARUS, http://jarus-rpas.org/. *See* Mike Lissone, fmr. Sec'y Gen., JARUS, Presentation at ICAO, DRONE ENABLE/2, in Chengdu, P.R.C. (Sept. 13-14, 2018), https://www.icao.int/Meetings/RPAS3/Documents/Presentations/1.2.1%20Mr.%20Mike%20Lissone,%20Secretary%20General,%20JARUS.pdf (providing an overview of JARUS work).
564. *See* Mike Lissone, *supra*, and JARUS, *Terms of Reference*, Ref. JARUS-ToR_v7.0.2020, Sect. 3.6, http://jarus-rpas.org/sites/jarus-rpas.org/files/imce/attachments/jarus_tor_v7.0.2020_and_annex_scb_tor_130818_released_20200304.pdf, presenting the WGs:
    WG 1: FLC – licensing and competencies
    WG 2: OPS – operations
    WG 3: Airworthiness
    WG 4: Detect and Avoid
    WG 5: Command Control and Communications
    WG 6: Safety and Risk Management
    WG 7: Concept of Operations (CONOPS)
565. *See generally* JARUS, *Publications*, http://jarus-rpas.org/publications; JARUS, Press Release, *New Chair, New Opportunities for JARUS in the New Era*, No. 2021/03 (March 23, 2021) (election of Giovanni Di Antonio as the new Chair), http://jarus-rpas.org/sites/jarus-rpas.org/files/jarus_chair_election_news_release_1.pdf.
566. At present JARUS work is organizised withing three permenent working groups, namely: (i) WG Operations, Pesonnel and Organizations (former WG 2), (ii) WG Airworthiness (former WG 3), (iii) WG Safety Risk Management (former WG 6), and a 1-year term working group for the development of the Autonomous Flight Concept of Operations, the AutoConOps WG (former WG 7).
567. JARUS, JAR doc 06 SORA (package) (Mar. 6, 2019), http://jarus-rpas.org/content/jar-doc-06-sora-package.
568. *See, e.g.*, FAA, Order 8040.6, Subj. *Unmanned Aircraft Systems Safety Risk Management Policy* (Oct. 4, 2019) https://www.faa.gov/regulations_policies/orders_notices/index.cfm/go/document.information/documentID/1036752; EU Specific Category, Art. 11 of the Regulation (EU) 2019/947 (Implementing Regulation) and related EASA AMC1 to Art. 11, https://www.easa.europa.eu/document-library/easy-access-rules/easy-access-rules-unmanned-aircraft-systems-regulation-eu; Transport Canada, *draft AC 903-001-X*, Subj. Remotely Piloted Aircraft Systems Operational Risk Assessment, https://www.unmannedsystems.ca/wp-content/uploads/2019/07/DRAFT-AC-903-001-RPAS-Operational-Risk-Assessment.pdf.
569. JARUS, JAR-DEL-WG6-D.04, Ed. 2.0 (Jan. 30, 2019), *ANNEX H –Unmanned Traffic Management (UTM) implications for SORA*, http://jarus-rpas.org/sites/jarus-rpas.org/files/jar_doc_06_jarus_sora_v2.0.pdf. This SORA annex remained unpublished at the time of publication of this book.
570. Email from Andy Thurling, CTO, NUAIR Alliance, and JARUS cyber group chair (Oct. 23, 2019 and Sept. 27, 2020) (on file with author) ("The cyber group is applying the SORA risk-based approach to establish appropriate cyber safety requirements for UAS operations in the 'Specific' category.").
571. Andy Thurling, CTO, NUAIR Alliance, Remarks at ICAO DRONE ENABLE Symposium 2021 (April 21, 2021) (recognizing challenges of securing the "middle part"—medium risk operations: "that's the difficult part," and underscoring need for proportionality with "a spectrum of mitigations"); and *id*. Las Vegas, Nev. (Sept. 5, 2019) (describing Annex E); JARUS, JARUS guidelines on SORA, Annex E, *Integrity and assurance levels for the Operational Safety Objectives (OSO)*, Doc ID. JAR-DEL-WG6-D-0.4, Ed. No. 1, (Jan 25, 2019), http://jarus-rpas.org/sites/jarus-rpas.org/files/jar_doc_06_jarus_sora_annex_e_v1.0_.pdf.
572. *See* NATO, *NATO Standardization Document Database*, https://nso.nato.int/nso/nsdd/listpromulg.html; Email from Simone de Manso, Press Officer, NATA (Sept. 15, 2020) (on file with author). The NATO Joint Capability Group Unmanned Aircraft Systems (JCGUAS) is a focal point for unmanned issues and programs.

573. Al Secen, VP, Avi. Tech. and Stds., RTCA, Presentation at the FAA UAS Symposium, Balt., Md. (June 5, 2019).
574. *See, e.g.,* [ANSI Roadmap], Sect. 7.1 ("RTCA: There is no activity" regarding UTM); Email from Al Secen, VP, Avi. Tech. and Stds., RTCA (July 25, 2019) (on file with author) (confirming ANSI Roadmap assertion). RTCA's role in UAS is frequently characterized as being for "heavy metal," and certified aircraft operating in higher altitudes in contrast to UTM for sUAS in VLL airspace.
575. Paul McDuffee, Exec./Ops. Analyst at Hyundai Urban Air Mobility, and Co-Chair, RTCA SC 228, Remarks (July 30, 2019) (further remarking that "there is still debate on what UTM is.").
576. RTCA, SC-228, https://www.rtca.org/content/sc-228; RTCA, ToR, SC 228, *Minimum Performance Standards for Unmanned Aircraft Systems*, RTCA Paper No. 099-18/PMC-1742 (March 22, 2018).
577. Both its Phase 2 *Terminal Area DAA Well Clear*, and *Non-Cooperative DAA Well Clear* have informed DAA development for UTM, including in ASTM F38. The Committees' membership shares many noted experts.
578. RTCA, SC-147, https://www.rtca.org/content/sc-147 (Its work has expanded into ACAS Xu, aircraft collision—working on a MOPS for ACAS sXu; has approved a MOPS for ACAS Xu for larger UAS, and is exploring another phase of work focusing on sUAS and AAM).
579. RTCA, DO-200B, https://my.rtca.org/NC__Product?id=a1B36000001IclyEAC. *See also* RTCA, DO-201 *User Requirements for Navigation Data*, https://my.rtca.org/NC__Product?id=a1B1R0000082opBUAQ.
580. *See infra* Chap. 4, Sect. 2 *USS Qualification*.
581. *See* www.rtca.org; FAA, *Advisory Circular AC 20-115D*, Subj: Airborne Software Development Assurance Using EUROCAE ED-12( ) and RTCA DO-178( ) (July 21, 2017), https://www.faa.gov/regulations_policies/advisory_circulars/index.cfm/go/document.information/documentID/1032046.
582. www.rtca.org.
583. *See* Paul McDuffee, Exec./Ops. Analyst at Hyundai Urban Air Mobility, and Co-Chair, RTCA SC-228, Presentation at GUTMA Annual Conference, Portland, Or. (June 2019); RTCA, Paper No. 163-20/PMC-2034, *Terms of Reference, SC-228, Minimum Performance Standards for Unmanned Aircraft Systems* (June 11, 2020), https://www.rtca.org/wp-content/uploads/2020/08/sc-228_tor_rev_10_approved_06-11-2020.pdf ["RTCA ToR SC-228"] (DO-368 "Contains Phase 1 Minimum Operational Performance Standards (MOPS) for DAA systems used in aircraft transitioning to and from Class A or special use airspace (higher than 500' Above Ground Level (AGL), traversing Class D, E, and G airspace in the NAS.").
584. *See* [RTCA ToR SC-228].
585. *See* [RTCA ToR SC-228] (ACAS X_u with completed MOPS will provide vertical and horizontal advisories; and interoperability with TCAS).
586. RTCA, DO-377, https://my.rtca.org/NC__Product?id=a1B1R000008bv4UUAQ.
587. RTCA, Change 1 (Sept. 16, 2019), https://my.rtca.org/NC__Product?id=a1B1R000009XgFoUAK.
588. RTCA, (Dec. 12, 2020), https://my.rtca.org/NC__Product?id=a1B1R00000LoYFjUAN.
589. Email from Al Secen, VP, AVi Tech. and Stds., RTCA (Aug. 26, 2020) (on file with author).
590. Al Secen, *id. See* Philip M. Kenul, Chairman, ASTM F38 Unmanned Aircraft Systems, Remarks (Aug. 19, 2019) ("Both RTCA and ASTM agree that in the interest of limited resources each SDO will not duplicate effort but will coordinate on those areas where there is overlap-transiltion from low altitude low risk to high altitude high risk.").
591. https://www.sae.org/; Email from Dorothy Lloyd and Mark P. DeAngelo, SAE Int'l (Aug. 5, 2020) (on file with author). *See infra* note 1146 (presenting SAE's levels of driving automation).

592. *See* [EASA Opinion 01/2018]; *See generally* EASA, *Civil drones (Unmanned aircraft)*, https://www.easa.europa.eu/easa-and-you/civil-drones-rpas; and EASA, *Drones - regulatory framework background* (webpage), https://www.easa.europa.eu/easa-and-you/civil-drones-rpas/drones-regulatory-framework-background. *See generally* EASA, *Opinions Information*, https://www.easa.europa.eu/document-library/opinions/information.
593. EASA, *SESAR launches call to establish U-space demonstrators across Europe* (Feb. 2, 2018), https://www.easa.europa.eu/newsroom-and-events/news/sesar-launches-call-establish-u-space-demonstrators-across-europe. *See infra* Chap. 5, Sect. 5.3 *U-space*.
594. ESCP, *The European Strategic Coordination Platform for Cybersecurity in Aviation, Charter,* Ver. 2.0 (Feb. 2019), https://www.easa.europa.eu/sites/default/files/dfu/ESCP%20Charter%20V2.0%20February%202019.pdf.
595. [EASA Reg. Framework U-space], Sect. 2.4. Main approach for the development of the U-space regulatory framework.
596. UAS standards development falls under AUS (the UAS Integration Office), and there is heavy involvement from AIR (Aircraft Certification Service) and AFS (Flight Standards Service).
597. Email from Jon Standley, Business Development Lead, Aviation Systems / L3Harris Technologies (Aug. 18, 2020) (on file with author). *See* Frank Wolfe, *MITRE Aviation Lab Looking At What National Airspace System Will Look Like in 2035,* AVIATION TODAY (Dec. 11, 2019), https://www.aviationtoday.com/2019/12/11/mitre-aviation-lab-looking-national-airspace-system-will-look-like-2035/; *The Future of Aerospace: Interconnected from Surface to Space*, MITRE (Jan. 15, 2020), https://www.mitre.org/news/in-the-news/the-future-of-aerospace-interconnected-from-surface-to-space.
598. *See generally* FAA, *What is NextGen*, https://www.faa.gov/nextgen/what_is_nextgen/; [FAA Roadmap]; [RTT 2017], Sect. 6, p. 13, NextGen, NASA, and Industry Evaluation and Demonstration Partnership (stating that research activities "are supported by the full range of NASA and FAA research and development capabilities"); *Statement of the Hon. Elaine L. Chao, Sec'y of Transportation before the Appropriations Subcommittee on Transportation, Housing, and Urban Development and Related Agencies, US Senate* (March 27, 2019), https://www.transportation.gov/testimony/appropriations-subcommittee-transportation-housing-and-urban-development-and-related (budget proposal for new Office of Innovation and $203 million for unmanned aircraft systems integration into the NAS) *See infra* Chap. 3, Sect. 3.4 *Further Standards Coordination* (NextGen and SESAR coordination).
599. Global UTM Association (website), https://gutma.org/.
600. Email from Eszter Kovacs, fmr. Acting Sec'y Gen, GUTMA (Sept. 2, 2020) (on file with author) ("With a clear mandate to further engage membership, understand member needs, and propose relevant initiatives and activities. GUTMA is currently expanding its portfolio of partners and is emerging as a credible facilitator for global UTM/ATM harmonization.").
601. *See* GUTMA, *Annual Report 2019/20* (April 2020), https://gutma.org/annual-reports/; *Global UTM Association*, GITHUB, https://github.com/gutma-org; GitHub, *Drone Registry Brokerage*, https://github.com/openskies-sh/aircraftregistry-broker/blob/master/documents/Registration-Brokerage-Specification.pdf (white paper). *See supra* Chap. 2, Sect. 2.3.2 *Registration*. GUTMA is also contributing to open source technical solutions and addressing harmonized concepts and standards for remote ID, geo-awareness, inter-USS communications, and "open FIMS."
602. *See, e.g.*, Global UTM Ass'n, *GUTMA and ASTM International Sign MOU on Drone Standard* (Nov. 27, 2019), https://gutma.org/blog/2019/11/27/gutma-and-astm-international-sign-mou-on-drone-standards/; ASTM, *ASTM International Signs Drone Standard MOU with GUTMA,* ASTM Standardization News (Jan./Feb. 2020), www.astm.org/sn.

603. https://gutma.org/blog/2019/08/15/map-of-international-utm-implementations-call-for-contribution/.
604. Email from Reinaldo Negron, Head of UTM, Wing and Co-President, GUTMA (Dec. 28, 2019) (on file with author).
605. NASA Technical Standards Program, https://www.nasa.gov/offices/oce/functions/standards/index.html.
606. Email from Jeffrey R. Homola, UTM Integration & Testing Lead, NASA Ames (Jan. 12, 2020) (on file with author) (noting "that they likely fed into standards discussions elsewhere with some level of influence."). *See supra* Chap. 3, Sect. 3.2.11 *JARUS.*
607. SESAR, the Single European Sky Air Traffic Management (ATM) Research project established per COUNCIL *REGULATION (EC) No 219/2007 of 27 Feb. 2007 on the establishment of a Joint Undertaking to develop the new generation European air traffic management system (SESAR)*, https://eur-lex.europa.eu/legal-content/EN/TXT/PDF/?uri=CELEX:32007R0219&from=EN.
608. SESAR JU, *Discover SESAR,* https://www.sesarju.eu/discover-sesar. *See* SESAR, U-space, *Supporting Safe and Secure Drone Operations in Europe, Consolidated Report on SESAR U-Space Research and Innovation Results* (Nov. 11, 2020), https://www.sesarju.eu/node/3691.
609. Presentation at the FAA UAS Symposium (July 8, 2020).
610. *See, e.g.,* Lorenzo Murzilli, *supra* (emphasizing that UTM and Swiss U-Space are "very much aligned as demonstrated by Swiss U-Space joining ASTM F38"). *See generally* Chap. 3, Sect. 3.1.1 *ANSI.*
611. NextGen – SESAR, *State of Harmonisation*, 3rd ed., Report prepared by the Coordination Committee (CCOM) & Deployment Coordination Committee (DCOM) for the US–EU MoC Annex 1 Executive Committee (EXCOM) (Sept. 2018), Sect. A5, https://www.faa.gov/nextgen/media/NextGen-SESAR_State_of_Harmonisation.pdf, https://op.europa.eu/en/publication-detail/-/publication/355688ca-2832-11e9-8d04-01aa75ed71a1. *See supra* Chap. 3, Sects. 3.3.2 *FAA NextGen*, and 3.3.5 *SESAR JU.*
612. [SESAR ATM Master Plan 2017], Fig. 4, p. 16 (Some items on this list have advanced, but its over-all content remains instructive.).
613. [ICAO UTM Framework Ed. 3], p. 14, and other requirements presented therein. *See* ASTM, WK75923, *New Specification for Positioning Assurance, Navigation, and Time Synchronization for Unmanned Aircraft Systems*, https://www.astm.org/DATABASE.CART/WORKITEMS/WK75923.htm?A&utm_source=tracker&utm_campaign=20210218&utm_medium=email&utm_content=standards.
614. *See* EUROCONTROL and EASA, *UAS ATM Flight Rules, Discussion Doc.,* Ed. 1.1, Sect. 5 (Nov. 27, 2018), *available at* https://www.eurocontrol.int/sites/default/files/publication/files/uas-atm-integration-operational-concept-v1.0-release%2020181128.pdf.
615. *See infra* Chap. 6, Sect. 6.6 *Stratospheric Operations.*
616. *See infra* Chap. 6, Sect. 6.5 *Advanced Air Mobility / Urban Air mobility.*
617. Other organizations contributing to standards coordination include, for example, AW Drones, https://www.aw-drones.eu/ (funded by the European Union's Horizon 2020 Research and Innovation Programme); and the 5GAA [5G Automotive Ass'n], https://5gaa.org/about-5gaa/vision-mission/ (in particular, "covering vehicle-to-vehicle (V2V) communication, vehicle-to-infrastructure (V2I) communication, vehicle-to-network (V2N)" communications). *See generally* Chap. 6, Sect. 6.3 *International Coordination.*
618. John Scull Walker, Sr. Partner, The Padina Group & Chair, ISO/TC 20/SC 16, Remarks at ICAO DRONE ENABLE Symposium 2021 (April 14, 2021), and Presentation at the FAA UAS Symposium (May 16, 2019).
619. Presentation at the FAA UAS Symposium, Balt., Md. (June 3, 2019).
620. UTM governance describes the assessment, approval, and oversight of the UTM ecosystem based on defined criteria such as applicable law, policy, standards, and agreements. It provides assurances that UTM serves its intended purposes.

621. For example, DSS architecture was fashioned to enforce a privacy policy. *See supra* Chap. 2, Sect. 2.4.5.2 *Discovery and Synchronization Service*, *infra* Chap. 4, Sect. 4.7.6 *Data Rights and Privacy Management*.
622. *See, e.g.*, Lawrence Lessig, *Code is Law, On Liberty in Cyber*space, Harvard Mag. (Jan. 1, 2000), https://www.harvardmagazine.com/2000/01/code-is-law-html (Code/APIs influence on governance - "This regulator is code—the software and hardware that make cyberspace as it is. This code, or architecture, sets the terms on which life in cyberspace is experienced.... In a host of ways that one cannot begin to see unless one begins to understand the nature of this code, the code of cyberspace regulates.").
623. *See supra* Chap. 2, Sect. 2.3 *Scope of Services*.
624. *See, e.g.*, Chap. 4, Sect. 4.6 *State, Regional, and Local* (e.g., identifying challenges in determining authority over airspace); Chap. 4, Sect. 4.7.4 *Liability*; Chap. 2, Sect. 2.2 *Participants*.
625. *See supra* Chap. 2, Sect. 2.4.7 *Cybersecurity*.
626. *See infra* Chap. 4, Sect. 4.7.3 *Business Models and Competition*.
627. *See infra* Chap. 4, Sect. 4.4.4 *Fair Access*.
628. *See supra* Chap. 2, Sect. 2.4.2 *Centralized vs. Federated Architecture*.
629. Founder and CEO, Murzilli Consulting, formerly Leader, Innovation and Digitization Unit, Swiss Federal Office of Civil Aviation (FOCA), Remarks, in Chicago, Ill. (May 1, 2019).
630. *See, e.g., infra* Chap. 4, Sect. 4.7.2 *Industry Consortia*, and Chap. 5, Sect. 5.3 *U-space*.
631. *See* Walter Schwenk, et al., *Aspects of International Co-operation in Air Traffic Management*, Martinus Nijhoff Publishers (1998), p. 3 (Offering a traditional view on delegation: "The functions of ATM may be delegated to private institutions, organized under private law, whereas the responsibility for the services remains with the State and is subject to public law. . . . In cases where ATM is of less importance to public safety, certain control functions may be let to private persons or institutions."); Gur Kimchi, GOEXA, *Progress on Remote ID* (Jan. 5, 2021), https://www.linkedin.com/pulse/progress-remote-id-gur-kimchi/?trackingId=eYM9LhwlSheEzG8z%2BLimMg%3D%3D (recounting output from Feb. 27th 2018 *DAC tenets working group* interim meetings: "*Industry* (e.g., operators, manufacturers, users) shall bear the responsibility for developing, operating and supporting the equipage and infrastructure (e.g., DAA capabilities, UTM, Cloud Services, etc.) required for UAS operations, ensuring compliance with standards, regulations and operating requirements. *Regulators* shall maintain authority over the safety, rule-making, standards-selection, ID/tracking, oversight, and compliance enforcement."). *See also* Ric Peri, VP, Gov't and Industry Affairs, Aircraft Electronics Ass'n, Remarks (Aug. 15, 2019) ("The major lesson learned was that safety standards should not be delegated. Where safety has been delegated, any deficiency in the safety standard doesn't show up until after an event.").

Email from Craig Bloch-Hansen, Prog. Mgr. - UAS Design Standards, Transport Canada (Dec. 19, 2019) (on file with author) ("The roles of the CAA... are not diminished in the UTM environment. A national CAA is ultimately responsible for the safety and security of flight operations both within their national boundaries, and, for signatories to the Chicago Convention, in international airspace as defined by ICAO SARPs. The ANSP is responsible for managing the operational safety of the airspace through the provision and operation of services to support the integration of aircraft into national and international airspace operations. Both of these entities form key pillars to the development and deployment of UTM: the CAA provides the safety requirements for UAS products, procedures, and personnel operating within UTM designated airspace, while the ANSP provides for service provision, flight dispatch, and coordination within the operational airspace." Further urging "[c]lear governance structures including well documented and legally enforceable roles, responsibilities, and reporting expectations will help to protect both equitable public interest, as well as private economic interest."); [Drone Alliance Europe, U-Space Whitepaper 2019], p. 6 (recognizing as most important roles to be retained by competent authorities: "provision of authoritative data such as registration data," setting the rules of operation and defining the USP approval process," and "setting EU-wide approval criteria." *Cf.* Email from Benoit Curdy, Digital Transformation, Swiss FOCA (Nov. 12, 2019) (on file with author) (Regarding *authorization*, "[w]e would not consider that we delegate anything to USS. Delegation only happens when an authority's duty is executed by a third party on behalf of the authority (not as a vendor). For instance, regulation requires FOCA to run an operator registry. We could decide to delegate it to skyguide, which would mean that skyguide is then fully responsible of fulfilling this duty. FOCA would be required to oversee its operation and to set its cost.").

632. *See, e.g.*, James Foltz, PE, Mgr., Strategic Policy Emerging Aircraft, Aircraft Certification Services, Remarks at the FAA UAS Symposium, Balt., Md. (June 4-5, 2019) ("UAS shattered certification constructs.").
633. *See, e.g.,* Chap. 2, Sect. 2.4.4 *System-Wide Information Management* (addressing SWIM governance and UTM). *See also* Stephen P. Creamer, Dir., Air Navigation Bureau, ICAO, Remarks at ICAO DRONE ENABLE 2021 (April 21, 2021) (For UTM, querying "Do we certify the cloud or the service?").
634. Presentation at ICAO, DRONE ENABLE/3, Montreal, CA (Nov. 13, 2019). *See* Chap. 5, Sect.5.3 *U-space.*
635. Remarks at the FAA UAS Symposium (July 8, 2020) (further remarking, "we're not calling it certification").
636. *See, e.g.,* [FAA ConOps UTM v2.0], Sect. 2.1 Overview ("Some services provided by USSs require *qualification* by the government...") (emphasis added); FAA, *USS Services Categorization*, UAS Traffic Management (UTM) Newsletter (Feb. 2019) (describing categorization of USS services, including, "Category 1 – Government-*Qualified* Services... Services to UAS operators that may be provided directly by the government or by an FAA-*qualified* third party that has been *qualified* to perform the service.") (emphasis added).
    Almost any term adopted for this purpose may be considered "loaded"—that is, having unintended meaning or association with unrelated procedures. Euphemistically, in practice, the only (informally) observed noncontroversial alternative to "qualification" was "blessing." Note, the operative term describing approvals by governance regimes vary. *See, e.g.*, [UAS Service Supplier Checkout], p. 18 ("register"); *infra* Chap. 5, Sect. 5.2 *Low Altitude Authorization and Notification Capability* ("accepted"); *infra* Chap. 5, Sect. 5.3 *U-space* (competent authority "certifies" USSP).
637. [ICAO UTM Framework Ed. 3], p. 15.
638. ICAO, *Unmanned Aircraft System Traffic Management (UTM), Request for Information* (2019), Sect. 2, p. 2, https://www.icao.int/Meetings/DRONEENABLE3/Documents/ICAO%20Request%20for%20Information%20(RFI)%202019.pdf. *See* Email from Benoit Curdy, Digital Transformation, Swiss FOCA (Nov. 12, 2019) (on file with author) ("There is a difference in approach between the US and us regarding the term 'authorization'. For us, it is reserved to competent authorities. USS can facilitate authorization by facilitating communication with authorities but they cannot, on their own, authorize flights.").
639. ICAO, *Unmanned Aircraft System Traffic Management (UTM), Request for Information* (2020), 3.b), p. 2, https://www.icao.int/safety/UA/Documents/RFI%20-%202020.pdf.
640. [SESAR ATM Master Plan 2017], Sect. 3.5.
641. *See infra* Chap. 6, Sect. 6.2.2 *Advanced Automation and Autonomy.* To the extent UTM becomes increasingly autonomous and will "learn" from its experience and environment to optimize performance and mitigate hazards, non-deterministic system characteristics may challenge the qualification process.
642. *See supra* Ch. 3 *Standards-Making for UTM.*
643. Presentation at ICAO, DRONE ENABLE/3, Montreal, CA (Nov. 13, 2019).
644. Keynote at ICAO, DRONE ENABLE/2, in Chengdu, P.R.C. (Sept. 13-14, 2018).

645. Email from Benoit Curdy, Digital Transformation, Swiss FOCA (Sept. 10, 2020) (on file with author).
646. Jarrett Larrow, UTM Prog. Mgr., FAA, Remarks at the ASTM UTM WG (June 19, 2020) (further stating "from a regulatory perspective, I do not have the means to approve a USS now."). Capability to assure clarity of responsibility has been a persistent goal throughout the standards development process.
647. Amit Ganjoo, Founder and CEO, ANRA Technologies, and Co-chair, ASTM F38 UTM WG, Remarks at the ASTM F38 UTM WG (June 19, 2020). *See infra* Chap. 4, Sect. 4.7.4 *Liability* (addressing, in part, apportionment of USS responsibilities).
648. [FAARA], Sect. 351 Unmanned aircraft system integration pilot program. *See* [FESSA], Sect. 2208(b) (establishment of a UTM system pilot program); *supra* Chap. 1, Sect. 1.5 *Research and Development.*
649. [FAARA], Sect. 376 *Plan for full operational capability of unmanned aircraft systems traffic management. See* FAA, *FY 2020 ANG Business Plan* (2020), p. 15, https://www.faa.gov/about/plans_reports/media/2019/ang_business_plan.pdf ("Activity: UAS Traffic Management (UTM) Data Exchange, Develop and implement... and *governance* for UAS Operations in the NAS") (emphasis added).
650. [FAARA], *id.* (emphasis added).
651. Although not providing for the general direct qualification of USS, Part 135 certifications relied, in part, on safety cases that included strategic mitigation of air risk via the applicants' proto-UTM capabilities—distinct from the FAA evaluating those USS against a comprehensive UTM-specific standard in a federated environment. Part 135 was not developed for UAS and AAM, and has required extensive waiver and customization for such uses. Human-carrying attributes of AAM and public acceptance / safety expectations of certification may prolong such practice. *See, e.g.,*

    Amazon Prime Air – FAA, *In the matter of the petition of Amazon Prime Air*, Exemption No. 18602, Reg. Docket No. FAA-2019-0622 (Part 135 exemption issued Aug. 27, 2020), https://beta.regulations.gov/document/FAA-2019-0622-0003; Amazon Prime Air–*Petition for Exemption Under 49 U.S.C. § 44807 and 14 C.F.R. Parts 61, 91, and 135* (July 16, 2019), https://www.federalregister.gov/documents/2019/08/08/2019-17010/petition-for-exemption-summary-of-petition-received-amazon-prime-air;

    UPS Flight Forward – FAA, *In the matter of the petition of UPS Flight Forward, Inc.*, Exemption No. 18338, Reg. Docket No. FAA-2019-0652, https://www.regulations.gov/document?D=FAA-2019-0628-0021 (Part 135 exemption issued Sept. 23, 2020); UPS, *UPS Flight Forward Attains FAA's First Full Approval For Drone Airline,* Press Release (Oct. 1, 2019), https://pressroom.ups.com/pressroom/ContentDetailsViewer.page?ConceptType=PressReleases&id=1569933965476-404;

    WING Aviation, LLC – FAA, *In the matter of the petition of Wing Aviation, LLC*, Exemption No. 18162, Reg. Docket No. FAA-2015-3344 (reissuing corrected copy April 2, 2019), https://www.regulations.gov/document?D=FAA-2018-0835-0024 ("UTM and the geo-fence system must be operational for flight operations" 33); FAA, Press Release–*U.S. Transportation Secretary Elaine L. Chao Announces FAA Certification of Commercial Package Delivery* (April 23, 2019), https://www.faa.gov/news/press_releases/news_story.cfm?newsId=23554 (first air carrier certification for commercial package delivery via UAS).

    Safety Imperative - *See* Daniel K. Elwell, formerly Deputy Admin'r, Presentation at the FAA UAS Symposium, Balt., Md. (June 3, 2019) (Underscoring, "[i]f it's not safe, it's not going to fly." Additionally stating, "[a]s we move toward integration . . . safety should remain fundamental to our collective foundation. Make no mistake, that foundation is safety."). *Cf.* Ali Bahrami, Asso. Admin. for Avi. Safety, FAA, Presentation at the FAA UAS Symposium (July 9, 2020) ("Safety is our great unifying passion."); Ric Peri, VP, Gov't and Industry Affairs, Aircraft Electronics Ass'n, Remarks (Aug. 15, 2019) (asserting that without safety requirements promulgated, that is, "until government sets UTM 'speed limits', industry must assure the highest standard of 'doing no harm'."); 49 U.S.C. Sect. 44701, General requirements (d)(1)(A), *available at* https://www.law.cornell.edu/U.S.C.ode/text/49/44701 (expressing "the duty of an air carrier to provide service with the *highest possible degree* of safety in the public interest"); 49 U.S.C. Sect. 40101, Policy, (a)(1), *available at* https://www.law.cornell.edu/U.S.C.ode/text/49/40101 (requiring "assigning and maintaining safety as the *highest priority* in air commerce," and (3) "recognizing the clear intent, encouragement, and dedication of Congress to further the *highest degree* of safety in air transportation…") (emphasis added).

652. [Comm Implementing Reg. (EU) 2021/664 - U-space], Art. 14, Application for a certificate, ¶ 1.
653. [Comm Implementing Reg. (EU) 2021/664 - U-space], Art. 18, Tasks of the competent authorities, ¶ (j). *See*, e.g., EASA, *Draft acceptable means of compliance (AMC) and Guidance Material (GM) to Opinion No 01/2020 on a high-level regulatory framework for the U-space, Issue 1 (XX Month 2020)*, https://www.easa.europa.eu/sites/default/files/dfu/Draft%20AMC%20%26%20GM%20to%20the%20U-space%20Regulation%20—%20for%20info%20only.pdf ["EASA Draft AMC & GM"]. *See infra* Chap. 4, Sect. 4.2.5 *Continuing Compliance and Safety Assurance.*
654. *See* Parimal Kopardekar, Ph.D., Dir., NARI, NASA, Presentation at the EUROCONTROL Stakeholder Forum on Urban Air Mobility (Feb. 4, 2021), https://www.eurocontrol.int/event/eurocontrol-stakeholder-forum-uam (NASA National Campaign data collection helping regulators and serving as a foundation for both aircraft and ATM certification).
    The adopted qualification method should be risk-based. [EDITOR: the text continues in the same paragraph with "The most rigorous method…"]
    The most rigorous method is typically "certification" by the CAA on some basis of conformity assessment to rules and standards. Less rigorous methods may include third party conformity assessment – typically by a recognized accredited body; or an applicant's self-declaration of conformity to standards – which may or may not be based on test reports from an independent entity.
655. *See* Chap. 4, Sect. 4.2.7 *Specific Operations Risk Assessment.* Industry consortia may also contribute to governance by: advocating how UTM should meet safety standards, ensuring harmonization among UTM service providers, and providing measured administration and support. *See also infra* Chap. 4, Sect. 4.7.2 *Industry Consortia.*
656. Presentation at ICAO, DRONE ENABLE/3, in Montreal, CA (Nov. 13, 2018) (and urging that the path to an approved provider can be done quickly). *See* Gur Kimchi, formerly VP, Amazon, Presentation at ICAO, DRONE ENABLE/3, Montreal, CA (Nov. 13, 2019) (asserting that "if everyone does standards the same way, then the approval process [addresses] the edge case and local concerns. Where 99 percent of complexity is per standards compliance, meeting international standards will be done. I think it is a huge cost savings.").
657. "Conformity" is more broadly defined and addressed in: ISO, *Certification and Conformity*, https://www.iso.org/conformity-assessment.html ("involves a set of processes that show your product, service or system meets the requirements of a standard"); ANSI, *U.S. Conformity Assessment System: Introduction*, https://www.standardsportal.org/usa_en/conformity_assessment/introduction.aspx ("a term used to describe steps taken by both manufacturers and other parties to evaluate whether products, processes, systems, or personnel adhere to the requirements identified in a specified standard. Conformity assessment activities such as testing, certification, and accreditation are closely associated with standards."). *See* ASTM Committee E36 on Accreditation and Certification, https://www.astm.org/COMMITTEE/E36.htm (addressing conformity assessment).
    James D. Foltz, PE, Mgr., Strategic Policy Emerging Aircraft, Aircraft Cert. Services, FAA, *UAS Type Certification, Durability & Reliability Means of Compliance*, Presentation at the ASTM F38 meeting, in Raleigh, N.C. (Nov. 5, 2019), https://www.astm.org/COMMIT/Nov2019.pdf. Foltz further stated:

We're trying to do something that we once utilized heavily, focusing on *reliability* for certification. For smaller UAS, durability and reliability (D&R) is appropriate. The inverse of failure is "reliability." Asking applicants to demonstrate reliability meets our intent—it's just a different way of getting there. D&R MoC utilizes verification & validation versus traditional 10^-X probabilities... and is used to show compliance with airworthiness criteria. The long-term plan is to publish airworthiness criteria MoC in Advisory Circulars that would recognize additional industry consensus standards.

See FAA, *Type Certification of Unmanned Aircraft Systems*, 85 Fed. Reg. 5905-6 (Feb. 3, 2020), https://www.federalregister.gov/documents/2020/02/03/2020-01877/type-certification-of-unmanned-aircraft-systems. *Cf.* FAA, *Aircraft Conformity*, https://www.faa.gov/licenses_certificates/airline_certification/conformity/ (addressing aircraft conformity inspection verifying "aircraft conforms to its type design and is configured/bridged to your approved program and operations.").

658. *See e.g.,* Email from Lane Hallenbeck, Exec. Dir., ANSI Nat'l Accreditation Board (Sept. 23, 2019) (on file with author) (underscoring that "[t]his is formally defined and referenced in international standards such as ISO/IEC 17050-1&2 [Conformity assessment—Supplier's declaration of conformity]... 'Self'-certification is considered an oxymoron given that certification by definition requires independence."); Canadian Aviation Regulations, Part IX - Remotely Piloted Aircraft Systems, Division VI—Advanced Operations—Requirements for Manufacturer, Sect. 901.77-79, https://lois-laws.justice.gc.ca/eng/regulations/SOR-96-433/FullText.html#s-901.76.
659. *See* Canadian Aviation Regulation, *id.*, Sect. 901.77-79, https://lois-laws.justice.gc.ca/eng/regulations/SOR-96-433/FullText.html#s-901.76 (self-certification also "allows an entity to voluntarily identify itself as subject to a regulatory environment" and correspondingly then meet certain regulatory obligations). *See supra* Email from Craig Bloch-Hansen (further suggesting that self-declaration may "lower costs to comply, increased opportunities for innovation, providing services which in areas which wouldn't otherwise be economically feasible (i.e., safety enhancing services), and acts as a 'steppingstone' to higher risk-regimes for new market entrants.").
660. That is, applied in cases of particularly low system risk (low introduction of risk to others) and a high degree of confidence in the ability of users to self-analyze and make conclusions upon which to base their declaration. In addition to ascertaining the veracity of declarations, the understanding of the people or organizations making them is important. Errors may be as likely to stem from problems of ignorance as they are to lack of integrity or forthrightness.
661. *See, e.g.,* FAA, *Aircraft Certification Process Review and Reform, FAA Response to FAA Modernization and Reform Act of 2012*, Pub. L. No. 112-95, Sect. 312, 126 Stat. 11 (Aug. 12, 2012), https://gama.aero/wp-content/uploads/FAA-ACPRR-Report-to-Congress-2012-08.pdf.
662. *Statement of Daniel K. Elwell*, formerly Acting Admin'r, FAA, Before the Comm. on Transp. and Infrastructure, U.S. House of Reps. (May 15, 2019), https://www.transportation.gov/content/status-boeing-737-max (addressing the Organization Designation Authorization (ODA) program) (emphasis added). *See* DoT, *Official Report of the Special Committee to Review the Federal Aviation Administration's Aircraft Certification Process* (Jan. 16, 2020), https://www.transportation.gov/sites/dot.gov/files/docs/briefing-room/362926/scc-final-report.pdf (concluding reforms must be adopted for the FAA's overall certification system to be effective); 14 C.F.R. Sect. 183.1, *et seq.* (addressing requirements for such delegees); *Cf.* DoT, Office of the Inspector General, *DOT's Fiscal Year 2020 Top Management Challenges,* Rpt. PT2020003 (Oct. 23, 2019), pp. 6-7, https://www.oig.dot.gov/sites/default/files/DOT%20Top%20Manage-

ment%20Challenges%20%20FY2020.pdf (asserting FAA "faces the significant oversight challenge of ensuring the ODA companies maintain high standards and comply with FAA safety regulations"); [FAARA], Sect. 44810(i) Non-Delegation (re: C-UAS).

663. *See* Ric Peri, VP, Gov't and Industry Affairs, Aircraft Electronics Ass'n, Remarks (Aug. 15, 2019) (asserting, "[n]ew and novel technologies do not lend themselves well to quantified assessment. As such, a risk-based assessment of level of involvement is not perfect.").

664. *See* Majority Staff of the Comm. on Transp. and Infrastructure, Final Comm. Rpt., *The Design, Development & Certification of the Boeing 737 MAX* (Sept. 20, 2020), Ch. 4 FAA Oversight and Delegation of Authority, https://transportation.house.gov/imo/media/doc/2020.09.15%20FINAL%20737%20MAX%20Report%20for%20Public%20Release.pdf (*inter alia*, "[e]xcessive FAA delegation of certification functions to Boeing on the 737 MAX eroded FAA's oversight effectiveness and the safety of the public."); Natalie Kitroeff, *The Roots of Boeing's 737 Max Crisis: A Regulator Relaxes Its Oversight,* N.Y. Times (July 27, 2019), https://www.nytimes.com/2019/07/27/business/boeing-737-max-faa.html?campaign_id=61&instance_id=0&segment_id=15620&user_id=7d6dc7a69ed16a75a2bf96de1983331f&regi_id=48239999&em-c=edit_ts_20190727ries (Describing "a broken regulatory process that effectively neutered the oversight authority of the agency.…*As planes become more technologically advanced, the rules, even when they are followed, may not be enough to ensure safety.*") (emphasis added); Joint Authorities Technical Review, *Boeing 737 MAX Flight Control System, Observations, Findings, and Recommendations* (Oct. 11, 2019), Sect. 8, https://www.faa.gov/news/media/attachments/Final_JATR_Submittal_to_FAA_Oct_2019.pdf?mod=article_inline (recommending, *inter alia*, need for analytical tools for complex systems; recognizing that "[d]esign and analysis techniques traditionally applied to deterministic risks or to conventional, non-complex systems may not provide adequate safety coverage for complex systems."); Eric Adams, *Boeing 737 Crashes May Make It Harder for Air Taxis to Take Off,* WIRED (April 16, 2019), https://preview.tinyurl.com/737-Regulation (considering adequate safety coverage for complex systems).

665. *See, e.g.,* [Remote ID Rule], Preamble, XIII. ("The FAA encourages consensus standards bodies to develop means of compliance and submit them to the FAA for acceptance [and] intends to rely increasingly on consensus standards as FAA-accepted means of compliance for UAS performance-based regulations for remote identification, consistent with FAA precedent for general aviation aircraft and other initiatives taken with respect to UAS."); Chap. 3, Sect. 3.1 *General—Standards-Making for UTM. See* Email from Adam Morrison, Owner, Streamline Designs LLC (Mar. 16, 2021) (on file with author) ("Operational rules and personnel certification (e.g., pilots, repairmen, etc.) have thus-far been closely held by the regulator which has created discontinuities in the healthy function of the overall market. UTM may be an opportunity in this regard.").

666. EUROCAE, *Technical Work Programme* (Public ver. 2019), Sect. 3.4, p. 8, https://eurocae.net/media/1567/eurocae-public-twp-2019.pdf.

667. FAA, Drone Advisory Committee, *DAC Member eBook*, Task Group #6 BVLOS, Autonomy Sub-Group, Wash., D.C. (Feb. 27, 2020), https://www.faa.gov/uas/programs_partnerships/drone_advisory_committee/media/Public_Ebook_v3a.pdf (further stating, "[c]ertain waivers could be automatically approved when using an approved UTM system.").

668. *See* White House, *Executive Order 12866 of Sept. 30, 1993*, Regulatory Planning and Review (Oct. 4, 1993), Sect. 1, https://www.archives.gov/files/federal-register/executive-orders/pdf/12866.pdf (emphasis added):

> (7) Each agency shall base its decisions on the best reasonably obtainable scientific, technical, economic, and other information concerning the need for, and consequences of, the intended regulation.

(8) Each agency shall identify and assess alternative forms of regulation and shall, to the extent feasible, *specify performance objectives, rather than specifying the behavior or manner of compliance* that regulated entities must adopt.

See also ICAO, *Global Air Traffic Management Operational Concept*, 1st ed., Doc. 9854 (2005), https://www.icao.int/Meetings/anconf12/Document%20Archive/9854_cons_en[1].pdf (identifying and structuring relevant performance areas); Gov't of Australia, CASR Part 139 – Aerodromes, https://www.casa.gov.au/standard-page/reviewing-rules-aerodromes-part-139?utm_source=phplist2273&utm_medium=e-mail&utm_content=HTML&utm_campaign=July+2020+Regulatory+wrap-up+%5B-SEC%3DOFFICIAL%5D (Certain aerodrome requirements may also inform UTM governance, e.g., "requirements will be scalable, dependent on the size and complexity of the aerodrome operations and associated risk.").

669. Remarks at the FAA UAS Symposium (July 9, 2020). *See* Reinaldo Negron, Head of UTM, Wing, Remarks at ICAO DRONE ENABLE Symposium 2021 (April 14, 2021) (Recognizing "performance-based is harder than prescriptive [rules] and requires regulators to think in ways they have not before .... Recognize that regulators have the responsibility and burden" and that "we need to help.").
670. Andy Thurling, CTO, NUAIR Alliance, Presentation at the GUTMA Annual Conference, Portland, Or. (June 20, 2019). *See* Email from Adam Morrison, Owner, Streamline Designs LLC (Mar. 16, 2021) (on file with author) (It might be helpful "to inform the standards development process in a data-driven manner instead of trying to predict the future and imposing life-long requirements up-front that may be needlessly expensive to comply with. Once the bar is set, it is very difficult to lower the bar even if it is warranted[. We also] need a proper framework that helps us to define what the appropriate performance measures even are. We can't set the performance requirements until we define how we're going to measure performance.").
671. [ICAO UTM Framework Ed. 3], pp. 8-9. *See* Email from Andrew Carter, Pres. and Co-founder, ResilienX, Inc. (Dec. 13. 2019) (on file with author) ("Additional consideration may be needed for…the performance level or accuracy of the services or data. However, this creates a paradigm where both the system and environment may be dynamic, greatly increasing the complexity of determining safety.").
672. [FAA ConOps UTM v2.0], Sect. 2.3.2. (emphasis added). *See* Steve Bradford, Chief Scientist for Architecture and NextGen Development, FAA, Remarks at the NASA, AAM Airspace WG (webinar) (Aug. 4, 2020) (stating that community-based rules will have performance requirements that the FAA reviews and approves, and asserting, "remember everything has to be approved by the FAA because it is *our* airspace.") (emphasis added).
673. [Comm Delegated Reg. (EU) 2019/945 - UAS], ¶ 29 — referencing Decision No. 768/2008/EC - on a common framework for the marketing of products. Additionally, the EU's structured use of conformity assessment procedures also reflects the Canadian approach.
674. FAA, *Advisory Circular AC 150/5220-16E CHG 1*, Subj: Automated Weather Observing Systems (AWOS) for Non-Federal Applications (Jan. 31, 2019), Ch. 2. Certification and Commissioning Process, https://www.faa.gov/documentLibrary/media/Advisory_Circular/AC_150_5220-16E_w-chg1.pdf (Further stating, "After the AWOS is installed and the MOA/OMM is signed, the FAA conducts a commissioning ground inspection to verify that the system is located, installed and operating correctly and that the owner has the resources to maintain the system in proper operating condition for the life of the system.…Finally, in addition to the annual revalidation inspection, there will be periodic visits to the operational AWOS by the FAA and other technical representatives to verify that the system continues to operate correctly." This process includes performance standards and an "ongoing validation program;" also restating that AWOS is defined as an "air navigation facility" in 49 U.S.C. Sect. 40102).

675. Jonathan Daniels, CEO, Praxis Aerospace Concepts Int'l, and Chair, ASTM F38.03, Remarks (Sept. 26, 2019). *See* 47 U.S.C. Part 87 – Aviation Services, *available at* https://www.law.cornell.edu/cfr/text/47/part-87. *But see* Adam Morrison, Owner, Streamline Designs LLC (Mar. 16, 2021) (on file with author) ("The concept of ground-based systems as a primary enabler for UTM is being explored. Certainly some infrastructure will be ground based, such as telecommunications. In some use cases, ground-based radar systems may have utility as well. However, the industry should consider carefully the strategic consequences that new, mandatory, purpose-built, ground-based infrastructure would have on the industry. It would serve to dramatically slow the adoption and widespread utilization of UAS in an even-handed manner. Unless a government pays to fund and construct such systems, the overhead associated with commercial BLOS operations would be increased to amortize these costs and serve to limit commercial use opportunities to only those that could afford the underlying infrastructure 'tax'. Thus far systems like this have also proven quite challenging to operate successfully and require significant custom tailoring and expertise.").
676. [RID NPRM 2019], Sect. I.C.5 *Remote ID USS*, 72442. *See, e.g.,* Email from Paul Dudley, VLC Managing Dir., Adv. Tech. Int'l, Vertical Lift Consortium (Apr. 9, 2021) (on file with author) ("Under an OTA, DoD is using the Vertical Lift Consortium (VLC) to support development of rotorcraft standards. The FAA's use of the VLC OTA could provide a recognized venue for engaging the vertical flight community in developing UTM governance."). *See also* 49 U.S.C. Sect. 106 Federal Aviation Administration, *available at* https://www.law.cornell.edu/uscode/text/49/106 (providing broad contracting authority); *See* [Remote ID Rule], Preamble, p. 4405 (reacknowledging "willingness to enter into no-cost contracts with the FAA"); Chap. 4, Sect. 4.7.2 *Industry Consortia.*
677. FAA, *Advisory Circular AC 20-153B*, Subj: Acceptance of Aeronautical Data Processes and Associated Databases (April 19, 2016), Sect. 5, https://www.faa.gov/regulations_policies/advisory_circulars/index.cfm/go/document.information/documentID/1029446 ["AC 20-153B"] (also defining *aeronautical data* as "data used for aeronautical applications such as navigation, flight planning, flight simulators, terrain awareness, and other purposes (e.g., navigation data, terrain and obstacle data, and airport mapping data) [and] new and novel aeronautical applications…" Compliance with an AC is one non-excusive way to satisfy certain regulatory requirements.). *Cf.* Canada: Civil Air Navigation Services Commercialization Act (S.C. 1996, c. 20), Sect. 10 Subsection 2(b), https://laws-lois.justice.gc.ca/eng/acts/C-29.7/page-2.html#docCont (NAV CANADA must provide written consent to an organization which will provide aviation services.).
678. Email from Andrew Carter, Pres. and Co-founder, ResilienX, Inc. (Aug. 19, 2020) (on file with author). *See supra* Chap. 3, Sect. 3.2.13 *RTCA* (introducing DO-278).
679. *See* [ICAO UTM Framework Ed. 3], Appn. F, UTM Service Providers (USP) Organizational Construct and Approval Processes, p. 32 ("[S]afety-critical services will have a higher degree of criticality [and] would need a greater degree of oversight…").
680. ASTM International, F3269-17, *Standard Practice for Methods to Safely Bound Flight Behavior of Unmanned Aircraft Systems Containing Complex Functions*, https://www.astm.org/Standards/F3269.htm; *See* Jay Merkle, Exec. Dir., UAS Integration office, FAA, UAS Symposium, *Innovation: Shaping the Future* (Aug. 19, 2020) (stating, "Who certifies autonomy? *Nobody certifies autonomy*… we actually certify designs or approve operations."); Chap. 6, Sect. 6.2.2 *Advanced Automation and Autonomy.*
681. Andy Thurling, CTO, NUAIR, *The need to define an appropriate set of UAS requirements for base UAS Traffic Management capabilities* (Feb. 2021), https://nuair.org/2021/02/04/establishing-and-validating-performance-requirements-for-utm/ (also describing such tranche as "local BVLOS" (L-BVLOS) – for drone-in-a-box type operations within an "appropriately constrained, well-understood environment"). *See* Adam Morrison, Owner, Streamline Designs LLC (Mar. 16, 2021) (on file with author) ("We need to avoid, however, that these explorations become embedded as regulations or become the long-term standard of care, lest we limit the long-term ability of UAS to deliver value-add to our lives and economy.").

682. [NASA USS Spec.], Sect. 16, p. 37; [Rios Strategic Deconfliction Performance], Table 4, p. 24 (Test implementations should include "targeted checkout tests" reflecting use cases serviced by a USS.); Joseph L. Rios, ATM-X Project Chief Eng'r, NASA, Remarks at ASTM F38 UTM WG meeting, Palo Alto, Cal. (Aug. 29, 2019) ("Every single constraint has to have a test."). *Compare* Irene Skupniewicz Smith, et al., NASA Research Center, *USS Service Supplier Checkout, How UTM Confirmed Readiness of Flight Tests with UAS Service Suppliers*, NASA/TM-2019-220456 (Dec. 1, 2019), p. 6, https://ntrs.nasa.gov/archive/nasa/casi.ntrs.nasa.gov/20190034170.pdf ["NASA USS Checkout"] (stating that even "the USS spec has hundreds of requirements, thousands of data model validations and interrelated protocols").
683. LAANC USS Performance Rules], Attachment D: Upgrade Onboarding Timeliness. *See* FAA, *USS Onboarding, Low Altitude Authorization and Notification Capability (LAANC),* v1.0 (Feb. 2018), Sect. 3.4, https://www.faa.gov/uas/programs_partnerships/data_exchange/laanc_for_industry/media/LAANC_UAS_Service_Supplier_onboarding_information.pdf ["LAANC USS Onboarding"] ("demonstrate successful execution of validation scenarios" as a precondition to the FAA's onboarding "decision," *id.* Sect. 4.9). The FAA views LAANC as not having functionally changed ATC responsibilities—rather, just enabling industry to help automate them. Since material conformance monitoring and feedback mechanism data supporting LAANC risk management were not necessarily captured robustly, the benefits of such data may not adequately support broader UTM development. LAANC validation scenarios are merely a logic check against an FAA data base. Nonetheless, many of its processes have matured. *See infra* Chap. 5, Sect. 5.2 *Low Altitude Authorization and Notification Capability*.
684. Email from Brandon Montellato, Solutions Mgr., Wing (Feb. 12, 2020) (on file with author). *See, e.g.*, Xuxi Yang, Maxim Egorov, et al., *Stress Testing of Unmanned Traffic Management Decision Making Systems* (June 8, 2020), *available at* https://arc.aiaa.org/doi/abs/10.2514/6.2020-2868 (perhaps such capability could assist testing of USS deconfliction capability via simulated conditions); Email from Maxim Egorov, Research Scientist, Airbus (Aug. 20, 2020) (on file with author) (automated deconfliction stress testing "can absolutely be used to stress test any interoperability functions as well by connecting multiple USS through a simulation environment. In fact, we do this already with multiple but unique instances of our USS when testing automated deconfliction and negotiation. From a governance perspective, I think the right approach to the UTM validation problem involves some simulation. The precedent for this has already been set in the certification of ACAS-X, where a simulated encounter model was used to validate large chunks of the logic in the software. The idea would be the same - an accredited (or regulatory) body could keep a set of simulated scenarios that USSs must checkout against (in theory this could be fully automated), and the scenario dataset is expanded over time by adding scenarios that are challenging during operation or are discovered through something like stress testing."); [Embraer UATM ConOps], Annex C.2.3, p. 80 (presenting "major issues" identified from simulation of UAM traffic management by ATCTs).
685. [UAS Service Supplier Checkout], p. 18. *See* ISO/IEC 17011:2017(en), *Conformity assessment — Requirements for accreditation bodies accrediting conformity assessment bodies,* https://www.iso.org/obp/ui/#iso:std:iso-iec:17011:ed-2:v1:en:term:3.1, Sect. 3.1 (addressing "remote assessment... using electronic means").
686. *See, e.g.,* [ASTM Remote ID], Annex A2 – Network Remote ID Interoperability Requirements, APIs and Testing.
687. *See infra* Chap. 4, Sect. 4.7.2 *Industry Consortia* (e.g., describing Nacha).
688. *See, e.g.,* [Baum FCA Liability] (surveying trusted infrastructure requirements in diverse industries).
689. Presentation at ICAO, DRONE ENABLE/3, Montreal, CA (Nov. 13, 2019).
690. *See e.g.,* ASTM, F3322-18, *Standard Specification for Small Unmanned Aircraft System (sUAS) Parachutes*, https://www.astm.org/Standards/F3322.htm (Sect. 6.3.1 requiring a third-party testing agency or CAA approved delegee to oversee testing).

691. *See, e.g.*, FAA, *LAANC Operating Rules*, Sect. 3.9.1, https://www.faa.gov/uas/programs_partnerships/data_exchange/laanc_for_industry/media/FAA_sUAS_LAANC_Ph1_USS_Rules.pdf (requiring "audit of service functionality on a periodic basis"); [LAANC MOA], Art. 21. AUDIT (requiring third-party auditor to be ISO/IEC 17020:2012 certified); [FAA ConOps UTM v2.0], Sect. 2.7.1.1 Strategic Management of Operations (operator accounts and records subject to FAA auditing at agency discretion); [India UTM Policy]. Sect. 7.1 Data Security and Privacy (requiring secure audit interface to DigitalSky for audit; and right of competent authority to conduct physical audit of IT system). *Compare* [Remote ID Rule], Preamble, XIV.F. Accountability (re RID design and production: "apply industry best practices to determine when and how often independent audits are needed.").
692. *See, e.g.*, Int'l Business Aviation Council, IS-BAO, https://www.ibac.org/is-bao; Sandy Murdock, *International Aviation Standards: government-to-government and government to non-governmental organization*, JDA J. (Aug. 22, 2018), https://jdasolutions.aero/blog/international-aviation-standards-govt-to-govt-government-ngo/ (EASA responding: "Operators are free to utilize on a voluntary basis any third-party SMS program of their choice, including the IS-BAO program. EASA will then perform its own assessment of the effective implementation of the operator's SMS, based on actual SMS deliverables."); ANSI National Accreditation Board (ANAB), https://anab.ansi.org/; ISO, ISO/IEC 17011:2017 *Conformity assessment - Requirements for accreditation bodies accrediting conformity assessment bodies,* ISO/IEC 17011:2017, https://www.iso.org/standard/67198.html; Am. Bar Ass'n, Sect. of Sci. & Tech., Info. Sec. Comm., *PKI Assessment Guidelines* (May 20, 2003), Sect. B.6 - PKI Assessment, and App. 4 - PKI Audit Methodology and Guidelines.
An international "toolbox" of conformity assessment standards is developed and maintained by the ISO Committee on Conformity Assessment (ISO/CASCO), https://www.iso.org/sites/cascoregulators/02_casco_toolbox.html. *See* ISO/CASCO, Committee on conformity assessment, https://www.iso.org/committee/54998.html (described as "relevant to a myriad of conformity assessment applications"); ISO, *Standards by ISO/CASCO*, Committee on conformity assessment, https://www.iso.org/committee/54998/x/catalogue/p/1/u/0/w/0/d/0. CASCO standards address the range of objects (products, people, services, processes, systems, etc.) of conformity and the spectrum of methods for assessing their conformity. However, it does not have direct involvement with 3rd party conformity assessment beyond defining benchmark requirements. ANSI members can participate in developing those requirements – on the ANSI mirror committee to ISO CASCO, called the International Conformity Assessment Committee (ICAC), https://www.ansi.org/iso/ansi-activities/ansi-administered-tags/icac.
693. ASTM, F3365 *Standard Practice for Compliance Audits to ASTM Standards on Unmanned Aircraft Systems*, https://www.astm.org/Standards/F3365.htm; ASTM, F3364, *Standard Practice for Independent Audit Program for Unmanned Aircraft Operators,* https://www.astm.org/Standards/F3364.htm.
694. Jonathan Daniels, CEO, Praxis Aerospace Concepts Int'l, and Chair, ASTM F38.03, Remarks (Sept. 26, 2019) (stating that for UTM audit, "I don't think we need anything new" beyond these ASTM standards).
695. *See, e.g.,* Letter from Kratos SecureInfo to Amazon Web Services, *NIST Cybersecurity Framework (CSF)*, Amazon Web Services (Jan. 2019), *App. B — Third Party Assessor Validation* (Sept. 19, 2018), https://d1.awsstatic.com/whitepapers/compliance/NIST_Cybersecurity_Framework_CSF.pdf; Email from Craig Bloch-Hansen, Prog. Mgr. - UAS Design Standards, Transport Canada (Sept. 29, 2020) (on file with author) ("What role do industry certifications play in the accreditations process? *See, e.g.,* ISO 9001 certification. In this framework the Governmental Responsible entity (e.g., ANSI in the US, SCC in Canada) certifies the accreditation and auditing bodies in accordance with published standards (e.g., ISO, ASTM) and these bodies act as regulated entities supporting the accreditation of other entities/products who then in turn become regulated entities (when 'Accredited'))."

696. [ICAO IATF 2019], Sect. 4.6.2 *Accreditation, Certification & Audit Process. See generally* ISO/IEC 27001, *Information Security Management*, https://www.iso.org/isoiec-27001-information-security.html (requiring post-assessment audit); ISO/IEC JTC 1/SC 72, *Information security, cybersecurity and privacy protection,* https://www.iso.org/committee/45306.html (addressing generic methods, techniques, and guidelines for infosec. and privacy conformance assessment, accreditation, and auditing); ISO/IEC 27007:2020, *Information security, cybersecurity and privacy protection – Guidelines for information security management systems auditing*, https://www.iso.org/standard/77802.html?browse=tc. *See supra* Chap. 2, Sect. 2.4.7.2 *International Aviation Trust Framework.*
697. Parimal Kopardekar, Ph.D., Dir., NARI, NASA, *National Unmanned Aerial System Standardized Performance Testing and Rating (NUSTAR)* (2016), https://ntrs.nasa.gov/archive/nasa/casi.ntrs.nasa.gov/20160012460.pdf. *See* NUAIR, *NUSTAR*, https://nuair.org/nustar/ (NUSTAR to "act as an independent third party to provide safety declarations for UAS, USS and SDSPs – providing manufacturers and suppliers an independent declaration of compliance.").
698. Andy Thurling, CTO, NUSTAR Alliance, *Why we are building a future system for third party verification of UAS & USS performance*, NUSTAR BLOG (March 7, 2019), https://nuair.org/2019/03/07/nustar-from-concept-to-gold-standard/ ("The need for third party statements of compliance has already been written into risk-based processes such as the Specific Operations Risk Assessment (SORA) process.... More recently, the European Union Aviation Safety Agency (EASA) framework for drones, methods for approving waivers by the FAA, and the new Modernization of Special Class Airworthiness Certification (MOSAIC) have all identified how third party validation of UAS and USS performance can improve safety and will be considered in granting operational approvals."). Andy Thurling, CTO, NUAIR Alliance, Remarks (Aug. 27, 2019) (underscoring that the NUAIR NUSTAR proposal is "very JARUS and SORA friendly," and responsive to manufacturer and integrator business cases).
699. NUAIR, *NUSTAR*, https://nuair.org/nustar/ ("In building NUSTAR, we expect to develop means of compliance with evolving standards and provide an environment for testing and verification that industry leaders will seek out for standards verification. NUSTAR will act as an independent third party to provide safety declarations for UAS, USS and SDSPs – providing manufacturers and suppliers an independent declaration of compliance.").
700. *See* Email from Craig Bloch-Hansen, Prog. Mgr. - UAS Design Standards, Transport Canada (Sept. 29, 2020) (on file with author) ("Declarations may also be used to gather information on products. An Accredited organization (as discussed above) may be given authorization to issue declarations which help identify services and equipment that meet a higher safety threshold than those self-declarations made under a pure declaration system (i.e., no organizational accreditation).").
701. Email from Benoit Curdy, Digital Transformation, Swiss FOCA (July 8, 2020). *See* [FAA ConOps UTM v2.0], Sect. 2.4.2.2 Obtaining a Performance Authorization (for "lower risk operations" permitting operator self-declaration of compliance to standards); [ASTM UTM Spec.], Sect. 10 Procedures (providing for self-declaring conformity with designated roles); [Remote ID Rule], Sect. 89.1 (Declaration of compliance submitted to FAA by RID producer attesting to satisfaction of subpart F requirements); [Remote ID Rule], Preamble, XIV.G Filing of Declaration of Compliance ("declaration of compliance process is not a self-certification process"). *Compare* (FAA *Advisory Circular AC 00-1.1B*, Subj. Public Aircraft Operations—Manned and Unmanned (Sept. 21, 2018), https://www.faa.gov/documentLibrary/media/Advisory_Circular/AC_00-1.1B.pdf (describing public aircraft operations (PAO), where the COA applicant self-certifies the airworthiness of its aircraft, the certification and qualifications of its pilot, and medical standards. Importantly, a PAO declaration—and compliance with Part 91 and other relevant Federal Aviation Regulations—represents a significant transfer of responsibility to the government entity and the FAA does not provide oversight for PAO flights).

702. [Comm Implementing Reg. (EU) 2021/664 - U-space], Art. 18, *Tasks of the competent authorities,* ¶ (j). *See* Stephen P. Creamer, Dir., Air Navigation Bureau, ICAO, Remarks at ICAO DRONE ENABLE 2021 (April 21, 2021) ("Certification is not the end of the process").
703. *Id.* (emphasis added).
704. *See, e.g.,* ICAO, *Safety Management Manual (SMM),* Doc. 9859, AN/474 (3rd. Ed. 2013), https://www.icao.int/SAM/Documents/2017-SSP-GUY/Doc%209859%20SMM%20Third%20edition%20en.pdf; FAA, *ORDER 8040.6*, Subj: Unmanned Aircraft Systems Safety Risk Management Policy (Oct. 4, 2019), https://www.faa.gov/regulations_policies/orders_notices/index.cfm/go/document.information/documentID/1036752; FAA ATO, *Safety Management Systems Manual* (April 2019), https://www.faa.gov/air_traffic/publications/media/ATO-SMS-Manual.pdf; FAA, *Advisory Circular AC 120-92B*, Subj: Safety Management Systems for Aviation Service Providers (Jan. 8, 2015), https://www.faa.gov/documentLibrary/media/Advisory_Circular/AC_120-92B.pdf; Amazon Prime Air, *Petition for Exemption Under 49 U.S.C. Sect. 44807 and 14 C.F.R. Parts 61, 91, and 135* (July 16, 2019), p. 16, *available at* https://www.aviationtoday.com/wp-content/uploads/2019/08/amazon_-_exemption_rulemaking.pdf; https://www.federalregister.gov/documents/2019/08/08/2019-17010/petition-for-exemption-summary-of-petition-received-amazon-prime-air ("Prime Air's SMS. . . provides the structure and tools necessary to establish safety policies, identify and mitigate risks, audit those risk mitigations, and promote safety and drive continuous improvement throughout the organization.").
705. Such systems should, *inter alia*, leverage conformance monitoring data (*see supra* Chap. 2, Sect. 2.3.6 *Conformance Monitoring* & Sect. 2.3.10 *Conflict Advisory and Alert*).
706. Isabel Del Pozo, Airbus & Mildred Troegeler, The Boeing Co., *A New Digital Era of Aviation, and the Path Forward for Airspace and Traffic Management,* ICAO, Uniting Aviation, ICAO ANC meeting (June 15, 2020), https://www.airbusutm.com/a-new-digital-era. *See* [SESAR ARIA], Sect. 3.4 U-space and urban air mobility (recognizing need for "[n]ew safety assurance modelling and assessment methodologies applicable to U-space"); [Comm Implementing Reg. (EU) 2021/664 - U-space], Art. 5 Common information services, ¶ 5(a) (*data quality requirements* per ATM/ANS.OR.A.085(b) of Comm Implementing Regulation (EU) 2017/373 – common requirements for providers of air traffic management/air navigation services); RTCA DO-200B, *Standards for Processing Aeronautical Data*, https://my.rtca.org/NC__Product?id=a1B36000001IclyEAC ("provides a level of assurance for maintaining data quality throughout all phases of the data handling process" via a quality management system (QMS)); [AC 20-153B], Sect. 3.1. ("provides a means for organizations to obtain FAA acceptance of their aeronautical data processes demonstrating compliance with RTCA/DO-200B").
707. Configuration Management (CM) in a federated UTM ecosystem is particularly challenging. *See, e.g.,* Email from Andrew Carter, Pres. and Co-founder, ResilienX, Inc. (Aug. 20, 2020) (on file with author) ("It is PAINFUL to change even a single line of code. In UTM, NASA literally had to make a rule during TCL-4 that USSs couldn't update their software while a drone was in the air during a test. It is so easy to push new web-based software that the UTM industry, in my opinion, actually needs protections from this. We need to verify that the version and configuration or the ecosystem components is the same version and configuration that was actually qualified."). *Compare* Email from Hunter Hudson, Head of Autonomy Technology, Northrop Grumman (Aug. 19, 2020) (on file with author) (emphasis added):

> The system complexity is increasingly being driven by the software that is in the systems. Defense systems have an imperative need to continue to operate in challenging environments. We need the capability to modify software to mitigate successful cybersecurity attacks. So what is the rate that software can be modified and pushed out to a fleet of air vehicles in a safe and compliant manner? This speaks to the

challenge that automated digital engineering needs to solve. All the airworthiness artifacts need to be produced as part of the automated software engineering build, test, and deployment process and depend less on manual human inspection. The reality is that we need the capability of responding to cyberattacks, including modification of software, in near real time and that really raises the bar. The mission is to have the ability to manage an airworthy system that is *certified and recertified as the software is changing on a daily basis.*

Consider Dr. Michael Tanner, SAF/AQ, DevStar, https://software.af.mil/dsop/dsop-devstar/ (urging "effective warfighting capabilities at the speed of relevance" characterized, in part, by "Speed. Tight cycle time enables user feedback; reduces integration risk").

708. [Comm Implementing Reg. (EU) 2021/664 - U-space], Art. 18 Tasks of the competent authorities, ¶ (m).
709. Email from Andrew Carter, *supra*. *Cf.* [Comm Implementing Reg. (EU) 2021/664 - U-space], Art. 18 Tasks of the competent authorities, ¶¶ (n) & (o) ("monitor and assess the levels of safety performance").
710. National Academies of Sciences, Engineering, and Medicine, *In-Time Aviation Safety Management: Challenges and Research for an Evolving Aviation System,* Nat'l Academies Press, DOI:10.17226/24962 (2018), p. 2, https://www.nap.edu/catalog/24962/in-time-aviation-safety-management-challenges-and-research-for-an. *See* Ersin Ancel, Ph.D., et al., Aerospace Eng'r, NASA, *Real-time Risk Assessment Framework for Unmanned Aircraft System (UAS) Traffic Management (UTM)* (2017), https://arc.aiaa.org/doi/10.2514/6.2017-3273 (introducing the UTM Risk Assessment Framework (URAF), a real-time risk assessment approach); Kyle Ellis, Ph.D., et al., NASA, *In-Time System-Wide Safety Assurance, Concept of Operations*, Presentation at the Autonomy Workshop, NASA Ames Research Center, Moffett Field, Cal. (Aug. 6, 2019) (on file with author); Kyle Ellis, et al., *In-Time System-Wide Safety Assurance (ISSA) Concept of Operations*, ARC-E-DAA-TN74516-1 (Oct. 31, 2019), https://ntrs.nasa.gov/archive/nasa/casi.ntrs.nasa.gov/20190032480.pdf; [NASA UAM Vision ConOps], Sect. 4.3.1.2 Processes ("Each SMS is supported by In-time System-wide Safety Assurance (ISSA) capabilities which are systems that monitor data, make assessments, and perform or inform a mitigation action").
711. Email from Andrew Carter, Pres. and Co-founder, Resilienx, Inc. (Nov. 14, 2019) (on file with author).
712. NTSB News Release, *NTSB Publishes Preliminary Report for Investigation of Mid-Air Collision, Calls for Greater Safety Measures for For-Hire Flights* (May 22, 2019), https://www.ntsb.gov/news/press-releases/Pages/mr-20190522.aspx. *See* Chris Van Buiten, HAI, *Designing Urban VTOL Safety* (2020 Winter), https://www.rotor.org/resource?ArtMID=493&ArticleID=5138 (comparing S-92 helicopter safety to expected risk tolerance for urban eVTOL operations; "Perhaps we could assume that, in a mature eVTOL market, people would accept one fatal accident a year. But this frequency would require an approximate target accident rate of one per 100 million flight hours—100 times better than that of the current state-of-the-art S-92. A 100-fold improvement in safety will be possible if the eVTOL community embraces the most exacting design, manufacturing, testing, and regulatory standards.").
713. Email from Anthony Rushton, Ph.D., UTM Technical Lead, NATS (March 3, 2020) (on file with author).
714. As UTM systems go online, with each USS making real-time performance test results available to the competent authority, some USS logging requirements may be relaxed for governance purposes. *See supra* Chap. 2, Sect. 2.4.9, *UTM Data Exchange*.
715. *See, e.g.,* [FAA & NASA UTM RTT WG #3], Sect. 1.1.6 (FIMS data exchange includes "archived UTM data"); GUTMA, *UAV/Operator Flight logging Exchange Protocol*, https://github.com/gutma-org/flight-logging-protocol/blob/master/Flight_logging_protocol.md (selected logged items informed by a reconciliation of: CAST/ICAO

Common Taxonomy Team, ICAO, *Definitions and Usage Notes* (Dec. 2017), http://www.intlaviationstandards.org/Documents/PhaseofFlightDefinitions.pdf, and CAST/ICAO Common Taxonomy Team, ICAO, *Phase of Flight* (April 2013), http://www.intlaviationstandards.org/Documents/PhaseofFlightDefinitions.pdf).

716. Email from Benjamin Pelletier, UTM Stds. Mgr., Wing, and ASTM F38 UTM WG member (June 30, 2020) (on file with author) (also positing five categories of data logging: non-utilization, blocking other users, compliance, performance, and data-mining abuse).

See [FAA USS Performance Rules], Attachment D: FAA Quality Assurance Process, Approach to Rules Violations, p. 8 (addressing self-reporting).

717. Presentation at the FAA UAS Symposium, Balt., Md. (June 3, 2019). Larrow rightly raises a critical point. Many risk management protocols start with hazard identification. However, before something can be classified as a hazard (a condition that could cause a mishap/accident), one has to understand the system and its operational environment and how elements of those systems and environments either singly or in their interactions, could cause performance variations that ultimately result in an accident. Although hazards may not be readily evident, they must be determined by system analysis. Interactions between system and environmental elements that in their nominal states do not appear problematic may interact in certain relationships. Prof. Erik Hollnagel terms this "stochastic resonance." Careful system analysis is, therefore, essential. *See* Erik Hollnagel, et al., *The Functional Resonance Accident Model* (2004), *available at* https://pdfs.semanticscholar.org/a538/fc4f7c5ebd7c083aa8474786f6886300251f.pdf.

718. *See, e.g.*, NIST, SP 800-30, *Guide to Conducting Risk Assessments* (Sept. 2012), https://nvlpubs.nist.gov/nistpubs/Legacy/SP/nistspecialpublication800-30r1.pdf; ISO, *ISO/IEC 27001:2013, Information technology — Security techniques — Information security management systems — Requirements*, https://www.iso.org/standard/54534.html, Sect. 8.2 Information security risk assessment.

719. Stephen Plishka, Human Factors Eng'r, UAS Integration Office, AUS-420, FAA, Remarks (June 27, 2019). In some cases, seeking to solve human factors problems via automation may have the unintended effect of merely emphasizing the human limitations of the design process, described by Professors Adamski and Westrum as *requisite imagination.* Anthony J. Adamski, et al., *Requisite Imagination: The Fine Art of Anticipating What Might Go Wrong,* in HANDBOOK OF COGNITIVE TASK DESIGN (Erik Hollnage, ed., 2003). Automation may also shift many human tasks of operation into monitoring tasks, something for which experience has shown humans do not excel.

720. Stephen Plishka, *id. See* [Rios TCL4], Sect. 5.A ("data exchanged between system components of UTM are not always displayed to the user").

721. [ICAO UTM Framework Ed. 3], Appn. H, UTM Risk Assessment and Contingency Planning, p. 42.

722. Letter from Glen A. Martin, VP, Safety and Technical Training, ATO, FAA to Michael Baum (June 22, 2020) (on file with author).

723. Regarding LAANC, *see* FAA, FAA Safety Risk Management Panel, Emerging Tech. Team, AJV-115, *Title 14 Code of Federal Regulations Part 107 Small Unmanned Aircraft System Rule: Conditions and Limitations for Allowing Operations in Class E Surface Area Safety Risk Management Document*, ver. 1.0 (Aug. 26, 2016) (copy on file with author).

Regarding a UTM Risk Model and Risk Message Capability, its demonstration is planned for "UTM Evaluation 4," a formal laboratory exercise planned for late 2021, led by FAA's ANG-C5 (Technology Development and Prototyping Division), in collaboration with other FAA entities and Embry-Riddle Aeronautical University, the Florida NextGen Testbed, and NASA.

724. Additionally, the need for "information from operators is critical and the result [in its absence] is no standard guidance for best practices. [Such information] would create

an incentive for UAS operators to report incidents." Peter Sachs, UTM Implementation Prog. Mgr., UAS Integration Office, FAA (formerly Safety and Risk Architect, Airbus), Remarks at the In-Time SWS Assurance ConOps Development webinar, System-Wide Safety Project, Aeronautics Research Directorate, NASA Langley Research Center (Sept. 26, 2019), https://www.nasa.gov/aeroresearch/programs/aosp/sws.

725. Nonetheless, the FAA is moving toward a reliability type of certification process based primarily on a review of past event rates via D&R. *See supra* Chap. 4, Sect. 4.2.2 *Who Assesses Conformity,* notes.
726. *See, e.g.,* James Reason, *Human Error,* Cambridge U. Press (1990).
727. Email (Dec. 27, 2019) (on file with author).
728. *See supra* Chap. 3, Sect. 3.2.11 *JARUS;* JARUS, *JARUS Guidelines on Specific Operations Risk Assessment (SORA),* Executive Summary, JAR doc 06 SORA *(package),* Ed. 1.0 (Jan. 30, 2019), http://jarus-rpas.org/content/jar-doc-06-sora-package (explaining the SORA approach and methodology). The SORA initiative has five active task forces: quantitative methods, cybersecurity, UTM, training, and standard scenarios. *See* Email from Giovanni Di Antonio, Chair, JARUS (April 23, 2021) (on file with author) (Explaining that in Europe, Art. 11 of Regulation (EU) 2019/947 (Implementing Regulation) lists the principles under which a holistic risk assessment must be undertaken to obtain an UAS operation authorizaton in the "specific" category. Art. 11 principles are precisely those implemented by JARUS SORA. Moreover the JARUS SORA has been endorsed by EASA, with some modifications, as an Acceptable Means of Compliance (AMC) to Art. 11).
729. JARUS, *JARUS Guidelines on Specific Operations Risk Assessment (SORA),* Ed. 2.0 (Jan. 30, 2019), JAR-DEL-WG6-D.04, http://jarus-rpas.org/sites/jarus-rpas.org/files/jar_doc_06_jarus_sora_v2.0.pdf ["JARUS SORA"]. *Cf.* ASTM, F3178-16, *Standard Practice for Operational Risk Assessment of Small Unmanned Aircraft Systems (sUAS),* https://www.astm.org/Standards/F3178.htm.
730. Recognize that "operator error" cannot be classified as a "root cause" of undesirable outcomes. This conclusion provides a description not an explanation of these outcomes. Further analysis of the system, including those elements that govern selection, training, qualification, and supervision of human operators, as well as other system and environmental factors must be considered to explain *why* operators erred.
731. Andy Thurling, CTO, NUAIR Alliance, *Assessing Risk with SORA and Andy Thurling of NUAIR,* InterDrone Podcast, Episode 24 (May 26, 2019), https://soundcloud.com/michael-pehel-690682997/epsiode-24-andy-thurling. The SORA approach of enfranchising operators to mitigate designated risks has found increasing endorsement.
732. *See* [JARUS SORA], List of Annexes & Sect. 1.5(h) (referencing ANNEX H—*Unmanned Traffic Management (UTM) implications for SORA*, and stating that certain UTM/U-space service provider "services may support an operator's compliance with their safety obligation and risk analysis"); Jarrett Larrow, UTM Prog. Mgr., FAA, and Dr. Marcus Johnson, NASA, *Annex H— SORA & UTM* (2018), https://ntrs.nasa.gov/archive/nasa/casi.ntrs.nasa.gov/20180004530.pdf.
733. Additionally, a separate Annex F under development quantifies the ground risk component of SORA. One of many advances, amidst the reams of math and simulation results, is normalizing the mitigation values. For example, "-1" must correspond to a demonstrated order-of-magnitude (10-fold) reduction in fatalities expectation (no longer defined in Annex F, as merely "level of risk"), and "-2" would correspond to a 100-fold reduction in fatality expectation. Getting "-2" for population data alone will be a high bar and require additional work to prove out how a UTM service can actually do so.
734. Peter Sachs, UTM Implementation Prog. Mgr., FAA (formerly, Safety and Risk Architect, Airbus), Presentation at the GUTMA Annual Conference, Portland, Or. (June 19, 2019) (e.g., resolving definitions such as what are "sparse" vs "populated" areas, and noting that to make an inference of 95 percent certainty you need 62 days of data, and a year of data in less busy areas—"a nontrivial amount of work").

735. Andy Thurling, CTO, NUAIR Alliance, *The Drone Market*, Enterprise Tech. Rev. (Aug. 1, 2019), p. 30, https://www.enterprisetechnologyreview.com/magazines/August2019/Display_Tech/#page=29. *Cf.* Giovanni Di Antonio, Chair, JARUS (April 23, 2021) (on file with author) ("[R]equirements and standards will have to be developed in order to define the level to which these services are effective, with the aim to quantify the amount of risk mitigation an Operator can claim when using a UTM/U-space service.").
736. This should be a critical element across UTM and ANSP providers, authorities, and individual UAS operators. *But see* Andrew Carter, Pres. and Co-founder, Resilienx, Inc. (Nov. 14, 2019) (on file with author) ("the problem with UTM is that none of the risk mitigations are quantifiable. UTM is just a risk mitigation service…nobody can assign a quantifiable value to them, so how do I know I get under the risk threshold? SORA is very specific. When you try to broaden it, it becomes more difficult.").
737. This may not only contribute to normalizing their use in this domain, but importantly, help formulate some key outstanding issues. Annex H includes an appendix dedicated to required SLA elements.
738. This is a foundational aspect of Annex H. "It contains a framework for dividing responsibility between the operator, competent authority and USP, where presently in SORA (and many other contexts) it presents a binary/exclusive relationship between operator and regulator." Email from Peter Sachs, UTM Implementation Prog. Mgr., FAA (formerly, Safety and Risk Architect, Airbus) (Aug. 3, 2020) (on file with author).
739. Peter Sachs, *id.*, Presentation at the GUTMA Annual Conference, Portland, Or. (June 19, 2019).
740. *See* Parimal Kopardekar, Ph.D., Dir., NARI, NASA, Remarks, in Moffett Field, Cal. (Aug. 16, 2019) (asserting that "[w]e can't just dump the hardest problems on the regulators [such as qualification]. To expect the FAA to keep to [an extremely fast] pace with the great safety issues [is unrealistic]. So, we need to find a way to shift the responsibility.").
741. Andreas Lamprecht, Ph.D., CTO, AIRMAP, Presentation at ICAO, DRONE ENABLE/3, Montreal, CA (Nov. 13, 2019). *Cf.* Email from Nathan Ruff, CEO, UASidekick (Mar. 22, 2021) (on file with author) ("If the LAANC system is not operating correctly and kicking out approvals in airspace volumes with high conflict potential, then safety would definitely be at risk and a dangerous manned/unmanned intersection could precipitate.").
742. *See supra* Peter Sachs (describing efforts to automate the SORA process); Chap. 4, Sect. 4.2.5 *Continuing Compliance and Safety Assurance* (regarding the NASA ISSA initiative). Separately, a need to transcend SORA's specific operations approach has been urged—to advance a more generalized approach to UAS/UTM flight rules akin to air carrier regulations. SORA version 3.0 may enhance quantitative methods for evaluation of UTM/U-space. *See also* AirHub (example of an early automated tool to determine SORA Air Risk Class in the Specific Category), https://sora.airhub.app/#/login. *Cf.* Email from Adam Morrison, Owner, Streamline Designs LLC (Mar. 16, 2021) ("I see some value in SORA as a tactical solution to some of today's near-term challenges across limited use-cases. I don't see it as a robust, long-term solution for this industry and I would hope to see it deprecated within the next 5 years.... Don't just keep overloading SORA trying to get it to solve all problems."); Email from Giovani Di Antonio, Chair, JARUS (April 23, 2023) (on file with author) ("I do not personally share Adam Morrison's opinion about the SORA. As matter of fact I believe SORA is a powerful and flexible holistic risk assessment able to face a lot of even complex operations carried out in the medium-risk "specific" category. For what I can gather from industry, they are generally quite happy with this tool (where design threat barriers are balanced with operational mitigations), that can be used instead of applying a traditional certification-like approach. After all, we still have the certified category to handle very high risk operation. Eventually I expect it'll survive. … Though, please notice, I do not know which type of operation or types of vehicle Adam Morrison was referring to.").

743. EUROCONTROL and EASA, *UAS ATM Flight Rules, Discussion Doc.,* Ed. 1.1 (Nov. 27, 2018), Sect. 3.3, *available at* https://www.eurocontrol.int/sites/default/files/publication/files/uas-atm-integration-operational-concept-v1.0-release%2020181128.pdf. Applicable EU flight rules are found in the *Standardised European Rules of the Air* (SERA), EASA, Commission Implementing Regulation (EU) 2016/1185, https://www.easa.europa.eu/document-library/regulations/commission-implementing-regulation-eu-20161185, https://www.easa.europa.eu/regulation-groups/sera-standardised-european-rules-air. Applicable U.S. flight rules are found largely in 14 C.F.R. Parts 91 and 107, *available at* https://www.law.cornell.edu/cfr/text/14/chapter-I/subchapter-F.
744. [FAA ConOps UTM v2.0], Sect. 2.7.1.2 Separation Provision/Conflict Management.
745. [Chicago Convention], *Annex 2,* www.icao.int, *available at* https://www.icao.int/Meetings/anconf12/Document%20Archive/an02_cons%5B1%5D.pdf (The *Rules of the air* are included in ICAO's SARPS and PANS).
746. Olivier Mrowicki, Network Mgr., Programme Manager, SASM/ATS/ATFCM Procedures, EUROCONTROL, Presentation at ICAO, DRONE ENABLE/3, Montreal, CA (Nov. 14, 2019).
747. Randy Willis, CANSO, Presentation at ICAO, DRONE ENABLE/3, Montreal, CA (Nov. 14, 2019) (also asserting the ICAO RPAS Panel's work will "lead the industry towards regulatory requirements").
748. [ICAO UTM Framework Ed. 3], Gaps, Issues and Challenges, *Rules of the Air,* p. 13 (emphasis added). *See* ICAO, Doc 10140, Assembly Resolutions in Force [as of 4 October 2019] (2020), A40-7: New entrants, https://www.icao.int/publications/Documents/10140_en.pdf (directing ICAO to review SARPs "relating to, inter alia, the rules of the air, air traffic services, certification, licencing, liability and the environment, for amendment or expansion as necessary, to facilitate the operation of New Entrants within a global, harmonized framework, taking into account regional frameworks and practices"); Joseph Rios, Ph.D., ATM-X Project Chief Eng'r, NASA Ames, Presentation at the FAA & NASA UTM Pilot Program Phase Two Industry Workshop, Moffett Field, Cal. (Dec. 9, 2019) (positing the following core operation principles: drones don't hit each other; "don't mess" with manned aviation; no anonymous access to airspace; common understanding of airspace; and provision for priority operations).
749. [EUROCAE OSED-DAA], Sect. 4.3.2.1., traffic. *Cf.* [CORUS ConOps], Sect. 3.2.2 Specific and General Flight rules in VLL ("VLOS and BVLOS drone operations are not compatible with VFR... Existing right of way rules are applicable not only to VFR traffic but also for VLL flights with drones.").
750. *See, e.g.,* [EUROCONTROL UAS ATM Integration], Sects. 2.2 & 2.3.
751. [India UTM Policy], Sect. 11.2 Flight Rules for Unmanned Aircraft Systems.
752. Parimal Kopardekar, Ph.D., Dir., NARI, NASA, Remarks at *UAM Virtual, Integrating Urban Air Mobility into the Airspace,* AVIATIONWEEK (Aug. 12, 2020). *See* Email from Dr. Kopardekar (Aug. 15, 2020) (on file with author) ("With digitization and connectivity everyone has the same situation awareness. We no longer require differences such as VFR and IFR since everyone can see everyone electronically.... Once we have the connectivity and situation awareness, fundamentally we can change the role of ATC."); NASA, David J. Wing, et al, *New Flight Rules to Enable the Era of Aerial Mobility in the National Airspace System,* NASA/TM–20205008308 (Nov. 2020), https://www.researchgate.net/publication/346321764_New_Flight_Rules_to_Enable_the_Era_of_Aerial_Mobility_in_the_National_Airspace_System (DFR augmenting IFR and VFR; introducing "a new digital layer between the environmental measurement and human actors" providing greater flexibility and access; and "capable of being introduced gradually.").
753. [FAA ConOps UTM v2.0], Sect. 2.7.1.2 Separation Provision/Conflict Management, note. 23. *See* Jay Merkle, Exec. Dir., UAS Integration Office, FAA, Presentation at *GUTMA, High Level Webinar Session - Ask the Experts* (June 9, 2020), https://gutma.org/ask-the-experts/ (considering expectations for operators: "if you're not participating in air traffic management you will be participating in UTM." Manned aircraft flying at VLL "are encouraged to participate" in the UTM system and help with separation from

unmanned aircraft by sharing their position cooperatively); Email from Jon Standley, Business Development Lead, Aviation Systems / L3Harris Technologies (Aug. 20, 2020) (on file with author) ("Although participation in UTM may not be required for all UAS operations, the sharing of state and intent data with UTM participants either directly or through ANSP supported mechanisms (i.e., UVR) will increase the overall awareness of all flight operations and create a safer and more efficient environment."); [Comm Implementing Reg. (EU) 2021/664 - U-space], Art. 11, Traffic information service, ¶¶ 1, 2 (requiring "information on any other conspicuous air traffic [including] manned aircraft…").

754. *See infra* Chap. 6, Sect. 6.2.2 *Advanced Automation and Autonomy*; David J. Wing, et al., Langley Research Center, *Autonomous Flight Rules A Concept for Self-Separation in U.S. Domestic Airspace*, NASA/TP–2011-217174 (Nov. 2011), https://ntrs.nasa.gov/archive/nasa/casi.ntrs.nasa.gov/20110023668.pdf; Email from David J. Wing, Langley Research Center (Aug. 7, 2020) (on file with author) ("We devised the AFR concept to be a practical implementation of the RTCA vision of Free Flight from the 1990's.… [M]ost of the ideas of AFR could well be applied to UAS and the emerging Urban Air Mobility vision for aircraft autonomy, with a key distinguishing feature being that AFR enables mixed operations of AFR, IFR, and VFR sharing the same airspace without segregation.").

755. Michael Glasgow, Technical Standards and UTM System Architect, Wing, Remarks at the ASTM F38 UTM WG meeting (Dec. 11, 2020) (proposing a standard abstracted priority mechanism that could accommodate various regulatory bodies' prioritization schemes for what kinds of flights could be permitted to co-exist in the same airspace and how those priorities interact when new flights are planned). *See* David Murphy, Chief Architect and Product Mgr., ANRA Technologies, Remarks at the ASTM F38 UTM WG meeting (Dec. 11, 2020) (urging a "regulatory overlay" to the UTM specification to provide the required response upon conflict detection).

756. [ICAO UTM Framework Ed. 3], App. D, UTM-ATM Boundaries and Transition, p. 26.

757. *See infra* Chap. 4, Sect. 4.6 *State, Regional, and Local* (re stakeholder engagement challenges).

758. Remarks at the NASA, AAM Airspace WG (webinar) (Aug. 4, 2020).

759. Cameron R. Cloar, Unmanned Aircraft in the National Airspace (Donna A. Dulo, ed., 2015), p. 100.

760. [ICAO UTM Framework Ed. 3], p. 13. *See* Ron Johnson, Assoc. Deputy Admin., ARMD, NASA, Presentation at the NASA-UTM Project Technical Interchange Meeting (TIM) (Feb. 23, 2021) ("I think we're in the initial stages of the airspace access question.").

761. EASA/EC, *EASA/EC workshop on the U-space regulatory framework Discussion with Member States and Industry,* TE.GEN.00404-003, Cologne (May 14-15, 2019), Sect. 2, p. 2, https://www.easa.europa.eu/sites/default/files/dfu/U-space%20workshop%20-%20discussion%20paper.pdf. *See infra* **Figure 4.1** - U-space VLL Volume types.

762. [ICAO UTM Framework Ed. 3], p. 13 ("*with the potential of establishing UAS-specific flight rules*"). *See* Jay Merkle, Exec. Dir., UAS Integration Office, FAA, Presentation at *GUTMA, High Level Webinar Session - Ask the Experts* (June 9, 2020), https://gutma.org/ask-the-experts/ ("Right now we view UTM closer to VFR than IFR.").

763. [EASA Reg. Framework U-space], Sect. 2.4 (the draft regulation therefore "does not contain any airspace classifications").

764. Air Line Pilots Ass'n, *Remotely Piloted Aircraft Systems, Challenges for Safe Integration into Civil Airspace,* ALPA White Paper (Dec. 2015), p. 14, http://www.alpa.org/-/media/ALPA/Files/pdfs/news-events/white-papers/uas-white-paper.pdf?la=en. *Cf.* [EASA Opinion 01/2020 - U-space], Sect. 3.5.8 Impact on General Aviation ("There is a new obligation for the manned aircraft to provide its position to the USSP…"); [Comm Implementing Reg. (EU) 2021/664 - U-space], Art. 7 Obligations for operators of manned aircraft operating in U-space airspace.

765. *See, e.g.,* [FAARA], Sect. 341, Definitions, Integration of civil unmanned aircraft systems into national airspace system. (includes, "(E) creation of a safe airspace designa-

tion for cooperative manned and unmanned flight operations in the national airspace system;").

766. *See, e.g.,* Alison Ferguson, *Pathfinder Focus Area 2, Phase III Report* (2018), https://tinyurl.com/Pathfinder-BVLOS (presenting noteworthy UAS research contributing to an understanding of airspace requirements; also recognizing need for "UTM-like systems" as BVLOS systems proliferate); [Volocopter Roadmap], Sect. 4.5.2 ("Simply stated, there can be no urban air taxi service without airspace integration.").
767. FAA, *Aviation Rulemaking Committee Charter*, Subj: Airspace Access Priorities Aviation Rulemaking Committee (Effective Nov. 20, 2017), 2, https://www.faa.gov/regulations_policies/rulemaking/committees/documents/media/Airspace%20Access%20Priorities%20ARC%20Charter%20(FINAL).pdf (emphasis added).
768. FAA, *Advisory and Rulemaking Committees, Unmanned Aircraft Systems (UAS) in Controlled Airspace ARC* (Effective June 6, 2017; Amended Nov. 20, 2017), https://www.faa.gov/regulations_policies/rulemaking/committees/documents/index.cfm/committee/browse/committeeID/617.
769. Telephone Interview with Heidi Williams, Dir. of Air Traffic Services, Nat'l Bus. Avi. Ass'n, Remarks (June 27, 2019). Jay Merkle, Exec. Dir., UAS Integration Office, FAA, Presentation at the World ATM Congress Virtual Panel: UTM & ATM: Integrate Now or Integrate Later? (Oct. 13, 2020) ("I think we have to assume that the right of way rules will evolve – and there will be a time when traditional traffic will yield to drone traffic."); Jay Merkle, *id.*, Presentation at the Electric VTOL Symposium, Vertical Flight Society (Jan. 26, 2021) ("The traditional operators in this airspace will not easily yield their positions, and there will have to be some detailed discussions on how to do scale operations, particularly in already densely dense aerospace and urban areas—and the ability to get aircraft in and around while the other operator's operations are ongoing.")
770. Tom Prevot, Ph.D., Dir., Air Taxi Product Lead, Joby Aviation, formerly Airspace Systems, Uber, Remarks, in Palo Alto, Cal. (Aug. 9, 2019) (also underscoring a need to prove that UAS can operate safety without needing verbal controller clearances and two-way communications). *See infra* Chap. 6, Sect. 6.2.1 *Communications Infrastructure* (addressing Voice over IP – VoIP).
771. [Comm Implementing Reg. (EU) 2021/664 - U-space], Art. 2 Definitions; Art. 3 U-Space airspace ("a UAS geographic zone designated by Member States, where UAS operations are only allowed to take place with the support of U-space services").
772. Robin Garrity, ATM Expert, SESAR JU, in Montreal, CA, Remarks, in Montreal, CA (Nov. 13, 2019). *See* Email from Anthony Rushton, Ph.D., UTM Technical Lead, NATS (June 30, 2020) (on file with author) ("Ultimately the draft [Opinion] says it will be up to the state, not the EU, to decided which blocks of airspace are designated as U-Space; in-turn each targeted U-Space volume (e.g., London maybe?) will have one or more designated USSP. This differs from the US approach of having Federally approved USS to provide e.g., LAANC and not establishing target blocks of UAS airspace requirements (I think the current FAA concept is to delegate UAS volume restrictions to the USS)."); Benoit Curdy, Digital Transformation, Swiss FOCA, Remarks at the ASTM F38 UTM WG telecon (April 8, 2020) ("there will be U-space bubbles over cities").
773. [Comm Implementing Reg. (EU) 2021/664 - U-space], Art. 4. Dynamic airspace reconfiguration.
774. [CORUS ConOps], Sect. 3.1.1 - The three *types of airspace* volume and the services provided in each; *id.* Sect. 3.1.2 - The services available in the different volumes. [CORUS ConOps], Sect. 3.1.1, p. 33, *Types of Airspace* (categorizing VLL airspace types: X, Y, and Z as a function of conflict resolution—with incrementally greater service availability). *See infra* Chap. 5, Sect. 5.3 *U-space.*
775. Source: [CORUS ConOps], Sect. 3.1.7.
776. *See, e.g.,* [SESAR SRIA], Sect. 3.3 Capacity-on-demand and dynamic airspace. *See* [India UTM Policy], Sect. 11.3 Airspace Classification of UTM Airspace (recommending "a different classification (say U-Airspace)" and consideration of "different sub-categories under U airspace.").

777. [Embraer[X] 2019], p. 15 (emphasis added). *See* 14 C.F.R. Parts 107.41 Operation in certain airspace, and 107.51 Operating limitations for small unmanned aircraft (addressing sUAS operations in controlled airspace).
778. Parimal Kopardekar, Ph.D., Dir., NARI, NSA, Remarks at NASA's Advanced Air Mobility (AAM) Ecosystem Airspace Working Group: Why Corridors? (Sept. 1, 2020).
779. Source: Derived from FAA, *Airspace 101-Rules of the Sky.*
780. *See, e.g.,* 14 C.F.R. Part 71 – Designation of Class A, B, C, D, and E Airspace Areas; Air Traffic Service Routes; and Reporting Points, *available at* https://www.law.cornell.edu/cfr/text/14/part-71; and 14 C.F.R. Part 91, Subpart B – Flight Rules, *available at* https://www.law.cornell.edu/cfr/text/14/part-91/subpart-B (presenting requirements for operating in specific classes of airspace). *See also* Chap. 4, Sect. 4.4.3 *Corridors* (addressing buffers and MOAs).
781. Presentation at NBAA BACE, Las Vegas, Nev. (Oct. 23, 2019) (addressing airspace; also asserting "[w]e have a long way to go yet, but we'll get there.").
782. Standards may support both polygons with associated min/max altitude (facilitating simpler computation of 3D intersections—effectively reducing them to a 2D intersection + height-check), and a circle (radius and a point—particularly useful to denote constraints).
783. *See infra* Chap. 4, Sect. 4.7.6 *Data Rights and Privacy Management* (addressing relevant privacy issues).
784. Email from Maxim Egorov, Research Scientist, Airbus (Jan. 16, 2019) (on file with author).
785. Designing operation volumes "tightly," that is, with the least necessary volume would improve UAS location accuracy and decrease airspace/route reservation conflict with other operators seeking UTM airspace/route authorizations. *See* Andrew R. Lacher, Sr. Principal, Aerospace Research and Autonomous Systems, Noblis, formerly Sr. Mgr. for Autonomous Systems Integration, Boeing NeXt, Remarks at the ASTM F38 WG meeting, in Brussels, Belg. (Feb. 25, 2020) ("We need small volumes otherwise we could create walls in the sky."). As UAS performance improves, the declared airspace volume will tend to decrease and better accommodate anticipated increased traffic density.
786. UAS operational volumes, buffers, corridors, and associated airspace constraints are being designed to be performance-based. *See, e.g.,* [ASTM UTM Spec.]; Marcus Johnson, Ph.D., NASA Ames, *Unmanned Aircraft Systems Traffic Management (UTM): Discussion on Operational Volume* (copy on file with author) ["Johnson NASA"]; Marcus Johnson, Ph.D., NASA Ames, Presentation to the ASTM F38 UTM WG (Dec. 5, 2019); Jaewoo Jung, et al., *Applying Required Navigation Performance Concept for Traffic Management of Small Unmanned Aircraft Systems*, 30th Cong. of the Int'l Council of the Aeronautical Sciences (Deejeon, Korea–Sept. 25-30, 2016), https://ntrs.nasa.gov/archive/nasa/casi.ntrs.nasa.gov/20160011496.pdf.
787. Violation of certain boundaries may invoke safety actions, including incremental exigent notifications (reflecting change of states, e.g., nonconforming or contingency status) and ultimately flight termination (transition to an ended state). *See supra* Chap. 2, Sect. 2.3.11 *Contingency Management.*
788. *See, e.g.,* [SESAR SRIA], Sect. 3.2 Air-ground integration and autonomy (describing "A common 4D trajectory, shared between every application" and "defin[ing] the TBO concept and requirements for drones to operate in U-space, interoperable with TBO in ATM" supporting convergence by 2030).
789. Source: NASA, derived from [Johnson NASA].
790. Certain ongoing developments such as ASTM UTM standards development, have done away with many volume representations for simplicity—combining operational intent, conformance volume, and flight geography into a single type of volume representation that serves the functions of all three simultaneously.
791. Terminology for airspace volumes varies, such as between the NASA/FAA and ASTM approaches to the "onion" of flight volume definitions. The former specifies an actual

"popcorn trail" for the flight—the "Flight Geometry"—where the latter conflates into nominal operational intent volumes.

792. *Cf.* [EASA Easy Access 2019/945], Fig. PDRA-01.1 – Graphical representation for the SORA semantic model (presenting "Contingency Volume" encapsulating "Flight Geography").
793. [FAA ConOps UTM v2.0], Sect. 2.4.2.2 Obtaining a Performance Authorization. *See* Jay Merkle, Exec. Dir., UAS Integration Office, FAA, Presentation at *GUTMA, High Level Webinar Session - Ask the Experts* (June 9, 2020), https://gutma.org/ask-the-experts/ ("We can think of [performance authorization] not flight-by-flight, but more meeting the requirements for Required Navigation Performance.").
794. Performance requirements will determine for how long a UAS can exceed the OIV before needing to declare itself in contingency/rogue—and needing to land immediately or take other designated action.
795. Email from Robin Garrity, ATM Expert – Airport and Airspace User Operations, SESAR Joint Undertaking (Dec. 18, 2019) (on file with author). *See* Mike Glasgow, Technical Standards and UTM System Architect, Wing, and Co-chair, ASTM F38 UTM WG, Remarks at the ASTM F38 UTM WG telecon (Nov. 5, 2019) (describing buffers as "the last big jump" in UTM development).
796. [FAA ConOps UAM v1.0], Sect. 3, ¶ 2.
797. Jeffrey Homola, Sr. Research Assoc., Human Sys. Integration Div., SJSU/NASA ARC, *Design and Evaluation of Corridors-in-the-sky Concept: The Benefits and Feasibility of Adding Highly Structured Routes to a Mixed Equipage Environment*, AIAA (2012), p. 12, https://humansystems.arc.nasa.gov/publications/Homola_AIAA_GNC_2012.pdf.
798. *See, e.g.*, EUROCONTROL, *European Route Network Improvement Plan – Part 1, European Airspace Design Methodology – Guidelines, European Network Operations Plan* 2019-2024, Sect. 6.5 *Free Route Airspace (FRA) Design* (Dec. 19, 2019), https://www.eurocontrol.int/concept/free-route-airspace (describing free reroute. Perhaps FRA routes could terminate at the entry point to an airport with frequent UAS operations, or a vertiport corridor).
799. *See* Email from Parimal H. Kopardekar, Ph.D., Dir., NARI, NASA (Sept. 13, 2019) (on file with author).
800. *See, e.g.*, T. V. Nguyen, *Dynamic Delegated Corridors and 4D Required Navigation Performance for Urban Air Mobility (UAM) Airspace Integration*, J. of Avi./Aerospace Edu. & Research, 29(2) (2020), https://doi.org/10.15394/jaaer.2020.1828 (considering "specialized airspace sector and trajectories"); Brock Lascara, et al., *Urban Air Mobility Airspace Integration Concepts*, MITRE, 19-00667-9 (June 2019), p. 8, https://www.mitre.org/publications/technical-papers/urban-air-mobility-airspace-integration-concepts ["MITRE 2019"] (proposing *Dynamic Delegated Corridors* "similar to the notion of VFR corridors and VFR flyways, except todays VFR corridors and flyways are rather static.").
801. [FAA ConOps UAM v1.0], Sect. 3, ¶ 2 (such "[s]tructure evolves from current helicopter routing to UAM-specific corridors…"). *See* Steve Bradford, Chief Scientist for Architecture and NextGen Development, FAA, Remarks at the NASA, AAM Airspace WG (webinar) (Aug. 4, 2020) ("There will clearly be new structures in the charts. … It may be like a VFR corridor. We'll have to see.") *Compare* [NASA UAM Vision ConOps], Sect. 2.0, The UAM Operating Environment (describing the "UAM Operating Environment (UOE) [as a] flexible airspace area encompassing the areas of high UAM flight activity"); Aeronautical Charting Forum, Rec. Doc., FAA Control. No 17-02-318, Subj. Charting for helicopter routes designated to meet RNP 0.3, https://www.faa.gov/air_traffic/flight_info/aeronav/acf/media/RDs/17-02-318_Charting_heli-routes-RNP03.pdf (addressing TK routes). *See* John Walker, Remarks at ICAO DRONE ENABLE Symposium 2021 (April 14, 20201) (describing FAA IFR/GPS Helicopter TK routes as "a delicious way to start thinking" about UTM using the same route structure at lower altitudes).
802. Source: FAA, *ORDER JO 7210.3BB CHG 1*, Subj: Facility Operation and Administration (Effective Date: Jan. 30, 2020), Sect. 12-4. Charted VFR Flyway Planning Chart

Program, https://www.faa.gov/documentLibrary/media/Order/7210.3BB_Chg_1_dtd_1-30-20.pdf; FAA, *VFR Flyway Planning Chart*, https://www.faa.gov/air_traffic/flight_info/aeronav/productcatalog/planningcharts/vfrflyway/; 14 C.F.R. Part 93, https://www.govinfo.gov/content/pkg/CFR-2012-title14-vol2/pdf/CFR-2012-title14-vol2-part93.pdf (Special Air Traffic Rules authorizing diverse corridors).

803. *See, e.g.,* [Airbus Blueprint], p. 18 (observing that "[o]ver time, corridors may be replaced by new constructs or eliminated entirely with more sophisticated, high assurance technology").
804. Interview with Tom Prevot, Ph.D., Air Taxi Product Lead, Joby Aviation, formerly Dir., Airspace Systems, Uber, Remarks, in Palo Alto, Cal. (Aug. 9, 2019). *See supra* Steve Bradford, ("I do think [corridors] need to be somewhat static... 'flexible' but not using the 'dynamic' word.").
805. Tom Prevot, *id.*, Presentation at Uber Elevate Summit 2018, Los Angeles, Cal. (2018), https://www.uber.com/it/it/elevate/summit/ ("Industry/operators manage aircraft within corridors and provide situation awareness to ATC as needed; ATC keeps manned aircraft outside."). *See Easy Access Rules for Unmanned Aircraft Systems (Regulation (EU) 2019/947 and Regulation (EU) 2019/945)* (March 3, 2020), https://www.easa.europa.eu/document-library/general-publications/easy-access-rules-unmanned-aircraft-systems-regulation-eu ["EASA Easy Access 2019/945"], Sect. C.5.2.2, Examples of mitigation by common airspace structure.
806. *See* [Embraer^X 2019], p. 17 (re "the urban airspace of the future will be structured with routes, corridors, and boundaries...").
807. *See, e.g.,* [EUROCAE OSED-DAA], Sect. 4.2 ("[c]orridors around the flight plan may be added to help maintain a safe distance from the different hazards, in accordance with the separation minima defined for the operation and expected to be larger than the separation minima used by the DAA RWC [remain well clear] capability.").
808. *See infra* Chap. 4, Sect. 4.6 *State, Regional, and Local; supra* Chap. 2, Sect. 2.2.6 *Other Participants and Stakeholders.*
809. John Scull Walker, Sr. Partner, The Padina Group & Chair, ISO/TC 20/SC 16, Remarks (May 16, 2019) ("Where are the merge points? What will the transition points be? What separation points?... This UTM thing will be operational then obsolete if not thinking about the connecting pieces."). *See* Steve Bradford, Chief Scientist for Architecture and NextGen Development, FAA, Remarks at the FAA UAS Symposium (July 8, 2020) ("Personally, there's got to be a mixing zone.").
810. *See, e.g.,* [FAA ConOps UAM v1.0], Sect. 1.4 ("Inside UAM Corridors:... UTM aircraft cross UAM Corridors").
811. Source: Derived from OneSky Systems.
812. [FAA ConOps UAM v1.0], Sect. 4.3.1.1.
813. [FAA ConOps UAM v1.0], Sect. 3.2.
814. Source: [FAA ConOps UAM v1.0], Sect. 1.4.
815. [DLR Blueprint], p. 8.
816. Tim McCarthy, et al., *Fundamental Elements of an Urban UTM*, Aerospace 2020 (June 27, 2020), Sect. 3.6.1, https://res.mdpi.com/d_attachment/aerospace/aerospace-07-00085/article_deploy/aerospace-07-00085.pdf.
817. John Scull Walker, Sr. Partner, The Padina Group & Chair, ISO/TC 20/SC 16, Remarks (July 22, 2020) (also stating, "I'm confident we'll have [corridor-like] substructure that may be grandfathered."). *See* John Scull Walker, Remarks at DRONE ENABLE Symposium 2021 (April 14, 2021) (urging consideration of TK routes for UAM).
818. Jeff Homola, UTM Integration and Testing Lead, NASA Ames, Presentation at the AeroClub of N. Cal. (Webinar) (July 23, 2020).
819. [Embraer UATM ConOps], Annex B.1.2, p. 56 (individual corridors may be open or closed as a function of active runway). Control of corridor airspace proximate to a vertiport could be delegated to the vertiport authority to best facilitate local operations. Nonetheless, this practice may raise fairness and competition issues to the extent scheduling, permits, fees or other indicia of selectivity are imposed by the vertiport.

See infra Chap. 4, Sect. 4.4.4 *Fair Access*; Chap. 4, Sect. 4.7.3 *Business Models and Competition.*

820. *See* Email from Tim McCarthy, Ph.D., Assoc. Prof., Maynooth University (July 4, 2020) (on file with author). Although the proposed drone safety tubes are characteristically dynamic, Dr. McCarthy posits:

> maybe a fixed model would work here i.e., one tube going in and one tube coming out - these could be stacked at 3 X levels - enabling 3 drones to arrive/depart at respective safety time/space intervals - general assumption might be that there might be only a handful of landing/take-off points on airport side i.e., no point in allowing drones to fly freely around active taxi/RWYs - but you would have thought that a space/time interval - similar to our "safety tube time" separation of 75sec would be adequate - this would allow (in our model) 144 (3 X 75 second spaced drones) drones to arrive/depart every hour. The control here would be the capacity of the urban UTM to handle the incoming/exiting drone - if the Urban UTM could handle more - then I see no reason to increase the number of fixed in/out tubes connecting urban UTM with airport. These "tubes" could be added to the existing landing/departing procedure plates for GA/Commercial traffic?

821. Heidi Williams, Dir., Air Traffic Services, NBAA, Remarks (June 27, 2019) (Further remarking that "I do think there is a happy medium and we certainly are not there. It will take bringing both sides of the table together. Probably means new airspace and structure.").
822. *See* [Baum 2019], Sect. 6.e. Display of Ad Hoc UTM Corridors, and Sect. 7.iv (urging "display of UTM corridors connecting airports"); [FAA ConOps UAM v1.0], Sect. 4.3.9 ("*Other NAS Airspace Users have the responsibility to know* about and meet the relevant performance and participation requirements to operate in, or cross, active UAM Corridors or avoid the active UAM Corridors. UAM Corridor definitions and availability will be publicly available.") (emphasis added). ["NASA UAM Vision ConOps"], Sect. 2.0 The UAM Operating Environment, ("UOE") ("The maximum possible extent of the UOE is static and *can be represented on traditional aeronautical charts.* The extent of this static, maximal UOE can be redefined and recharted over time following accepted methods." (emphasis added). UOE is a UTM "inspired construct."); *id.* Sect 4.1.1 Airspace Design ("a fixed maximal size that is tailored and charted based on the unique characteristics and needs of specific metropolitan areas [e.g., changeable in response to] the flow pattern at a nearby major airport…"). *Cf.* [EUROCONTROL UAS ATM Integration], Sect. A6.4 ("The different UAS operations and application of geo-fencing to specific areas should be described and *promulgated to all airspace users* to ensure a common understanding of airspace use and access requirements.") (emphasis added).
823. Email from Robert Champagne, ASTM F38 UTM WG member (July 13, 2020) (on file with author). *See* Jay Merkle, Exec. Dir., UAS Integration Office, FAA, Presentation at the Electric VTOL Symposium (Jan. 26, 2021) ("Corridors are great for aircraft that need to transit airspace infrequently, understanding that the traffic in that area typically gets priority. Those constructs just don't work for highly-scalable, high-density UAM or AAM.").
824. *See*, e.g., [FAA ConOps UTM v2.0], Sect. 2.5.6 Separation (focused on UAS operator remaining within the bounds of his/her flight volumes), Sect. 2.5.1 Participation (manned aircraft need not participate), Sect. 1.1 ("[s]olutions that extend beyond the current paradigm for manned aircraft operations, to those that promote shared situations awareness among Operators are needed.").
825. *Cf.* [FAA & NASA UTM RTT WG #3], Sect. 2.1, notes 6 & 7 (recognizing an alternative to external buffers as Area Based Operation Volumes (ABOVs) and Transit-Based Operation Volumes (TBOVs) expanded to compensate where "the UAS has minimal navigation capabilities (cannot handle trajectories with small margins of in-flight

error)" and "includes any geographic buffer required to account for the UAS' ability to maintain flight along the centerline").

826. *See, e.g., infra* Chap. 5, Sect. 5.2.3 *LAANC Safety Challenges*; [Baum 2019], Appx. 1, *Example 3: Marginal LAANC Airspace Buffer Inside FAF*, and *Example 4: Final Approach Path Without LAANC Buffer*. Separately, the size of minimum external buffers (or similar constructs) is presumably an airspace matter and thus regulatory, and may, in part, be guided by surveillance and airworthiness equipage requirements, manned aircraft airspace risk, human factors, navigational performance, and other factors.
827. FAA, *ORDER 7110.65X*, Subj: Air Traffic Control, Unmanned Aircraft System (UAS) Activity Information (Oct. 12, 2017), Sect. 9-3-2 Separation Minima, https://www.faa.gov/documentLibrary/media/Order/7110.65X_w_CHG_1_3-29-18.pdf (ATC provides explicit separation minima of at least 500 ft above/below the upper/lower limits of a MOA and certain other special activity airspace.); FAA, *ORDER JO 7210.37H*, Subj: En Route Instrument Flight Rules (IFR) Minimum IFR Altitude (MIA) Sector Charts (Jan. 3, 2019), Sect. 6.e.2(a), https://www.faa.gov/documentLibrary/media/Order/FAA_Order_JO_7210.37H__En_Route_Instrument_Flight_Rules_(IFR)_Minimum_IFR_Altitude_(MIA)_Sector_Charts.pdf; CAA of the UK, *Special Use Airspace – Safety Buffer Policy for Airspace Design Purposes*, Sect. 2.4 (Aug. 22, 2014), https://publicapps.caa.co.uk/docs/33/20140822PolicyStatementSafetyBufferPolicy.pdf ("an additional safety buffer would be required between the edge of the SUA containing such activity and adjacent airspace structures"); [Baum 2019], App. 1, note 56 (re MOA vs. UASFM separation).
828. Lorenzo Murzilli, Leader, Innovation and Digitization Unit, FOCA, Remarks, in Chicago, Ill. (May 1, 2019) (emphasis added).
829. Constraints on internal buffer size for AAM purposes may also heighten support for external buffers. *See* Steve Bradford, Chief Scientist for Architecture and NextGen Development, FAA, Remarks at the NASA AAM WG (Aug. 4, 2020) ("I don't think corridors should be that thick."). Email from Christopher T. Kucera, Head of Strategic Partnerships, OneSky Systems (April 15, 2019) (on file with author) ("Contingency isn't part of Total System Error, but perhaps it should be part of the corridor sizing.... Our buffers are just meant to act like a yellow light for tactical deconfliction.... It's nice to have a buffer."). *Cf.* Tim McCarthy, Ph.D., Sr. Lecturer, Maynooth U., Presentation at the Global ATM Congress, Madrid, ES (March 13, 2019) (describing "a 'safety bubble' as buffer of sorts"); [India UTM Policy], Sect. 11.4.1 Concept of Flight Bubble.
830. Remarks at the NASA, AAM Airspace WG (webinar) (Aug. 4, 2020).
831. Peter Sachs, et al., *Evaluating Fairness in UTM Architecture and Operations,* Airbus UTM, Ver. 1.1, TR-010 (Feb. 2020), p. 4, https://storage.googleapis.com/blueprint/UTM_Fairness_Tech_Report-v1.1.pdf ["Airbus UTM Fairness 2020"] (also stating that "[t]he price of fairness is significant because it quantifies the degradation of other metrics..." *id.* p. 21). *See* [Terrence Martin Interview] ("there is not yet consensus on what equitable access really means.").
832. [Airspace Access Priorities ARC]. Because of criticism of the term "equitable" purportedly implicating "equal" (that is, interpreted as synonymous with "identical") access, there is movement in the UTM industry to embrace the term "fair" access. *Cf.* [CORUS ConOps], Sect. 2.7.4 Equitable access; [Airbus UTM Fairness 2020], Sect. 2.2, p. 6 (defining fairness as "the state in which each stakeholder's welfare is increased to the extent possible, given limited resources, after taking proper account of disparate claims and individual circumstances" and clarifying that "'equity' refers to the special case of fairness when stakeholders with similar characteristics are treated the same").
833. [SESAR U-space Blueprint], pp. 1-2; [Comm Implementing Reg. (EU) 2021/664 - U-space], Art. 5 Common information services ("Access to common information services shall be granted to relevant authorities, air traffic service providers, U-space service providers and UAS operators on a non-discriminatory basis, including with the same data quality, latency and protection levels.").

834. *See* [Rios TCL4 Sprint 2], p. 5; [NASA USS Spec.], Sect. 12.3 Strategic Deconfliction (prioritization scheme includes: "MUST allow for preemption of operations with lower priority by those with higher priority." [UTM-CM.23], and "MUST be equivalently calculable by each USS given the same operation data." [UTM-CM.30]); Philip M. Kenul, Chair, ASTM F38 on Unmanned Aircraft Systems, *ASTM Int'l F38 Committee UTM Overview*, Presentation at the GUTMA Annual Conference, Portland, Or. (June 19, 2019) ("…expect to need updated versions of the core standards that incorporate…performance requirements for…*fair usage* of VLL (very low level) airspace.") (emphasis added). *Cf.* [EASA Reg. Framework U-space], Art. 7, *U-space priority rules* (requiring U-space service providers that grant UAS flight authorizations to UA operators to adhere to the following priority rules in the specified order):
   1. when conducting special operations, in the meaning of Article 4 of Commission Implementing Regulation (EU) No 923/2012, manned aircraft shall have priority over unmanned aircraft;
   2. when conducting special operations, in the meaning of Article 4 of Commission Implementing Regulation (EU) No 923/2012, unmanned aircraft shall have priority over any other air traffic;
   3. aircraft carrying passengers shall have priority over aircraft without passengers on board;
   4. manned aircraft shall have priority over unmanned aircraft;
   5. BVLOS operations shall have priority over VLOS operations.

Contention over the control and prioritization of airspace, particularly in areas of dense traffic, will likely accelerate. One industry observer stated:

> Everyone is going to want to "control" LA/Chicago/NY/MIA etc.…Airspace. Who is going to want to spend the money to set up package delivery in South Dakota or worse Montana (where the only customer is Ted Turner :). That's going to be a problem and will likely push multiple providers into one airspace sector forcing the harder to implement Federated System sooner…now back to SD.…How do those folks get served? Will the regulator (government) REQUIRE services in those areas in order to participate in say Denver? Makes sense to me, but of course would drive the cost up for everyone…kinda like the post office now…and we see where that is headed (shame).

Email from Jeffrey Richards, NATCA (Sept. 13, 2018) (on file with author).

835. [Airbus UTM Fairness 2020], Sect. 2.1, p. 5. *See* Email from Christopher T. Kucera, Head of Strategic Partnerships, OneSky Systems (Dec. 31, 2019) (on file with author) ("The concepts behind USS to USS interaction using ASTM standards enable sharing of flight plans and corridors for strategic planning and deconfliction. Negotiation for airspace is a part of the strategic deconfliction process when two users want the same airspace. Without checks and balances on airspace usage, including prioritization, we can't have equitable access. We need to understand how to regulate fair access to airspace in a UTM construct.").
836. [NASA USS Spec.], Sect. 10, UAS Volume Reservations. *See supra* Chap. 2, Sect. 2.3.11 *Contingency Management.*
837. Tom Prevot, Ph.D., Air Taxi Product Lead, Joby Aviation, formerly Dir., Airspace Systems, Uber, Remarks, in Palo Alto, Cal. (Aug. 9, 2019).
838. Benoit Curdy, Digital Transformation, Swiss FOCA, Presentation at the ASTM F38 meeting, Raleigh, N.C. (Nov. 6, 2019).
839. *See* Anthony Evans, PhD, Maxim Egorov, et al., Airbus UTM, *Fairness in Decentralized Strategic Deconfliction in UTM* (2020), p. 15, https://storage.googleapis.com/blueprint/Fairness_in_Decentralized_Strategic_Deconfliction_in_UTM.pdf, and https://doi.org/10.2514/6.2020-2203 (in part, simulating and observing "that there could be significant inequity if the file-ahead time is not constrained in a FCFS [first-come first-serve) allocation, but primarily at high demand levels"); [Comm Implementing Reg. (EU) 2021/664 - U-space], Art. 10 UAS flight authorization service, ¶ 10 (UAS flight

authorizations with same priority "processed on a first come first serve basis."). *See also* Peter Crampton, et al., *Combinatorial Auctions* (MIT Press, 2010).

840. Robert Roth, Head of Software Eng'g, Uber ATG (formerly, Eng'g Dir., Prime Air at Amazon), Presentation at the GUTMA Annual Conference, Portland, Or. (June 20, 2019).
841. *See supra* Chap. 2, Sect. 2.4.5.2 *Discovery and Synchronization Service* (describing a *Bad Actor List*).
842. Jessie Mooberry, formerly Head of Deployment, Airbus, Presentation at the GUTMA Annual Conference, Portland, Or. (June 20, 2019) (emphasis added). *Cf.* Email from Christopher T. Kucera, Head of Strategic Partnerships, OneSky Systems (Dec. 31, 2019) ("Perhaps it's time to consider airspace as a licensed resource, just as we do with spectrum? Railroads wouldn't operate as a share resource. They are managed by their owning organization and track time is leased to other operators. The same analogy takes place with the cellular industry owning the use of spectrum in certain bands that is optimized and managed for their customers. Maybe for new entrants like Urban Aerial Mobility (UAM) operators, we see the need for licensed airspace around vertiports that is owned by the vertiport and licensed by the regulator.").
843. Joseph L. Rios, ATM-X Project Chief Eng'r, NASA Ames Research Center, *Strategic Deconfliction: System Requirements, Final Report* (July 31, 2018), https://utm.arc.nasa.gov/docs/2018-UTM-Strategic-Deconfliction-Final-Report.pdf (referencing UTM-CM.30).
844. Marcus Johnson, Ph.D., UAM Grand Challenge Deputy Lead, NASA Ames, *Unmanned Aircraft Systems Traffic Management (UTM): Discussion on Operational Volume* (Dec. 5, 2019) (copy on file with author). *See* Andrew R. Lacher, Sr. Principal, Aerospace Research and Autonomous Systems, Noblis, formerly, Sr. Mgr. for Autonomous Systems Integration, Boeing NeXt, Remarks at the ASTM F38 meeting, in Raleigh, N.C. (Nov. 6, 2019) ("Someone could just flood the system with intent. So you might need some sort of anti-gaming approach."). Additionally, the ASTM UTM specification has proposed various performance and usage metrics to be logged for fairness audit purposes. The exact definitions of the metrics or parameters may need refinement or regulatory designation, but the concept is in place.
845. [FAA ConOps UTM v2.0], Sect. 2.7.3 Equity.
846. Email from Greg Deeds, CEO, Technology Exploration Group, Inc. (Aug. 21, 2019) (on file with author). Perhaps such race conditions could be mitigated by adopting a "Reasonable Time To Act" (RTTA). *See, e.g.,* SESAR JU, CORUS, *U-Space Concept of Operation,* Vol. 2, Ed. 03.00.02 (Oct. 25, 2019), Sect. 3.3.7 Reasonable time to act, https://www.sesarju.eu/node/3411 ["CORUS ConOps"]; [Airbus UTM Fairness 2020], Sect. 4.3, p. 14 (designating a time period during which all non-exigent flights would be given equal priority).
847. [FAA ConOps UTM v2.0], Sect. 2.7 Airspace Management.
848. Shinji Nakadai, Principal Researcher, NEC Corp., *Field Test in Japan and R&D on Coordinated Negotiation*, Presentation at the GUTMA Annual Conference, Portland, Or. (June 20, 2019), *citing* Amy R. Pritchett, et al., *Negotiated Decentralized Aircraft Conflict Resolution* (May 3, 2017), https://ieeexplore.ieee.org/abstract/document/7918615 (Nakadai suggests a neutral mediator would "[p]ropose mutually good solution for conflicting operators"; "AI to predict Operator's preferences"; and that it is "just advisory" with rejections fed-back for learning.).
849. *See, e.g.,* Shinji Nakadai, *id.*
850. Tom Prevot, Ph.D., Air Taxi Product Lead, Joby Aviation, formerly Dir., Airspace Systems, Uber, Remarks, in Palo Alto, Cal. (Aug. 9, 2019). *See* [Airbus UTM Fairness 2020], Sects. 5.2 and 7.1 (recognizing that air transport "is orders of magnitude more complex than [other domains and exhibits] technical, stochastic, political, legal, dynamic, and safety aspects" that demand consideration; and proposing a "Six-Step Evaluation Process": define the domain, determine operator utility and value criteria, determine allocation method, consider behavior incentives, measure and analyze fairness before and after allocation, and iterate steps 1-5, as needed).

851. Remarks at the ASTM F38 WG meeting (Jan. 9, 2020).
852. Presentation at AUVSI Xponential (Oct. 8, 2020).
853. *See generally* FAA, *Spectrum Engineering & Policy – Radio Frequency Bands Supporting Aviation*, https://www.faa.gov/about/office_org/headquarters_offices/ato/service_units/techops/safety_ops_support/spec_management/engineering_office/rfb.cfm (presenting a break-down of aeronautical RF spectrum); FCC, *FCC Online Table of Frequency Allocations*, 42 C.F.R. Sect. 2.106 (Rev. June 18, 2020), https://transition.fcc.gov/oet/spectrum/table/fcctable.pdf.
854. [Davis 2019], p. 17 (further identifying factors affecting spectral needs to include "peak data rate to be supported by a radio system, spectral efficiency, average data rate, [and] expected device density.").
855. *See, e.g.*, Communication from the Commission to the European Parliament, the Council, the European Economic and Social Committee and the Committee of the Regions, *Promoting the Shared Use of Radio Spectrum Resources in the Internal Market* (2012), *available at* https://ec.europa.eu/digital-single-market/sites/digital-agenda/files/com-ssa.pdf; Jennifer M. McCarthy, VP, Legal Advocacy, Federated Wireless, Inc. to Ian Atkins, FAA Spectrum Strategy and Policy, Comment in response to FAA draft pre-decisional overview regarding the Reauthorization Act of 2018, Sect. 374 Spectrum (Nov. 18, 2019) (urging consideration of Spectrum Access System (SAS) cloud-based dynamic spectrum management and sharing framework) (on file with author); Bob Brown, *FAQ: What the wireless world is CBRS?*, NETWORK WORLD (Feb. 26, 2020), https://www.networkworld.com/article/3180615/faq-what-in-the-wireless-world-is-cbrs.html (explaining the Citizens Broadband Radio Service (CBRS), a SAS in the 3.5GHz band, and its anticipated support for IoT devices; considering if this approach could enable diverse vertical applications of LTE in other CBRS bands supporting regional drone use cases off the traditional cellular networks); Hon. Ajit Pai, Chairman, FCC, Remarks at the Consumer Electronics Show, Las Vegas, Nev. (Jan. 7, 2020) (urging that to continue to deliver value for the American consumer, spectrum sharing models are essential; that "tech has already evolved for sharing"; and that "massive MIMO-multiple input multiple output could deliver a lot of value.").
856. *See e.g.*, Drone Alliance Europe, *Drones, UTM and Spectrum–A review* (2016), http://dronealliance.eu/wp-content/uploads/2016/06/Spectrum-Allocation-White-Paper-Drone-Alliance-Europe-fin.pdf.
857. Jennifer Richter, JD, Partner, Akin Gump, Presentation at AUVSI Xponential, in Chicago, Ill. (April 29, 2019).
858. [FAARA], Sect. 374 (Congressional study to include whether to permit sUAS to operate on L-band, 960-1164 MHz, and C-band, 5030-5091 MHz, including reallocation of 30 MHz of spectrum for non-federal and shared aviation use). These frequencies service the AM(R)S affecting safety and regularity of flight. The inquiry considers, in part, altitude limits, infrastructure, interference and its mitigation. *See* FAA, *Wireless Telecom. Bureau and Office of Eng'g & Tech. Seek Comment on UAS Ops in the 960-1164 MHz and 5030-5091 MHz Bands Pursuant to Sect. 374 of the FAA Reauth. Act of 2018*, DA 19-1207 (Nov. 25, 2019), https://www.fcc.gov/document/wtb-and-oet-seek-comment-section-374-faa-reauthorization-act.
859. FCC, *Report on Section 374 of the FAA Reauthorization Act of 2018*, prepared by the Wireless Telecom. Bureau, Office of Eng'g and Tech, FCC (Aug. 20, 2020), https://docs.fcc.gov/public/attachments/DOC-366460A1.pdf.
860. https://www.ntia.doc.gov/category/csmac. *See, e.g.*, CSMAC, *Unmanned Aircraft Spectrum Briefing* (April 22, 2020), https://www.ntia.doc.gov/other-publication/2020/unmanned-aircraft-spectrum-briefing-04222020.
861. AIA, *Petition To Adopt Service Rules for Unmanned Aircraft Systems ("UAS") Command and Control in the 5030-5091 MHz Band,* RM-11798 (Feb. 8, 2018), https://www.aia-aerospace.org/wp-content/uploads/2018/02/AIA-Petition-for-Rulemaking-on-UAS-2018-02-08-FILED.pdf (to extend use of spectrum designated for AM(R)S for LOS control links for UA).

862. *See supra* Chap. 3, Sect. 2.2.10 *ITU*.
863. *See, e.g.,* FAA, *Spectrum Efficient National Surveillance Radar (SENSR)*, https://www.faa.gov/air_traffic/technology/sensr/ (unlikely a factor until at least the mid-2020s); [FAARA], Sect. 571, Spectrum Availability ("(b) SENSE OF CONGRESS… that the SENSR Program of the FAA should continue its assessment of the feasibility of making the 1300–1350 megahertz band of electromagnetic spectrum available for non-Federal use."); and DoT, Office of the Inspector General, *DOT's Fiscal Year 2020 Top Management Challenges,* Rpt. PT2020003 (Oct. 23, 2019), p. 16, https://www.oig.dot.gov/sites/default/files/DOT%20Top%20Management%20Challenges%20%20FY2020.pdf (asserting FAA "faces a number of high risks and challenges in advancing SENSR" including revenue generation uncertainty from auction).
864. *But see* Email from Barbara Pareglio, Exec. Dir. for Connectivity for Aviation and Drones, GSMA (Aug. 13, 2020) (on file with author) ("I believe it will be more a case of hybrid systems than sharing.").
865. Email from Oliver Chapman, Policy Director, GSMA (Aug. 13, 2020) (on file with author), *citing* Alex Douglas, *Skyports collaborates with Vodafone and Deloitte on NHS drone deliveries*, Commercial Drone Professional (July 10, 2020), https://www.commercialdroneprofessional.com/skyports-collaborates-with-vodafone-and-deloitte-on-nhs-drone-deliveries/.
866. Matt Fanelli, Dir. of Strategy, Skyward, Presentation at the FAA UAS Symposium, in Balt., Md. (June 4, 2019) (For example, by reducing latency. "If a vehicle is traveling at 60 mph, it takes 4.4 feet to stop. In 5G it stops in 1 inch."). *See* Hon. Ajit Pai, Chairman, Federal Communications Commission, Remarks at the Consumer Electronics Show, Las Vegas, Nev. (Jan. 7, 2020) (characterizing "spectrum the lifeblood of 5G").
867. *See* Mark Davis, Pres., Crossbar Inc. (formerly Intel), et al., *Aerial Cellular: Tutorial for UAV operators considering cellular-based C2 links* (2019), slides 59-70, *available at* https://s3.amazonaws.com/unodel/assets/6946/FOyYv867TROUfzHYqPIS_Davis_Mark_20190905.pptx (presenting the converged tower concept).
868. *See* Chap. 6, Sect. 6.2.1 *Communications Infrastructure*.
869. Jay Merkle, Exec. Dir., UAS Integration Office, FAA, Keynote Presentation at the FAA UAS Symposium (July 8, 2020).
870. *See* Email from Hrishikesh Ballal, Ph.D., Founder and CEO, Openskies Aerial Technology Ltd (Sept. 24, 2019) (on file with author) ("This is actually a very broad topic. For example, in cities under multiple jurisdictions such as the Zurich metro area (with two / three separate cantons), if each canton runs their own bus services and they don't encroach on one another how must jurisdiction be redefined for effective urban mobility").
871. *See, e.g.,* Kate Fraser, Head of Public Policy, Uber Aviation, Presentation at Uber Elevate Summit 2019, Wash., D.C. (June 12, "Regulating ridesharing services is very much a local issue.").
872. State, regional, and local investment in UTM is material and growing. *See, e.g.,* NUAIR, *Governor* Cuomo *Announces First Segment of New York's 50-Mile Drone Corridor Receives Authority from FAA to Fly Beyond Visual Line of* Sight (Nov. 8, 2019), https://nuair.org/2019/11/08/first-segment-of-ny-50-mile-uas-corridor-receives-bvlos-authority/ (building on a 2016 $30 million NY State investment); followed by additional investment— New York State, *Governor Cuomo Unveils 23rd Proposal of 2020 State of the State: Investing $9 Million In Unmanned Aerial System Experimentation and Test Facility at Griffiss International Airport* (Jan. 5, 2020), https://www.governor.ny.gov/news/governor-cuomo-unveils-23rd-proposal-2020-state-state-investing-9-million-unmanned-aerial; and *North Dakota Investing $33M In UAS Infrastructure*, Business Facilities (May 1, 2019), https://businessfacilities.com/2019/05/north-dakota-investing-33m-uas-infrastructure/.
873. *See* [Embraer^{X} 2019], p. 13. Certain operational characteristic of UTM raise local government sovereignty and property owner rights; Ken Hanly, *Wing to bring drone delivery to Helsinki Finland,* Digital J. (May 18, 2019), http://www.digitaljournal.com/tech-and-science/technology/wing-to-bring-drone-delivery-to-helsinki-finland/article/

550075#ixzz5taZqV0W2 (recounting noise from neighborhood drone deliveries—yet failing to prevent regulatory approval for such deliveries):

> In the Bonython area of Canberra some residents are up in arms over the noise of delivery drones. Bonython resident Irena claimed the daily life of her family was severely impacted, and admitted she took her children away from the house for hours each weekend just to escape the sound of drones. Irena said: [video: https://www.abc.net.au/news/2018-11-09/noise-from-drone-delivery-service-divides-canberra-residents/10484044] "We're worried about the noise issue, the issue about privacy, we're worried about the wildlife that seems to have disappeared from the area — there aren't as many birds as before. From a quarter past seven in the morning we'd hear our first drone flyover and you won't be able to sleep for the rest of the day because the drones are flying from 7am to 4pm."

See generally Pavan Yenavalii, et al., Airbus UTM: Defining Future Skies, *An Assessment of Public Perception of Urban Air Mobility (UAM)* (2018), p. 10, *available at* https://drive.google.com/file/d/15YTiXudG_eb5IlkqqCMcE2SGvDBJA7_L/view (noise a major factor of public concern); Joseph Rios, Ph.D., ATM-X Project Chief Eng'r, NASA Ames, NASA in Silicon Valley Live - *Air Taxis and the Future of Flight,* Moffett Field, Cal. ("Noise is an issue [and] is something considered in the design of all the pieces of the system."). *See also* John Walker, Sr. Partner, The Padina Group & Chair, ISO/TC 20/SC 16, Remarks at ICAO DRONE ENABLE Symposium 2021 (April 14, 2021) (urging that "visual pollution—not just noise" must be considered).

874. *See, e.g.,* The Drone Integration and Zoning Act of 2019, H.R. __, 116th Cong. (2019), *available at* https://www.scribd.com/document/430621355/Drone-Integration-and-Zoning-Act#download, and S.2607, https://www.congress.gov/bill/116th-congress/senate-bill/2607/text#toc-idfd0c7e760acb4b20a0f299574143d901 (seeking to advance local governance and cooperative federalism).
875. *See supra* Chap. 3, Sect. 3.2.3 *Cellular Industry*; and telecommunication requirements for BVLOS corridors.
876. *Supra* Chap. 4, Sect. 4.1 *General—UTM Governance* (addressing environmental impact).
877. Email from Anna Mracek Dietrich, Co-Exec. Dir., Community Air Mobility Initiative (CAMI) (Dec. 13, 2019) (on file with author) (Urging that each of these things are part of the underlying mission of state government. Further observing that "[t]he combination may result in having close cooperation between different offices, but that's true of a number of transportation-related projects.... the focus is really on transportation planning (read: congestion reduction) and economic development.").
878. *See, e.g.,* Reinaldo Negron, Head of UTM, Wing, Presentation at the Global ATM Congress, Madrid, ES (March 13, 2019) (underscoring the need for deliberate, rigorous engagement and consultation with communities and affected persons as a best practice to advance public acceptance and implementation for delivery operations in the community). *See also* Parimal Kopardekar, Ph.D., Dir., NARI, NASA, Remarks, in Moffett Field, Cal. (Aug. 26, 2019) (emphasizing the need for a regional focus on modeling and simulation, and NASA's development of a responsive toolkit for regional and state use).
879. [FAARA], Sect. 373, *Federal and Local Authorities.*
880. *See* FAA, *UAS Integration Pilot Program*, https://www.faa.gov/uas/programs_partnerships/integration_pilot_program/ (involving local, state, and tribal entities); *supra* note 79 (describing IPPs).
881. *See, e.g.,* Uniform Law Commission, *Uniform Tort Law Relating to Drones Act, Introduction, Annual Meeting Draft* 2019, https://www.uniformlaws.org/committees/community-home?communitykey=2cb85e0d-0a32-4182-adee-ee15c7e1eb20&tab=groupdetails ("With the United States Congress and the FAA asserting jurisdiction over many aspects of unmanned aircraft operations, and states and local governments asserting jurisdiction over others, a patchwork quilt of regulatory and legal requirements is developing.").

882. Email from Brittney Kohler, Prog. Dir., Transp. & Infrastructure, Federal Advocacy, National League of Cities (NLC) (Dec. 10, 2019) (on file with author).
883. *See supra* The Uniform Law Commission, *Tort Law Relating to Drones Committee.*
884. Henry E. Smith, Fessenden Professor of Law, and Reporter for the American Law Institute's Restatement Fourth of the Law, Property, to the Nat'l Conf. of Comm'rs on Uniform State Laws, *Memorandum*, re: Comments on May 30, 2019 draft "Uniform Tort Law Relating to Drones Act" (June 20, 2019), https://www.uniformlaws.org/HigherLogic/System/DownloadDocumentFile.ashx?DocumentFileKey=44cb6696-a733-81f8-c9da-fb5c47852d59&forceDialog=0; Jason Snead, the HeritageFoundation(July8,2019),https://www.uniformlaws.org/committees/community-home?communitykey=2cb85e0d-0a32-4182-adee-ee15c7e1eb20&tab=groupdetails (asserting, *inter alia*, that the FAA "has made a regulatory determination that the federal legal term 'aircraft' includes drones; this decision, however, is not determinative of how states should address, for purposes of trespass law, extremely low-level drone flights."). *Cf.* Uniform Law Commission, *Comments of the unmanned aircraft industry to the Tort Law Relating to Drones Committee* (June 14, 2019), https://www.uniformlaws.org/HigherLogic/System/DownloadDocumentFile.ashx?DocumentFileKey=05b054ab-f5de-27e1-a87d-239501372755&forceDialog=0. During its annual plenary (Anchorage, July 12-18, 2019), the ULC tabled the draft for at least another year of drafting).
885. [FAA ConOps UAM v1.0], Sect. 7. *See* Steve Bradford, Chief Scientist for Architecture and NextGen Development, FAA, Remarks at the NASA, AAM Airspace WG (webinar) (Aug. 4, 2020) (suggesting CBRs will be developed through the standards bodies, such as for UTM.)
886. EASA/EC, *EASA/EC workshop on the U-space regulatory framework Discussion with Member States and industry,* TE.GEN.00404-003 (May 14-15, 2019), p. 3, https://www.easa.europa.eu/sites/default/files/dfu/U-space%20workshop%20-%20discussion%20paper.pdf. *But see*, Nat'l Conf. of State Legislatures, *Current Unmanned Aircraft State Law Landscape* (Sept. 10, 2018), http://www.ncsl.org/research/transportation/current-unmanned-aircraft-state-law-landscape.aspx (presenting diverse approaches to state law addressing UAS); Email from Anna Mracek Dietrich, Co-Exec. Dir., Community Air Mobility Initiative (CAMI) (Dec. 13, 2019) (on file with author) ("For UAS, this fragmentation has in many ways already occurred, with thirty-eight different states having laws covering more than a dozen different issues relating to drone operation within their jurisdiction.").
887. Rebecca Venis, Dir. Neighborhood Services, City of Reno, Nev., Remarks at the FAA UAS Symposium (Aug. 18, 2020).
888. GUTMA, *GUTMA General Assembly & new Board of Directors* (June 22, 2019), https://gutma.org/blog/2019/06/22/2019-gutma-general-assembly-new-board-of-directors/.
889. Lorenzo Murzilli, Murzilli Consulting, formerly Leader, Innovation and Digitization Unit, Swiss Federal Office of Civil Aviation, Remarks, in Chicago, Ill. (May 1, 2019).
890. *See e.g.*, [FAA ConOps UAM], Appn. C (Community Based Rules development for UAM); *Together in Safety*, https://togetherinsafety.info/ ("non-regulatory industry consortium…working together to improve safety performance"); Joe Barkai, *Collaboration is the New Competition: Industry Consortia and Why You Should Join One!* (Dec. 17, 2019), http://joebarkai.com/collaboration-is-the-new-competition-industry-consortia/. *See generally* Chap. 4, Sect. 4.4.4 *Fair Access.*
891. *See* RPAS CIVOPS 2019 – European Civil RPAS Operators & Operations Forum, *The Madrid Declaration of Intent* (Madrid – Jan. 23-24, 2019), https://rpas-conference.com/wp-content/uploads/2019/03/Madrid-Declaration-of-Intent_Final_190124_B_TR.pdf; *supra* Chap. 4, Sect. 4.6 *State, Regional, and Local.*
892. *See, e.g.*, Alnoor Peermohamed, *13 consortia get aviation ministry approval to operate drones,* The Economic Times (May 7, 2020), https://economictimes.indiatimes.com/industry/transportation/airlines-/-aviation/13-consortia-get-ministry-approval-to-operate-drones/articleshow/75585314.cms (approval by India's Ministry of Civil Aviation);); *BT & Altitude Angel Lead Consortium to Deliver 'Future Flight'*

Drones Project to Revolutionise Airspace, sUAS NEWS (Dec. 15, 2020), https://www.suasnews.com/2020/12/bt-altitude-angel-lead-consortium-to-deliver-future-flight-drones-project-to-revolutionise-airspace/; *infra* Chap. 5, Sect. 5.4 *Selected Country-Specific Implementations*; *ASV Global-Led Consortium Develops 'Long Endurance, Multi-Vehicle, Autonomous Survey Solutions'*, AUVSI NEWS (July 16, 2018), https://www.auvsi.org/industry-news/asv-global-led-consortium-develops-long-endurance-multi-vehicle-autonomous-survey (consortium including ASV Global (ASV), Sonardyne International Ltd., the National Oceanography Centre (NOC)).

893. Email from Peter van Blyenburgh, formerly Pres., UVS International (July 4, 2020) (on file with author) (also noting that industry consortia "have a very specific meaning in Europe [as being] temporary [and] formed to bid on calls for proposal[s] issued and funded by the European Commission or its agencies." Additionally, "industry consortia are referred to as 'industry stakeholder representative organisations' [with the purpose] to defend the interests of their members with the European Commission and its agencies."). *See* Email from Andy Updegrove, Esq., Gesmer Updergrove, LLC (July 7, 2020) (on file with author) ("This isn't an area where there is wide agreement on specific terms.").
894. GUTMA, https://gutma.org/.
895. Interview with Fanni Lukácsy, formerly Sec'y Gen., GUTMA, *in* UNMANNED AIRSPACE (June 7, 2019), https://www.unmannedairspace.info/latest-news-and-information/the-utm-market-should-be-open-competitive-where-service-providers-just-plug-in-and-do-their-business-fanni-lukacsy-gutma/.
896. Nacha, *About us*, www.nacha.org/about ("Nacha brings together diverse stakeholders to develop rules and standards that foster compatibility and integration across a range of payment systems.").
897. Email from Jane Larimer, Pres. and CEO, Nacha (Sept. 29, 2020) (on file with author).
898. *See* [ICAO UTM Framework Ed. 3], Appn. F, p. 37 ("ANSPs and even existing aviation regulators may not yet have the experience and capability to define such a dynamic validation mechanism. As such *it may be necessary to learn from non-aviation domains,* and to determine if alternative mechanisms are suitable for aviation purposes.") (emphasis added).
899. *For example, addressing the UTM Pilot Program (UPP)* (consortia via sponsored government research); and SESAR (research projects and demonstrations).
900. Examples include the Aerospace Industries Association (AIA), https://www.aia-aerospace.org/; American Institute of Aeronautics and Astronautics, www.aiaa.org; International Air Transport Ass'n (IATA), https://www.iata.org/ (*see, e.g.,* ICAO, A40-WP/342, EX/143 (Aug. 8, 2019), Agenda Item 26, The Safe and Efficient Integration of UAS into Airspace, Presented by CANSO, IATA, and IFALPA, https://www.iata.org/contentassets/e45e5219cc8c4277a0e80562590793da/safe-efficient-integration-uas-airspace.pdf), and the Vertical Flight Society, https://hover.vtol.org/home (a cacophony of constituent legacy aerospace associations offer expertise, institutional experience, and advocacy). Formerly, UVS International, https://uvs-international.org, terminated operations in early 2021 (initiatives included the Drone REGIM (REGulation IMplementation), a European drone community action "to produce consensually agreed guidance documents, contribute to existing standards efforts, and, in coordination with the NAARIC [National Aviation Authority Regulation Implementation Coordination Group], define consensually agreed recommendations on [urgent] topics" and accelerate European harmonization under EU drone regulations.). *See* Chap. 4, Sect. 4.2.3 *Basis of Conformity Assessment* (mentioning the FAA's Other Transaction Authority and OTA's facilitation of consortia).
901. Andrew Updegrove, *The Essential Guide to Standards* (2017), ConsortiumInfo.org, https://www.consortiuminfo.org/essentialguide/whatisansso.php#section1. *See supra* Email from Andrew Updegrove (stating, "can an SDO be a consortium?…

definitionally I'd say no. That said, some SDOs, like IEEE, have taken on many consortium-like attributes."); Chap. 3 *Standards-Making for UTM* (identifying UTM SDOs).

902. *See supra* Chap. 3, Sect. 3.2.6 *ICAO*; ICAO, *Organizations able to be invited to ICAO Meetings,* https://www.icao.int/about-icao/Pages/Invited-Organizations.aspx (presenting a list of aviation NGOs).
903. *See, e.g.,* The World Economic Forum (WEF) (funded by the World Bank and assisting developing nations with UAS); WEF, *Drones, Airspace Management and Infrastructure,* https://www.suasnews.com/2020/05/european-aviation-community-calls-for-cooperation-to-ensure-the-safe-integration-of-drones-in-european-airspace/; Scott H. Kimpel, *World Economic Forum Announces Global Consortium for Digital Currency Governance*, Hunton (Jan. 20, 2020), https://www.blockchainlegalresource.com/2020/01/world-economic-forum-announces-global-consortium-for-digital-currency-governance/; WEF, *Drone Innovators Network,* https://www.weforum.org/projects/drone-innovators-network.
904. *See, e.g.,* [RID NPRM 2019] (re economic models); [India UTM Policy], Sect. 13 UTM Service Charges (addressing various pricing models).
905. *See* Koen Meuleman, Co-founder, Unifly, Remarks, ICAO TV, *Aeronautical Information Management for Unmanned Traffic Management (AIM for UTM)* webinar (Mar. 11, 2021), https://www.icao.tv/videos/ep03-aeronautical-information-management-for-unmanned-traffic-management (the preliminary guidance document is expected to include recommendations to ICAO) ("The whole business model around UTM … free doesn't exist."); Email from Scott Blum, Ph.D., Dir. of Int'l Relations, Jeppesen (Aug. 21, 2019) (on file with author) ("Sources and availability of data are yet to be resolved, some of the data is either new or at a higher resolution than existing ATM data. States are not generally prepared for the new data requirements, and existing agreements (ICAO) do not cover it either. The providers of new data (i.e., microweather, or high-resolution cultural data) are not required to provide it by those agreements. Governance criteria will need to address this gap."); GAO, *Unmanned Aircraft Systems, FAA Should Improve Drone-Related Cost Information and Consider Options to Recover Costs*, GAO-20-136 (Dec. 17, 2019), https://www.gao.gov/products/GAO-20-136 (considering various cost-recovery models for UAS-related regulation, including UTM; and stating that "policy makers may decide not to recover these costs based on "the goal of promulgating UAS-related regulations…related to the general safety of the airspace [and thus] funded with general revenues." pp. 31-32); [FAARA], Sect. 360 Study on Financing of Unmanned Aircraft Services.
906. *See e.g.,* [EASA Reg. Framework U-space], Art. 5,¶5 ("The organization in charge of the common information function shall not be related or connected in any manner or form to any U-space service provider and shall not provide any U-space services itself."—raising industry concerns that certain providers, such as ANSPs, might be precluded from offering U-space services). *Cf.* EC, Comm. Staff Working Document, *A fresh look at the Single European Sky*, 4.3.3. Common information services for unmanned aircraft (i.e. drones), SWD(2020) 187 final (Sept. 22, 2020) https://eur-lex.europa.eu/legal-content/EN/TXT/HTML/?uri=CELEX:52020SC0187&rid=45 (addressing role of the ANSP and responsibility to "make data available at marginal cost. In addition, if an ANSP wishes to become a CIS provider, and in the interest of transparency and to avoid discrimination and cross-subsidisation, it should have separate accounts.").
907. *See, e.g.,* EASA/EC, *EASA/EC workshop on the U-space regulatory framework Discussion with Member States and industry,* TE.GEN.00404-003 (May 14-15, 2019), pp. 5-6, https://www.easa.europa.eu/sites/default/files/dfu/U-space%20workshop%20-%20discussion%20paper.pdf ("Who will pay for U-space? * The user.… The U-space market is subject to competition between various U-space service providers…important private companies are ready to invest as soon as the rules of the game will become clear.").

908. [SESAR ATM Master Plan 2017], Sect. 4.4 The ATM business model and the need for incentivization.
909. *See, e.g.*, Chap. 3, Sect. 3.2.3 *Cellular Industry*. Query how UTM undertaken primarily by MNOs (rather than ANSPs) impact sustainable business models?
910. *See, e.g.*, [RID NPRM 2019], Sect. XIV.B ("Remote ID USS may have a variety of business models."); Email from Scott Blum, Ph.D., Dir. of Int'l Relations, Jeppesen (Dec. 4, 2019) (on file with author) ("I specifically think of the ENAIRE [demonstrations that spawned the] concept of centralized, confederated, and federated USS models.").
911. EHang, *The Future of Transportation: White Paper on Urban Air Mobility Systems* (Jan. 15, 2020), Ch. 2, https://www.ehang.com/app/en/EHang%20White%20Paper%20on%20Urban%20Air%20Mobility%20Systems.pdf ["EHang 2020"]. *But see* Wolfpack, *EHANG: A Stock Promotion Destined to Crash and Burn* (Feb. 16, 2021), https://wolfpackresearch.com/research/ehang/. *See* Lorenzo Murzilli, Founder and CEO, Murzilli Consulting, formerly Leader, Innovation and Digitalization Unit at Federal Office of Civil Aviation Switzerland (FOCA), *DroneTalks* (Podcast) (Feb. 24, 2021), https://www.youtube.com/watch?v=gJ4ijfoFMSQ (Asserting "Consolidation is inevitable.").
912. *See infra* Chap. 5, Sect. 5.2 *Low Altitude Notification and Authorization Capability*; John Walker, Sr. Partner, The Padina Group. Remarks at ICAO DRONE ENABLE Symposium 2021 (April 15, 2021) ("We better start putting UTM/ATM in the business model"—referring to organizations that plan to incorporate emerging technologies, including UAS and other transportation robotics, as part of their operating system).
913. *See, e.g.*, [Swiss U-Space ConOps v1.1], Sect. 3.4.2.1 *Legal considerations and open platform* ("Providers of U-Space services should not impose restrictions on the use of equipment connecting to the U-Space system [and] the USPs should treat all traffic equally, without discrimination, restriction or interference, independently of its sender or receiver, content, application or service, or terminal equipment.... Any traffic management practices which go beyond such reasonable traffic management measures, by blocking, slowing down, altering, restricting, interfering with, degrading or discriminating between Operators should be prohibited, subject to justified and defined exceptions."). *See supra* Chap. 4, Sect. 4.4.4 *Fair Access;* Chap. 2, Sect. 2.4.5.2 *Discovery and Synchronization Service.*
914. *See* [FAA ConOps UAM], Appn. C – Glossary (CBRs – a "[c]ollaborative set of UAM operational business rules developed by the stakeholder community."). *See infra* Chap. 6, Sect. 6.5 *Advanced Air Mobility / Urban Air Mobility.*
915. Such law includes, for example, 49 U.S.C. Sect. 40101. Policy (a) Economic Regulation:

> ... the Secretary of Transportation shall consider the following matters, among others, as being in the public interest and consistent with public convenience and necessity:
>
>> (10) avoiding unreasonable industry concentration, excessive market domination, monopoly powers, and other conditions that would tend to allow at least one air carrier or foreign air carrier unreasonably to increase prices, reduce services, or exclude competition in air transportation.

and 49 U.S.C. Sect. 40101. Policy (f) Strengthening Competition. (strengthen competition to prevent unreasonable concentration in the air carrier industry), *available at* https://www.law.cornell.edu/uscode/text/49/40101.

Since the UTM ecosystem is led predominantly by some of the largest tech companies in the world, antitrust policy is relevant. *See, e.g.*, U.S. Dept. of Justice, News, *Justice Department Sues Monopolist Google For Violating Antitrust Laws* N.Y. Times (Oct. 20, 2020), https://www.justice.gov/opa/pr/justice-department-sues-monopolist-google-violating-antitrust-laws; Katie Benner, et al., *Justice Dept. Plans to File Antitrust Charges Against Google in Coming Weeks*, N.Y. Times (Sept. 3, 2020), https://www.nytimes.com/2020/09/03/us/politics/google-antitrust-justice-department.html?action=click&module=Top%20Stories&pgtype=Homepage (mentioning "lawyers in the antitrust division...describe[ing] it as the case of the century, on par with the breakup of Standard Oil after the Gilded Age"); David McLaughlin, *FTC Chief Says He's Willing to Break Up Big Tech Companies,* Bloomberg (Aug. 13, 2019), https://www.bloomberg.com/news/articles/2019-08-13/ftc-chief-says-willing-to-break-up-companies-amid-big-tech-probe; Hon. Joseph Simons, Chairman, FTC, Remarks at the Consumer Electronics Show, Las Vegas, Nev. (Jan. 7, 2020) (stating that "we're worried about the large tech entrenched players" because of evidence that GDPR is having the effect of creating "a barrier to entry in favor of the big companies."). *See infra* Chap. 4, Sect. 4.7.6 *Data Rights and Privacy Management.*

916. ICAO, *Air Navigation Services Provider (ANSPs) Governance and Performance,* ATConf/6-WP/73 (March 4, 2013), https://www.icao.int/Meetings/atconf6/Documents/WorkingPapers/ATConf.6.WP.073.2.en.pdf. *See* ICAO, *Anticompetitive Behaviors,* https://www.icao.int/sustainability/Compendium/Pages/1-2-Anticompetitive-Activities.aspx.

917. *See, e.g.*, [ICAO UTM Framework Ed. 3], p. 13 (stating, "challenges that must be addresses [include] Liability and Insurance implications for USPs in relation to UAS operators..."); EBAA, *We are all one in the sky Initiative,* European Business Aviation Assn (EBAA) (Nov. 13, 2019), Sect. 2, *available at* https://www.ebaa.org/news/joint-statement-we-are-all-one-in-the-sky/, stating, in part:

    2. Clarify the responsibilities and liabilities to be borne by different actors

    The responsibility and liability of all actors using and managing the airspace must be clearly defined. For instance, when considering the creation of volumes of 'U-Space' airspace in a flight information region (FIR) where ANSPs have full responsibility, the regulatory framework needs to make clear where responsibilities and liabilities lie.

    Separately, where U-space service providers "avail themselves of services of another service provider, they [must] have the agreements concluded to that effect, specifying the allocation of liability between them" [Comm Implementing Reg. (EU) 2021/664 - U-space], Art. 15 Conditions for obtaining a certificate, subsection (j).

918. Consider the applicability of certain established liability frameworks such as those governing international air transport (e.g., considering future large UAS delivering cargo that originated in another state, or UAM transport associated with international travel). *See, e.g., Convention for the Unification of Certain Rules for International Carriage by Air,* done at Montreal on 28 May 1999 ("the Montreal Convention of 1999" - ICAO Doc 9740), *available at* https://www.jus.uio.no/lm/air.carriage.unification.convention.montreal.1999/ (limiting airline liability for death, injury, and damage or delay of international cargo).

    Separately, *the Convention on the Suppression of Unlawful Acts Relating to International Civil Aviation,* done at Beijing on 10 Sept. 2010 ("The Beijing Convention of 2010"), https://www.icao.int/secretariat/legal/Docs/beijing_convention_multi.pdf. *See* Email from William Voss, Special Advisor to ICAO (Nov. 26, 2019) (on file with author) ("This convention is critical to small UAS operations since it deals with the legal liabilities associated with the 'weaponization of aircraft.' It updates previous conventions on unlawful interreference in response to 9/11, but it accidentally addresses some of the most pressing issues following a drone attack that was organized or executed across borders. Something I find interesting is Article 5. It seems clear to me that the convention would only be applicable to a UAV that was registered as an aircraft in a State. This puts an entirely new spin on drone registration. [I]t is the only document that establishes that an international network (cyber) attack on an aircraft or aviation infrastructure is an act of illegal interreference. [I]f a UTM system were controlling drones registered as aircraft in a State, then it would be infrastructure that would fall under the protections of this document.").

919. *See, e.g.,* **Autonomy:** [NASA UAM Vision Conops], Sect. 3.0, Table 2 NASA UAM Framework Barriers ("Challenges in ... developing a framework for the determination

of liability associated with the development and operation of increasingly . . . autonomous systems."); **Data Rights and Privacy:** [RID NPRM 2019], Sect. IX.C ("The MOA signed by Remote ID USS would require it to agree to privacy protections of any data . . . ."), and Chap. 4, Sect. 4.7.6 *Data Rights and Privacy Management*; **Competition:** Chap. 4, Sect. 4.7.3 *Business Models and Competition.*

920. *See, e.g.,* The Federal Torts Claims Act [FTCA] (28 U.S.C. Sects. 1346(b), 1401, 1402, 2402, 2411, 2412 & 2671-80), *available at* https://www.law.cornell.edu/uscode/text/28/1346 (precluding tort claims against federal agencies and employees for personal or property loss or injury due to negligence, subject to exceptions); and Douglas M. Marshall, *Drone versus Manned Aircraft: An Analysis of the Application of the Discretionary Function Exception to the Federal Tort Claims Act to Accidents Caused by a Collision Between a Drone and a Manned Aircraft*, Issues in Avi. Law and Policy, Vol. 19:2 (Spring 2020), pp. 237-268, https://las.depaul.edu/centers-and-institutes/chaddick-institute-for-metropolitan-development/research-and-publications/Documents/IALPLatestIssueSpring2020b.pdf (analysis of the FTCA and the discretionary function).
921. Since people on the ground are nonparticipants, having not consented to take any flight risk, they should receive commensurate rights/remedies. Judges and juries faced with laws that do not especially protect bystanders and those in the wrong place at the wrong time may still view such parties as "innocent" (at least unconsciously) and want to maximize their protection.
922. Email from Benoit Curdy, Digital Transformation, Swiss FOCA (Sept. 14, 2020) (on file with author) ("[I]n reality, just to set up Net-RID, you might have four distinct companies involved: a DSS provider, a Net-RID provider, a Display Provider and a Display Client Provider. I expect a similar amount of complexity for all services.... responsibility will not be trivial to determine in case of an incident."). *See* [ICAO UTM Framework Ed. 3], Appen. F, p. 33 ("Policies need to be produced that clearly show who is responsible for each UTM service [and] include performance-based requirements that enable an effective USP approval and accreditation process . . ."); [India UTM Policy], Sect. 4.5.4 Service Responsibility Matrix ("Availability of the UTM service at the UTMSP level does not necessarily mean that the core responsibilities of providing the service will be the responsibility of the UTMSP.").
923. Richard Parker, Founder and CEO, Altitude Angel, Remarks at *Above & Beyond: The Role of UTM in Enabling BVLOS Flight* (webinar) (Aug. 11, 2020), https://www.altitudeangel.com/news/posts/2020/august/webinar-11-august-2020/.
924. Reinaldo Negron, Head of UTM, Wing, Remarks at the ASTM F38 UTM WG (June 19, 2020).
925. John Walker, Sr. Partner, The Padina Group & Chair, ISO/TC 20/SC 16, Remarks (July 22, 2020).
926. Andy Thurling, CTO, NUAIR Alliance, Remarks (Aug. 27, 2019). *See* [Drone Alliance Europe, U-Space Whitepaper 2019], p. 4 ("Empower USPs to take on as much responsibility as possible"); Chap. 5, Sect. 5.2 *Low Altitude Authorization and Notification Capability.*
927. [Rios Strategic Deconfliction Performance], Table 4, p. 24; [FAA ConOps UTM v2.0], Sect. 2.4.6 *Separation. Cf.* [RID NPRM 2019], Sect. IV.B. Unmanned Aircraft Systems Traffic Management (UTM) ("where *UAS operators have the responsibility* for the coordination, execution, and management of a safe operating environment.") (emphasis added).
928. Email from Richard Parker, Founder and CEO, Altitude Angel (Sept. 29, 2020) (on file with author).
929. [Comm Implementing Reg. (EU) 2021/664 - U-space], recital, ¶ (7).
930. [Comm Implementing Reg. (EU) 2021/664 - U-space], Art. 15 Conditions for obtaining a certificate, 1(b) & (j) (emphasis added). *See infra* Chap. 4, Sect. 4.7.5 *Insurance.*
931. *See* [ICAO UTM Framework Ed. 3], Appn. F, p. 34 ("In order to ensure a consistent level of performance of service provisions, service-level agreements (SLA) will need

to be established between UAS operators and USPs."); ICAO, *Manual of Standards – Aeronautical Information Management*, Sect. 2.3 Service Level Agreement, Ver. 0.1 (Oct. 31, 2017), https://www.icao.int/NACC/Documents/RegionalGroups/ANIWG/AIM/AIMManualofStandards-MoS.PDF (explaining SLA objectives, including to "[p]rotect against excessive expectations"); Simon Johnson, fmr. Acting Sec'y Gen., GUTMA, Interview by Unmanned Airspace (Feb. 10, 2020), https://www.unmannedairspace.info/news-first/utm-and-atm-are-going-to-be-one-and-much-more-simon-johnson-gutma (addressing USS outage and redundancy response analogous to the mobile phone industry, stating "[s]imilar agreements will happen with UTM."); ICAO, *Service Level Agreement, Aeronautical Information Service (AIS)*, https://www.icao.int/NACC/Documents/RegionalGroups/ANIWG/AIM/AIMServiceLevelAgreement-SLA.PDF; [Comm Implementing Reg. (EU) 2021/664 - U-space], Annex V ("exchange of information shall be ensured through an agreement on a service level").

932. [LAANC MOA].
933. SUSI, *SWISS U-space Implementation* (Oct. 2020), http://susi.swiss (in particular, MASTER AGREEMENT, Sect. 8.2(iii) restricting default to gross negligence or willful conduct, Sect. 15 INDEMNITY, and Sect. 17 LIMITATION OF LIABILITY).
934. FAA, *FAA Data Disclaimer* (2018), https://udds-faa.opendata.arcgis.com/pages/uas-disclaimer.
935. *See, e.g.*, Jeppesen, *Jeppesen Receives Industry's First FAA Letter of Acceptance Endorsement for Aviation Navigation Database* (Oct. 26, 2011), https://news.jeppesen.com/news-room/jeppesen-receives-industrys-first-faa-letter-acceptance-endorsement-aviation-navigation-database/; Jeppesen, *Jeppesen Distribution Manager and Jeppesen Services Update Manager End User License Agreement*, https://ww2.jeppesen.com/legal/jeppesen-distribution-manager-and-jeppesen-services-update-manager/.
936. *See* FAA, *Advisory Circular AC 00-63A, CHG 1*, Subj: Use of Flight Deck Displays of Digital Weather and Aeronautical Information, App. 1 (Jan. 6, 2017), https://www.faa.gov/regulations_policies/advisory_circulars/index.cfm/go/document.information/documentID/1024126 ["AC 00-63A"]. In both current and anticipated data structures, the provider and user of the data mutually agree to limitations and governance specified in the contract signed between them that are in practice very similar to those established by the FAA. Liability apportionment schemes in other government-private partnerships for IT infrastructures, such as public key certificate infrastructure may also inform UTM since both share trust infrastructure, highly distributed IT, and private-public sector attributes and responsibilities. *See, e.g.*, [Baum FCA Liability] (addressing, in part, liability constructs for secure distributed systems).
937. Email from Scott Blum, Ph.D., Dir. of Int'l Relations, Jeppesen (Aug. 21, 2019) (on file with author).
938. *See* [ICAO IATF 2019], Sect. 3.3.1 Business Value & Liability Flows; ICAO, *IATF Master Trust Framework Service Agreement (MTFSA)*; the ICAO, *X.509 Certificate Policy for the International Aviation Trust Framework (IATF) Certification Authority*, Ver. 0.2 (Nov. 2019) (copy on file with author). *See supra* Chap. 2, Sect. 2.4.7.2 *International Aviation Trust Framework.*
939. *See, e.g.*, RTCA DO-200 *Standards for Processing Aeronautical Data; supra* Chap. 3, Sect. 3.2.13 *RTCA*.
940. *See, e.g.*, OpenSky, *Opensky Terms of Service* (June 24, 2019), https://opensky.wing.com/visitor/map?lat=-28.418053&lng=133.454211&zoom=4:
    a. Your use of OpenSky Services, including to generate or automate a flight path for any drone, or connecting your drone to a device while using OpenSky Services, is at your sole risk. You assume the entire risk related to the operation of your drone and as to the quality and performance of OpenSky Services.
    b. Any information or tools offered through OpenSky Services, including information related to flight restrictions or the automation of a flight path, are offered for convenience purposes only…

c. Wing and its licensors make no representations or warranties regarding the accuracy or completeness of OpenSky Services or any contract. To the fullest extent permitted by law, OpenSky Services, including any content therein, are provided "as is" and "as available" without warranties of any kind, either express or implied. We expressly disclaim all implied warranties, including warranties of merchantability, fitness for a particular purpose, title, absence of viruses or latent or other defects, and non-infringement.

AIRMAP, *Terms of Service,* Sect. 18, Limitation of Liability (May 29, 2019), https://www.airmap.com/terms-service/; AIRXOS, *Terms* (copy on file with author) (in part, stating, "You shall exercise all reasonable diligence and implement all reasonable measures while using the Services as may be necessary to ensure that your use thereof does not threaten national security, public security, public safety or the health and safety of any person [and] [a]ny reliance you place on information you obtain from or through the Services *is strictly at your own risk.*") (emphasis added); NAV CANADA, *Terms of Use*, http://www.navcanada.ca/en/pages/terms-of-use.aspx (disclaimers).

941. *See* Chrystel Erotokritou, *The Legal Liability of Air Traffic Controllers*, INQUIRIES JOURNAL/STUDENTPULSE, Vol. 4(02) (2019), http://www.inquiriesjournal.com/a?id=613 (recognizing lack of a universal convention regulating ATC liability; also addressing ANSP liability, observing, e.g., "that skyguide in Switzerland can be sued [and that the] government of the Swiss Confederation will substitute the agency only as a last resort. In the United Kingdom, NATS... bears the exclusive responsibility to compensate victims in case of an accident. The UK government does not have any legal duty to provide assistance to NATS....") ["Erotokritou 2019"]. *See also* ICAO, Legal Committee, *Remotely Piloted Aircraft Systems Legal Survey*, LC/37-WP/2-1 (July 26, 2018), https://www.icao.int/Meetings/LC37/Documents/LC37%20WP%202-1%20EN%20Remotely%20Piloted%20Aircraft.pdf (Notably, the Secretariat concluded that "it appears... that there are currently no international legal issues that urgently need to be addressed through the development of new treaties or protocols." *id.* Sect. 6).
942. *See, e.g.,* Bart Elias, *Air Traffic Inc.: Considerations Regarding the Corporatization of Air Traffic Control*, Congressional Research Service (May 16, 2017), *available at* https://fas.org/sgp/crs/misc/R43844.pdf.
943. [Erotokritou 2019], p. 2. Separately, *see* FAA, *Aeronautical Information Manual*, Sect. 5-5-1(e) (2020), https://www.faa.gov/air_traffic/publications/atpubs/aim_html/chap5_section_5.html (intentionally overlapping pilot and controller responsibilities "expected to compensate, in many cases, for failures that may affect safety.").
944. To the extent UTM-based operations will be primarily BVLOS, they may be more akin to IFR operations to which ATC generally has greater responsibilities; and the primary beneficiary of separation are proximate manned aircraft—whether operating VFR or IFR.
945. Daniel W. Woods, et al., *Does insurance have a future in governing cybersecurity*, IEEE SECURITY & PRIVACY (2019), p. 1, https://tylermoore.utulsa.edu/govins20.pdf (also stating, "[p]rivate governance influences how responsibilities and liabilities are aligned for organisations." *id.*) ["Woods"].
946. [ICAO UTM Framework Ed. 3], *Gaps, Issues and Challenges*, p. 13.
947. Daniel W. Woods, et al., *Policy measures and cyber insurance: a framework*, J OF CYBER POLICY, Vol. 2, Issue 2 (Aug. 2017), pp. 209-226, *available at* https://www.tandfonline.com/doi/full/10.1080/23738871.2017.1360927. *See generally* Nat'l Cyber Security Center, UK, *Cyber Insurance Guidance* (Aug. 6, 2020), https://www.ncsc.gov.uk/guidance/cyber-insurance-guidance.
948. Email from Scott Blum, Ph.D., Dir. of Int'l Relations, Jeppesen (Sept. 19, 2019) (on file with author) (also asserting that "[u]ntil governments define this relationship and associated requirements, there will be little standardization in industry as each stakeholder develops procedures to optimize their competitive advantages.").
949. [LAANC MOA] (Note that Ver. 1 of the LAANC MOA (Sept. 18, 2017) required the USS "to arrange by insurance or otherwise the *full protection* of itself from and against all liability to third parties arising out of, or related to, its performance of this

Agreement.") (emphasis added), *available at* https://www.suasnews.com/wp-content/uploads/2018/10/LAANC_MOA_Skyward.pdf.

950. Swiss FOCA, *Memorandum of Cooperation established between The Federal Office of Civil Aviation and skyguide* (Dec. 20, 2018), Sect. 9, https://www.bazl.admin.ch/dam/bazl/en/dokumente/Gut_zu_wissen/Drohnen_und_Flugmodelle/moc_susi.pdf.download.pdf/Memorandum%20of%20Cooperation.pdf.
951. *See* [Comm Implementing Reg. (EU) 2021/664 - U-space], Art. 15 Condition for obtaining a certificate, ¶ 1.(i) (have in place arrangements to cover liabilities related to the execution of their tasks appropriate to the potential loss and damage). Separately, in scoping future UTM insurance requirements, note EU minimum accident coverage requirements for air carriers and aircraft operators in *Regulation (EC) No 785/2004 of the European Parliament and of the Council of 21 April 2004*, L 138/1, https://eur-lex.europa.eu/legal-content/EN/TXT/HTML/?uri=CELEX:32004R0785&from=EN (Its Art. 6 mandates minimum insurance coverage of 250,000 SDRs/passenger; and its Art. 7 insurance regarding third party liability—although excluding aircraft with a MTOM <500 kg and microlights used for non-commercial purposes.).
952. *See, e.g.*, Unmanned Risk Management, *UTMAssure*, http://unmannedrisk.com/uas-uav/utm-assure/ (offering an insurance product for operators described as "Assurance For The Unmanned Aerial System Traffic Management System – UTM").
953. *See* ISO/IEC 72001, *Information technology — Security techniques — Information security management systems — Overview and vocabulary*, https://www.iso.org/standard/73906.html; NIST, *Enhanced Security Requirements for Protecting Controlled Unclassified Information: A Supplement to NIST Special Publication 800-172* (Feb. 2021), https://csrc.nist.gov/publications/detail/sp/800-172/final; Tracy, *Could NIST SP 800-171 Be A Model for the Cyber Insurance Industry?*, TELOS.VISION (July 10, 2019), https://multimedia.telos.com/blog/nist-800-171b-cyber-insurance/.
954. [Woods], p. 1. Indeed, traditional manned aviation insurance underwriting imposes extensive requirements addressing pilot qualifications, proficiency, and equipage.
955. [Woods], p. 3 (also stating that "evidence suggests today's cyber insurance market is not fully delivering on its predicted governance functions, with security obligations in contracts particularly lacking"). This takes on heightened importance as safety-critical operations increase within UTM).
956. In addition to "proprietary information" – commercial trade secret-type information (such as the routes and flight parameters chosen by a new entrant testing UAs), governments too may seek to keep some of this information confidential. For example, would knowledge that a UA is in the area of a mobster's location potentially tip off the mobster? Or reveal information about defense research from a national security perspective? *See* [FAA & NASA UTM RTT WG #3], Sect. 3.1.5.3 *Shared Information Across Actors* (presenting a detailed table of types of information, providers, and entities with access). Furthermore, UTM's potential surveillance capability is noted. *See, e.g.*, EPIC, *Comments of the Electronic Privacy Information Center to the FAA, Notice: Petition for Exemption; Summary of Petition Received; Causey Aviation Unmanned, Inc.*, Docket No. FAA-2020-42 (Aug. 4, 2020), https://epic.org/apa/comments/EPIC-Comments-FAA-Drone-Delivery-Exemption-Aug2020.pdf.
957. *See, e.g.*, [NASA USS Spec.], Sect. 9 Operator Support (presenting requirement: "[UTM-USS-030] A USS MUST protect an operator's Personally Identifiable Information (PII) from unlawful/or unintended disclosure."). Consider applicability of "UTM-specific data" to National Institute of Standards and Technology (NIST), Special Publication 800.122, *Guide to Protecting the Confidentiality of Personally Identifiable Information (PII)*, https://nvlpubs.nist.gov/nistpubs/Legacy/SP/nistspecialpublication800-122.pdf. *Cf.* FAA, *ADS-B Privacy*, https://www.faa.gov/nextgen/equipadsb/privacy/ (presenting the Privacy ICAO Address (PIA) program – permitting "alternate, temporary ICAO Aircraft Address, which will not be assigned to the owner in the Civil Aviation Registry (CAR)"); FAA, *Limiting Aircraft Data Displayed*

(LADD), https://ladd.faa.gov/ (replacing the Block Aircraft Registry Request (BARR) program). GA privacy rights under this program should harmonize with those of UAS remote pilots.

958. *E.g.*, [ASTM Remote ID], Sect. 5.5.5.7 (intentional obfuscation of specific UA position data). Also, the scope of RID broadcast data communications is restricted.
959. [ASTM Remote ID], Sect. 4.6.9.
960. This may include the source of data needed by other system players. Also, litigants may want to issue subpoenas to a DSS to obtain data relevant to civil or criminal proceedings.
961. Email from Mike Glasgow, Technical Standards and UTM System Architect, Wing, and Co-chair, ASTM F38 UTM WG (Aug. 23, 2020) (on file with author) ("The ASTM UTM specification's logging is being designed such that the detailed information needed for event reconstruction is accomplished via logged messages, not retained data in the DSS. One can derive the state of the DSS from examination of the inputs and outputs.").
962. For example, USS will eventually decommission, and their data may become unavailable. Therefore, to assure accountability, a long-term data archival program may be necessary. System resilience limitations may require at least a temporal data store. Also, any USS that can reasonably anticipate that certain data may be relevant to dispute resolution proceedings may have preservation obligations under applicable law. *See, e.g.*, [RID NPRM 2019], Sect. 89.135 *Record retention* (requiring any "Remote ID USS to retain any remote identification message elements…for 6 months from the date" received or first possessed); [Comm Implementing Reg. (EU) 2021/664 - U-space], Art. 15 Conditions for obtaining a certificate, ¶1(g) (USSP must demonstrate they can: "retain for a period of at least 30 days recorded operational information and data or longer, where the recordings are pertinent to accident and incident investigations until it is evident that they will no longer be required"); *id.* recital ¶ (30) ("should establish a system of record keeping that allows adequate storage of the records and reliable traceability"). Counsel for USS should ensure USS prepare litigation hold procedures that may be triggered by such preservation obligations.
963. Email from Reinaldo Negron, Head of UTM, Wing (Dec. 28, 2019) (on file with author) ("In the evolving space…many regulators would prefer if the DSS met the USS paradigm and not something new."). This is in addition to national security requirements that may have independent trust requirements. Responsive DSS procedural/organizational controls deserve further consideration.
964. [ASTM Remote ID], Sect. 5.5.4.2 *(3)*.
965. *See* Faine Greenwood, *Drone Pilots Deserve Privacy Too*, Slate (April 23, 2019), https://slate.com/technology/2019/04/drone-tracking-registry-faa-unmanned-traffic-management.html; Presentation by Mike Glasgow, Technical Standards and UTM System Architect, Wing, and Co-chair, ASTM F38 UTM WG (July 10, 2019) ("Once you get the license plate [number] due to privacy considerations, the access issues for law enforcement are very challenging.").
966. Hrishi Ballal, Ph.D., Founder and CEO, Openskies Aerial Technology Ltd, *Registry Identity and Authentication*, GitHub, https://github.com/openskies-sh/aircraftregistry/blob/master/documents/registration-identity-authentication.md#information-security-assessment ("The registry by its nature will store personally identifiable information (PII) and the database will come under the local or national privacy and data protection laws. In many cases, this means that the data has to be stored in different servers and/or relevant security and isolation procedures must be followed."). *See supra* Chap. 2, Sect. 2.3.2 *Registration*; [Remote ID Rule], Preamble, X. Privacy Concerns on the Broadcast of Remote identification Information ("As with all other information maintained within the registry, the FAA has implemented the required privacy and security measures to protect data maintained in the registry system." Also asserting protection by the Privacy Act, 5 U.S.C. Sect. 552a.).

967. [FAARA], Sect. 357, *Unmanned Aircraft Systems Privacy Policy* (Also, [FAARA], Sect. 358 tasked the Controller General of the US to review privacy issues and concerns associated with UAS operations in the NAS; Sect. 375 made privacy policy violation by a commercial UAS user in the NAS a violation of the *Federal Trade Commission Act*, 15 U.S.C. 45(a)), Sect. 5(a). *See* [RID NPRM 2019], XXII. Privacy (FAA privacy impact statement finding RID "NPRM requirements... affect privacy"). *Cf.* [Remote RID Rule], Preamble, XI. Government and Law Enforcement Access to Remote Identification Information ("The FAA emphasizes that any use of remote identification data by law enforcement agencies is bound by all Constitutional restrictions and any other applicable legal restrictions.").
968. [EASA Reg. Framework U-space], Art. 10 *U-space service providers*, ¶ 2. *See* European Parliament, Regulation (EU) 2016/679 of the European Parliament and of the Council of 27 April 2016 on the protection of natural persons with regard to the processing of personal data and on the free movement of such data, and repealing Directive 95/46/EC (General Data Protection Regulation), 2016 O.J. (L 119) 1 (April 27, 2016), https://eur-lex.europa.eu/legal-content/EN/TXT/HTML/?uri=CELEX:32016R0679&from=EN, and https://gdpr-info.eu/ ["GDPR"]; [Comm Implementing Reg. (EU) 2021/664 - U-space], recitals ¶¶ (2), (4), (5), and Art. 3 U-space airspace (authorizing U-space airspace for privacy reasons).
969. FAA, *Privacy Statement Regarding LAANC and USS Providers Collection of Information in Accordance with 14 C.F.R. Part 107* (last modified July 23, 2019), https://www.faa.gov/uas/programs_partnerships/data_exchange/privacy_statement/.
970. FAA, *Memorandum of Agreement for Low Altitude Authorization and Notification Capability (LAANC) Between FEDERAL AVIATION ADMINISTRATION (FAA) And _______*, v2.3 (April, 2020), *Art. 22. DATA PROCEDURAL PROTECTIONS,* https://www.faa.gov/uas/programs_partnerships/data_exchange/laanc_for_industry/media/Memorandum_of_Agreement.pdf ["LAANC MOA"].
971. *See* FAA, Drone Advisory Committee, *Public eBook* (June 19, 2020), TG7 Recommendation on data protection, p. 14, https://www.faa.gov/uas/programs_partnerships/drone_advisory_committee/media/Public_eBook_06192020_v5.pdf ("For those services that are required due to FAA regulation, USSs should meet internationally accepted data protection standards in order to ensure that customer, government, and peer services are secured for the continued and safe operation of the UTM network.").
972. Ancillary privacy issues may include warrantless searches implicated by GPS-based monitoring. *See, e.g.,* U.S. v Jones, 565 U.S. 400 (2012), *available at* https://supreme.justia.com/cases/federal/us/565/400/ (Gov't obtained search warrant permitting GPS tracking device installation constituted 4th Amendment search).
973. *See* Hon. Joseph Simons, Chairman, Federal Trade Commission, Remarks at the Consumer Electronics Show, Las Vegas, Nev. (Jan. 7, 2020) (urging a federal privacy law; noting "our primary statute is a 100-year-old act and that it's time we think of doing something modern... We think it's time.").
974. Cal. Civil Code 1798.100-1798.199, https://leginfo.legislature.ca.gov/faces/codes_displayText.xhtml?division=3.&part=4.&lawCode=CIV&title=1.81.5 (Geolocation data).
975. *See, e.g.,* AIRMAP, *Privacy Notice for California Residents* (effective Jan. 1, 2020), https://www.airmap.com/privacy-notice-for-california-residents/.
976. Cal. Civil Code 1798.140(o)(1)(G).
977. [GDPR], Art. 3.
978. *See* Privacy Shield Framework, https://www.privacyshield.gov/Program-Overview. The International Trade Administration within the FTC administers the Privacy Shield Program for US-based organizations, including for LAANC providers. *See, e.g.,* Wing, *Privacy Policy* (Effective Nov. 5, 2018), https://storage.googleapis.com/wing-opensky-files/privacy.html (asserting, "[w]e also adhere to several self regulatory frameworks"), and Jeppesen, *EU-U.S. and Swiss-U.S. privacy shield notice* (Aug. 2, 2019), https://ww2.jeppesen.com/legal/eu-u-s-and-swiss-u-s-privacy-shield-notice/.

979. Court of Justice of the European Union, Press Release No 91/20 (July 16, 2020), *Judgement in Case C-311/18, Data Protection Commissioner v Facebook Ireland and Maximillian Schrems,* https://curia.europa.eu/jcms/upload/docs/application/pdf/2020-07/cp200091en.pdf ("The Court of Justice invalidates Decision 2016/1250 on the adequacy of the protection provided by the EU-US Data Protection Shield. However, it considers that Commission Decision 2010/87 on standard contractual clauses for the transfer of personal data to processors established in third countries is valid").
980. European Commission, *Standard Contractual Clauses (SCC)*, https://ec.europa.eu/info/law/law-topic/data-protection/international-dimension-data-protection/standard-contractual-clauses-scc_en (UTM providers such as AIRMAP point to the SCC for such data transfers).
981. For a general discussion of privacy and security requirements under law and contracts for advanced technologies such as UAS, *see* Stephen S. Wu, *Privacy and Security Challenges of Advanced Technologies*, PLI CURRENT: THE JOURNAL OF PLI PRESS, vol. 3, no. 3 (Summer 2019), p. 549, https://legacy.pli.edu/Content/Journal/PLI_Current_The_Journal_of_PLI_Press_Vol/_/N-bqZ1z0zkxf?ID=374249.
982. Presentation at the FAA UAS Symposium (July 8, 2020). *See* Ali Bahrami, Assoc. Admin. for Avi. Safety, FAA, Presentation at ICAO DRONE ENABLE Symposium 2021 (April 14, 2021) (LAANC, "an initial UTM").
983. [FAA ConOps UTM v2.0], Sect. 2.1 Overview. *See* [FAARA], Sect. 376 Plan for Full Operational Capability of Unmanned Aircraft Systems Traffic Management, (c)(2) ("establishing UTM services... and implementation of the Low Altitude Authorization and Notification Capability and future expanded UTM services").
984. *See* [RID NPRM 2019], Sect. XIV.A (reiterating that "Section 376 of Pub. L. 115-254 recommended that the FAA use the LAANC model of private sector participation in implementing future expanded UTM services.").
985. Michele Merkle, Dir. of ATO Operations Planning and Integration, FAA, Remarks at the FAA UAS Symposium (July 8, 2020). *See* Ali Bahrami, *supra* (LAANC: "a big win for us").
986. FAA, *ORDER JO 7210.3BB CHG 1*, Subj: Facility Operation and Administration (Effective Date: Jan. 30, 2020), Sect. 12-10-1.a. Program Description, https://www.faa.gov/documentLibrary/media/Order/7210.3BB_Chg_1_dtd_1-30-20.pdf. *See* FAA, *ORDER N JO 7210.914*, Subj: Low Altitude Authorization and Notification Capability - LAANC (July 23, 2019), https://www.faa.gov/documentLibrary/media/Notice/N_JO_7210.914_Low_Altitude_Authorization_and_Notification_Capability_-_LAANC.pdf; [FAARA], Sect. 349, Exception for Limited Recreational Operations of Unmanned Aircraft (authorization for inclusion of model aircraft operators — now referred to as recreational UA or limited recreational operations of UAS), FAA, *Exception for Limited Recreational Operations of Unmanned Aircraft*, 84 Fed. Reg. 22552, *Notice implementing the exception for limited recreational operations of unmanned aircraft* (May 17, 2019), https://www.federalregister.gov/documents/2019/05/17/2019-10169/exception-for-limited-recreational-operations-of-unmanned-aircraft.
987. FAA ATO, *Low Altitude Authorization and Notification Capability (LAANC) Concept of Operations*, Ver. 2.1 (March 20, 2020), Sect. 6.1 Scenario 1: Part 107 Authorization Below Automatically Approved Altitude, https://www.faa.gov/uas/programs_partnerships/data_exchange/laanc_for_industry/media/FAA_LAANC_CONOPS.pdf ["LAANC ConOps 2.1"].
988. Email from Parimal Kopardekar, Ph.D., Dir., NARI, NASA (Sept. 13, 2019) (on file with author). LAANC sets no aircraft density limits.
989. FAA, *UAS Data Exchange (LAANC)*, https://www.faa.gov/uas/programs_partnerships/data_exchange/.
990. FAA, *FAA Air Traffic Facilities Participating in LAANC*, https://www.faa.gov/uas/programs_partnerships/data_exchange/laanc_facilities/.
991. *See id.*

992. *See* FAA, *LAANC*, www.faa.gov; Teri Bristol, COO, ATC, FAA, Keynote: *Airspace Integration Update*, Presentation at the FAA UAS Symposium (July 8, 2020) (citing 320,000 LAANC operations thru April 2020).
993. Email from Jon Hegranes, Founder and CEO, Kittyhawk.io (Sept. 20, 2020) (on file with author); Jon Hegranes, *All-Time Activity Across the Kittyhawk Platform* (July 13, 2020), https://kittyhawk.io/blog/all-time-activity-across-the-kittyhawk-platform/.
994. In this regard, it has served an important data collection function. Before LAANC, primarily due to workload, the ATO was issuing blanket authorizations for 6-months for an entire UASFM. The FAA neither knew how many operations were occurring in an individual grid, nor in the entire UASFM. This lack of operational data made conducting safety analysis and panels extremely difficult. LAANC provides actual, individual operation specific data/counts to assist in refining the process and continued integration.
995. FAA, *ORDER JO 7200.23A*, Subj: Unmanned Aircraft Systems (Aug. 1, 2017), https://www.faa.gov/documentlibrary/media/order/jo_7200.23a_unmanned_aircraft_systems_(uas).pdf ["JO 7200.23A"].
996. *See* FAA, *FAA UAS Data Delivery System*, https://udds-faa.opendata.arcgis.com/ (hosting published UASFM) ["UAS Data Delivery System"]; FAA, *UAS Facility Maps*, https://www.faa.gov/uas/commercial_operators/uas_facility_maps/ (presenting the UASFM); FAA, [untitled], https://faa.maps.arcgis.com/apps/webappviewer/index.html?id=9c2e4406710048e19806ebf6a06754ad, and FAA, *B4UFLY Mobile App,* https://www.faa.gov/uas/recreational_fliers/where_can_i_fly/b4ufly/.
997. [LAANC ConOps 2.1], Sect. 5.2.2 Further Coordination.
998. Email from Jon Hegranes, Founder and CEO, Kittyhawk.io (Sept. 20, 2020) (on file with author).
999. *See* [LAANC USS Onboarding], *Attachment* A: USS-FAA High-Level Exchange Model; [LAANC USS Performance Rules], Sect. 3.2.1 API-Based Interface Between USS and FAA ("The USS must [] conform to the 'USS-FAA Authorizations and Notifications Interface Control Document' (ICD) version in effect. The ICD includes details on connecting to the FAA's LAANC system via the Internet."). *See* Chap. 4, Sect. 4.2.4.1 *Checkout and Onboarding.*
1000. The Mission Support Network provides access and connectivity for the FAA's non safety-critical planning and coordination tools. The network is included in the FAA's Enterprise Network Services Program (FENS), https://www.faa.gov/air_traffic/technology/cinp/fens/.
1001. *See supra* Chap. 2, Sect. 2.4.7 *Cybersecurity* (noting LAANC security control limitations); Maria A. DiPasquantonio, Deputy Dir., UAS Integration Pilot Program (IPP), FAA UAS Integration Office, *Memorandum to Lead Participants*, Subj.: Data Security (Dec. 5, 2019), *available at* https://www.suasnews.com/2019/12/does-laanc-have-security-issues/ (in part, "advising interagency partners **not to use LAANC for sensitive UAS missions**" and that the FAA is establishing "new data security and sovereignty requirements and policy" for LAANC USS); DoT, Office of the Inspector Gen., *FAA Lacks Sufficient Security Controls and Contingency Planning for Its DroneZone System*, FAA, Rpt. No. IT2020027 (April 15, 2020), p. 6, https://www.oig.dot.gov/sites/default/files/FAA%20DroneZone%20Security%20Controls%20Final%20Report.pdf?utm_medium=email&utm_source=govdelivery ("FAA did not effectively assess (e.g., select, test, implement, monitor, or develop a risk mitigation strategy for)…LAANC security controls before authorizing the systems to operate. In addition, FAA's inadequate monitoring of security controls increases the risk of the systems being compromised. Furthermore, FAA's use of unauthorized cloud systems increases the likelihood of security vulnerabilities. Finally, FAA did not adequately assess privacy and security controls for protecting PII.").
1002. Rather, airspace is precisely categorized as Class A, B, C, D, E, or G in the United States with specific purposes, definitions and geometry (in some jurisdictions Class "F" substitutes for "G"). *See supra* **Figure 4.2** – U.S. airspace volumes. *Cf.* [Comm Implementing Reg. (EU) 2021/664 - U-space], Art. 3 *U-space airspace.*

1003. That is, it would be considered an unsafe operation for manned and unmanned aircraft to operate simultaneously in the same proximate LAANC-enable airspace; and best practice for manned aircraft is generally avoidance.
1004. *See* FAA, *FAA Safety Briefing* (2018), https://www.faa.gov/news/safety_briefing/2018/media/SE_Topic_18_08.pdf (more than 25% of general aviation fatal accidents result from maneuvering); Peter Sachs, UTM Implementation Prog. Mgr., FAA (formerly, Safety and Risk Architect, Airbus), *A Quantitative Framework for UAV Risk Assessment*, Altiscope, Vol. 1, Rpt. TR-008 (Sept. 13, 2018), https://drive.google.com/file/d/1KNk8eHkvlRUpbxtyUsGi4deGoxstIkqJ/view ("**Operations in proximity to manned aircraft:**... are subject to more stringent risk thresholds to comply with higher target levels of safety associated with manned aircraft (that is, the risk calculation must take into account lethality as a primary harm...).")
1005. *See* FAA, *ORDER 8260.56*, Subj: Diverse Vector Area (DVA) Evaluation (Aug. 2, 2011), https://www.faa.gov/documentLibrary/media/Order/Order%208260.56.pdf.
1006. For example, from a precision approach, descending below glideslope to a corresponding lower non-precision minimum altitude abutting LAANC-enabled airspace; cancelling IFR and descending further; and for traffic conflict, exercising PIC authority to avoid an accident.
1007. *See* [AC 107-2], Sect. 5.8.1.2 ("the FAA expects that most remote PICs will avoid operating in the vicinity of airports because their aircraft generally do not require airport infrastructure, and the concentration of other aircraft increases in the vicinity of airports."); FAA, *NOTICE N 8900.529*, SUBJ: Extended Unmanned Aircraft Systems Oversight (Nov. 13, 2019), https://www.faa.gov/documentLibrary/media/Notice/N_8900.529.pdf (citing FAA analysis: "indicat[ing] an increase in UAS sightings that pose potential risks to air transport due to UAS sightings in communities *bordering airport approach and departure paths*.") (emphasis added). *Cf.* FAA, *Operation and Certification of Small Unmanned Aircraft Systems,* 81 Fed. Reg. 42064 (June 28, 2016), *Discussion of the Final Rule*, Sect. III.E.2.a, https://www.federalregister.gov/documents/2016/06/28/2016-15079/operation-and-certification-of-small-unmanned-aircraft-systems ("Because of the limits on their access to airspace that is controlled... small unmanned aircraft will avoid busy flight paths and are unlikely to encounter high-speed aircraft that would be difficult for the remote pilot to see-and-avoid."); 14 C.F.R. Sect. 107.43 Operation in the vicinity of airports (prohibits operation of a sUA "in a manner that interferes with operations and traffic patterns at any airport, heliport, or seaplane base.").
1008. Nonetheless, controller awareness of *all* hazards in their airspace is an important goal, and there are accelerating industry developments that signal this view. *See, e.g.,* CANSO, *AirMap and Raytheon virtually demo drone monitoring tools on next-generation air traffic control workstation* (Oct. 21, 2019), https://www.prnewswire.com/news-releases/raytheon-airmap-virtually-demo-drone-monitoring-tools-on-next-generation-air-traffic-control-workstation-300941971.html ("Air traffic controllers need real-time airspace awareness and alerts to unusual drone activity; it's a key step on the path toward safe drone integration into the national airspace system." Matt Gilligan, VP, Raytheon Intelligence, Information and Services); Amit Ganjoo, Founder and CEO, ANRA Technologies, and Co-chair, ASTM F38 UTM WG, *quoted in, Talking UTM with Amit Ganjoo, Founder of ANRA Technologies*, Avionics Int'l (Feb. 19, 2020), https://www.aviationtoday.com/2020/02/19/talking-utm-amit-ganjoo-founder-anra-technologies/?oly_enc_id=5780G0396467F4Y ("In true UTM sense, air traffic controllers need to be able to track as well, to see that traffic if it's going to impact anything.").
1009. *See* [FAA ConOps UTM v2.0], Sect. 2.7.1 Safety (re multiple layers of separation assurances); Paul McDuffee, Exec./Ops. Analyst at Hyundai Urban Air Mobility (formerly, Boeing Horizon X), *quoted in,* Bill Carey, *Enter The Drones, The FAA and UAVs in America* (Schiffer, 2016), p. 120 ("We don't believe there's going to be a one-size-fits-all solution to any of this.").
1010. *See, e.g.,* [Baum 2019] (presenting one such proposal by the Aviators Code Initiative).
1011. Remarks by Jay Merkle, Exec. Dir., UAS Integration Office, FAA *GUTMA, High Level Webinar Session - Ask the Experts* (June 9, 2020), https://gutma.org/ask-the-experts/.

1012. Such information should be integrated into ATCT displays in a manner that does not distract controllers from their primary duties. It should also extend to airport personnel: "The [Blue Ribbon Task Force on UAS Mitigation at Airports] encourages the FAA to partner with the LAANC-authorized USS through the InterUSS Platform as soon as possible to enable another level of known information in the airport environment, including…airport operators." Blue Ribbon Task Force on UAS Mitigation at Airports, Interim Report (July 2019), p. 25, https://uasmitigationatairports.org/blue-ribbon-task-force-on-uas-mitigation-at-airports-interim-report/.
1013. *See* Michele Merkle, Presentation at the FAA UAS Symposium (July 8, 2020) (stating plans for "using a more structured and objective approach to determine the max altitudes").
1014. Bundesamt für Zivilluftfahrt BAZL OFAC UFAC FOCA, *Swiss-wide U-space* (July 3, 2018), https://youtu.be/-dRj428sJfw.
1015. *See* Maria Algar Ruiz, EASA Drone Programme Mgr., EASA, Presentation at *GUTMA, High Level Webinar Session - Ask the Experts* (June 9, 2020), https://gutma.org/ask-the-experts/ (introducing U-space); EASA/EC, *EASA/EC workshop on the U-space regulatory framework Discussion with Member States and industry,* TE.GEN.00404-003 (May 14-15, 2019), p. 1, https://www.easa.europa.eu/sites/default/files/dfu/U-space%20workshop%20-%20discussion%20paper.pdf.
1016. [GUTMA, 2017], Sect. 3.
1017. [SESAR U-space Blueprint], p. 4.
1018. EC, *EASA/EC workshop on the U-space regulatory framework Discussion with Member States and industry*, Cologne (May 14-15, 2019), p. 1, https://www.easa.europa.eu/sites/default/files/dfu/U-space%20workshop%20-%20discussion%20paper.pdf. *See* [EASA Reg. Framework U-space], Sect. 2.1 Why we need U-space (describing U-space as "the enabler to manage more complex and longer-distance operations, and to ensure…seamless exchange of and operators intent, operational constraints, and other data critical for safety and security purposes.").
1019. Guido Manfridi, *U-SPACE: THE BLUEPRINT*, ENAC (June 19, 2017), http://drone-chair.enac.fr/rpas-news/u-space-the-blueprint/. *See* Email from Benoit Curdy, Digital Transformation, Swiss FOCA (Nov. 12, 2019) (on file with author) ("There is a consensus at the European Commission level that U-Space, by nature, is not limited to VLL."). *Cf.* Florian Guillermet, Exec. Dir., SESAR JU, Presentation at ICAO, DRONE ENABLE/3, Montreal, CA (Nov. 13, 2019) ("U-space is not an airspace.…U-space is not carved in stone.").
1020. [Comm Implementing Reg. (EU) 2021/664 - U-space], Art. 3 U-space airspace ¶ 2(a)-(d), Art. 8, Network identification service, Art. 9 Geo-awareness service, Art. 10 UAS flight authorization service, and Art. 11 Traffic information service. *See infra* Figure 5.5 Structure of the *U-space regulation* and subsequent text (addressing the regulation); Email from Benoit Curdy, Swiss FOCA (April 19, 2021) (The term *UAS flight authorization service* "does not provide an authorization in the regulatory sense (like an ANSP would provide for operations in the 5km radius of an airport)…the name of the service is creating a lot of confusion, as expected.").
1021. [Comm Implementing Reg. (EU) 2021/664 - U-space], Art. 12 Weather information service, and Art. 13 Conformance Monitoring service. *See* Benoit Curdy, *id.*, Remarks at ICAO DRONE ENABLE Symposium 2021 (April 16, 2021) ("I don't see a case where conformance monitoring will be optional. It's such a useful service once you have everything in place." To reduce confusion and "provide stability, a set of certified services" should include conformance monitoring.); Email from Benoit Curdy, id., (April 19, 2021) (on file with author) ("Our opinion was that all services should be optional and that a set of 'mandatory' service would emerge naturally as part of the learnings linked to going through a series of risk assessment processes. We thought that, except for weather, the likelihood that all other five services would always be used was high. The one that we had in mind for potentially being optional was traffic information in controlled airspace because segregation is ensured by ATS units. In any case, we will wait until having designated our first couple of U-Space airspaces to take

a position. For now, we don't have enough concrete arguments to justify a change in the regulation.").

1022. *See* SESAR JU, U-space, Overview, https://www.sesarju.eu/U-space.
1023. [SESAR U-space Blueprint], p. 5.
1024. Source: [SESAR ATM Master Plan 2020], p. 99 (updated to reflect geo-awareness). *See generally* EUROCONTROL, *U-SPACE Services, Implementation Monitoring Report* (Nov. 2020), https://www.eurocontrol.int/sites/default/files/2020-11/uspace-services-implementation-monitoring-report-2020-1-2.pdf.
1025. Email from Stefano Giovannini, Staff CEO, d-Flight S.p.A. (Aug. 17, 2019) (on file with author).
1026. [Comm Implementing Reg. (EU) 2021/664 - U-space], Art. 5 Common information services. *Cf.* [EASA Reg. Framework U-space], Sect. 2.4. Main approach for the development of the U-space regulatory framework (presenting requirements for "the functioning of U-space in a centralised manner and made available through one single gateway. The most important objective is to ensure that the necessary information comes from trusted sources and is of sufficient quality, integrity, and accuracy to ensure safe operations in the U-space airspace."); Lorenzo Murzilli, Founder and CEO, Murzilli Consulting, formerly Leader, Innovation and Digitization Unit, Swiss Federal Office of Civil Aviation, Remarks, in Chicago, Ill. (May 1, 2019) (describing an initial, comparatively centralized model); Email from Benoit Curdy, Digital Transformation, SWISS FOCA (Aug. 24, 2020) (on file with author) ("FOCA has since trimmed down the FIMS quite substantially. FIMS now only provides three items: NOTAM, traffic information, and U-Space Facility Maps (the data used for our LAANC-like function). All those are existing mandates of the Swiss ANSP, skyguide. There are no data going back to the ANSP via the FIMS anymore.").
1027. SESAR, Exploratory Research Projects, https://www.sesarju.eu/exploratoryresearch; SESAR, *U-Space, Supporting Safe and Secure Drone Operations in Europe, A preliminary summary of SESAR U-space research and innovation results (2017-2019)* (2020), https://www.sesarju.eu/sites/default/files/documents/u-space/U-space%20Drone%20Operations%20Europe.pdf.
1028. A helpful resource to locate these and related projects and their respective reports is available via EC, *CORDIS, EU research results,* https://cordis.europa.eu/.
1029. SESAR, *AIRPASS*, https://www.sesarju.eu/node/3327.
1030. SESAR, *CLASS*, https://www.sesarju.eu/projects/class.
1031. [CORUS ConOps]; *see* **Figure 4.1** *U-space VLL volume types.*
1032. SESAR, *DACUS*, https://www.sesarju.eu/projects/dacus.
1033. SESAR, *DIODE*, https://www.sesarju.eu/node/3200, Email from Stefano Giovannini, Staff CEO, d-Flight S.p.A. (Aug. 20, 2019) (on file with author). *See* EUROUSC, *Project Diode*, https://www.eurousc-italia.it/en/research-and-development/diode/.
1034. Email from Stefano Giovannini, Staff CEO, d-Flight S.p.A. (Aug. 17, 2019) (on file with author).
1035. SESAR, *DREAMS*, https://www.sesarju.eu/projects/dreams.
1036. SESAR, *DroC2Om*, https://www.sesarju.eu/projects/droc2om.
1037. SESAR, *EuroDrone*, https://www.sesarju.eu/node/3202; Email from Stefano Giovannini, *supra.*
1038. SESAR, *GEOSAFE*, https://www.sesarju.eu/projects/geosafe.
1039. SESAR, *GOF USPACE*, https://www.sesarju.eu/node/3203; *infra* Chap. 5, Sect. 5.4 *Selected Country-Specific Implementations.*
1040. SESAR, *ICARUS*, https://www.sesarju.eu/projects/icarus; Alberto Mennella, *ICARUS: A Proposal for a Common Altitude Reference*, SESAR (Nov. 20, 2020), https://www.eurocontrol.int/sites/default/files/2020-11/tim-2020-day-2-icarus-mennella.pdf. *See generally* [Corus ConOps], Sect, 2.5.2., Figure 2. An overview of the various vertical datums and vertical measurements (proposing service to convert between different altitude systems). *Cf.* ASTM, WK75923, *New Specification for Positioning Assurance, Navigation, and Time Synchronization for Unmanned Aircraft Systems* (2021), https://www.astm.

org/DATABASE.CART/WORKITEMS/WK75923.htm?A&utm_source=tracker&utm_campaign=20210218&utm_medium=email&utm_content=standards; Stephen P. Creamer, Dir., Air Navigation Bureau, ICAO, Remarks at ICAO DRONE ENABLE Symposium 2021 (April 21, 2021) (Regarding altitude and other common measurement, "no one size fits all—that's a new paradigm.").

1041. SESAR, *IMPETUS*, http://impetus-research.eu/.
1042. LABYRINTH, http://labyrinth2020.eu/ (launched June 2020).
1043. SESAR, *PercEvite,* http://www.percevite.org/.
1044. SESAR, *PODIUM*, https://www.sesarju.eu/projects/podium. *See PODIUM events aim to prove drone operations with initial UTM,* AIR TRAFFIC MANAGEMENT.NET (April 15, 2019), https://airtrafficmanagement.keypublishing.com/2019/04/15/podium-event-aim-to-prove-drone-operations-with-initial-utm/; Florian Guillermet, Exec. Dir., SESAR JU, Presentation at ICAO, DRONE ENABLE/3, Montreal, CA (Nov. 13, 2019) (underscoring PODIUM's key element is interoperability with manned aircraft).
1045. SESAR, *SAFEDRONE*, https://www.sesarju.eu/node/3199.
1046. SESAR, *SAFIR*, https://www.sesarju.eu/projects/safir. The SAFIR Consortium includes Unifly and Amazon Prime Air.
1047. SESAR, *SECOPS*, https://www.sesarju.eu/projects/secops.
1048. SESAR, *TERRA*, https://www.sesarju.eu/projects/terra.
1049. SESAR, *USIS*, https://www.sesarju.eu/projects/usis (emphasis removed). *See* Florian Guillermet, Exec. Dir., SESAR JU, Presentation at ICAO, DRONE ENABLE/3, Montreal, CA (Nov. 13, 2019) (asserting USIS's "main demonstration objective is to show the services are capable to support BVLOS Ops, all EASA's categories of drones, and all population densities.").
1050. SESAR, *VUTURA*, https://www.sesarju.eu/projects/vutura; Email from Stefano Giovanni, Staff CEO, d-Flight S.p.A. (Aug. 24, 2019) (on file with author).
1051. Email from Andrew Hately, UTM Concept Expert, EUROCONTROL (Sept. 3, 2020 and April 9, 2021) (on file with author):

> SESAR Exploratory Research call 3 (ER3) resulted in 9 projects: AIRPASS, CLASS, CORUS, DREAMS, Droc2om, IMPETUS, PercEvite, SECOPS, TERRA. Contemporaneously, SESAR funded 10 demonstration projects . . . DIODE, DOMUS, EURODRONE, GEOSAFE, GOF USPACE, PODIUM, SAFEDRONE, SAFIR, USIS, VUTURA. All have now closed. These 19 projects are covered in SESAR, *U-space, Supporting Safe and Secure Drone Operations in Europe, A report of the consolidated SESAR U-space research and innovation results* (2020) (describing nine projects of the ER3 call and the ten contemporary demonstration projects), https://www.sesarju.eu/sites/default/files/documents/reports/U-space%20research%20innovation%20results.pdf, and https://www.sesarju.eu/U-space ("Publications").
>
> SESAR Exploratory Research call 4 (ER4) has been made, answered, awarded, and work has started. In addition to the projects presented in the text of Chap. 5, Sect. 5.3 *U-space, see* BUBBLES—*Defining the BUilding Basic BLocks for a U-Space SEparation Management Service*, https://bubbles-project.eu/), FACT, INVIRCAT, SAFELAND, and URClearED. *See also* Metropolis2—*A unified approach to airspace design and separation management for U-space*, https://www.sesarju.eu/projects/; USE PE—*U-space Separation in Europe*, https://www.sesarju.eu/projects/USEPE.
>
> SESAR has also launched another call for demonstrations and is expected to award 4 or so in the coming months.... Outside SESAR, at least three other drone projects are being funded by Horizon 2020: 5D-Aerosafe, Drone4Safety (D4S) and LABYRINTH. Simultaneously SESAR has launched Pj34 AURA—*ATM-U-space interface*, https://www.sesarju.eu/projects/aura.
>
> There are also six demonstration projects addressing how U-space will support UAM, including TINDAIR—*Tactical Instrumental Deconfliction And in flight Resolution*, https://www.sesarju.eu/node/3759; and Uspace4UAM—*UAM in Every Days*

Traffic, to include a UTM provider, https://cordis.europa.eu/project/id/101017643; as well as CORUS-XUAM, the CORUS eXtension for Urban Air Mobility https://corus-xuam.eu/.

1052. SESAR, *Strategic Research and Innovation Agenda, Digital European Sky*, (Sept. 2020), https://www.sesarju.eu/sites/default/files/documents/reports/SRIA%20Final.pdf ["SESAR SRIA"]; Email from Robin Garrity, ATM Expert – Airport and Airspace User Operations, SESAR Joint Undertaking (Dec. 16, 2019) (on file with author).
1053. *See* [SESAR SRIA]. *See generally* EC *Horizon 2020*, https://ec.europa.eu/programmes/horizon2020/what-horizon-2020 ("the biggest EU Research and Innovation program ever," provides contract vehicles through multiple EC agencies). Among the non-SESAR JU-funded initiatives are: *EGNSS4RPAS*, funded though the European Global Navigation Satellite Systems Agency (GSA), https://www.gsa.europa.eu/newsroom/news/ec-project-showcases-benefits-egnss-drones; SUGUS - *Solution for E-GNSS U-Space Service*—to accelerate GNSS and Galileo for UAVs, funded via Horizon 2020; GMV, *SUGUS Kicks Off, A European Project for Integrating Drones into the Airspace* (March 25, 2020), https://www.gmv.com/en/Company/Communication/PressReleases/2020/NP_006_SUGUS.html; CORDIS, Horizon 2020, *Defining the BUilding Basic BLocks for a U-Space SEparation Management Service* (BUBBLES), Grant Agt. ID: 893206 (May 1, 2020 – Oct. 31, 2022), https://cordis.europa.eu/project/id/893206 (validation of separation services and drafting performance specifications). *See also* SESAR, *Sustainability, airspace optimisation and urban air mobility – focus of latest very large-scale demonstrations* (Feb. 23, 2021), https://www.sesarju.eu/news/sustainability-airspace-optimisation-and-urban-air-mobility-focus-latest-very-large-scale (introducing large-scale demonstration projects including CORUS-XUAM, relevant to U-space).
1054. [EASA Opinion 01/2020 - U-space].
1055. EASA, A-NPA 2015-10, *Introduction of a regulatory framework for the operation of drones* (July 31, 2015), https://www.easa.europa.eu/sites/default/files/dfu/A-NPA%202015-10.pdf; EASA, *Notice of Proposed Amendment (NPA)* (May 5, 2017), https://www.easa.europa.eu/newsroom-and-events/press-releases/easa-publishes-proposal-operate-small-drones-europe.
1056. EASA, NPA 2017-05 (A), *Introduction of a regulatory framework for the operation of unmanned aircraft systems in the 'open' and 'specific' categories*, https://www.easa.europa.eu/sites/default/files/dfu/NPA%202017-05%20%28A%29_0.pdf (also asserting, "... there is a strong link between the U-Space concept and the Regulation (EU) 201X/XXX....," Sect. 2.3.1.9 *Link with the U-Space*).
1057. EASA, *Drones Amsterdam Declaration* (Nov. 28, 2018), https://www.easa.europa.eu/sites/default/files/dfu/Drones%20Amsterdam%20Declaration%2028%20Nov%202018%20final.pdf.
1058. [Comm Delegated Reg. (EU) 2019/945 - UA].
1059. [Comm Implementing Reg. (EU) 2019/947 - UA].
1060. EASA, *Proposed Special Condition for Light UAS*, Doc. No. SC Light-UAS 01, Issue: 1 (July 20, 2020), https://www.easa.europa.eu/document-library/product-certification-consultations/proposed-special-condition-light-uas ["EASA SC Light UAS"]. *See supra* [Comm Implementing Reg. (EU) 2019/947 - UA].
1061. [Comm Implementing Reg. (EU) 2019/947 - UA], ¶ 26.
1062. EASA, *U-space workshop*, Köln (May 14-15, 2019), https://www.easa.europa.eu/sites/default/files/dfu/U-space%20workshop%2014-15%20May%20v2.0.pdf (considering "[p]riority on air traffic, airspace classes"; applicability to both manned and unmanned operations; and "SERA to be used [as] a source of inspiration for U-space," Slide 27).
1063. EASA, U-space Regulatory Framework Workshop, *Summary of Conclusions*, Cologne (May 14-15, 2019), https://www.easa.europa.eu/sites/default/files/dfu/U-space%20workshop%20-%20summary%20of%20conclusions%20%28final%29.pdf.
1064. [EASA Opinion 01/2020 - U-space] ("The final decision will be published by EASA once the European Commission has adopted the regulation and once the necessary

consultation with the affected stakeholders has been performed." Sect. 1.2. The next steps.).

1065. *See* EASA, *U-space workshop,* Köln (May 14-15, 2019), https://www.easa.europa.eu/sites/default/files/dfu/U-space%20workshop%2014-15%20May%20v2.0.pdf. *See generally* EASA, *Civil drones (Unmanned aircraft)*, https://www.easa.europa.eu/easa-and-you/civil-drones-rpas (providing a summary of U-space regulatory developments).

1066. [EASA Opinion 01/2020 - U-space], Exec. Summary.

1067. *Id.*

1068. [EASA Opinion 01/2020 - U-space], Sect. 2.4 Overview of the proposals.

1069. See EC, Mobility and Transport, Air, *Drones: Commission adopts new rules and conditions for safe, secure and green drone operations* (April 22, 2021), https://ec.europa.eu/transport/modes/air/news/2021-04-22-drones_en (EC "today adopting the U-space package"):
    - EC, *Commission Implementing Regulation (EU) 2021/664 of 22 April 2021 on a regulatory framework for the U-space*, https://eur-lex.europa.eu/legal-content/EN/TXT/?uri=CELEX%3A32021R0664 ["Comm Implementing Reg. (EU) 2021/664 - U-space"]
    - EC, *Commission Implementing Regulation (EU) 2021/665 of 22 April 2021 amending Implementing Regulation (EU) 2017/373 as regards requirements for providers of air traffic management/air navigation services and other air traffic management network functions in the U-space airspace designated in controlled airspace*, https://eur-lex.europa.eu/legal-content/EN/TXT/?uri=CELEX%3A32021R0665
    - EC, *Commission Implementing Regulation (EU) 2021/666 of 22 April 2021 amending Regulation (EU) No 923/2012 as regards requirements for manned aviation operating in U-space airspace*, https://eur-lex.europa.eu/legal-content/EN/TXT/?uri=CELEX%3A32021R0666 ["Comm Implementing Reg. (EU) 2021/666 - U-space"]

    See also https://www.easa.europa.eu/regulations. As a practical matter, the U-space regulation will become effective when U-space airspace has been designated.

1070. "Significant milestone" — Patrick Ky, Exec. Dir, EASA, European Aviation Safety, Remarks at the ICAO DRONE ENABLE Symposium 2021 (April 16, 2021) (further stating the regulation was done in 2 years "compared to the usual 4-5 years [and this] is quite a success I have to say"); "Not yet the end state…" — Maria Algar Ruiz, EASA Drone Programme Manager, EASA, Presentation at *GUTMA, High Level Webinar Session - Ask the Experts* (June 9, 2020), https://gutma.org/ask-the-experts/.

1071. *See* [SESAR ATM Master Plan 2020], p. XII (acknowledging that "achieving the SESAR vision by 2040 will be challenging"); SESAR, *U-Space Implementation Map Tool*, https://www.atmmasterplan.eu/depl/u-space (indicating certain U-space research and demonstration project delays).

1072. Interview, *Interdrone Podcast,* at 29:00 (May 21, 2020), https://interdrone.com/news/episode-39-dji-loses-patent-suit-charges-filed-after-shot-down-drone-interview-with-amit-ganjoo-of-anra-technologies/. *See* Email from Maxim Egorov, Research Scientist, Airbus (Dec. 4, 2020) (on file with author) ("The future of UTM ecosystem generally depends on how quickly and effectively UTM services can be operationalized in the near term. The first iteration of UTM architectures must be operationalized first before it can be evolved to accommodate future technologies, visions, and concepts.").

1073. For comprehensive, updated implementations, *see* GUTMA, *Map of International UTM Implementations*, https://gutma.org/?s=international+implementations (presenting diverse initiatives within a dynamic World map); Vertical Flight Society, *eVTOL Timeline*, https://evtol.news/evtol-timeline (presenting an updated timeline of eVTOL developments, some of which will affect UTM).

1074. Wing, *Australia | Canberra*, https://wing.com/australia/canberra/. *See* Reinaldo Negron, Head of UTM, Wing, Presentation at ICAO, DRONE ENABLE/2, in Chengdu, P.R.C. (Sept. 13-14, 2018) (describing "on demand drone delivery & UTM platform" and in close collaboration with CASA); OpenSky (App), https://opensky.wing.com/visitor/map?lat=-28.418053&lng=133.454211&zoom=4 (Terms of Service).

1075. Uber Newsroom, *Uber announces Melbourne, Australia as first International Uber Air pilot city* (June 12, 2019), https://www.uber.com/en-AU/newsroom/uberelevatemelbourne/. Separately, *see* the Australian Association for Unmanned Systems (AAUS), https://aaus.org.au/ (listing other Australian UTM initiatives).
1076. *See* Airbus, Zephyr, https://www.airbus.com/defence/uav/zephyr.html.
1077. Unifly, *Unifly supports 7 simultaneous UTM demonstrations across Europe*, https://www.unifly.aero/news/unifly-supports-simultaneous-utm-demonstrations-across-europe ("Supported by the Unifly UTM system, people all over Europe brought U-space one step closer to reality.").
1078. Email from Craig Bloch-Hansen, Prog. Mgr. - UAS Design Standards, Transport Canada (April 9, 2020) (on file with author). *See* Transport Canada, *RPAS Traffic Management (RTM) Services Trials, CALL FOR PROPOSALS, Phase 1 – Round 1* (May 11, 2020), https://www.tc.gc.ca/en/services/aviation/drone-safety/drone-innovation-collaboration/remotely-piloted-aircraft-systems-rpas-traffic-management-services-testing-call-proposals.html (presenting the RTM's 29 services).
1079. NAV CANADA, *NAV CANADA signs strategic agreement with Unifly* (Feb. 26, 2020), https://www.navcanada.ca/EN/Pages/NR-14-2020.aspx.
1080. *See* Jianping Zhang, *UOMS in China* (Shenzhen, June 6-8, 2018), https://rpas-regulations.com/wp-content/uploads/2018/06/1.2-Day1_0910-1010_CAAC-SRI_Zhang-Jianping_UOMS-_EN.pdf ("Both UOMS and current ATM system receive real-time flight data from RPAS, and perform collision risk calculations to ensure safety.") (on file with author).
1081. Presentation at ICAO, DRONE ENABLE/2, in Chengdu, P.R.C. (Sept. 13-14, 2018).
1082. Charles Alcock, *EHang Makes First Passenger-carrying Autonomous Flight*, AINonline (Sept. 6, 2019), https://tinyurl.com/EHang-Pax (demonstrated during the 2019 Northeast Asia Expo, Changchun, CN (Aug. 23-27, 2019)). *See* EHANG, *EHang to Provide UAM Services in Hengqin New Area in Zhuhai, China*, Globe Newswire (Jan. 12, 2021), https://ir.ehang.com/node/7456/pdf (initiate UAM operations for aerial sightseeing and other services).
1083. SESAR JU, Media Release, *SESAR JU GOF U-space project: First demos successfully completed* (Jan. 12, 2021), https://www.frequentis.com/sites/default/files/pr/2019-07/PR_GOF_USPACE_072019.pdf; SESAR, *SESAR GOF U-space project: Start of demonstrations in June 2019* (May 15, 2019), https://www.sesarju.eu/news/ready-steady-go-gulf-finland-u-space-begin-demonstrations-june. *See supra* Chap. 5, Sect. 5.3 *U-space* (presenting GOF U-space).
1084. SESAR JU, *SESAR U-space demonstration in the Gulf of Finland (GOF) ready for take-off!* (July 2, 2019), https://www.sesarju.eu/news/sesar-u-space-demonstration-gulf-finland-gof-ready-take. *See* Bill Carey, *New Airspace Concepts Floated For 'Nontraditional' Entrants*, Aviation Week and Space Tech. (May 21, 2019), https://tinyurl.com/Cary-AviationWeek (includes Estonia; and testing urban, package delivery, and mixed manned/unmanned operations).
1085. Wing, *Finland | Helsinki*, https://wing.com/finland/helsinki/.
1086. La Direction générale de l'Aviation civile (DGAC), https://www.ecologique-solidaire.gouv.fr/en/french-civil-aviation-authority. *See generally* DSNA, *Direction des Services de la Navigation Aérienne*, RAPPORT D'ACTIVITÉ, https://www.ecologique-solidaire.gouv.fr/sites/default/files/DSNA-DGAC-RA-2017-FR.pdf; MINISTÈRE DE LA TRANSITION ÉCOLOGIQUE ET SOLIDAIRE, Direction générale de l'Aviation civile Paris, le 10 decembre 2018, *Demande d'information (RFI) sur l'opportunité et les conditions de mise en place de partenariats concernant la mise en oeuvre des services U-space avec la direction des services de la Navigation aérienne (DSNA).*

1087. *Journal officiel de l'Union européenne, Bulletin officiel des annonces de marchés publics and PLACE, Plate-forme Des Achats De L'Etat,* www.marches-publics.gouv.fr:

> seeks to better understand the market, operational and technical drivers that prevail now around the management of drone traffic, and their evolution over the next 10 years [and] to assess the value and feasibility of establishing one or more long-term structured partnerships for the design, testing and operation of U-space services in geographical volumes of the French airspace (referred to as U-subspaces) and delivered with the support of SWIM services....

Inquiry regarding the RFP's federated (versus centralized) character precipitated the following response: "It is indeed certainly possible that we will establish a more federative architecture. However, the ATC/UTM interface itself (UAS authorizations from ATC in controlled airspace) might be unique and/or centralised (even if it is distributed from a technical point of view)." Email from Antoine Martin, New ATM Services, Programme Dir., DGAC/DSNA (July 4, 2019) (on file with author). *See* DGAC, *U-space Together, Fast-Tracking Drone Integration in a Safe Sky* (2020), https://www.ecologique-solidaire.gouv.fr/sites/default/files/dsna_WAC_USPACE.pdf.

1088. Claude Le Tallec, et al., *Low Level RPAS Traffic Management Potential systems solutions* (Nov. 5-6, 2015), https://www.onera.fr/sites/default/files/ressources_documentaires/cours-exposes-conf/ASTECH_Session5_LeTallec-1.pdf (ONERA). *See supra* Chap. 5, Sect. 5.3 *U-space* (introducing PODIUM and GEOSAFE).

1089. https://droniq.de/en/utm.

1090. *See* Happiest Minds, https://www.happiestminds.com/, and ANRA, www.anratechnologies.com. Indian regulations require electronic flight plan approval via UTM. Gov't of India, Office of the Dir. Gen. of Civil Avi., Civil Aviation Requirements, Sect. 3 - Air Transport, Series X Part 1, Issue I (Aug. 27, 2018, Effective Dec. 1, 2018), *id.* F. No. 05-13/2014-AED Vol. IV, http://bathindapolice.in/civil.pdf (Sect. "12.4 The operator [except Nano intending to operate up to 50 ft (15 m) AGL in uncontrolled airspace / enclosed premises] shall obtain permission before undertaking flight through 'Digital Sky Platform'." Nano is defined as "less than or equal to 250 grams" *id.* Sect. 3.1).

1091. Email from Stefano Giovannini, Staff CEO, d-Flight S.p.A. (Nov. 13, 2019) (on file with author). *See* d-Flight Portal, https://www.d-flight.it/portal/en/.

1092. *See* Hiroyuki Ushijima, METI, *UTM Project in Japan*, Presentation at the Global UTM Conference 2017, GUTMA, Montreal, CA (June 26, 2017), *available at* https://bit.ly/2UMzDyw (UTM a critical part of this vision).

1093. Jeffrey Homola, UTM Integration and Testing Lead, UAS Traffic Management, et al., NASA Ames, *UTM and D-NET: NASA and JAXA's Collaborative Research on Integrating Small UAS with Disaster Response Efforts* (2019), https://utm.arc.nasa.gov/docs/2018-Homola-Aviation2018-Jun.pdf.

1094. Private sector participants included the major ATM system suppliers in Japan, and others. Academic institutions and the public sector also participate. Flight tests of tens of simultaneous operations in the same air space, all connected to UTM systems, conducted successfully at the Fukushima Robot and Drone Test Field test site. The NEDO project is now being extended to address remote ID and other important UTM-related topics. The envisioned architecture centralizes many of the functions that are more distributed among USS in some other implementations and may evolve when deployed.

1095. Email from Fred Borda, COO, Aerial Innovation (Aug. 31, 2019) (on file with author).

1096. *See* Tom Dent-Spargo, *Korea's First UTM System*, Robotics Law J. (June 11, 2018), https://roboticslawjournal.com/global/koreas-first-utm-system-86846618.

1097. *LMT successfully conducts their first cross-border drone flight on the mobile network*, sUAS News (Sept. 4, 2020), https://www.suasnews.com/2020/09/lmt-successfully-conducts-their-first-cross-border-drone-flight-on-the-mobile-network/ (facilitated by leading Baltic mobile operator LMT).

1098. Swoop Aero, *The First International Piloted Drone Delivery* (March 24, 2020), https://swoop.aero/.
1099. *See Altitude Angel awarded nationwide UTM contract by Luchtverkeersleiding Nederland (LVNL)*, CANSO (Jan. 3, 2020), https://canso.org/altitude-angel-awarded-nationwide-utm-contract-by-luchtverkeersleiding-nederland-lvnl/.
1100. *See* NZ Gov't, *NZ Government establishes innovative, industry-focused Airspace Integration Trials Programme* (Oct. 4, 2019), https://www.beehive.govt.nz/release/nz-government-establishes-innovative-industry-focused-airspace-integration-trials-programme.
1101. AirShare Ltd., *UAV Traffic Management Trial Launched in New Zealand* (March 4, 2018), https://www.airshare.co.nz/news-hub#ufh-i-390582501-uav-traffic-management-trial-launched-in-new-zealand.
1102. *See UTM system to be implemented for the first time in Nordic region*, INT'L AIRPORT REV. (Jan. 16, 2020), https://www.internationalairportreview.com/news/110464/utm-system-implemented-nordic-region/; www.altitudeangel.com; www.frequentis.com (allowing Norway's ANSP, Avinor Air Navigation Services to commence drone integration at 18 airports).
1103. PANSA, *PansaUTM*, https://www.pansa.pl/en/pansautm/.
1104. *See* NOVA Systems, *UTM Research and Development in Singapore*, https://novasystems.com/experiences/utm-research-development-singapore/.
1105. *See* https://onesky.blog (collaborators also included Nanyang Technological University's Air Traffic Management Research Institute (ATMRI), M1 Limited, and Scout Aerial).
1106. Future Flight, *About,* https://futureflight.sg/about/.
1107. Ministry of Transport, CAAS, *Connected Urban Airspace Management for Unmanned Aircraft,* Future Flight Consortium Press Release (July 24, 2018), *available at* http://www.acornint.com/future-flight-consortium/.
1108. LFV, *UTM Solution*, Ref. No. 19/8 (June 20, 2019), https://ted.europa.eu/TED/notice/udl?uri=TED:NOTICE:287048-2019:TEXT:EN:HTML&src=0.
1109. skyguide, https://www.skyguide.ch/en/company/innovation/drones-at-skyguide/. *See Swiss U-Space Implementation (SUSI) platform* (April 1, 2019), https://airscope.ae/creative/2019/04/01/swiss-u-space-implementation-susi-platform/ (noting that a memorandum of cooperation for U-Space in Switzerland includes: Skyguide, Swisscom, Involi, Sensefly, Wing and Auterion). The SUSI platform includes support from ANRA's UTM DroneUSS. *See* ANRA Technologies, *ANRA Technologies Partners with Civil Aviation Authority FOCA for Swiss U-Space Initiative* (Aug. 9, 2019), http://www.anratechnologies.com/home/news/anra-technologies-parters-with-foca-for-swiss-u-space/.
1110. Swiss Confederation, *Swiss U-space* (Feb. 20, 2019), https://www.bazl.admin.ch/dam/bazl/en/dokumente/Gut_zu_wissen/Drohnen_und_Flugmodelle/Swiss_U-space_Implementation.pdf.download.pdf/Swiss%20U-Space%20Implementation.pdf.
1111. skyguide, *Swiss U-space Deploys National Flight Information Management System for Drones (FIMS) to Enable a Safe and Open Drone Economy*, Media Release (Aug. 6, 2019), https://www.skyguide.ch/en/events-media-board/news/#p96744-96749-96754.
1112. Matternet, https://mttr.net/.
1113. Thomas Black, et al., *UPS Seeks Airline-Type Status for Long-Distance Drone Deliveries*, BLOOMBERG (July 23, 2019), https://www.bloomberg.com/news/articles/2019-07-23/ups-seeks-airline-type-status-for-long-distance-drone-deliveries.
1114. [Swiss U-Space ConOps v1.0].
1115. World Economic Forum, *Drones are saving lives in Tanzania's remote communities* (April 11, 2019), https://www.weforum.org/agenda/2019/04/drones-are-saving-lives-in-tanzania-remote-communities/. *See* Margaret Eichleay, et al., *Using Unmanned Aerial Vehicles for Development: Perspectives from Citizens and Government Officials in Tanzania* (Feb. 2016), https://www.researchgate.net/publication/295547373; *Deliver Future: DHL Parcelcopter 4.0 in Tanzania, Africa* – Trailer, https://www.youtube.com/watch?v=n-v6xIcQa03Q. *See also* Tanzania Civil Aviation Authority, *Circular on Unmanned Aircraft Systems*, Doc. No. TCAA/FRM/ANS/AIS-30 (Jan. 1, 2017), https://my-road.de/downloads/Tansania_AIC-05-2017_Unmanned_Aircraft_Systems.pdf (the sophistication of this circular, particularly its Sect. 9 on autonomous UAS is noteworthy).

1116. *Lake Victoria Challenge pioneers UTM operations in East Africa*, Unmanned airspace (Oct. 30, 2018), https://www.unmannedairspace.info/uncategorized/lake-victoria-challenge-pioneers-utm-operations-east-africa/.
1117. NATS (Nov. 23, 2018), https://www.nats.aero/news/uk-first-drones-fly-safely-controlled-airspace/.
1118. Altitude Angel, *Altitude Angel powers access to the UK's skies as Airspace User Portal goes live* (April 11, 2019), https://www.altitudeangel.com/blog/altitude-angel-powers-access-to-the-uks-skies-as-airspace-user-portal-goes-live/; Altitude Angel, *Strategic Conflict Resolution, Strategic Flight Deconfliction API*, https://docs.altitudeangel.com/docs/conflict-resolution-service.
1119. UK Civil Aviation Authority, *From air taxis to artificial intelligence in air traffic control – UK Civil Aviation Authority announces first six participants for new innovation work* (May 20, 1019), https://www.caa.co.uk/News/From-air-taxis-to-artificial-intelligence-in-air-traffic-control-%E2%80%93-UK-Civil-Aviation-Authority-announces-first-six-participants-for-new-innovation-work/.
1120. *Connected Places Catapult, Towards a UTM System for the UK, Preparing the UK for the Commercial Drone Industry* (2019), https://s3-eu-west-1.amazonaws.com/media.cp.catapult/wp-content/uploads/2019/09/30150855/Towards-a-UTM-System-for-the-UK.pdf (architecture for an open access UTM system); Connected Places Catapult, *Enabling UTM in the UK* (May 2020), https://s3-eu-west-1.amazonaws.com/media.cp.catapult/wp-content/uploads/2020/05/22110912/01296_Open-Access-UTM-Report-V4.pdf.
1121. The important roles of state DoTs, UAS Test Sites, and both private and public sector entities are acknowledged in fielding UTM systems and infrastructure in the US. Issuance of Part 135 operator certificates to companies such as Amazon Prime Air, UPS, and Wing, intended for package delivery are further examples. *See supra* note 651 (addressing these Part 135 exemptions).
1122. Remarks at ICAO, DRONE ENABLE/3, Montreal, CA (Nov. 12, 2019) (further stating, "we still have a tremendous amount of work ahead of us"). *See* Robert Pierce, Deputy Assoc. Admin., ARMD, NASA, Presentation at the NASA-UTM Project Technical Interchange Meeting (TIM) (Feb. 23, 2021) (UTM: "Inspiration for future airspace innovation").
1123. Presentation at ICAO, DRONE ENABLE/3, Montreal, CA (Nov. 12, 2019).
1124. Including authorizations that require neither the use of chase aircraft nor visual observers.
1125. *See supra* note 655 (re: Amazon, UPS, and Wing air carrier certification). *Cf.* Andy Thurling, CTO, NUAIR Alliance, *The Drone Market*, Enterprise Technology Review (Aug. 1, 2019), p. 30, https://www.enterprisetechnologyreview.com/magazines/August2019/Display_Tech/#page=29 ("[W]e have failed to take a step back from our demonstrations of new capability and features to establish a fieldable tranche of UTM capability with which we may show the value of BVLOS applications.").
1126. Presentation at TEDxChandigarh, *Future of Drones and Airborne Autonomy* (Aug. 14, 2020), https://www.youtube.com/watch?v=SpEZcL8z8fU&app=desktop.
1127. Steve Bradford, *Welcome to the Future: The Next State of UAS Traffic Management*, FAA (Dec. 9, 2020), https://www.suasnews.com/2020/12/welcome-to-the-future-the-next-stage-of-uas-traffic-management/.
1128. Dani Grant, et al., *The Myth of the Infrastructure Phase*, USV (Oct. 1, 2018), https://www.usv.com/writing/2018/10/the-myth-of-the-infrastructure-phase/.
1129. Dallas Brooks, Aviation Regulatory Lead, Wing, formerly Dir. of Miss. State U. Raspet Flight Research Laboratory, and Asso. Dir. of the FAA's Center of Excellence for UAS Research, Testimony before the Senate Commerce Committee (May 8, 2019), *quoted in*, Nick Zuzulia, *FAA to Debut Remote ID Rule in July*, Avionics Int'l (May 10, 2019), https://www.aviationtoday.com/2019/05/10/faa-remote-id-drones/.
1130. Technical Officer, ICAO, Presentation at ICAO, DRONE ENABLE/2, Chengdu, P.R.C. (Sept. 13-14, 2018).

1131. Michael Gadd, Head of Int'l Regulatory Affairs, Altitude Angel, Presentation at the Technical Workshop on U-space and ATM Aspects - Stream 2A - EASA High Level Conference on Drones 2019, Amsterdam, NL (Dec. 10, 2019), *available at* https://tinyurl.com/Amsterdam-U-spaceServ.

1132. *See* Email from Tim McCarthy, Ph.D., Professor Maynooth Univ. (Aug. 22, 2019) (on file with author) ("It is likely that 100s if not 1000s of drones will be required to provide various urban services; data gathering, package delivery, air-taxi. This in turn can result in tens of thousands of missions over the course of a day across the city - resulting in adverse Capacity... within the UTM.").

1133. *See* Paige Smith, *The move to IP-based communications in aviation,* AEROSPACE TESTING INT'L (Dec. 19, 2019), https://www.aerospacetestinginternational.com/features/the-move-to-ip-based-communications-in-aviation.html ("...a new telecommunications system that would then bring the IP core to the FAA. We're just getting on that road now because the whole telecommunications system needs to go IP as well as all the edge equipment.").

VoIP—Until substantial ATM/UTM integration, Voice-over-IP (VoIP) integration between ATM and UAS/UTM may accommodate certain safety mitigation requirements. While the planned UTM is being designed to operate without direct human intervention, UTM specifications may include a voice requirement (at least) for contingency operations until the reliability of the system is demonstrated in operation over time. To the extent that ATM is quickly moving to VoIP for its voice-based communications, VoIP can be expected to also serve UTM purposes. *See* Email from Alexander Engel, Sr. Expert Standardization, ASTERIX Mgr. and EUROCAE Liaison, DECMA/PCS/SCS, EUROCONTROL (Dec. 3, 2019) (on file with author) ("whenever voice is used for applications within UTM (for instance for coordination with ATC) I would not see why [ED-137] could not be applied."); EUROCAE, ED-137, www.eurocae.net (defining rules for VoIP implementations supporting ATM communications and interoperability for radio, voice communications systems, recording and supervision); *Voice over Internet Protocol – going global for aviation*, sUASNEWS (Nov. 28, 2019), https://www.suasnews.com/2019/11/voice-over-Internet-protocol-going-global-for-aviation/ (integration of the EUROCAE ED-137 Standard into ICAO, *Manual on the Aeronautical Telecommunication Network (ATN) using Internet Protocol Suite (IPS) Standards and Protocol* (Doc 9896)). *But see* Bruce Eckstein, Sr. Sys. Eng'r, FirebirdSE, LLC, Remarks (Dec. 2, 2019) (asserting "This [ATM VoIP integration] is ten years out.").

1134. *See, e.g.*, FCC Application, File No. SAT-LOA-20161115-00118, https://licensing.fcc.gov/myibfs/download.do?attachment_key=1364689; FCC, *In the Matter of Space Exploration Holdings, LLC Application For Approval for Orbital Deployment and Operating Authority for the SpaceX NGSO Satellite System, Memorandum, Order and Authorization*, IBFS File No. SAT LOA-20161115-00118, Call Sign S2983, et al. (adopted Mar. 28, 2018), https://licensing.fcc.gov/cgi-bin/ws.exe/prod/ib/forms/reports/swr031b.hts?q_set=V_SITE_ANTENNA_FREQ.file_numberC/File+Number/%3D/SATLOA2016111500118&prepare=&column=V_SITE_ANTENNA_FREQ.file_numberC/File+Number&utm_content=bufferda647 (FCC grant of proposal for "4,425 satellites in 83 orbital planes, at an approximate altitude of 1,110 to 1,325 kilometers.... SpaceX proposes to operate in the 10.7-12.7 GHz, 13.85-14.5 GHz, 17.8-18.6 GHz, 18.8-19.3 GHz, 27.5-29.1 GHz, and 29.5-30 GHz bands," as supplemented); Sissi Cao, *SpaceX Expands Starlink Project to 42,000 Satellites, 'Drowns' ITU in Filing Paperwork*, OBSERVER (Oct. 21, 2019), FCC, *Order and Authorization*, DA 19-1294 [in aforementioned SpaceX proposal] (Dec. 19, 2019), https://observer.com/2019/10/spacex-elon-musk-starlink-satellite-Internet-itu-fcc-filing/; https://docs.fcc.gov/public/attachments/DA-19-1294A1.pdf (approving 72 orbital planes at 550 Km altitude); Todd Shields, *Chasing SpaceX, Amazon Seeks to Launch 3,236 Internet Satellites*, BLOOMBERG (July 8, 2019), https://www.bloomberg.com/news/articles/2019-07-05/amazon-asks-to-join-broadband-space-race-with-elon-musk-s-spacex. *Cf.* NAV

CANADA, *Space-based Automatic Dependent Surveillance-Broadcast (ADS-B)*, https://www.navcanada.ca/en/search.aspx?search-words=aireon (Aireon's space-based global air traffic service; extending to 70% of the world's airspace), but this initiative's 2021 equipage mandate has been put in abeyance in response to stakeholder feedback pending further regulatory action.).

1135. *See infra* Chap. 6, Sect. 6.6 *Stratospheric Operations*; [EUROCAE OSED-DAA], Sect. 2.4.4.2 ("Exposure to C2 link loss situations is higher in VLL than it is when operations are conducted at higher altitudes."); AeroVironment, *High-Altitude Pseudo-Satellites*, https://www.avinc.com/about/haps; HAPS MOBILE, https://www.hapsmobile.com/en/. Tony Spouncer, Sr. Dir., Inmarsat, Remarks at the *Out of Sight: Deploying Pop-Up UTM* (webinar) Altitude Angel (Feb. 16, 2021) (describing successful use of satcom for C2 and a 200KB link "sufficient").

1136. *See, e.g.,* Vaughn Maillo, Technical Officer, Airspace Mgt. and Optimization Sect., ICAO, Presentation at ICAO, DRONE ENABLE/3, Montreal, CA (Nov. 12, 2019) (overcoming communication limitations, including of 1090 MHz and 24-bit aircraft addressing, "This system has served us well *way back in the 1950s when developed…*") (emphasis added).

1137. Presentation at the Enabling Autonomous Flight & Operations in the National Airspace System Workshop 2, NASA Ames Research Center, Moffett Field, Cal. (Aug. 6, 2019).

1138. Presentation at the FAA UAS Symposium, Balt., Md. (June 5, 2019) (further asserting, "Human error is the greatest enemy.").

1139. Reinaldo Negron, Head of UTM, Wing, Presentation at the FAA UAS Symposium (July 9, 2020).

See Edward "Ted" Lester, formerly Chief Technologist, AiRXOS, Presentation at *Drone Services at the Edge - Business Implications*, GUTMA (May 6, 2020). Proposals to introduce AI into the UTM ecosystem include applying an AI agent to broker negotiations between USS (*see supra* Chap. 4, Sect. 4.4.4 *Fair Access*), and to provide autonomous conflict avoidance (*see supra* [Sheng Li DRL 2019]). Additionally, "In cases where the technical systems have been demonstrated to be reliable and capable of providing information to support independent operations, allowing for autonomous decision making in the automated UTM procedures (e.g., contingency/emergency management) will give airspace managers a useful tool in supporting the integration of routine large scale operations." Email from Craig Bloch-Hansen, *supra.*

1140. Remarks at the EAA Spirit of Aviation (webinar) (July 27, 2020).

1141. *See generally* ASTM, Stephen Cook, Chair, ASTM AC377, et al., Technical Report, *Autonomy Design and Operations in Aviation: Terminology and Requirements Framework,* TR1-EB (July 2019), p. 1, https://www.astm.org/DIGITAL_LIBRARY/TECHNICAL_REPORTS/PAGES/1fe1b67c-5ff0-488e-ab61-cb0d329bc63c.htm and https://www.astm.org/DIGITAL_LIBRARY/TECHNICAL_REPORTS/index.html ["ASTM Autonomy Framework"] ("Increased automation moving towards autonomy has shown great promise to improve the safety of aviation and transform the industry."); ASTM, Stephen Cook, Chair, ASTM AC377, et al., Technical Report, *Developmental Pillars of Increased Autonomy for Aircraft Systems*, TR2-EB (2020), https://www.astm.org/DIGITAL_LIBRARY/TECHNICAL_REPORTS/PAGES/fc2da6dc-733d-4d86-bb09-75a555b85330.htm; ICAO, A40-WP/3271EX/137, *New Operational Concepts Involving Autonomous Systems*, Presented by the Int'l Coordinating Council of Aerospace Industries Associations (ICCAIA) (Aug. 19, 2019), ¶ 2.7, https://www.icao.int/Meetings/a40/Documents/WP/wp_327_en.pdf ("inviting the [ICAO] Assembly to acknowledge…the increasing automation in aircraft and ATC systems, with a strong trend towards autonomous operations"); Craig Bloch-Hansen, Prog. Mgr. - UAS Design Standards, Transport Canada, Presentation at the ASTM F38 meeting, in Raleigh, N.C. (Nov. 5, 2019) (In Canada, "some degree of autonomy is required for all operations. There isn't going to be infrastructure out there to support every possible operation because of the size of the country—so really what we're looking for are solutions to the autonomy question and finding ways to create predictable and certifiable autonomous systems.").

1142. *See* P. D. Vascik, et al., *Assessment of air traffic control for urban air mobility and unmanned systems,*" Report No. ICAT-2018-03, MIT, Int'l Center for Air Transp. (June 2018), Sect. 3.D, https://dspace.mit.edu/bitstream/handle/1721.1/117686/ICAT-2018-03_Vascik_2018a%20Vascik%20ICRAT%20UAM%20and%20UAS%20ATC.pdf?sequence=1 ("The wide variability of autonomy…will create a mixed-performance challenge for ATC."); [MITRE 2019], p. 6 ("Integrating autonomous systems into the [NAS] is challenging but not impossible").

1143. Tom Prevot, Ph.D., Air Taxi Product Lead, Joby Aviation, formerly Dir., Airspace Systems, Uber, Presentation at the FAA UAS Symposium, Balt., Md. (June 5, 2019) (asserting that humans in the system are a "tricky balance"). *See* Email from Bruce Eckstein, Sr. Sys. Eng'r., FirebirdSE LLC (Dec. 2, 2019) (on file with author) ("Integration of autonomous ops for uncontrolled airspace is likely in my opinion and within the foreseeable future (but not easy). Autonomous ops in controlled airspace is not likely in my life time and according to my wife, I am supposed to live until she reaches 100 which is quite a ways off."); Anjan Chakrabarty, et al., NASA Ames, *Autonomous flight for Multi-copters flying in UTM -TCL4+sharing common airspace* (Jan. 2020), https://utm.arc.nasa.gov/docs/2020-Chakrabarty_SciTech_2020-0881_UTM.pdf (asserting that the last fifty feet of a UAS operation is "the most difficult phase of autonomous operations").

1144. *See* Brian Garrett-Glaser, *Why Automation Shouldn't Push Pilots Out of Air Taxis,* Rotor & Wing Int'l (May 23, 2019), https://www.rotorandwing.com/2019/05/23/automation-shouldnt-push-pilots-air-taxis/ (For example, "First, while a pilot can make decisions in the moment about how to respond to a situation, risk-averse automation will likely choose to 'bail out.' Over time, those operational decisions will add to costs making the automated flights more expensive because they are flown in a less-optimized fashion. Second, the development of that technology will have a significant upfront cost.").

1145. Parimal Kopardekar, Ph.D., Sr. Technologist, Air Transportation System, and Director, NASA NARI, *Enabling Future Airspace Operations*, J. of Air Traffic Control, Vol. 62:1 (Spring 2020), pp. 25, 27. *See* Joseph Rios, ATM-X Project Chief Eng'r, NASA, *Strategic Deconfliction: System Requirements, Final Report* (July 31, 2018), https://utm.arc.nasa.gov/docs/2018-UTM-Strategic-Deconfliction-Final-Report.pdf ("We agree there will be special cases where humans need to be involved and this need may persist long into UTM's existence."); Alfredo Giuliano, Intelligent Systems Mgr., Aurora Flight Sciences, *Presentation at Enabling Autonomous Flight & Operations in the National Airspace System Workshop 2*, NASA Ames Research Center, Moffett Field, Cal. (Aug. 6, 2019) (underscoring that "humans will have to surveil the autonomy").

1146. *See, e.g.*, Society of Automotive Engineers Int'l (SAE) SAE J3016, Levels of Driving automation, https://www.sae.org/news/2019/01/sae-updates-j3016-automated-driving-graphic. SAE J3016, *Surface Vehicle Recommended Practice – Taxonomy and Definitions for Terms Related to Driving Automation Systems for On-Road Motor Vehicles* (Sept. 2016), Table 1 – Summary of levels of driving automation, https://www.sae.org/standards/content/j3016_201609/ (incorporating level labels in brackets, as follows):

–Level 0: [No Driving Automation] System issues warnings, but cannot sustain control

–Level 1: [Driver Assistance] "Hands On" driver and system share control such as adaptive cruise control & parking assistance

–Level 2: [Partial Driving Automation] "Hands Off" system takes full control but driver must be prepared to intervene

–Level 3: [Conditional Driving Automation] "Eyes Off" driver can turn attention away from driving tasks, but must be prepared to intervene when called upon, by the system, to do so

–Level 4: [High Driving Automation] "Mind Off" driver may sleep – self driving only in geofenced areas or under special circumstances

–Level 5: [Full Driving Automation] "Steering Wheel Optional" no human intervention required, such as a robotic taxi

See also [ASTM Autonomy Framework], App. A: Summary of Various Existing Levels of Automation Taxonomies and Frameworks; [EUROCAE OSED-DAA], Sect. 2.1.1 & App. A. (levels of automation regarding UTM); Stephen Cook, Ph.D., et al., *Promoting Autonomy Design and Operations in Aviation*, 38th DASC, San Diego, Cal. (2019), https://ieeexplore.ieee.org/document/9081809; ASTM, WK76044, *New Practice for Exercising a Contextual Framework for Increasingly Autonomous Aviation Systems*, https://www.astm.org/DATABASE.CART/WORKITEMS/WK76044.htm.

1147. [ASTM Autonomy Framework], p. 2. *See* [Nat'l Academies AAM], Sect. 5-3 ("lack of harmonized view on acceptable autonomy levels is producing a negative impact on advancing overall development of advanced aerial mobility").

1148. [ASTM Autonomy Framework], pp. 8–9.

1149. Suresh K. Kannan, Ph.D., CEO, Nodein Autonomy Corp., Remarks at NBAA BACE, Las Vegas, Nev. (Oct. 23, 2019) (Furthermore, "[e]ventually we want to take the band-aid off (pilot), but it must be done slowly and systematically (evolve) to avoid adverse interactions between humans and autonomy."). *Cf.* Michael McNair, Innovation Mgr., Bell, Presentation at *Enabling Autonomous Flight & Operations in the National Airspace System Workshop 2*, NASA Ames Research Center, Moffett Field, Cal. (Aug. 6, 2019) ("Rather than asking for full autonomy, it has been urged that the better questions are "what is the right amount of autonomy [and] when can I get the autonomy I need.").

1150. *See* Email from Craig Bloch-Hansen, Program Mgr. - UAS Design Standards, Transport Canada (Dec. 19, 2019) (on file with author) ("Various regulatory and research initiatives (e.g., the Nat'l Research Council, Canada) are actively invested in working to define architectures to support the development and deployment of autonomous operations. While this work is still in its early development decades of experience in automation, human-machine teaming, and thoughtful lessons learned from aviation tragedies have provided a strong basis to build the core principles surrounding safe and reliable automation in support of autonomous operations. This work will need to be complemented by both academic and industrial means of compliance to effectively deploy technology to meet the high demands of safety critical systems."); Nat'l Research Council Canada), *Integrated Autonomous Mobility* (Aug. 2019), https://criaq.aero/wp-content/uploads/2019/08/Technology-focus-for-NRCs-IAM-initiative.pdf.

1151. Email from Suresh K. Kannan, Ph.D., CEO, Nodein Autonomy Corp. (Jan. 4, 2020) (on file with author).

1152. Adapted from email from Wes Ryan, Prog. Mgr., NASA Aeronautical Research Institute (Mar. 4, 2021) (on file with author). *See* Biruk Abraham, ICAO and Global Initiatives, FAA, Remarks at the ICAO DRONE ENABLE Symposium 2021 (April 15, 2021) (urging that "autonomy is not going to be binary … it is continuing to evolve.").

1153. *See* [SESAR ARIA], Sect. 3.8 Artificial Intelligence (AI) for aviation (to include machine learning, deep learning and big data analytics); Email with Maxim Egorov, Research Scientist, Airbus (Oct. 19, 2019) (on file with author). *See, e.g.,* [Sheng Li DRL 2019]; Email from Maxim Egorov, Research Scientist, Airbus (Dec. 23, 2019) (on file with author):

> I think we are still fairly far away from a world where we can use deep reinforcement learning in any part of aviation. As much as I think approaches like deep reinforcement learning can revolutionize complex autonomous systems, we still have a long way to go in ensuring that the neural networks that are foundational to these approaches can be properly verified for safety-critical systems. There is a lot of research on this topic, but the nice thing about the simpler and traditional methods that are rule based or driven by a well-defined logic table, they can be verified to be predictable, understood, and thus safe to the desired degree. In short, I would say that new approaches are needed to solve autonomy and traffic management problems for VTOL/drone

delivery ops, etc., BUT alongside we must develop new approaches to verify and validate the systems powered by these approaches that would allow them to be safely integrated into the airspace at safety levels we've come to expect from aviation.

1154. *See* ASTM International, F3269-17, *Standard Practice for Methods to Safely Bound Flight Behavior of Unmanned Aircraft Systems Containing Complex Functions*, https://www.astm.org/Standards/F3269.htm (providing run-time assurance for non-deterministic complex functions that may become increasingly the norm for UTM services in the future, and reducing C2 requirements/reliance to establish trust that a vehicle will complete its operations safety; a pending update to this standard practice is entitled, *Methods to Safely Bound Behavior of Aircraft Systems Containing Complex Functions Using Run-Time Assurance*); Francis X. Grovers II, Bell, Presentation at Enabling Autonomous Flight & Operations in the National Airspace System Workshop 2, NASA Ames Research Center, Moffett Field, Cal. (Aug. 6, 2019) (proposing to "surround non-deterministic with a deterministic system for a deterministic outcome").

1155. *See* SAE G-34 / WG-114, Artificial Intelligence in Aviation, https://www.sae.org/works/committeeHome.do?comtID=TEAG34.

1156. Jai Xu, Sr. Dir. of Strategy, Urban Air Mobility and Unmanned Aerial Systems, Honeywell, Remarks at the FAA UAS Symposium, Innovation Shaping the Future (Aug. 19, 2020). *See generally* ASTM Int'l, Committee F45 on Robotics, Automation, and Autonomous Systems, https://www.astm.org/COMMITTEE/F45.htm.

1157. Remarks at the NASA, AAM Airspace WG (webinar) (Aug. 4, 2020). *See* Email from Steve Bradford (Dec. 29, 2020) (on file with author) (explaining this "reflects the fleet mix and the need to support vehicles which exhibit limited levels of autonomy. Some think that we can automate ATC, but since aircraft coming online today and for the next 15 years will last 30 plus years, it is not likely that the fleet will have homogenously high levels of autonomy in 2050 and will need to be served in a manner much as today."); [Volocopter Roadmap], Sect. 4.7.2 ("Implementing UAM autonomy is … a gradual and evolutionary process…."). As a practical matter, any timeline should qualify what constitutes "autonomous", the required level of safety, volume of vehicles, where they operate, etc. *See American Robotics Becomes First Company Approved by the FAA To Operate Automated Drones Without Human Operators On-Site*, Media Release (Jan. 15, 2021), available at https://www.american-robotics.com/media-coverage (Arguably, the FAA first approved autonomous BVLOS operations: "to operate automated drones without human operators on-site" and characterizing the waiver to "represent[] a pivotal inflection point in the commercial drone industry").

1158. *See, e.g.*, EC, *Proposal for a Regulation on a European approach for Artificial Intelligence* (April 21, 2021), https://digital-strategy.ec.europa.eu/en/library/proposal-regulation-european-approach-artificial-intelligence (includes Annex II mentioning unmanned aircraft); EC, *Ethics guidelines for trustworthy AI* (April 8, 2019), https://ec.europa.eu/digital-single-market/en/news/ethics-guidelines-trustworthy-ai (urging, *inter alia*, AI systems to be: lawful, ethical, and robust); John Markoff, *A Case for Cooperation Between Machines and Humans*, N.Y. TIMES (May 21, 2020), https://www.nytimes.com/2020/05/21/technology/ben-shneiderman-automation-humans.html?referringSource=articleShare (urging robot collaboration with humans; and avoiding "absolving humans of ethical responsibility"); Int'l Comm. For Robot Arms Control, www.icrac.net; Mark Rosekind, Ph.D., Chief Safety Innovation Officer, Zoox, and prior Admin'r of the Nat'l Highway Traffic Safety Admin., *quoted in*, Eric A. Taub, *How Jaywalking Could Jam Up the Era of Self-Driving Cars*, N.Y. Times (Aug. 1, 2019), https://www.nytimes.com/2019/08/01/business/self-driving-cars-jaywalking.html ("[w]ith autonomous vehicles, the technical stuff will get worked out. It's the societal part that's the most challenging.").

Regarding regulatory constraints, *see* Chris Anderson, CEO, 3DR, InterDrone Podcast, (June 18, 2020), https://podcasts.apple.com/us/podcast/interdrone-podcast/id1376928140?i=1000478498371 ("AI: you can't use it for any kind of regulated use. [The] FAA won't accept the letters AI …. Drones are supposed to be autonomous … that was the dream and the fact that it's still one-to-one essentially means we've failed … what's stopping us is regulation.");

Email from Wes Ryan, Prog. Mgr., NASA Aeronautical Research Institute (Mar. 4, 2021) ("- Current traffic management is very human centric, and is under a stronghold of FAA Air Traffic policy and ATC Controller Union coverage. This will make it tough to move forward at the pace some are hoping. - FAA Flight Standards and ATO rarely do any changes in technology without substantial data and trials of new equipment in test environments").

1159. Chief, Avi. Safety Mgt. Sys. Div., U.S. DOT Volpe National Transportation Systems Center, *quoted in* DoT, *Blockchain for Unmanned Aircraft Systems* (April 2020), p. 1, https://rosap.ntl.bts.gov/view/dot/48789.

1160. Bronwyn E. Howell, Victoria U. of Wellington, et al., *Governance of Blockchain Distributed Ledger Technology Projects: a Common-Pool Resource View* (June 11, 2019), https://dlc.dlib.indiana.edu/dlc/bitstream/handle/10535/10527/20190611_pm_Ostrom_DLTs_Polycentric.pdf?sequence=1&isAllowed=y; Rory Houston, Co-founder, Skyy Network, *UTM is a Social Problem, More than a Technical One* (Aug. 1, 2019), https://medium.com/skyynetwork/utm-is-a-social-problem-more-than-a-technical-one-751b8b82072c (describing blockchain).

1161. *See, e.g.,* Ronald J. Reisman, Aero Computer Eng'r, NASA Ames, *Air Traffic Management Blockchain Infrastructure for Security, Authentication, and Privacy* (2018), https://ntrs.nasa.gov/archive/nasa/casi.ntrs.nasa.gov/20190000022.pdf.
See generally Maryanne Murray, *A Reuters Visual Guide, Blockchain explained*, Reuters Graphics (June 15, 2018), http://graphics.reuters.com/TECHNOLOGY-BLOCKCHAIN/010070P11GN/index.html.
The blockchain is a database structure that is read and write only. One that is required to be accessed by multiple stakeholders within an ecosystem as a means to collectively access or modify the data stored within, without the disadvantages injected by (or need for) a trusted 3rd party intermediary. The stakeholders either do not, or need not trust one another, as the blockchain provides powerful "trust minimization" over the data contained within, typically through verification mechanisms that mathematically prove the data to be true. The mechanism behind it is known as asymmetric key cryptography, whereby two "keys" known as the public and private keys are generated by the sender (who wishes to modify the database state). The private key is kept confidential by the sender and is used to "sign" transactions broadcast to the network, yet the public key (which is generated from the private key) is shared at will and can be used by anyone in possession to verify that the transaction did originate from the intended source.

1162. *See* Satoshi Nakamoto, *Bitcoin: A Peer-to-Peer Electronic Cash System* (Oct. 31, 2008), https://bitcoin.org/bitcoin.pdf.

1163. Email from Rory Houston, Co-founder, Skyy Network (Dec. 12, 2019) (on file with author). *See generally* Rory Houston, *Why NASA's Blockchain Paper is significant for Aviation* (April 26, 2019) https://medium.com/skyynetwork/why-nasas-blockchain-paper-is-significant-for-aviation-b0649fa535c6.

1164. This is known as *economic security*, in direct contrast to the more commonly understood *trust-based security* touted as robust by centralized institutions globally. Customers cannot necessarily ascertain and trust what UTM does with customer private data. Trust in blockchain ecosystems requires only trust in the underlying math and cryptography, thereby enabling the power to *verify* truth. *See generally* John-Paul Thorbjornsen, et al., *Crytoeconomics of Enterprise Blockchains* (Dec. 27, 2018), https://www.youtube.com/watch?v=ODbAW7dQFiQ&feature=youtu.be.

1165. *See, e.g.,* Puyang, *Blockchain-based Decentralized Autonomous Organization Is Future!*, Stanford Mgt. Sci. & Eng'g (July 8, 2018), https://mse238blog.stanford.edu/2018/07/puyang/blockchain-based-decentralized-autonomous-organization-is-future/.

1166. The Bitcoin block reward is a financial incentive *increasing* the security of the network by incentivizing "miners" to commit expensive computational power toward the difficult challenge of solving for and committing blocks to the ledger. Diverting this computer power elsewhere, or attempting to commit an invalid block would be met with no block reward (economic loss). Thus, miners are disincentivized to behave in contravention of the consensus rules or otherwise devalue the network.

1167. Only a better system might make it obsolete, by stakeholders' loss of interest and abandonment.
1168. *See* Robin Hanson, *Futarchy: Vote Values, But Bet Beliefs* (undated), http://mason.gmu.edu/~rhanson/futarchy.html; coindesk, *The Father of Futarchy Has an Idea to Reshape DAO Governance* (May 23, 2016), https://www.coindesk.com/futarchy-dao-governance. Self-organization of developers forge informal hierarchical networks who sit exposed under constant audit, where personal integrity, honestly, and engineering talent of the highest order are required. Further combinations of formal hard-coded (on-chain) and informal extra-protocol (off-chain) decision making exists.
1169. Email from Rory Houston, *supra. Compare* the Bitcoin Network with tens of thousands of nodes, each enfranchised with an equitable voice in network governance. *See* Ralph C. Merkle, *DAOs, Democracy and Governance*, Ver. 1.9 (May 31, 2016), https://merkle.com/papers/DAOdemocracyDraft.pdf (characterizing DAO as a "new form of democracy").
1170. *See, e.g.,* Nathaniel Popper, *Twitter and Facebook Want to Shift Power to Users. Or Do They?*, N.Y. Times (Dec. 18, 2019), https://www.nytimes.com/2019/12/18/technology/facebook-twitter-bitcoin-blockchain.html; Dan Goodin, *Fire (and lots of it): Berkeley researcher on the only way to fix cryptocurrency*, ARS Technica (Feb. 4, 2019), https://arstechnica.com/information-technology/2019/02/researcher-counts-the-reasons-he-wants-cryptocurrency-burned-with-fire/ ["Goodin Blockchain"].
1171. *See supra* Chap. 2, Sects. 2.4.2 *Centralized vs. Federated Architecture*; Chap. 4 *UTM Governance*.
1172. https://canso.org/member/cocesna/.
1173. https://asecnaonline.asecna.aero/index.php/en/.
1174. *See supra* Chap. 3, Sect. 3.3.1 *EASA*.
1175. *See supra* Chap. 2, Sect. 2.4.5.3 *InterUSS Platform* (Google tech now hosted by the Linux Foundation), and Chap. 4, Sect. 4.7.2 *Industry Consortia*.
1176. *See* Toshendra Kumar Sharma, *Permissioned and Permissionless Blockchains: A Comprehensive Guide*, https://www.blockchain-council.org/blockchain/permissioned-and-permissionless-blockchains-a-comprehensive-guide/.
1177. Email from Rory Houston, *supra. Cf.* [Goodin Blockchain] ("The problem…is that these chains have existed for decades in the form of hash chains and have already been used for just about anything that could benefit from it.").
1178. Email from Rory Houston, *supra*.
1179. Unlike Bitcoin latency (challenged by its vast networks and probabilistic finality requirements), UTM participating network nodes are smaller/constrained, and can use separate consensus algorithms (e.g., the practical byzantine fault tolerance algorithm (PBFT)) providing near-instant finality. *See* Email from Rory Houston, Co-Founder, Skyy Network (Dec. 21, 2019) (on file with author) ("We argue that centralised path planners/network state gods aren't going to know of that rogue drone with sufficient time just as much as a blockchain network would.…That is where alternate systems of redundancy kick in to keep the sky safe (USS, DAA/SAA, mesh networks, Computer Vision, etc.").
1180. The consensus rules determine when the airspace state can be changed and coupled with private key cryptography and multi-party signature schemes (built into the aircraft themselves) assuring ultra-secure, tiered rule-based airspace access that prohibits non-conforming flight—the rules cannot be feasibly broken.
1181. *See generally* Iuon-Chang Lin, Asia Univ. & Nat'l Chung Hsing Univ., et al., *A Survey of Blockchain Security Issues and Challenges*, Int'l J of Network Security, Vol. 19, No. 5 (Sept. 2017), *available at* https://pdfs.semanticscholar.org/f61e/db500c023c-4c4ef665bd7ed2423170773340.pdf; Jai Xu, Sr. Dir. of Strategy, Urban Air Mobility and Unmanned Aerial Systems, Honeywell, Remarks at the FAA UAS Symposium, Innovation Shaping the Future (Aug. 19, 2020) ("[D]istributed ledger is useful for certain things, maybe tracking parts and components…but it's a tricky thing because we have a tendency to say that distributed ledger is applicable to everything.").

1182. Collectively, blockchain is asserted to solve the following UTM challenges: "Social challenges, airspace equity (market mechanism, no negotiation required), USS interoperability, single source shared airspace representation, privacy/pseudonymity, airspace state/vehicle/payload *verification.*" Email from Rory Houston, *supra* (Dec. 13, 2019) (on file with author). *See generally,* US DoT, Volpe Center, *Blockchain for Unmanned Aircraft Systems* (April 15, 2020), https://rosap.ntl.bts.gov/view/dot/48789; SAE, *Determination of Cost Benefits from Implementing a Blockchain Solution*, ARP6984 (Sept. 3, 2019), https://www.sae.org/standards/content/arp6984/; Skygrid, *The Power of Blockchain in Unmanned Aviation* (June 2020), https://www.skygrid.com/whitepaper/blockchain-in-unmanned-aviation.pdf.

1183. Jay Merkle, Exec. Dir., UAS Integration Office, FAA, Keynote Presentation at the FAA UAS Symposium (July 8, 2020).

1184. *See, e.g.,* ICAO, *ICAO signs new agreements with RTCA, EUROCAE, SAE and ARINC to better align international aviation standardization* (Dec. 12, 2017), https://www.icao.int/Newsroom/Pages/ICAO-signs-new-agreements-with-RTCA,-EUROCAE,-SAE-and-ARINC-to-better-align-international-aviation-standardization-.aspx (Enhancing interoperability to improve sectorial "safety, sustainability, and efficiency.").

1185. *See, e.g., NextGen-SESAR Memorandum of Cooperation,* NAT-I-9406, as amended (2014, 2017), OFFICIAL J. OF THE EU, L 90/3 (April 6, 2018), https://eur-lex.europa.eu/legal-content/EN/TXT/PDF/?uri=CELEX:22018A0406(01)&from=EN (bilateral agreement between FAA and EU [SESAR] to coordinate and harmonize all aspects of ATM – includes "Integration into ATM of new air vehicles, including Unmanned Aircraft Systems (UAS)"); NextGen - SESAR, *State of Harmonization*, 3rd ed., Report prepared by the Coordination Committee (CCOM) for the US–EU MoC Annex 1 High-Level Committee, 2nd Ed. (2016), https://www.faa.gov/nextgen/media/nextgen_sesar_harmonisation.pdf (noted for historical purposes).

1186. Maria Algar Ruiz, EASA Drone Programme Mgr., EASA, Remarks at *GUTMA, High Level Webinar Session - Ask the Experts* (June 9, 2020), https://gutma.org/ask-the-experts/ (with Jay Merkle "in absolute agreement").

1187. *See, e.g.*, FAA, *Bilateral Agreements Overview*, https://www.faa.gov/aircraft/air_cert/international/bilateral_agreements/overview/.

1188. *Declaration of Intent Between the FAA and Federal Office of Civil Aviation, Dept of the Environment, Transportation, Energy and Communications, Swiss Confederation* (May 2020), https://www.faa.gov/news/media/attachments/Declaration_Intent_Swiss_FAA.pdf. *Cf.* NextGen – SESAR, *State of Harmonisation*, 3rd ed., Report prepared by the Coordination Committee (CCOM) & Deployment Coordination Committee (DCOM) for the US–EU MoC Annex 1 Executive Committee (EXCOM) (Sept. 2018), https://www.faa.gov/nextgen/media/NextGen-SESAR_State_of_Harmonisation.pdf.

1189. E.U. and P.R.C., *Agreement on civil aviation safety between the European Union and the Government of the People's Republic of China,* 2020 (O.J.) (L 240) 4, https://eur-lex.europa.eu/legal-content/EN/TXT/PDF/?uri=OJ%3AL%3A2020%3A240%3AFULL&from=en.

1190. Maria Algar Ruiz, *supra.*

1191. Dale Sheridan, Assistant Dir. - Airspace and Future Technology - Aviation and Airports, Dept. of Infrastructure, Transport, Reg'l Dev. and Comm., Australian Gov't, Presentation at the World ATM Congress Virtual Panel: *UTM & ATM: Integrate Now or Integrate Later?* (Oct. 13, 2020) (further urging that "international consensus [should not be] too reflective of any one jurisdiction"). *See, e.g.*, International Coordinating Council of Aerospace Industries Associations, https://iccaia.org (CNS/ATM Committee standing up *ICCAIA Advisory Group* to advance, *inter alia*, airspace integration and ConOps incorporating current and emerging airspace user needs).

1192. Presentation at the EASA High Level Conference on Drones 2019, Amsterdam, NL (Dec. 4-6, 2019), *available at* https://tinyurl.com/Amsterdam-U-spaceServ.

1193. Email from Mark Wuennenberg, Technical Officer, ICAO (Dec. 18, 2019) (on file with author) (noting that ICAO does not want to develop regulations "to early" as regulators need data to develop effective regulations that will not hinder or restrict industry advancement). *See supra* Chap. 3, Sect. 3.2.6 *ICAO*; [ICAO UTM Framework Ed. 3]; Ali Bahrami, Asso. Admin. for Avi. Safety, FAA, Presentation at the FAA UAS Symposium (July 9, 2020) ("It is not always easy to fit new things into the same regulatory box."); Dan Dalton, VP, Global Partnerships, Wisk, Remarks at ICAO DRONE ENABLE Symposium 2021 (April 21, 2021) (UTM a "change in the mindset for us bureaucrats.").

1194. William Stanton, Sr. Advisor – Policy, FAA, Remarks at *UAM Virtual, Integrating Urban Air Mobility into the Airspace*, Aviation Week (Aug. 12, 2020).

1195. *See, e.g.*, [SESAR ATM Master Plan 2017], Sect. 3.5. (safety and regulatory needs); [CORUS ConOps], Annex J, Current, and night operations; [India UTM Policy], Sect. 5.3.2 Mandatory and Voluntary Participation; *JARUS, Regulations*, http://jarus-rpas.org/regulations (providing comparison of selected national regulations); *supra* Chap. 3, Sect. 3.2.6 *ICAO* (considering needed reform to its Annexes); Chap. 4 *UTM Governance*.

1196. *See supra* Chap. 4, Sect. 4.7.4 *Liability*.

1197. Murzilli Consulting, Presentation at the FAA UAS Symposium (July 8, 2020). *See* Jay Merkle, *supra* ("UTM: now we're understanding how it interacts with the regulatory side").

1198. [FAARA], Sects. 376 & 377 (ordering implementation plan and interim UTM services). *See* [GAO UAS], p. 30 ("Until FAA release its statutorily mandated UTM implementation plan, providing additional information through existing communications channels on the timing and substance of next steps could help FAA and stakeholders continue to make progress on UTM activities.").

1199. *See* DJI, *Elevating Safety: Protecting The Skies In The Drone Era* (May 2019), Sect. 5.III, https://terra-1-g.djicdn.com/851d20f7b9f64838a34cd02351370894/Fly-safe/190521_US-Letter_Policy-White-Paper_web.pdf ("[g]iven the pace of government rule-making, implementation [of a UAS remote ID] is likely to be years away."). Nonetheless, the Remote ID Rule has issued; MoC are in development; and the FAA indicated it may give some form of "credit" for early voluntary compliance.

1200. Brian Wynne, Pres. and CEO, AUVSI, Keynote presentation at the FAA UAS Symposium (July 8, 2020). Email from Michael Robbins, EVP, AUVSI (Sept. 24, 2020) (on file with author). ("At its core, AUVSI is an advocacy association, and therefore working to advance UTM through the support of public policy." Also underscoring AUVSI's focus to advance the FAA's RID rule).

1201. [FAA ConOps UAM v1.0.], Sect. 3, ¶ 3 (*E.g.*, "UAM driven regulatory changes" are envisioned "to address the needs for UAM operations' structure and performance."); [Volocopter Roadmap], Sect. 4.6 ("Merely translating old regulations will not work since there is a multitude of new considerations including new types of vehicles, unprecedented ways of operation, unique environmental situations within an urban environment, and new operators."). *See NATA, Urban Air Mobility: Considerations for Vertiport Operation* (2019), p. 9, https://www.nata.aero/assets/Site_18/files/GIA/NATA%20UAM%20White%20Paper%20-%20FINAL%20cb.pdf ["NATA UAM"] (Asserting a "regulatory void" regarding UAM vertiport operations); Chap. 6, Sect. 6.6 *Stratospheric Operations*.

1202. *See, e.g.*, [Comm Delegated Reg. (EU) 2019/945 - UAS]; [Comm Implementing Reg. (EU) 2019/947 - UA], ¶ 29.

1203. [Comm Implementing Reg. (EU) 2021/664 - U-space]; [SESAR SRIA], Sect. 3.4 U-space and urban air mobility (recognizing "U-space will have to overcome extraordinary challenges" and develop "[a] new regulatory framework").

1204. *See supra* Chap. 5, Sect. 5.3 *U-space*.

1205. William Stanton, *supra*.

1206. Interview by NASA in Silicon Valley Live - *Air Taxis and the Future of Flight*, Moffett Field, Cal. (Dec. 19, 2019), *available at* https://tinyurl.com/Rios-SVL. *See* [EHang 2020], Ch. 1 ("UAM as a revolutionary idea that can be implemented now.").

1207. Presentation at the Transp. Research Board (TRB) Annual Mtg., Orlando, Fl. (Jan. 13, 2020), https://www.c-span.org/video/?468065-1/automation-technology-transportation

(also noting the "emergence of UAM [as] filling in the last 30-300 miles which will then meet commercial aviation and complete the whole aviation supply chain").

1208. *See* Steve Bradford, Chief Scientist for Architecture and NextGen Development, FAA, Presentation at ICNS, Herndon, Va. (April 9, 2019) (describing UAM vehicles as "drone taxis" or "flying Teslas").

1209. [MITRE 2019], p. 10 (Also, "[t]he architecture for service provision may be similar to and possibly share elements of... (UTM) architecture."). *See generally* NASA, *UAM Overview*, https://www.nasa.gov/uam-overview/; FAA, *Urban Air Mobility and Advanced Air Mobility*, https://www.faa.gov/uas/advanced_operations/urban_air_mobility/.

1210. [FAA ConOps UAM v1.0], Sect. 1.3.1.

1211. [NATA UAM], p. 4; [FAA ConOps UAM v1.0], Sect. 1.2.1 (also defining it as "part of intermodal transportation links"). UAM also includes hybrid electric powered eVTOLs. *See* Alfredo Giuliano, Intelligent Systems Mgr., Aurora Flight Sciences, Presentation at Enabling Autonomous Flight & Operations in the National Airspace System Workshop 2, NASA Ames Research Center, Moffett Field, Cal. (Aug. 6, 2019) (characterizing UAM requirements to include: flights averaging 20 miles, payload of 2-4 passengers, thousands of trips daily, and hundreds of vehicles within a city infrastructure of 1-20 vertiports); NEXA Advisors and NBAA, *Business Aviation Embraces Electric Flight, How Urban Air Mobility Creates Enterprise Value* (Oct. 21, 2019), p. 11, https://nbaa.org/wp-content/uploads/aircraft-operations/uas/NEXA-Study-2019-Business-Aviation-Embraces-Electric-Flight.pdf (eVTOL to deliver increased *enterprise value* "similar to those generated by business aircraft, and contribute directly to shareholder value creation at multiple levels...").

1212. Presentation at NBAA BACE, Las Vegas, Nev. (Oct. 23, 2019).

1213. [FAA ConOps UAM v1.0], Sect. 1.1. *See* Paul McDuffee, Exec./Ops. Analyst at Hyundai Urban Air Mobility (formerly, Boeing Horizon X), Presentation at NBAA BACE, Las Vegas, Nev. (Oct. 23, 2019) (suggesting the term *UAM* is too limiting); [Nat'l Academies AAM], Sect. 3-1; p. vii ("UAM, is but one subset in a much broader field of advanced aerial mobility" [and] "increasingly used."). The terms AAM and UAM are sometimes used interchangeably. *Cf.* [EmbraerX 2019], p. 5 (UAM characterized as *urban air traffic management* (UATM)).

1214. Source: NASA Ames. *See generally* [NASA UAM Vision ConOps]; NASA, *Advanced Air Mobility Mission Studies / Reports / Presentations*, https://www.nasa.gov/aam-studies-reports/.

1215. [FAA ConOps UAM v1.0], Sect. 4.1. *See, e.g.*, Ch. 2, Sect. 2.2.1 *UAS Service Suppliers (USS)* (introducing the "PSU"—provider of services for UAM—the UAM analog to the USS); [NASA UAM Vision ConOps], Sect. 4.1.2.1 Airspace Rules and Procedures (describing "The PSU Network [that] governs and deconflicts PSU operations.").

1216. [EUROCAE Programme 2019], Sect. 2.3.6.

1217. [Embraer^X 2019], p. 13.

1218. Tom Prevot, Ph.D., Air Taxi Product Lead, Joby Aviation, formerly Dir., Airspace Systems, Uber, Presentation at ASTM F38 meeting, Seattle, WA (April 18, 2019) (adding, "[with] a lot more to be solved"). *See* [MITRE 2019], p. 6 (proposing four "operational concept components [to] help enable effective integration of automated UAM operators into the NAS": augmented visual flight rules; dynamic delegated corridors; automated decision support services; and performance-based operations).

1219. [Nat'l Academies AAM], Sect. 3-2.

1220. [FAA ConOps UAM v1.0], p. iii. *See generally* [NASA UAM Vision ConOps]. *Cf.* Mark Huber, *FAA's New Urban Air Mobility ConOps Raises Questions*, Avi. Int'l News (July 15, 2020), https://tinyurl.com/Huber-UAM-ConOps (addressing integration issues, such as protecting noncooperative aircraft).

1221. Joseph Rios, ATM-X Project Chief Eng'r, NASA Ames, Presentation at ASTM F38 meeting, Seattle, WA (April 17, 2019). *See* Michael McNair, Innovation Mgr., Bell, Presentation at Enabling Autonomous Flight & Operations in the National Airspace System Workshop 2, NASA Ames Research Center, Moffett Field, Cal. (Aug. 6, 2019) (underscoring ODM [on demand mobility] challenges for the NAS: modality, scalability, traffic density, awareness, and systems integration; and urging the importance of multi-vehicle cooperative formation management). *See generally* Eric Mueller, Parimal Kopardekar, et. al., NASA Ames Research Center, *Enabling Airspace Integration for High-Density On-Demand Mobility* Operations (June 2017), https://utm.arc.nasa.gov/docs/2017-Mueller_Aviation_ATIO.pdf.

1222. Tom Prevot, Ph.D., Air Taxi Product Lead, Joby Aviation, formerly Dir., Airspace Systems, Uber, Presentation at Uber Elevate Summit 2018, L.A., Cal. (May 8, 2018), https://www.youtube.com/watch?v=FQVqisv2rYE. *Cf.* [EHang 2020], Ch. 1 (proposing a "centralized platform"). *See supra* Chap. 2, Sect. 2.4.4 (*SWIM*).

1223. Tom Prevot, Ph.D., *id. See* Sanjiv Singh, CEO, Near Earth Autonomy, *Lessons Learned from Autonomous sUAS*, Presentation at Enabling Autonomous Flight & Operations in the National Airspace System Workshop 2, NASA Ames Research Center, Moffett Field, Cal. (Aug. 6, 2019) (asserting that operational UAM does not exist today for want of solutions to "two big issues": contingencies in case of failure, and countering willful misuse).

1224. *See* [Embraer UATM ConOps], Sect. 2.4. (UAM will operate "primarily below 1,500 ft" AGL, but also at higher altitudes); [NASA UAM Vision ConOps], Sect. 2.0 The UAM Operating Environment ("The UOE exists adjacent to actively controlled airspace rather than as a separate airspace class."); Drone Alliance Europe, *U-space Whitepaper, Ver. 2.0* (Feb. 2020), https://resourcecenter.dronealliance.eu/wp-content/uploads/2019/05/DAE-UTM-U-Space-whitepaper-2.0-final-1.pdf, pp. 1, 6-7 (urging UAM accommodation "in both low and intermediate altitudes" and "between uncontrolled and controlled airspace").

See also supra Chap. 6, Sect. 6.2.2 *Advanced Automation and Autonomy. But see* Brian Garrett-Glaser, *Why Automation Shouldn't Push Pilots Out of Air Taxis,* Rotor & Wing Int'l (May 23, 2019), http://go.rotorandwing.com/yMcZawin0H0nTTk6St00030 ("Many entrants to the air taxi game expect autonomy to solve the key challenge, providing safer, more reliable and more cost-effective flight without the need to pay (or develop) a workforce or use up a seat on a non-paying body.").

1225. Hervé Martins-rivas, Vehicle Partners Engineering Lead, Uber Elevate, Remarks at the NASA, NARI, AAM Aircraft WG Kickoff (May 28, 2020), https://youtu.be/13gGX-YGMHNE. *See* Email from Rex Alexander, Five-Alpha LLC (Mar. 15, 2021) (on file with author) (re unknown aircraft designs).

1226. Email from Andrew Carter, Pres. and Co-founder, ResilienX, Inc. (Dec. 12, 2019) (on file with author). *See* Tom Prevot, Ph.D., Air Taxi Product Lead, Joby Aviation, formerly Dir. of Eng'g, Airspace Systems, Uber Elevate (Dec. 12, 2019) (on file with author) ("It's also not quite clear whether the process should be design assurance or a different process with an equivalent or better level of safety, but more adequate for complex systems."); Email from James Licata, Bus. Dev. Mgr., Hidden Level, Inc. (Jan. 20, 2020) (on file with author) ("Problem is no one has defined these things, and backing into them is a hard (and expensive)."). *See supra* Ch. 4, Sect. 4.1 *USS Qualification*.

1227. *See* [Nat'l Academies AAM], Sect. 4 Safety, Security, and Contingency Management, p. 43 (recognizing the "unique nature of software," urging use of safety modeling and analysis tools rather than only testing or simulation, extending such tools for UAM, and "research on new, more powerful safety analysis tools that are widely used today that can be applied to software-intensive advanced system."); *see supra* Chap. 4, Sect. 4.2.3 *Basis of Conformity Assessment*; Chap. 6, Sect. 6.2.2 *Advanced Automation and Autonomy*.

1228. *See, e.g.*, Jeffrey R. Homola, et al., NASA, *UAS Traffic Management (UTM) Simulation Capabilities and Laboratory Environment* (Sept. 25-29, 2016), https://utm.arc.nasa.gov/docs/Homola_DASC_157026369.pdf; Email from John Scull Walker, Sr. Partner, The Padina Group & Chair, ISO/TC 20/SC 16 (May 16, 2019) (underscoring that there are many airspace separation issues that require resolution).

1229. *See, e.g.*, [EASA Reg. Framework U-space], Sect. 2.1. Why we need U-space (urging "society expects more than safety [and this] is why the U-space regulatory approach must go beyond traditional aviation safety issues and integrate the urban dimension of aerial mobility."); François Sillion, Dir. Advanced Technologies Centre Paris, Uber, *quoted in, Flying taxis are taking off to whisk people around cities*, THE ECONOMIST (Sept. 12, 2019), https://www.economist.com/science-and-technology/2019/09/12/flying-taxis-are-taking-off-to-whisk-people-around-cities ("No one really knows exactly how it is going to happen.").

1230. David Silver, VP for Civil Aviation, AIA, Remarks at the NASA, NARI, AAM Aircraft WG Kickoff (May 28, 2020), https://youtu.be/13gGXYGMHNE.

1231. *See* Nikhil Goel, Head of Product, Aviation at Uber, Presentation at Uber Elevate Summit 2019, Wash., D.C. (June 11, 2019) (testing with, and underscoring the exclusive use of "twin engine, twin piloted" helicopters, and the paramount need for safety); Bala Ganesh, VP, UPS Adv. Tech. Group, *quoted in CNBC, UPS agrees to buy electric vertical aircraft to speed up package delivery in small markets* (April 7, 2021) https://www.cnbc.com/2021/04/07/ups-to-buy-evtol-aircraft-to-speed-up-package-delivery-in-small-markets.html ("The new type of aircraft … "unlocks new business models that don't exist today…"); NASA, *Advanced Air Mobility National Campaign,* https://www.nasa.gov/aamnationalcampaign (previously entitled the *Grand Challenge*, key effort integrates and tests vehicles and airspace concepts).

1232. Tom Prevot, Ph.D., Air Taxi Product Lead, Joby Aviation, formerly Dir., Airspace Systems, Uber, Remarks, in Palo Alto, Cal. (Aug. 16, 2019).

1233. [Nat'l Academies AAM], Sect. 3-9.

1234. *See* [NATA UAM], p. 9 ("At present there is no comprehensive canon of policy guidance or regulatory mandates governing vertiport operation…no mandatory design standards…that speak to eVTOL infrastructure…to be considered 'safe' by an objective standard."); Deborah J. Peisen, *Analysis of Vertiport Studies Funded by the Airport Improvement Program (AIP)*, DOT/FAA/RD-93/37 (May 1994), https://apps.dtic.mil/dtic/tr/fulltext/u2/a283249.pdf (identifying vertiport and vertistop challenges); *Cf.* ASTM Int'l, F_ [WK59317], *New Specification for Vertiport Design*, https://www.astm.org/DATABASE.CART/WORKITEMS/WK59317.htm (providing early support and guidance for civil vertiport and vertistop design, and associated operational best practices).

1235. Jennifer Richter, JD, Partner, Aiken Gump, Presentation at AUVSI Xponential, Chicago, Ill. (April 29, 2019) (underscoring that edge computing, fiber optics, and "the benefits of features offered by smart city infrastructure" can advance UAM). *See generally* GCTC Smart Buildings Super Cluster, *Smart Buildings: A Foundation for Safe, Healthy & Resilient Cities* (Aug. 2020), https://pages.nist.gov/GCTC/uploads/blueprints/2020-SBSC-blueprint.pdf (SBSC Blueprint).

1236. *See supra* Chap. 4, Sect. 4.6 *State, Regional, and Local. See also* Kevin DeGood, *Flying Cars Will Undermine Democracy and the Environment*, Center for Am. Progress (May 28, 2020), https://www.americanprogress.org/issues/economy/reports/2020/05/28/481148/flying-cars-will-undermine-democracy-environment/ (raising potential community and societal opposition issues; challenging the positive impact of flying cars on traffic congestion; and "represent[ing] the apotheosis of sprawl…a spatial manifestation of deepening inequality [that] will amplify this separation and exacerbate the worst tendencies of wealth, power, and privilege with deleterious long-term effects on democracy.").

1237. Stan Swaintek, formerly Head of Aviation, Uber, Presentation at Uber Elevate Summit 2019, Wash., D.C. (June 11, 2019) (urging collaboration with smart city initiatives).

1238. To some extent UAM implements *mobility as a service* (MaaS)—the shift from person/owned to shared transportation resources—thus, UAM is affected by human choice/behavior.

1239. Dan Dalton, VP of Global Partnerships, Wisk, Remarks at *UAM Virtual, Integrating Urban Air Mobility into the Airspace*, AVIATION WEEK (webinar) (Aug. 12, 2020); Editorial, *BLADE Tests Urban Air Mobility Pilot Program in The Bay Area*, Transport UP (March 19, 2019), https://transportup.com/editorials/blade-in-the-bay/ (transition path for UAM from manned to unmanned status unresolved due, in part, to public perception and regulatory issues).

1240. Presentation at ICAO, DRONE ENABLE/2, in Chengdu, P.R.C. (Sept. 13-14, 2018), https://www.youtube.com/watch?v=Uz2cyq0khQY. *See generally* Uniting Aviation, *High fliers: high-altitude long-endurance aircraft are seeking new operational flight levels* (Aug. 28, 2019), https://tinyurl.com/FL600.
1241. FAA, *Aviation Rulemaking Committee Charter*, Subj: UAS in Controlled Airspace Aviation Rulemaking Committee (Amend. Nov. 20, 2017), p. 1, https://www.faa.gov/regulations_policies/rulemaking/committees/documents/media/UAS%20Controlled%20Airspace%20Charter%20Amendment%20(signed%2011-20-17).pdf, and https://www.faa.gov/regulations_policies/rulemaking/committees/documents/index.cfm/document/information/documentID/3222. *See* Robin Garrity, ATM Expert, Airports and Airspace User Operations, SESAR, Presentation at ICAO, DRONE ENABLE/3, Montreal, CA (Nov. 14, 2019) (noting that at high altitudes air traffic is low density – mitigating traffic conflict).
1242. FAA NextGen, *Concept of Operations, Upper Class E Traffic Management*, v1.0 (April, 22, 2020), Sect. 1, https://nari.arc.nasa.gov/sites/default/files/attachments/ETM_ConOps_V1.0.pdf ["FAA High E ConOps v1.0"].
1243. Other challenges include that flight characteristics of balloons and certain other lighter-than-air vehicles are not well-suited to the ATM system, in part, because their behavior is unconventional, their navigation requires changing altitudes to exploit favorable wind currents/direction—including nondeterministically, and vary by day or night. *See* Léonard Bouygues, Head of Aviation Strategy, Loon, and Wajahat Beg, Head of Overflights, Loon, Presentation at ICAO TV, Innovation, *Stratospheric Operations* (webinar) (Sept. 2, 2020). Perhaps the greatest challenge may be to "get costs low enough to build a long-term sustainable business." Alaster Westgarth, CEO, Loon, *Saying Goodbye to Loon* (Jan. 21, 2021), https://medium.com/loon-for-all/loon-draft-c3fcebc11f3f (announcing demise of Loon).
1244. *See, e.g.*, ICAO, Gary Christiansen, Tech. Advisor, FAA, *ATC Services Above FL600*, Twenty-Fourth Meeting of the Cross Polar Trans East Air Traffic Management Providers' Work Group (CPWG/24), IP/03 (Dec. 12, 2017), https://www.faa.gov/about/office_org/headquarters_offices/ato/service_units/mission_support/ato_intl/cross_polar/.
1245. John Scull Walker, Sr. Partner, The Padina Group & Chair, ISO/TC 20/SC 16, Remarks (March 23, 2019).
1246. Nancy Graham, Graham Aerospace Int'l, Presentation at ICAO, DRONE ENABLE/2, in Chengdu, P.R.C. (Sept. 13-14, 2018), https://www.youtube.com/watch?v=Uz2cyq0khQY.
1247. Source: [NASA High E ConOps v1.0], Fig. 2-7.
1248. Andy Tailby, Head of Flight Operations, Zephyr, Presentation at ICAO, DRONE ENABLE/2, in Chengdu, P.R.C. (Sept. 13-14, 2018), https://www.youtube.com/watch?v=Uz2cyq0khQY. *See* [Rios 2020], p. 31 ("NASA plans to continue research on UTM. This includes graduating concepts and architectures to other aviation domains such as Urban Air Mobility (UAM*), high altitude operations (over 60,000 ft)*, and space traffic management.") (emphasis added).
1249. *See, e.g.*, Email from Philip M. Kenul, Chair, ASTM F38 on Unmanned Aircraft Systems, (Jan. 18, 2021) (High altitude airspace operations have been "added to the TOR [terms of reference] for UTM as a future requirement."); Telephone interview with John Scull Walker, Sr. Partner, The Padina Group & Chair, ISO/TC 20/SC 16, remarks (March 23, 2019) (noting, "If the same [VLL] UTM principles are used above 60,000' then you have the 'sandwich effect'.").
1250. Presentation at ICAO, DRONE ENABLE/3, Montreal, CA (Nov. 14, 2019) (Recognizing that each aircraft knows its position, "[w]hy should air traffic get involved if operators have the information? I think this is fairly an easy concept if we go back to our first principles.").
1251. Leo Bouygues, Head of Flight Operations, Loon, Presentation at ICAO, DRONE ENABLE/3, Montreal, CA (Nov. 14, 2019). *See* Abdi Latif Dahir, *A Bird? A Plane? No, It's a*

Google Balloon Beaming the Internet, N.Y. Times (July 7, 2020), https://www.nytimes.com/2020/07/07/world/africa/google-loon-balloon-kenya.html (high altitude delivery of 4G LTE service to central and western Kenya).

1252. [FAA High E ConOps v1.0], Sect. 2.2, n. 3.
1253. Parimal Kopardekar, Ph.D., Dir., NARI, NASA Senior Technologist for Air Transportation System, *Airspace operations*, Tele-presentation at GUTMA Annual Conference, Portland, Or. (July 2019). *Cf.* Email from Christopher T. Kucera, Head of Strategic Partnerships, OneSky Systems (Dec. 24, 2019) (on file with author) ("I would argue that UTM is a model for space traffic management (STM), but that it really only extends to ETM, not space and STM. Concepts from STM have been shared with UTM and vice versa. The domain of space and air are completely different from a physics perspective and require different methods."); Brittany Sauser, *A Better Network for Outer Space, Why Vint Cerf wants to put Internet-style networking in space,* MIT Tech. Rev. (Oct. 27, 2008), https://www.technologyreview.com/s/411092/a-better-network-for-outer-space/ (inventor of the *Interplanetary Internet* –"Ultimately, the network could interconnect manned and robotic spacecraft, forming the backbone of a communications system that reaches across the solar system."). Separately, UTM services (e.g., automated traffic deconfliction) could possibly be used to manage the growing problem of space debris (i.e., noncollaborative spacecraft).
1254. Presentation at ICAO, DRONE ENABLE/2, in Chengdu, P.R.C. (Sept. 13-14, 2018), https://www.youtube.com/watch?v=Uz2cyq0khQY.
1255. Presentation at the *ICAO UTM Framework - Core Principles for Global Harmonization* (webinar) (Aug. 25, 2020).
1256. Peter F. Dumont, CEO, ATCA, *quoted in,* Kristina Knott, ATCA, *Airbus UTM and ATCA at World ATM Congress: Defining Future Skies, The Evolution of ATM and UTM*, ATCA Bulletin, Issue 2 (2019), p. 5, https://www.atca.org/atca-bulletin (emphasis added).
1257. [Embraer^{X} 2019], p. 15. UTM's impact on future ATM has also been characterized as the *UTMification* of ATM. *See, e.g.,* US DOT, *Budget Highlights*, Fiscal year 2020, https://www.transportation.gov/sites/dot.gov/files/docs/mission/budget/333126/budgethighlights030719final518pm25082.pdf (request includes $202.6 million for: "The Unmanned Air Traffic Management System, which will *pair* with the traditional Air Traffic Management System..." Emphasis added).
1258. GSMA, *Using Mobile Networks to Coordinate Unmanned Aircraft Traffic* (2018), p. 32, https://www.gsma.com/iot/wp-content/uploads/2018/11/Mobile-Networks-enabling-UTM-v5NG.pdf.
1259. ATCA Bulletin, *Preparing for our future ATM state*, Issue 2 (2019), p. 3, https://www.atca.org/atca-bulletin.
1260. d-flight, ITALY submission on ICAO Unmanned Aircraft System Traffic Management (UTM, Request for Information (DRONE ENABLE/3) (2020) (copy on file with author).
1261. [SESAR SRIA], Sect. 2.1 Vision leading to the Digital European Sky.
1262. Lorenzo Murzilli, Founder and CEO, Murzilli Consulting, formerly Leader, Innovation and Digitization Unit, Presentation at the FAA UAS Symposium (July 8, 2020). NASA has characterized a UTM-inspired environment "ATM-X". *See* NASA, *Air Traffic Management – eXploration (ATM-X),* https://www.nasa.gov/aeroresearch/programs/aosp/atm-x.
1263. [EUROCONTROL UAS ATM Integration], p. 7. *See* Frank Matus, Remarks at the ICAO DRONE ENABLE Symposium 2021 (April 20, 2021) ("The term UTM will cease to exist ... in a number of years.").
1264. Leslie Cary, Chief, Remotely Piloted Aircraft Systems (RPAS) Section, ICAO, *quoted in,* Unmanned Airspace (Nov. 25, 2018), https://www.unmannedairspace.info/utm-industry-leader-interview/connecting-atm-utm-presents-technical-regulatory-cultural-conflicts-leslie-cary-icao/. *See* [Comm Implementing Reg. (EU) 2021/664 -

U-space], Art. 7 U-space service providers, 3. ("shall establish arrangements with the air traffic services providers to ensure adequate coordination of activities, as well as the exchange of relevant operational data and information").

1265. Paul Albuquerque, Aerospace Eng'r, FAA Flight Standards Service, Presentation at the FAA UAS Symposium, Balt., Md. (June 3, 2019).

1266. *See, e.g.*, SESAR JU, *Exploring the boundaries of air traffic management, A summary of SESAR exploratory research results* 2016-2018 (2018), https://www.sesarju.eu/sites/default/files/documents/reports/ER_Results_2016_2018.pdf; *supra* NASA, ATM-X.

1267. Remarks at the FAA UAS Symposium (July 9, 2020).

1268. [Kopardekar 2016], p. 16.

Selected References/ Bibliography

6 U.S.C. Sect. 214n. – Protection of certain facilities and assets from unmanned aircraft, *available at* https://www.law.cornell.edu/uscode/text/6/124n

14 C.F.R. Part 107 – Small Unmanned Aircraft Systems, *available at* https://www.law.cornell.edu/cfr/text/14/part-107

49 U.S.C. Ch. 448 – Unmanned Aircraft Systems, *available at* https://uscode.house.gov/view.xhtml?path=/prelim@title49/subtitle7/partA/subpart3/chapter448&edition=prelim

49 U.S.C. Part 87 – Aviation Services, *available at* https://www.law.cornell.edu/cfr/text/47/part-87

49 U.S.C. Sect. 44809 – Exception for limited recreational operations of unmanned aircraft, *available at* https://www.law.cornell.edu/uscode/text/49/44809

5G Slicing Association, *5G Network Slicing for Cross Industry Digitization: Position Paper* (2018), p. 20, https://cdn0.scrvt.com/fokus/b77ede33ea8d7dc8/1819ab58384e/5G-Network-Slicing-for-Cross-Industry-Digitization-Position-Paper–Digital.pdf

A Deep Dive into UTM and the Flight Information Management System for Drones, DRONELIFE (Aug. 22, 2019), https://dronelife.com/2019/08/22/a-deep-dive-into-utm-and-the-flight-information-management-system-for-drones-long-form/

Access 5 (archival website), https://web.archive.org/web/20060627203947/http://www.access5.aero/site_content/index.html

ACJA, *LTE Aerial Profile* v1.00 (Nov. 2020), https://gutma.org/acja/wp-content/uploads/sites/10/2020/11/ACJA-WT3-LTE-Aerial-Profile_v1.00.pdf

Adamski, Anthony J., et al., *Requisite Imagination: The Fine Art of Anticipating What Might Go Wrong,* IN HANDBOOK OF COGNITIVE TASK DESIGN (ERIK HOLLNAGE, ED., 2003), DOI: 10.1201/9781410607775, https://www.taylorfrancis.com/books/9780429228216

Aerospace Industries Association of America, Inc. (AIA), *Civil Aviation Cybersecurity Industry Assessment & Recommendations, Report to the AIA Civil Aviation Council, Civil Aviation Regulatory & Safety Committee* (Aug. 2019), https://www.aia-aerospace.org/wp-content/uploads/2019/10/AIA-Civil-Aviation-Cybersecurity-Recommendations-Report-2019-Final-1.pdf

Agreement on civil aviation safety between the European Union and the Government of the People's Republic of China, 2020 (O.J.) (L 240) 4, https://eur-lex.europa.eu/legal-content/EN/TXT/PDF/?uri=OJ%3AL%3A2020%3A240%3AFULL&from=en

Air Line Pilots Ass'n, *Remotely Piloted Aircraft Systems, Challenges for Safe Integration into Civil Airspace,* ALPA White Paper (Dec. 2015), http://www.alpa.org/-/media/ALPA/Files/pdfs/news-events/white-papers/uas-white-paper.pdf?la=en

Airbus & Boeing, *A New Digital Era of Aviation: The Path Forward for Airspace and Traffic Management* (Sept. 2020) (on file with author)

Airbus, *Blueprint For The Sky* (2018), https://storage.googleapis.com/blueprint/Airbus_UTM_Blueprint.pdf, and https://www.airbusutm.com/ ["Airbus Blueprint"]

Airbus, *Technical Whitepapers*, https://www.airbusutm.com/uam-resources-technical-reports

AIRMAP, *Privacy Notice for California Residents* (effective Jan. 1, 2020), https://www.airmap.com/privacy-notice-for-california-residents/

Am. Bar Ass'n, Sect. of Sci. & Tech., Info. Sec. Comm., *PKI Assessment Guidelines* (PAG) (May 20, 2003), https://www.americanbar.org/content/dam/aba/events/science_technology/2013/pki_guidelines.pdf

Amazon Prime Air, *Petition for Exemption Under 49 U.S.C. Sect. 44807 and 14 C.F.R. Parts 61, 91, and 135* (July 16, 2019), https://www.aviationtoday.com/wp-content/uploads/2019/08/amazon_-_exemption_rulemaking.pdf; https://www.federalregister.gov/documents/2019/08/08/2019-17010/petition-for-exemption-summary-of-petition-received-amazon-prime-air

Amazon, *Revising the Airspace Model for the Safe Integration of Small Unmanned Aircraft Systems* (July 2015), *available at* https://images-na.ssl-images-amazon.com/images/G/01/112715/download/Amazon_Revising_the_Airspace_Model_for_the_Safe_Integration_of_sUAS.pdf

An Introduction to Unmanned Aircraft Systems (Douglas Marshall et al. eds., CRC 3rd ed. 2021)

Ancel, Ersin, Ph.D., Aerospace Eng'r, NASA, et al., *In-Time Non-Participant Casualty Risk Assessment to Support Onboard Decision Making for Autonomous Unmanned Aircraft* (June 14, 2019), https://arc.aiaa.org/doi/abs/10.2514/6.2019-3053

Ancel, Ersin, Ph.D., Aerospace Eng'r, NASA, *Ground Risk Assessment Service Provider (GRASP) Development Effort as a Supplemental Data Service Provider (SDSP) for Urban Unmanned Aircraft System (UAS) Operations,* DASC (Sept. 1, 2019), https://ieeexplore.ieee.org/document/9081659

Ancel, Ersin, Ph.D., et al., Aerospace Eng'r, NASA, *Real-time Risk Assessment Framework for Unmanned Aircraft System (UAS) Traffic Management (UTM)* (2017), https://arc.aiaa.org/doi/10.2514/6.2017-3273

ANRA UTM powers NASA TCL4 campaigns in Nevada and Texas, sUAS News (May 15, 2019), https://www.suasnews.com/2019/05/anra-utm-powers-nasa-tcl4-campaigns-in-nevada-and-texas/?mc_cid=415259365a&mc_eid=a9df228121

ANSI, *Standardization Roadmap for Unmanned Aircraft Systems*, Ver. 2.0 (June 2020), https://share.ansi.org/Shared%20Documents/Standards%20Activities/UASSC/ANSI_UASSC_Roadmap_V2_June_2020.pdf ["ANSI Roadmap"]

ANSI/CTA-2063-A, *Small Unmanned Aerial Systems Serial Numbers* (Sept. 2019), https://shop.cta.tech/collections/standards/products/small-unmanned-aerial-systems-serial-numbers

ASSURE, https://assureuas.org/

ASTM International, F_ [WK59317], *New Specification for Vertiport Design*, https://www.astm.org/DATABASE.CART/WORKITEMS/WK59317.htm

ASTM International, F_ [WK62344], Revision of F3196, *Standard Practice for Seeking Approval for Extended Visual Line of Sight (EVLOS) or Beyond Visual Line of Sight (BVLOS) Small Unmanned Aircraft System* (sUAS) Operations), https://www.astm.org/DATABASE.CART/WORKITEMS/WK62344.htm

ASTM International, F_ [WK63418], *New Specification for UAS Traffic Management (UTM) UAS Service Supplier (USS) Interoperability*, https://www.astm.org/DATABASE.CART/WORKITEMS/WK63418.htm ["ASTM UTM Spec."]

ASTM International, F_ [WK69690], *New Specification for Surveillance UTM Supplemental Data Service Provider (SDSP) Performance,* https://www.astm.org/DATABASE.CART/WORKITEMS/WK69690.htm ["ASTM Surveillance SDSP"]

ASTM International, F_ [WK70877], *Standard Practice for Development of a Durability and Reliability Flight Demonstration Program for Low-Risk Unmanned Aircraft Systems (UAS) under FAA Oversight*, www.astm.org

ASTM International, F_ [WK75923], *New Specification for Positioning Assurance, Navigation, and Time Synchronization for Unmanned Aircraft Systems*, https://www.astm.org/DATABASE.CART/WORKITEMS/WK75923.htm?A&utm_source=tracker&utm_campaign=20210218&utm_medium=email&utm_content=standards

ASTM International, F_[WK75981], *New Specification for Vertiport Automation Supplemental Data Services Provider (SDSP)*, https://www.astm.org/DATABASE.CART/WORKITEMS/WK75981.htm?A&utm_source=tracker&utm_campaign=20210223&utm_medium=email&utm_content=standards

ASTM International, F3196-18, *Standard Practice for Seeking Approval for Beyond Visual Line of Sight (BVLOS) Small Unmanned Aircraft System (sUAS) Operations*, https://www.astm.org/Standards/F3196.htm

ASTM International, F3269-17, *Standard Practice for Methods to Safely Bound Flight Behavior of Unmanned Aircraft Systems Containing Complex Functions*, https://www.astm.org/Standards/F3269.htm

ASTM International, F3322-18, *Standard Specification for Small Unmanned Aircraft System (sUAS) Parachutes*, https://www.astm.org/Standards/F3322.htm

ASTM International, F3411-19, *Standard Specification for Remote ID and Tracking* (July 2, 2019), DOI: 10.1520/F3411-19, https://www.astm.org/Standards/F3411.htm ["ASTM Remote ID"]

ASTM International, F3442/F3442M-20, *Standard Specification for Detect and Avoid System Performance Requirements,* DOI: 10.1520/F3442_F3442M-20, https://www.astm.org/Standards/F3442.htm

ASTM International, Stephen Cook, Chair, ASTM AC377, et al., Technical Report, *Autonomy Design and Operations in Aviation: Terminology and Requirements Framework,* TR1-EB (July 2019), DOI: 10.1520/tr1-eb, https://www.astm.org/DIGITAL_LIBRARY/TECHNICAL_REPORTS/PAGES/1fe1b67c-5ff0-488e-ab61-cb0d329bc63c.htm, and https://www.astm.org/DIGITAL_LIBRARY/TECHNICAL_REPORTS/index.html ["ASTM Autonomy Framework"]

ASTM International, Stephen Cook, Chair, ASTM AC377, et al., Technical Report, *Developmental Pillars of Increased Autonomy for Aircraft Systems*, TR2-EB (2020), https://www.astm.org/DIGITAL_LIBRARY/TECHNICAL_REPORTS/PAGES/fc2da6dc-733d-4d86-bb09-75a555b85330.htm

Atherton, Kelsey, *Amazon's Delivery Plans aren't Realistic*, Popular Science (Dec. 4, 2013), *available at* https://www.businessinsider.com/amazons-delivery-drones-arent-realistic-2013-12

ATIS, *Support for UAV Communications in 3GPP Cellular Standards*, ATIS-1-0000069 (Oct. 2018), https://access.atis.org/apps/group_public/download.php/42855/ATIS-I-0000069.pdf

ATIS, *Unmanned Aerial Vehicle (UAV) Utilization of Cellular Services, Enabling Scalable and Safe Operations*, ATIS-I-0000060 (2017), https://access.atis.org/apps/group_public/download.php/36134/ATIS-I-0000060.pdf

Australian Gov't, Airservices Australia, *FIMS Request for Information (RFI)*, ASA RFI 394578597 (Aug. 19, 2020), https://engage.airservicesaustralia.com/50159/widgets/263980/documents/177901

Australian Gov't, CASA, Advisory Circular AC 101-01, v3.0, *Remotely piloted aircraft systems – licensing and operations*, Ref. D19/460393 (Dec. 2019), https://www.casa.gov.au/sites/default/files/101c01.pdf

Australian Gov't, Dept. of Infrastructure, Transport, Reg'l Dev. and Comm., *Emerging Aviation Technologies, Nat'l Aviation Policy Issues Paper* (Sept. 2020), https://www.infrastructure.gov.au/aviation/drones/files/drone-discussion-paper.pdf

Aweiss, Arwa S., et al., NASA Ames, *Unmanned Aircraft Systems (UAS) Traffic Management (UTM) National Campaign II* (2018), https://ntrs.nasa.gov/archive/nasa/casi.ntrs.nasa.gov/20180000682.pdf

Baum, Michael S., et al., Aviators Code Initiative, *Improving Cockpit Awareness of Unmanned Aircraft Systems Near Airports* (March 8, 2019), http://www.secureav.com/UAS-Awareness-Listings-Page.html ["Baum 2019"]

Baum, Michael S., et al., Aviators Code Initiative, *UAS Pilots Code* (2018), https://www.secureav.com/UAS-Listings-Page.html; *id.*, annotated ver., http://www.secureav.com/UASPC-annotated-v1.0.pdf ["UAS Pilots Code"]

Baum, Michael S., *Federal Certification Authority Liability and Policy*, NIST, by MITRE Corp. under Contract #50SBN1C6732 (1992), DOI: 10.1016/s0267-3649(00)80027-8, *available at* https://tinyurl.com/PKI-Baum ["Baum FCA Liability"]

Beam, Tim, CEO, Fortem, *quoted in*, Nick Zazulia, *Fortem and Unifly Plan to Solve Unmanned Airspace*, ROTOR & WING (APRIL 30, 2019), https://preview.tinyurl.com/Tim-Beam

Bender, Walter, Johns Hopkins U. Applied Physics Lab., *Airborne Collision Avoidance System Xu for Smaller UAS (ACAS sXu), Position White Paper*, ACAS_RPS_20_001_V2R0, Ver. 2, Rev. 0, FAA TCAS Program Office (Feb. 26, 2020)

BIS Research, *Global UAV Sense-and-Avoid Systems Market - Analysis & Forecast, 2017-2022* (2018), https://bisresearch.com/industry-report/global-uav-sense-avoid-systems-market-2021.html

Booz, Allen, Hamilton, *Urban Air Mobility Market Study* (Oct. 5, 2018), https://www.nasa.gov/sites/default/files/atoms/files/bah_uam_executive_briefing_181005_tagged.pdf

Bowden, John, *NASA begins testing system to manage drone traffic in cities,* THE HILL (MAY 24, 2019), https://thehill.com/policy/technology/445393-nasa-launches-tests-of-system-to-manage-drone-traffic-in-cities

Bradford, Steve, FAA, & Kopardekar, Parimal, NASA, *Unmanned Aircraft Systems (UAS) Traffic Management (UTM) UTM Pilot Program (UPP), UPP Summary Report* (Oct. 2019)

Brooks, Dallas, Dir. of Miss. State U. Raspet Flight Research Laboratory, and Asso. Dir. of the FAA's Center of Excellence for UAS Research, Testimony before the Senate Commerce Committee (May 8, 2019), *quoted in, Nick Zuzulia, FAA to Debut Remote ID Rule in July,* AVIONICS INT'L (MAY 10, 2019), https://www.aviationtoday.com/2019/05/10/faa-remote-id-drones/

Burleson, Carl, FAA Acting Deputy Admin'r, Presentation at AUVSI Xponential, Chicago, Ill. (May 2, 2019), *quoted in, Dronelife*, https://dronelife.com/2019/05/01/faa-acting-deputy-administrator-carl-burleson-our-job-is-to-find-a-way-forward/

Butterworth-Hayes, Phillip, et al., *The Market for UAV Traffic Management Services –2019-2023,* Ed. 1.04 (Dec. 2018), https://www.unmannedairspace.info/wp-content/uploads/2019/01/unmanned-airspace-forecast-report.-Edition-1.04.sample.pdf, https://www.unmannedairspace.info/uav-traffic-management-services/, and https://www.unmannedairspace.info/category/uncategorized/

California Consumer Privacy Act of 2018, AB 375, https://leginfo.legislature.ca.gov/faces/billTextClient.xhtml?bill_id=201720180AB375

Campbell, S.E., et al., *Preliminary Weather Information Gap Analysis for UAS Operations*, Project Rpt. ATC-437, Rev. 1, Lincoln Laboratory (Oct. 2017), https://www.ll.mit.edu/sites/default/files/publication/doc/2018-05/Campbell_2017_ATC-437.pdf

CANSO, *ANSP Considerations for Unmanned Aircraft Systems (UAS) Operations*, v 1.1 (2016), https://canso.fra1.digitaloceanspaces.com/uploads/2020/02/ANSP-Considerations-for-UAS-Operations.pdf

CASR Part 139 – Aerodromes, https://www.casa.gov.au/standard-page/reviewing-rules-aerodromes-part-139?utm_source=phplist2273&utm_medium=email&utm_content=HTML&utm_campaign=July+2020+Regulatory+wrap-up+%5BSEC%3DOFFICIAL%5D

CAST/ICAO Common Taxonomy Team, ICAO, *Phase of Flight* (April 2013), http://www.intlaviationstandards.org/Documents/PhaseofFlightDefinitions.pdf

CBS This Morning, *Amazon CEO Unveils Drone Delivery Concept* (Dec. 2, 2013), *available at* https://www.youtube.com/watch?v=-qOBm3Iwlzo

Chakrabarty, Anjan, et al., NASA Ames, *Autonomous flight for Multi-copters flying in UTM -TCL4+sharing common airspace* (Jan. 2020), DOI: 10.2514/6.2020-0881, https://utm.arc.nasa.gov/docs/2020-Chakrabarty_SciTech_2020-0881_UTM.pdf

Chan, William N., Proj. Mgr., et al., *Overview of NASA's Air Traffic Management – eXploration (ATM-X) Project* (2018), https://ntrs.nasa.gov/archive/nasa/casi.ntrs.nasa.gov/20180005224.pdf

Cloar, Cameron R., UNMANNED AIRCRAFT IN THE NATIONAL AIRSPACE (DONNA A. DULO, ED., 2015)

Connected Places Catapult, *Enabling UTM in the UK* (May 2020), https://s3-eu-west-1.amazonaws.com/media.cp.catapult/wp-content/uploads/2020/05/22110912/01296_Open-Access-UTM-Report-V4.pdf ["Connected Places Catapult"]

Connected Places Catapult, *Towards a UTM System for the UK* (Sept. 30, 2019), https://s3-eu-west-1.amazonaws.com/media.cp.catapult/wp-content/uploads/2019/09/30150855/Towards-a-UTM-System-for-the-UK.pdf

Convention on International Civil Aviation, *see* ICAO, Convention on Civil Aviation

Cook, Stephen, Ph.D., et al, *Promoting Autonomy Design and Operations in Aviation*, 38th DASC (2019), https://www.researchgate.net/publication/341075826_Promoting_Autonomy_Design_and_Operations_in_Aviation

CORDIS, Horizon 2020, *Defining the Building Basic Blocks for a U-Space Separation Management Service* (BUBBLES), Grant Agt. ID: 893206, (May 1, 2020 – Oct. 31, 2022), https://cordis.europa.eu/project/id/893206

Court of Justice of the European Union, Press Release No 91/20 (July 16, 2020), *Judgement in Case C-311/18, Data Protection Commissioner v Facebook Ireland and Maximillian Schrems,* https://curia.europa.eu/jcms/upload/docs/application/pdf/2020-07/cp200091en.pdf

Davis, Mark, et al., *Aerial Cellular: What can Cellular do for UAVS with and without changes to present standards and regulations*, Presented at AUVSI Xponential, in Chicago, Ill. (May 2, 2019) ["Davis 2019"]

DeGood, Kevin, *Flying Cars Will Undermine Democracy and the Environment*, Center for Am. Progress (May 28, 2020), https://www.americanprogress.org/issues/economy/reports/2020/05/28/481148/flying-cars-will-undermine-democracy-environment/

Del Pozo, Isabel, Airbus & Troegeler, Mildred, The Boeing Co., *A New Digital Era of Aviation, The Path Forward for Airspace and Traffic Management*, ICAO, Uniting Aviation, ICAO ANC meeting (June 15, 2020), https://www.airbusutm.com/a-new-digital-era

Deloitte, *Managing the evolving skies, Unmanned aircraft system traffic management (UTM), the key enabler* (2018), https://www2.deloitte.com/content/dam/Deloitte/global/Images/infographics/gx-eri-managing-the-evolving-skies.pdf

DGAC, *U-space Together, Fast-Tracking Drone Integration in a Safe Sky* (2020), https://www.ecologique-solidaire.gouv.fr/sites/default/files/dsna_WAC_USPACE.pdf

DJI, *Elevating Safety: Protecting The Skies In The Drone Era* (May 2019), Section 5.III, https://terra-1-g.djicdn.com/851d20f7b9f64838a34cd02351370894/Flysafe/190521_US-Letter_Policy-White-Paper_web.pdf

DLR, *DLR Blueprint, Concept for Urban Airspace Integration* (Dec. 2017), https://www.dlr.de/fl/Portaldata/14/Resources/dokumente/veroeffentlichungen/Concept_for_Urban_Airspace_Integration.pdf ["DLR Blueprint"]

DoT, *Blockchain for Unmanned Aircraft Systems* (April 2020), https://rosap.ntl.bts.gov/view/dot/48789

DoT, *Budget Highlights*, Fiscal year 2020, https://www.transportation.gov/sites/dot.gov/files/docs/mission/budget/333126/budgethighlights030719final518pm25082.pdf

DoT, Office of the Inspector Gen., *FAA Lacks Sufficient Security Controls and Contingency Planning for Its DroneZone System*, FAA, Rpt. No. IT2020027 (April 15, 2020), https://www.oig.dot.gov/sites/default/files/FAA%20DroneZone%20Security%20Controls%20Final%20Report.pdf?utm_medium=email&utm_source=govdelivery

DoT, Office of the Inspector General, *DOT's Fiscal Year 2020 Top Management Challenges*, Rpt. PT2020003 (Oct. 23, 2019), https://www.oig.dot.gov/sites/default/files/DOT%20Top%20Management%20Challenges%20%20FY2020.pdf

DoT, Office of the Inspector General, *FAA Has Made Progress but Additional Actions Remain To Implement Congressionally Mandated Cyber Initiatives*, Rpt. AV2019021 (Mar. 20, 2019), https://www.oig.dot.gov/sites/default/files/FAA%20Cybersecurity%20Program%20Final%20Report%5E03.20.19.pdf

DoT, *Official Report of the Special Committee to Review the Federal Aviation Administration's Aircraft Certification Process* (Jan. 16, 2020), https://www.transportation.gov/sites/dot.gov/files/docs/briefing-room/362926/scc-final-report.pdf

DoT, Volpe Center, *Blockchain for Unmanned Aircraft Systems* (April 15, 2020), https://rosap.ntl.bts.gov/view/dot/48789

Douglas, Alex, *Skyports collaborates with Vodafone and Deloitte on NHS drone deliveries*, Commercial Drone Professional (July 10, 2020), https://www.commercialdrone-professional.com/skyports-collaborates-with-vodafone-and-deloitte-on-nhs-drone-deliveries/

Drone Alliance Europe, *U-space Whitepaper, Ver. 2.0* (Feb. 2020), https://resourcecenter.dronealliance.eu/wp-content/uploads/2019/05/DAE-UTM-U-Space-whitepaper-2.0-final-1.pdf ["Drone Alliance Europe, U-Space Whitepaper 2020"]

Drone Alliance Europe, *U-space Whitepaper* (July 2019), http://dronealliance.eu/wp-content/uploads/2019/07/Drone-Alliance-Europe-U-Space-Whitepaper-2-July-2019.pdf ["Drone Alliance Europe, U-Space Whitepaper 2019"]

DSNA, *Direction des Services de la Navigation Aérienne, RAPPORT D'ACTIVITÉ*, https://www.ecologique-solidaire.gouv.fr/sites/default/files/DSNA-DGAC-RA-2017-FR.pdf

E-Government Act of 2002, Pub. L. No. 107-374 (Title III), 44 U.S.C. 101 note, https://www.govinfo.gov/content/pkg/PLAW-107publ347/pdf/PLAW-107publ347.pdf

EASA and European Commission, *Initial draft Regulation on U-space* (issued July 4, 2019), *available at* https://www.dropbox.com/s/7braj95hti4b5go/EASA-U-Space-Proposed%20Reg-July%204-2019.docx?dl=0

EASA, A-NPA 2015-10, *Introduction of a regulatory framework for the operation of drones* (July 31, 2015), https://www.easa.europa.eu/sites/default/files/dfu/A-NPA%202015-10.pdf

EASA, Draft Opinion, *High-level regulatory framework for the U-space*, RMT.0230 (Oct. 2019), *available at* https://rpas-regulations.com/wp-content/uploads/2019/10/EASA_Draft-Opinion-on-U-space.pdf ["EASA Reg. Framework U-space"]

EASA, *Drones Amsterdam Declaration* (Nov. 28, 2018), Sect. 6.1, https://www.easa.europa.eu/sites/default/files/dfu/Drones%20Amsterdam%20Declaration%2028%20Nov%202018%20final.pdf

EASA, *Easy Access Rules for ATM-ANS (Regulation (EU) 2017/373*, https://www.easa.europa.eu/sites/default/files/dfu/Easy_Access_Rules_for_ATM-ANS.pdf

EASA, *Easy Access Rules for Unmanned Aircraft Systems* (Regulation (EU) 2019/947 and Regulation (EU) 2019/945) (March 3, 2020), https://www.easa.europa.eu/document-library/general-publications/easy-access-rules-unmanned-aircraft-systems-regulation-eu ["EASA Easy Access 2019/945"]

EASA, NPA 2017-05 (A), *Introduction of a regulatory framework for the operation of unmanned aircraft systems in the 'open' and 'specific' categories*, https://www.easa.europa.eu/sites/default/files/dfu/NPA%202017-05%20%28A%29_0.pdf

EASA, Opinion No 01/2018, *Unmanned aircraft system (UAS) operations in the 'open' and 'specific' categories*, https://www.easa.europa.eu/sites/default/files/dfu/Opinion%20No%2001-2018.pdf ["EASA Opinion 01/2018"]

EASA, Opinion No 01/2020, *High-level regulatory framework for the U-space*, RMT.0230 (March 13, 2020), https://www.easa.europa.eu/document-library/opinions/opinion-012020 ["EASA Opinion 01/2020 - U-space"]

Appendix to Opinion No 01/2020, Comment-response document (CRD) to draft Opinion (March 2020), https://www.easa.europa.eu/sites/default/files/dfu/Appendix%20to%20Opinion%20No%2001-2020.pdf

Draft acceptable means of compliance (AMC) and Guidance Material (GM) to Opinion No 01/2020 on a high-level regulatory framework for the U-space, issue 1 (XX Month 2020), https://www.easa.europa.eu/sites/default/files/dfu/Draft%20AMC%20%26%20GM%20to%20the%20U-space%20Regulation%20—%20for%20info%20only.pdf ["EASA Draft AMC & GM"]

EASA, *Proposed Special Condition for Light UAS*, Doc. No. SC Light-UAS 01, Issue: 1 (July 20, 2020), https://www.easa.europa.eu/document-library/product-certification-consultations/proposed-special-condition-light-uas ["EASA SC Light UAS"]

EASA, *Standardised European Rules of the Air* (SERA), *contained in* Commission Implementing Regulation (EU) 2016/1185, https://www.easa.europa.eu/document-library/regulations/commission-implementing-regulation-eu-20161185, and https://www.easa.europa.eu/regulation-groups/sera-standardised-european-rules-air

EASA, *U-space workshop*, Köln (May 14-15, 2019), https://www.easa.europa.eu/sites/default/files/dfu/U-space%20workshop%2014-15%20May%20v2.0.pdf

EASA/EC, *EASA/EC workshop on the U-space regulatory framework Discussion with Member States and industry*, TE.GEN.00404-003 (May 14-15, 2019), Sect. 7, https://www.easa.europa.eu/sites/default/files/dfu/U-space%20workshop%20-%20discussion%20paper.pdf

Egorov, Maxim, et al., Airbus, *Encounter Aware Flight Planning in the Unmanned Airspace*, ICNS Conference, Herndon, Va. (2019), https://storage.googleapis.com/blueprint/icns2019.pdf

EHang, *The Future of Transportation: White Paper on Urban Air Mobility Systems* (Jan. 15, 2020), https://www.ehang.com/app/en/EHang%20White%20Paper%20on%20Urban%20Air%20Mobility%20Systems.pdf

Eichleay, Margaret, et al., *Using Unmanned Aerial Vehicles for Development: Perspectives from Citizens and Government Officials in Tanzania* (Feb. 2016), DOI: 10.13140/RG.2.1.1737.7048, https://www.researchgate.net/publication/295547373

Elias, Bart, *Air Traffic Inc.: Considerations Regarding the Corporatization of Air Traffic Control*, Congressional Research Service (May 16, 2017), *available at* https://fas.org/sgp/crs/misc/R43844.pdf

Ellis, Kyle, et al., *In-Time System-Wide Safety Assurance (ISSA) Concept of Operations*, ARC-E-DAA-TN74516-1 (Oct. 31, 2019), https://ntrs.nasa.gov/archive/nasa/casi.ntrs.nasa.gov/20190032480.pdf

EmbraerX and Airservices, *Urban Air Traffic Management, Concept of Operations*, Ver. 1 (Dec. 2020), https://embraerx.embraer.com/global/en/uatm ["Embraer UATM ConOps"]

Embraer^{X}, et al., *Flight Plan 2030, An Air Traffic Management Concept for Urban Air Mobility* (May 2019), https://daflwcl3bnxyt.cloudfront.net/m/4e5924f5de45fd3a/original/embraerx-whitepaper-flightplan2030.pdf ["Embraer^{X} 2019"]

EPIC, *Comments of the Electronic Privacy Information Center to the FAA, Notice: Petition for Exemption; Summary of Petition Received; Causey Aviation Unmanned, Inc.*, Docket No. FAA-2020-42 (Aug. 4, 2020), https://epic.org/apa/comments/EPIC-Comments-FAA-Drone-Delivery-Exemption-Aug2020.pdf

Erotokritou, Chrystel, *The Legal Liability of Air Traffic Controllers*, INQUIRIES JOURNAL/ STUDENT PULSE, Vol. 4(02) (2019), http://www.inquiriesjournal.com/a?id=613 ["Erotokritou 2019"]

EUROCAE, *EUROCAE Technical Work Programme* (Public ver. 2019), https://eurocae.net/media/1567/eurocae-public-twp-2019.pdf ["EUROCAE Programme 2019"]

EUROCAE, *Operational Services and Environment Definition for Detect & Avoid in Very Low Level Operations*, Draft ED-269 for Open Consultation (June 2019), https://eurocae.net/news/posts/2019/june/eurocae-open-consultation-ed-267/ ["EUROCAE OSED-DAA"]

EUROCONTROL and EASA, *UAS ATM Airspace Assessment*, Ed. 1.2 (Nov. 27, 2018), https://www.eurocontrol.int/sites/default/files/publication/files/uas-atm-airspace-assessment-v1.2-release-20181127.pdf

EUROCONTROL and EASA, *UAS ATM Flight Rules, Discussion Doc.*, Ed. 1.1 (Nov. 27, 2018), *available at* https://www.eurocontrol.int/sites/default/files/publication/files/uas-atm-integration-operational-concept-v1.0-release%2020181128.pdf

EUROCONTROL and EASA, *UAS ATM Integration, Operational Concept*, Ed. 1 (Nov. 27, 2018), *available at* https://www.eurocontrol.int/sites/default/files/publication/files/uas-atm-integration-operational-concept-v1.0-release%2020181128.pdf ["EUROCONTROL UAS ATM Integration"]

EUROCONTROL, *European Route Network Improvement Plan – Part 1, European Airspace Design Methodology – Guidelines, European Network Operations Plan* 2019-2024 (Dec. 19, 2019), https://www.eurocontrol.int/concept/free-route-airspace

EUROCONTROL, *U-SPACE Service, Implementation Monitoring Report* (Nov. 2020), https://www.eurocontrol.int/sites/default/files/2020-11/uspace-services-implementation-monitoring-report-2020-1-2.pdf

European Commission, *Annex to EASA Opinion No. 01/2020, Commission Implementing Regulation (EU) …/… of XXX on a high-level regulatory framework for the U-space*, https://www.easa.europa.eu/sites/default/files/dfu/Draft%20COMMISSION%20IMPLEMENTING%20REGULATION%20on%20a%20high-level%20regulatory%20fram….pdf ["Comm Implementing Reg. (EU) …/… - U-space"]

European Commission, *Commission Delegated Regulation (EU) 2019/945 of 12 March 2019 on unmanned aircraft systems and on third-country operators of unmanned aircraft systems*, Annex 1, Part 2, Sect. 12, 2019 (O.J.) (L 152/1) (June 11, 2019), https://eur-lex.europa.eu/legal-content/EN/TXT/PDF/?uri=CELEX:32019R0945&from=EN ["Comm Delegated Reg. (EU) 2019/945 – UAS"]

European Commission, *Commission Implementing Regulation (EU) 2017/373 of 1 March 2017, laying down common requirements for providers of air traffic management/air navigation services and other air traffic management network functions and their oversight …*, https://eur-lex.europa.eu/legal-content/EN/TXT/PDF/?uri=CELEX%3A32017R0373&from=EN

European Commission, *Commission Implementing Regulation (EU) 2019/947 of 24 May 2019 on the rules and procedures for the operation of unmanned aircraft* (O.J.) (L 152/1) (July 11, 2019), https://eur-lex.europa.eu/eli/reg_impl/2019/947/oj ["Comm Implementing Reg. (EU) 2019/947 - UA"]

European Commission, Commission Implementing Regulation (EU) 2021/664 of 22 April 2021 on a regulatory framework for the U-space, https://eur-lex.europa.eu/legal-content/EN/TXT/?uri=CELEX%3A32021R0664 ["Comm Implementing Reg. (EU) 2021/664 - U-space"]

European Commission, Commission Implementing Regulation (EU) 2021/665 of 22 April 2021 amending Implementing Regulation (EU) 2017/373 as regards requirements for providers of air traffic management/air navigation services and other air traffic management network functions in the U-space airspace designated in controlled airspace, https://eur-lex.europa.eu/legal-content/EN/TXT/?uri=CELEX%3A32021R0665

European Commission, Commission Implementing Regulation (EU) 2021/666 of 22 April 2021 amending Regulation (EU) No 923/2012 as regards requirements for manned aviation operating in U-space airspace, https://eur-lex.europa.eu/legal-content/EN/TXT/?uri=CELEX%3A32021R0666

European Commission, Communication from the Commission to the European Parliament, the Council, the European Economic and Social Committee and the Committee of the Regions, *Promoting the Shared Use of Radio Spectrum Resources in the Internal Market* (2012), *available at* https://ec.europa.eu/digital-single-market/sites/digital-agenda/files/com-ssa.pdf

European Commission, *Standard Contract Clauses (SCC)*, https://ec.europa.eu/info/law/law-topic/data-protection/international-dimension-data-protection/standard-contractual-clauses-scc_en

European Parliament, *Regulation (EC) 785/2004 of the European Parliament and of the Council of 21 April 2004*, L 138/1, https://eur-lex.europa.eu/legal-content/EN/TXT/HTML/?uri=CELEX:32004R0785&from=EN

European Parliament, *Regulation (EU) 2016/679 of the European Parliament and of the Council of 27 April 2016 on the protection of natural persons with regard to the processing of personal data and on the free movement of such data, and repealing Directive 95/46/EC (General Data Protection Regulation)*, 2016 O.J. (L 119) 1 (April 27, 2016), https://eur-lex.europa.eu/legal-content/EN/TXT/HTML/?uri=CELEX:32016R0679&from=EN, and https://gdpr-info.eu/ ["GDPR"]

European Parliament, *Regulation (EU) 2018/1139 of the European Parliament and of the Council* (July 4, 2018), https://eur-lex.europa.eu/legal-content/EN/TXT/?uri=CELEX:32018R1139

Evans, Anthony, Egorov, Maxim, et al., Airbus UTM, *Fairness in Decentralized Strategic Deconfliction in UTM* (2020), https://storage.googleapis.com/blueprint/Fairness_in_Decentralized_Strategic_Deconfliction_in_UTM.pdf, and https://doi.org/10.2514/6.2020-2203

Experimental Aircraft Ass'n, *Comment on Amazon Prime Air Petition for Exemption, Regulatory Docket No. FAA-2019-0573* (Aug. 28, 2019), https://www.regulations.gov/document?D=FAA-2019-0573-0049

FAA & NASA, *UAS Traffic Management (UTM), Research Transition Team (RTT) Plan*, V1.0 (Jan. 31, 2017), https://www.faa.gov/uas/research_development/traffic_management/media/FAA_NASA_UAS_Traffic_Management_Research_Plan.pdf ["RTT 2017"]

FAA & NASA, *Uncrewed Aircraft Systems (UAS) Traffic Management (UTM), UTM Pilot Program (UPP), UPP Phase 2 Final Report*, V. 1.0 (July 29, 2021), https://www.faa.gov/uas/research_development/traffic_management/utm_pilot_program/media/FY20_UPP2_Final_Report.pdf

FAA & NASA, Unmanned Aircraft System Traffic Management (UTM) Research Transition Team (RTT), Concept Working Group, *Concept & Use Cases Package #2 Addendum: Technical Capability Level 3*, Ver. 1.0, Doc. No. 20180007223 (July 2018), https://ntrs.nasa.gov/search.jsp?R=20180007223 ["TCL 3"]

FAA & NASA, Unmanned Aircraft System Traffic Management (UTM) Research Transition Team (RTT), Concept Working Group, *Concept & Use Cases Package #3: Technical Capability Level 4*, Ver. 1.0 (March 2019) ["FAA & NASA UTM RTT WG #3"]

FAA & NASA, *Unmanned Aircraft Systems (UAS) Traffic Management (UTM) Pilot Program (UPP) Phase 2, Industry Workshop* (Dec. 2019), https://www.faa.gov/uas/research_development/traffic_management/utm_pilot_program/media/UPP2_Industry_Workshop_Briefing.pdf ["UPP Phase 2"]

FAA & NASA, *Unmanned Aircraft Systems (UAS) Traffic Management (UTM), UTM Pilot Program (UPP) Phase Two (2) Progress Report, V. 1.0* (March 2021), https://www.faa.gov/uas/research_development/traffic_management/utm_pilot_program/media/UTM_Pilot_Program_Phase_2_Progress_Report.pdf

FAA & NASA, *UTM Pilot Program Brochure,* https://www.faa.gov/uas/research_development/traffic_management/utm_pilot_program/media/UTM_Pilot_Program_Smart_Sheet.pdf

FAA & SWISS Confederation, *Declaration of Intent Between the FAA and Federal Office of Civil Aviation, Dept of the Environment, Transportation, Energy and Communications, Swiss Confederation* (May 2020), https://www.faa.gov/news/media/attachments/Declaration_Intent_Swiss_FAA.pdf

FAA ATO, *Low Altitude Authorization and Notification Capability (LAANC) Concept of Operations*, Ver. 2.1 (March 20, 2020), https://www.faa.gov/uas/programs_partnerships/data_exchange/laanc_for_industry/media/FAA_LAANC_CONOPS.pdf ["LAANC ConOps 2.1"]

FAA ATO, *Low Altitude Authorization and Notification Capability (LAANC), USS Onboarding Test Procedure and Report*, Ver. 5.0 (Feb. 19, 2021), https://www.faa.gov/uas/programs_partnerships/data_exchange/laanc_for_industry/media/LAANC_USS_Onboarding_Test_Procedure_and_Report.pdf

FAA ATO, *Low Altitude Authorization and Notification Capability (LAANC), USS Operating Rules*, Ver. 1.3 (Dec. 14, 2018), https://www.faa.gov/uas/programs_partnerships/data_exchange/laanc_for_industry/media/FAA_sUAS_LAANC_Ph1_USS_Rules.pdf ["LAANC USS Operating Rules"]

FAA ATO, *Low Altitude Authorization and Notification Capability (LAANC), USS Performance Rules*, Ver. 5.0 (Feb. 5, 2021), https://www.faa.gov/uas/programs_partnerships/data_exchange/laanc_for_industry/media/LAANC_USS_Performance_Rules.pdf ["LAANC USS Performance Rules"]

FAA ATO, *Remote Identification (Remote ID) of Unmanned Aircraft System, Concept of Use (ConUse): FAA Data Exchanges with UAS Service Suppliers (USS)*, v. 1.0 (Jan. 17, 2019), *available at* https://jrupprechtlaw.com/wp-content/uploads/2021/01/FAA-UAS-Remote-ID-Data-Exchange-ConUse-v1.0-highlighted.pdf

FAA ATO, *Safety Management Systems Manual* (April 2019), https://www.faa.gov/air_traffic/publications/media/ATO-SMS-Manual.pdf

FAA Extension, Safety, and Security Act of 2016, Pub. L. No. 114-190, 130 Stat. 615 (2016), codified at 49 U.S.C. 40101 note, https://www.congress.gov/114/plaws/publ190/PLAW-114publ190.pdf ["FESSA"]

FAA Modernization and Reform Act of 2012, Pub. L. No. 112-95, Sect. 332(c), 126 Stat. 11 (Feb. 14, 2012), https://www.congress.gov/bill/112th-congress/house-bill/658/text

FAA NextGen, *Concept of Operations for Urban Air Mobility (UAM)*, v1.0 (June 26, 2020), https://nari.arc.nasa.gov/sites/default/files/attachments/UAM_ConOps_v1.0.pdf ["FAA ConOps UAM v1.0"]

FAA NextGen, *Concept of Operations, Unmanned Aircraft System (UAS) Traffic Management* (UTM), v1.0 (May 2018), https://utm.arc.nasa.gov/docs/2018-UTM-ConOps-v1.0.pdf ["FAA ConOps UTM v1.0"]

FAA NextGen, *Concept of Operations, Unmanned Aircraft System (UAS) Traffic Management* (UTM), v2.0 (March 2, 2020), https://www.faa.gov/uas/research_development/traffic_management/media/UTM_ConOps_v2.pdf ["FAA ConOps UTM v2.0"]

FAA NextGen, *Concept of Operations, Upper Class E Traffic Management*, v1.0 (April 22, 2020), https://nari.arc.nasa.gov/sites/default/files/attachments/ETM_ConOps_V1.0.pdf ["FAA High E ConOps v1.0"]

FAA Reauthorization Act of 2018, Pub. L. No. 115-254, 132 Stat. 3186 (Oct. 5, 2018), *available at* https://www.congress.gov/115/bills/hr302/BILLS-115hr302enr.pdf ["FAARA"]

FAA UAS Symposium (July 8-9; Aug. 18-19, 2020), https://www.faa.gov/uas/resources/events_calendar/archive/

FAA, Advisory and Rulemaking Committees, Unmanned Aircraft Systems (UAS) in Controlled Airspace ARC, https://www.faa.gov/regulations_policies/rulemaking/committees/documents/index.cfm/committee/browse/committeeID/617

FAA, *Advisory Circular AC 00-1.1B*, Subj. Public Aircraft Operations—Manned and Unmanned (Sept. 21, 2018), https://www.faa.gov/documentLibrary/media/Advisory_Circular/AC_00-1.1B.pdf

FAA, *Advisory Circular AC 00-63A, CHG 1*, Subj: Use of Flight Deck Displays of Digital Weather and Aeronautical Information, App. 1 (Jan. 6, 2017), https://www.faa.gov/regulations_policies/advisory_circulars/index.cfm/go/document.information/documentID/1024126 ["AC 00-63A"]

FAA, *Advisory Circular AC 107-2A*, Subj: Small Unmanned Aircraft Systems (Small UAS) (Feb. 1, 2021), https://www.faa.gov/regulations_policies/advisory_circulars/index.cfm/go/document.information/documentID/1038977

FAA, *Advisory Circular AC 120-92B*, Subj: Safety Management Systems for Aviation Service Providers (Jan. 8, 2015), https://www.faa.gov/documentLibrary/media/Advisory_Circular/AC_120-92B.pdf

FAA, *Advisory Circular AC 150/5220-16E CHG 1*, Subj. Automated Weather Observing Systems (AWOS) for Non-Federal Applications (Jan. 31, 2019), https://www.faa.gov/documentLibrary/media/Advisory_Circular/AC_150_5220-16E_w-chg1.pdf

FAA, *Advisory Circular AC 20-153B*, Subj: Acceptance of Aeronautical Data Processes and Associated Databases (April 19, 2016), https://www.faa.gov/regulations_policies/advisory_circulars/index.cfm/go/document.information/documentID/1029446 ["AC 20-153B"]

FAA, *Aeronautical Information Manual* (2019), https://www.faa.gov/air_traffic/publications/atpubs/aim_html/chap5_section_5.html

FAA, *Aircraft Certification Process Review and Reform, FAA Response to FAA Modernization and Reform Act of 2012*, Pub. L. No. 112-95, Sect. 312, 126 Stat. 11 (2012), https://gama.aero/wp-content/uploads/FAA-ACPRR-Report-to-Congress-2012-08.pdf

FAA, ANPRM, *Safe and Secure Operations of Small Unmanned Aircraft Systems*, 84 Fed. Reg. 3732-33 (2019) (to be codified at 14 C.F.R. pt. 107), https://www.federalregister.gov/documents/2019/02/13/2019-00758/safe-and-secure-operations-of-small-unmanned-aircraft-systems ["FAA ANPRM 2019"]

FAA, *Aviation Rulemaking Committee Charter,* Subj: Airspace Access Priorities Aviation Rulemaking Committee (Effective, Nov. 20, 2017), https://www.faa.gov/regulations_policies/rulemaking/committees/documents/index.cfm/committee/browse/committeeID/677 ["Airspace Access Priorities ARC"]

FAA, *Aviation Rulemaking Committee Charter*, Subj: UAS in Controlled Airspace Aviation Rulemaking Committee (Amend. Nov. 20, 2017), https://www.faa.gov/regulations_policies/rulemaking/committees/documents/media/UAS%20Controlled%20Airspace%20Charter%20Amendment%20(signed%2011-20-17).pdf, and https://www.faa.gov/regulations_policies/rulemaking/committees/documents/index.cfm/document/information/documentID/3222

FAA, DOJ, FCC & DHS, *Advisory on the Application of Federal Laws to the Acquisition and Use of Technology to Detect and Mitigate Unmanned Aircraft Systems,* 9.95.300-UAS (Aug. 2020), https://www.justice.gov/file/1304841/download

FAA, Drone Advisory Committee (DAC), *Drone Access to Airspace, Final Report* (2017), https://www.faa.gov/uas/programs_partnerships/drone_advisory_committee/rtca_dac/media/dac_tg2_final_reccomendations_11-17_update.pdf

FAA, Drone Advisory Committee, *eBook, DAC Member (Public) Information for the Oct. 17, 2019 DAC Meeting,* Wash., D.C. (Oct. 17, 2019), https://www.faa.gov/uas/programs_partnerships/drone_advisory_committee/media/eBook_10-17-2019_DAC_Meeting.pdf

FAA, Drone Advisory Committee, *Public eBook, DAC Task Group #7, UTM Report* (June 19, 2020), https://www.faa.gov/uas/programs_partnerships/drone_advisory_committee/media/Public_eBook_06192020_v7.pdf

FAA, *Executive Summary Final Rule on Remote Identification of Unmanned Aircraft (Part 89)* (Dec. 28, 2020), https://tinyurl.com/FAA-RID-Summary

FAA, *FAA Aerospace Forecast Fiscal Years 2020–2040* (2020), https://www.faa.gov/data_research/aviation/aerospace_forecasts/media/FY2020-40_FAA_Aerospace_Forecast.pdf ["FAA Aerospace Forecast"]

FAA, *FAA Air Transportation Centers of Excellence*, https://www.faa.gov/about/office_org/headquarters_offices/ang/grants/coe/

FAA, *FAA Continues Drone Integration Initiatives*, https://www.faa.gov/news/updates/?newsId=95371

FAA, FAA Safety Risk Management Panel, Emerging Tech. Team, AJV-115, *Title 14 Code of Federal Regulations Part 107 Small Unmanned Aircraft System Rule: Conditions and Limitations for Allowing Operations in Class E Surface Area Safety Risk Management Document*, ver. 1.0 (Aug. 26, 2016)

FAA, *Fact Sheet – FAA UAS Test Site Program* (Dec. 30, 2013), https://www.faa.gov/news/fact_sheets/news_story.cfm?newsId=15575

FAA, *In the matter of the petition of Amazon Prime Air*, Exemption No. 18602, Reg. Docket No. FAA-2019-0622 (issued Part 135 exemption, Aug. 26, 2020), https://beta.regulations.gov/document/FAA-2019-0622-0003

FAA, *Integration of Civil Unmanned Aircraft Systems (UAS) in the National Airspace System (NAS) Roadmap*, Second Edition (July 2018), https://www.faa.gov/uas/resources/policy_library/media/Second_Edition_Integration_of_Civil_UAS_NAS_Roadmap_July%202018.pdf ["FAA Roadmap"]

FAA, *Integration of Unmanned Aircraft Systems into the National Airspace System, Concept of Operations,* v2.0 (Sept. 28, 2012), *available at* http://www.sarahnilsson.org/app/download/965094433/FAA-UAS-Conops-Version-2-0-1.pdf

FAA, *LAANC Drone Program Expansion Continues*, https://www.faa.gov/news/updates/?newsId=94750

FAA, *Memorandum of Agreement for Low Altitude Authorization and Notification Capability (LAANC) Between FEDERAL AVIATION ADMINISTRATION (FAA) And _______*, v2.3 (April, 2020), https://www.faa.gov/uas/programs_partnerships/data_exchange/laanc_for_industry/media/Memorandum_of_Agreement.pdf ["LAANC MOA"]

FAA, *NextGen Implementation Plan 2018-19*, https://www.faa.gov/nextgen/media/NextGen_Implementation_Plan-2018-19.pdf

FAA, *Notice implementing the exception for limited recreational operations of unmanned aircraft,* 84 Fed. Reg. 22552 (May 17, 2019), https://www.federalregister.gov/documents/2019/05/17/2019-10169/exception-for-limited-recreational-operations-of-unmanned-aircraft

FAA, *NOTICE N 8900.529*, SUBJ: Extended Unmanned Aircraft Systems Oversight (Nov. 13, 2019), https://www.faa.gov/documentLibrary/media/Notice/N_8900.529.pdf

FAA, NPRM, *Remote Identification of Unmanned Aircraft Systems*, 48 Fed. Reg. 72438-72524 (Dec. 31, 2019) (to be codified at 14 C.F.R. pts. 1, 47, 48, 89, 91, and 107), https://www.federalregister.gov/documents/2019/12/31/2019-28100/remote-identification-of-unmanned-aircraft-systems ["RID NPRM 2019"]

FAA, *Operation and Certification of Small Unmanned Aircraft Systems*, 81 Fed. Reg. 42064 (June 28, 2016), https://www.federalregister.gov/documents/2016/06/28/2016-15079/operation-and-certification-of-small-unmanned-aircraft-systems

FAA, *ORDER 7110.65X, CHG 1*, Subj: Air Traffic Control, Unmanned Aircraft System (UAS) Activity Information (Oct. 12, 2017), https://www.faa.gov/documentLibrary/media/Order/7110.65X_w_CHG_1_3-29-18.pdf

FAA, *ORDER 8040.6*, Subj: Unmanned Aircraft Systems Safety Risk Management (Oct. 4, 2019), https://www.faa.gov/regulations_policies/orders_notices/index.cfm/go/document.information/documentID/1036752

FAA, *ORDER 8260.56*, Subj: Diverse Vector Area (DVA) Evaluation (Aug. 2, 2011), https://www.faa.gov/documentLibrary/media/Order/Order%208260.56.pdf

FAA, *ORDER JO 7200.23A*, Subj: Unmanned Aircraft Systems (Aug. 1, 2017), https://www.faa.gov/documentlibrary/media/order/jo_7200.23a_unmanned_aircraft_systems_(uas).pdf ["JO 7200.23A"]

FAA, *ORDER JO 7210.3BB CHG 1*, Subj: Facility Operation and Administration (Effective Date: Jan. 30, 2020), https://www.faa.gov/documentLibrary/media/Order/7210.3BB_Chg_1_dtd_1-30-20.pdf

FAA, *ORDER N JO 7210.914*, Subj: Low Altitude Authorization Notification Capability - LAANC (July 23, 2019), https://www.faa.gov/documentLibrary/media/Notice/N_JO_7210.914_Low_Altitude_Authorization_and_Notification_Capability_-_LAANC.pdf

FAA, *Preliminary Hazard Analysis (PHA) for Low Altitude Authorization and Notification Capability (LAANC)* - Safety Risk Management Document, Ver. 1.0, p. 3 (Oct. 23, 2017) ["LAANC-PHA"]

FAA, *Preparation of Interface Documentation*, FAA-STD-025f (Nov. 30, 2007), http://www.tc.faa.gov/its/worldpac/standards/faa-std-025f.pdf

FAA, *Privacy Statement Regarding LAANC and USS Providers Collection of Information in Accordance with 14 CFR Part 107* (last modified Dec. 19, 2018), https://www.faa.gov/uas/programs_partnerships/data_exchange/privacy_statement/

FAA, *Remote Identification of Unmanned Aircraft,* Final rule (issued Dec. 28, 2020), 86 Fed. Reg. 4390 (Jan. 15, 2021), https://www.federalregister.gov/d/2020-28948 ["Remote ID Rule"]

FAA, *Statement of Objectives for Test Site Selection to Participate in UTM Pilot Program Demo Program,* Rev#3 (July 20, 2018), https://www.faa.gov/uas/research_development/traffic_management/utm_pilot_program/media/UTM_Pilot_Program_Smart_Sheet.pdf ["UPP Objectives"]

FAA, *System Wide Information Management (SWIM), Governance Policies*, Ver. 3.1 (Feb. 6, 2020), https://www.faa.gov/air_traffic/technology/swim/governance/standards/media/SWIM%20Governance%20Policies%20v3.1_20200206_Final.pdf ["FAA SWIM Governance"]

FAA, *The Future of the NAS* (2016), https://www.faa.gov/nextgen/media/futureOfTheNAS.pdf

FAA, *Type Certification of Unmanned Aircraft Systems*, 85 Fed. Reg. 5905-6 (Feb. 3, 2020), https://www.federalregister.gov/documents/2020/02/03/2020-01877/type-certification-of-unmanned-aircraft-systems

FAA, *UAS Identification and Tracking (UAS ID) Aviation Rulemaking Committee (ARC), ARC Recommendations, Final Report* (Sept. 30, 2017), https://www.faa.gov/regulations_policies/rulemaking/committees/documents/media/UAS%20ID%20ARC%20Final%20Report%20with%20Appendices.pdf

FAA, *UAS Integration Pilot Program*, https://www.faa.gov/uas/programs_partnerships/integration_pilot_program/

FAA, *Unmanned Aircraft System Traffic Management (UTM)* (webpage), https://www.faa.gov/uas/research_development/traffic_management/

FAA, *Unmanned Aircraft Systems (UAS) Traffic Management (UTM) Pilot Program Phase Two, Concept of Use V1.0* (Dec. 2019) ["FAA UPP Phase 2 ConUse"]

FAA, *Unmanned Traffic Management System Demonstrations* (Sept. 4, 2019), https://www.youtube.com/watch?v=zpc4aoJKefA

FAA, *USS Onboarding, Low Altitude Authorization and Notification Capability (LAANC)*, v1.0 (Feb. 2018), https://www.faa.gov/uas/programs_partnerships/data_exchange/laanc_for_industry/media/LAANC_UAS_Service_Supplier_onboarding_information.pdf ["LAANC USS Onboarding"]

FAA, *UTM Pilot Program*, https://www.faa.gov/uas/research_development/traffic_management/utm_pilot_program/

FCC Application, File No. SAT-LOA-20161115-00118, https://licensing.fcc.gov/myibfs/download.do?attachment_key=1364689

FCC, *FCC Online Table of Frequency Allocations* (Rev. May 7, 2019), https://transition.fcc.gov/oet/spectrum/table/fcctable.pdf

FCC, *In the Matter of Expanding Flexible Use in the 3.7-4.2 GHz Band*, GN Docket No. 18-122, 35 FCC Rcd 2343 (2020), https://www.fcc.gov/document/fcc-expands-flexible-use-c-band-5g-0

FCC, *In the Matter of Expanding Flexible use of the 3.7 to 4.2 GHz Band, Petition for Partial Reconsideration of the 3.7 GHz Band Report and Order*, GN Docket No. 18-122 (May 26, 2020), *Petition for Partial Reconsideration of the 3.7 GHz Band Report and Order*, GN Docket No. 18-122 (May 26, 2020), https://ecfsapi.fcc.gov/file/10527379225572/C-BAND%20Petition%20for%20Recon.pdf

FCC, *Report on Section 374 of the FAA Reauthorization Act of 2018*, prepared by the Wireless Telecom. Bureau, Office of Eng'g & Tech., FCC (Aug. 20, 2020), https://docs.fcc.gov/public/attachments/DOC-366460A1.pdf

Federal Information Security Management Act (FISMA) of 2014, Pub. L. No. 113-283 (Dec. 18, 2014), https://csrc.nist.gov/Topics/Laws-and-Regulations/laws/FISMA

Ferguson, Alison, *Pathfinder Focus Area 2, Phase III Report* (2018), https://tinyurl.com/Pathfinder-BVLOS

Ford, Warwick and Baum, Michael S., *Secure Electronic Commerce* (Prentice Hall 2nd. ed., 2000), DOI: 10.1016/s0267-3649(97)89803-2, https://tinyurl.com/Secure-ECommerce

Frost & Sullivan, *Analysis of Urban Air Mobility and the Evolving Air Taxi Landscape*, 2019 (Oct. 25, 2019), https://store.frost.com/analysis-of-urban-air-mobility-and-the-evolving-air-taxi-landscape-2019.html

GAO, *Unmanned Aircraft Systems, FAA Could Better Leverage Test Site Program to Advance Drone Integration*, Rpt. to the Ranking Member, Comm. on Transport. and Infrastructure, House of Representatives, GAO-20-97 (Jan. 9, 2020), https://www.gao.gov/assets/710/703726.pdf

GAO, *Unmanned Aircraft Systems, FAA Could Strengthen Its Implementation of a Drone Traffic Management System by Improving, Communication and Measuring Performance*, GAO-21-165 (Jan. 2021), https://www.gao.gov/assets/720/712037.pdf ["GAO UAS"]

GAO, *Unmanned Aircraft Systems, FAA Should Improve Drone-Related Cost Information and Consider Options to Recover Costs*, GAO-20-136 (Dec. 17, 2019), https://www.gao.gov/products/GAO-20-136

GAO, *Unmanned Aircraft Systems, FAA's Compliance and Enforcement Approach for Drones Could benefit from Improved Communication and Data*, GAO-20-29 (Oct. 2019), https://www.gao.gov/assets/710/702137.pdf

Garrett-Glaser, Brian, *Why Automation Shouldn't Push Pilots Out of Air Taxis*, ROTOR & WING INT'L (MAY 23, 2019), https://www.rotorandwing.com/2019/05/23/automation-shouldnt-push-pilots-air-taxis/

GSA, *Federal Public Key Infrastructure Guides*, HTTPS://FPKI.IDMANAGEMENT.GOV/

GSMA, *Accelerating the Commercial Drone Market using Cellular Technology* (Nov. 15, 2017), https://www.gsma.com/iot/wp-content/uploads/2017/11/Drones-Webinar-Slides-FINAL.pdf

GSMA, *Cellular Vehicle-to-Everything (C-V2X), Enabling Intelligent Transport* (2018), https://www.gsma.com/iot/wp-content/uploads/2017/12/C-2VX-Enabling-Intelligent-Transport_2.pdf

GSMA, *THE 5G GUIDE, A Reference for Operators* (April 2019), https://www.gsma.com/wp-content/uploads/2019/04/The-5G-Guide_GSMA_2019_04_29_compressed.pdf ["5G GUIDE"]

GSMA, *Using Mobile Networks to Coordinate Unmanned Aircraft Traffic* (2019), https://www.gsma.com/iot/wp-content/uploads/2018/11/Mobile-Networks-enabling-UTM-v5NG.pdf

GUTMA, *Designing UTM for Global Success* (Nov. 2020), https://gutma.org/designing-utm-for-global-success/

GUTMA, *ICAO Global Aviation Trust Framework* (Sept. 20, 2018), https://gutma.org/blog/2018/09/20/icao-global-aviation-trust-framework/

GUTMA, *Map of International UTM Implementations*, https://gutma.org/?s=international+implementations

GUTMA, *Meet Our Associate Members – Dr. Terrence Martin* (June 29, 2020), https://gutma.org/blog/2020/06/29/meet-our-members-dr-terrence-martin-nova-systems/

GUTMA, *UAS Traffic Management Architecture,* v1 (2017), https://www.gutma.org/docs/Global_UTM_Architecture_V1.pdf ["GUTMA Architecture 2017"]

Hass, Douglas A., ImageStream Internet Solutions, Inc., *The Never-Was-Neutral Net and Why Informed End Users Can End the Net Neutrality Debates*, BEPRESS LEGAL SERIES (2008), http://www.kentlaw.edu/faculty/rwarner/classes/ecommerce/materials/privacy/materials/network_neutrality/hess_neverwasneutralnet.pdf

Hollnagel, Erik, et al., *The Functional Resonance Accident Model* (2004), *available at* https://pdfs.semanticscholar.org/a538/fc4f7c5ebd7c083aa8474786f6886300251f.pdf

Homola, Jeffrey R., et al., NASA Ames, *UAS Traffic Management (UTM) Simulation Capabilities and Laboratory Environment* (Sept. 25-29, 2016), DOI: 10.1109/dasc.2016.7778078, https://utm.arc.nasa.gov/docs/Homola_DASC_157026369.pdf

Homola, Jeffrey R., et al., NASA Ames, *UTM Integration and Testing Lead, UAS Traffic Management, UTM and D-NET: NASA and JAXA's Collaborative Research on Integrating Small UAS with Disaster Response Efforts,* AIAA Aviation, Atlanta, GA (June 25-29, 2018), DOI: 10.2514/6.2018-3987, https://utm.arc.nasa.gov/docs/2018-Homola-Aviation2018-Jun.pdf

Homola, Jeffrey, Sr. Research Asso., Human Sys. Integration Div., SJSU/NASA ARC, *Design and Evaluation of Corridors-in-the-sky Concept: The Benefits and Feasibility of Adding Highly Structured Routes to a Mixed Equipage Environment*, AIAA (2012), https://humansystems.arc.nasa.gov/publications/Homola_AIAA_GNC_2012.pdf

Howell, Bronwyn E., Victoria U. of Wellington, et al., *Governance of Blockchain Distributed Ledger Technology Projects: a Common-Pool Resource View* (June 11, 2019), https://dlc.dlib.indiana.edu/dlc/bitstream/handle/10535/10527/20190611_pm_Ostrom_DLTs_Polycentric.pdf?sequence=1&isAllowed=y

ICAO and ICAN, *ICAO-ICANN MOU* (Feb. 5, 2019), *available at* https://www.icann.org/en/system/files/files/mou-icao-icann-05feb19-en.pdf

ICAO, A40-WP/3271EX/137, *New Operational Concepts Involving Autonomous Systems*, Presented by the Int'l Coordinating Council of Aerospace Industries Ass'ns (ICCAIA) (Aug. 19, 2019), https://www.icao.int/Meetings/a40/Documents/WP/wp_327_en.pdf

ICAO, A40-WP/342, EX/143 (Aug. 8, 2019), Agenda Item 26, *The Safe and Efficient Integration of UAS into Airspace*, Presented by CANSO, IATA, and IFALPA, https://www.iata.org/contentassets/e45e5219cc8c4277a0e80562590793da/safe-efficient-integration-uas-airspace.pdf

ICAO, *Addressing Cybersecurity in Civil Aviation,* Working Paper presented to the Council of ICAO, Agenda Item 16: Aviation Security – Policy, A39-WP/17, EX/5 (May 30, 2016), https://www.icao.int/Meetings/a39/Documents/WP/wp_017_en.pdf

ICAO, *Air Navigation Services Providers (ANSPs) Governance and Performance*, Working Paper, CEANS-WP/36, presented by CANSO (Aug. 28, 2008), https://www.icao.int/Meetings/ceans/Documents/Ceans_Wp_036_en.pdf

ICAO, *Annex 2*, ver. 10 of the Convention on International Civil Aviation (2005), *available at* https://www.icao.int/Meetings/anconf12/Document%20Archive/an02_cons%5B1%5D.pdf; current version, www.icao.int

ICAO, *Annex 13 Aircraft Accident and Incident Investigation, available at* http://www.emsa.europa.eu/retro/Docs/marine_casualties/annex_13.pdf

ICAO, *Annex 17 Security, to the Convention on International Civil Aviation* (10th Ed. 2017), https://www.icao.int/Security/SFP/Pages/Annex17.aspx

ICAO, *Assembly Resolutions in Force [as of 4 October 2019],* Doc 10140 (2020), https://www.icao.int/publications/Documents/10140_en.pdf

ICAO, *Convention on International Civil Aviation*, Doc 7300/9 (9th Ed. 2006), https://www.icao.int/publications/Documents/7300_9ed.pdf ["Chicago Convention"]

ICAO, *Convention on the Suppression of Unlawful Acts Relating to International Civil Aviation*, Done at Beijing 2020, https://www.icao.int/secretariat/legal/Docs/beijing_convention_multi.pdf

ICAO, *Drone Enable* (2017), https://www.youtube.com/watch?v=Aidkt8RDock

ICAO, Gary Christiansen, Tech. Advisor, FAA, *ATC Services Above FL600*, Twenty-Fourth Meeting of the Cross Polar Trans East Air Traffic Management Providers' Work Group (CPWG/24), IP/03 (Dec. 12, 2017), https://www.faa.gov/about/office_org/headquarters_offices/ato/service_units/mission_support/ato_intl/cross_polar/

ICAO, *Global Air Traffic Management Operational Concept*, 1st ed., Doc 9854 (2005), https://www.icao.int/Meetings/anconf12/Document%20Archive/9854_cons_en%5B1%5D.pdf

ICAO, *Global Resilient Aviation Network Concept of Operations, For a secure and trusted exchange of information* (2018), https://www4.icao.int/ganpportal/trustframework ["ICAO Resilient ConOps"]

ICAO, *ICAO Model UAS Regulations, Part 101 and Part 102* (June 23, 2020), https://www.icao.int/safety/UA/Documents/Final%20Model%20UAS%20Regulations3%20-%20Parts%20101%20and%20102.pdf

ICAO, *ICAO signs new agreements with RTCA, EUROCAE, SAE and ARINC to better align international aviation standardization* (Dec. 12, 2017), https://www.icao.int/Newsroom/Pages/ICAO-signs-new-agreements-with-RTCA,-EUROCAE,-SAE-and-ARINC-to-better-align-international-aviation-standardization-.aspx

ICAO, *ICAO Unmanned Aircraft System Advisory Group (UAS-AG) Update*, APUAS/TF/3-IP/02, 04, Bangkok (March 7, 2019), https://www.icao.int/APAC/Meetings/2019%20APUASTF3/IP02%20ICAO%20Unmanned%20Aircraft%20System%20Advisory%20Group%20Update.pdf#search=UAS%20Advisory%20Group

ICAO, Legal Committee, *Remotely Piloted Aircraft Systems Legal Survey*, LC/37-WP/2-1 (July 26, 2018), https://www.icao.int/Meetings/LC37/Documents/LC37%20WP%202-1%20EN%20Remotely%20Piloted%20Aircraft.pdf

ICAO, *Manual of Standards – Aeronautical Information Management*, Ver. 0.1 (Oct. 31, 2017), https://www.icao.int/NACC/Documents/RegionalGroups/ANIWG/AIM/AIMManualofStandards-MoS.PDF

ICAO, *Manual on Flight and Flow—Information for a Collaborative Environment (FF-ICE)*, Doc 9965, AN/483, 1st ed. (2012), https://www.icao.int/Meetings/anconf12/Documents/9965_cons_en.pdf

ICAO, *Manual on System-Wide Information Management (SWIM) Concept* (Doc 10039) AN/511 (20xx) (Adv. ed., unedited), https://www.icao.int/airnavigation/IMP/Documents/SWIM%20Concept%20V2%20Draft%20with%20DISCLAIMER.pdf ["ICAO SWIM Concept"]

ICAO, *Manual on System-Wide Information Management (SWIM) Concept* (2015), https://www.icao.int/safety/acp/ACPWGF/CP%20WG-I%2019/10039_SWIM%20Manual.pdf#search=Doc%2010039

ICAO, *Operational Use Cases to Validate The Trust Framework*, Agenda Item 5.6: FAA Operational Scenarios, Trust Framework Study Group (TFSG), First Meeting of the Working Group on Current and Future Operational Needs, TFSG-1-WP/19 (May 5, 2019)

ICAO, *Regulations and Procedures for the International Registry*, Doc 9864 (8th ed. 2019), https://www.icao.int/publications/Documents/9864_8ed.pdf

ICAO, *Remotely Piloted Aircraft System (RPAS) Concept of Operations (CONOPS) for International IFR Operations* (2018), https://www.icao.int/safety/UA/Documents/RPAS%20CONOPS.pdf

ICAO, *Safety Management Manual (SMM),* Doc. 8959, AN/474 (3rd. Ed. 2013), https://www.icao.int/safety/fsix/Library/DOC_9859_FULL_EN.pdf

ICAO, *Service Level Agreement, Aeronautical Information Service (AIS)*, https://www.icao.int/NACC/Documents/RegionalGroups/ANIWG/AIM/AIMServiceLevelAgreement-SLA.PDF

ICAO, *Summary*, Third Meeting of the Asia/Pacific Unmanned Aircraft Systems Task Force, APUAS/TF/3 – WP/02 (March 7, 2019), https://www.icao.int/APAC/Meetings/2019%20APUASTF3/WP02%20Related%20Meetings%20Outcomes.pdf#search=Search%2E%2E%2Eutm

ICAO, *SWIM as a Foundation for UTM/ATM Integration*, presented by Frequentis, SWIM TF/3–IP/11, Agenda Item 3(h), The Third Meeting of System Wide Information Management Task Force (SWIM TF/3), Bangkok (May 7-10, 2019), https://www.icao.int/APAC/Meetings/2019SWIMTF3/IP11_Frequentis%20AI3h%20-%20Task%201-8_SWIM%20Foundation%20UTM%20ATM%20Integration%20Rev1.pdf#search=Search%2E%2E%2Eutm

ICAO, *The Report of the Thirteenth Air Navigation Conference (AN-Conf/13)* (Doc 10115), *available (restricted) at* https://portal.icao.int/icao-net/Pages/Doc10115.aspx2; Supplement No. 1 to the Report of the Thirteenth Air Navigation Conference (AN-Conf/13) (Doc 10115), www.icao.int

ICAO, *Thirteenth Air Navigation Conference (AN-CONF/13) Outcome*, Agenda Item 2: Global Development in Aviation, MIDANPIRG/17 & RASG-MID/7 Meeting (Cairo, Egypt, April 15-18, 2019), WP/4 (April 4, 2019), https://www.icao.int/MID/MIDANPIRG/Documents/MID17%20and%20RASG7/WP4.pdf

ICAO, *Treaty Collection*, https://www.icao.int/Secretariat/Legal/Pages/TreatyCollection.aspx

ICAO, *Unmanned Aircraft System Traffic Management (UTM), Request for Information* (2019), https://www.icao.int/Meetings/DRONEENABLE3/Documents/ICAO%20Request%20for%20Information%20(RFI)%202019.pdf

ICAO, *Unmanned Aircraft System Traffic Management (UTM), Request for Information* (2020), https://www.icao.int/safety/UA/Documents/RFI%20-%202020.pdf

ICAO, *Unmanned Aircraft Systems Traffic Management (UTM) – A Common Framework with Core Principles for Global Harmonization,* Ed. 1 (April 3, 2019), https://www.icao.int/safety/UA/Pages/UTM-Guidance.aspx, and https://www.icao.int/safety/UA/Documents/UTM-Framework.en.alltext.pdf

ICAO, *Unmanned Aircraft Systems Traffic Management (UTM) – A Common Framework with Core Principles for Global Harmonization*, Ed. 2 (Nov. 2019), https://www.icao.int/safety/UA/Documents/UTM-Framework%20Edition%202.pdf

ICAO, *Unmanned Aircraft Systems Traffic Management (UTM) – A Common Framework with Core Principles for Global Harmonization*, Ed. 3 (Sept. 2020), https://www.icao.int/safety/UA/Pages/UTM-Guidance.aspx ["ICAO UTM Framework Ed. 3"]

ICAO, *Working Outline for International Civil Aviation Trust Framework (ICATF),* Draft Ver. 2 (Oct. 2019) ["ICAO ICATF 2019"]

ICAO, *X.509 Certificate Policy for the International Aviation Trust Framework (IATF) Certification Authority,* Ver. 0.2 (Nov. 2019)

IEFT, *Drone Remote ID Protocol (drip)* [documents], https://datatracker.ietf.org/wg/drip/documents/

IEFT, *RFC 6698, The DNS-Based Authentication of Named Entities (DANE) Transport Layer Security (TLS) Protocol: TLSA* (July 29, 2020), https://datatracker.ietf.org/doc/rfc6698/

IETF, *RFC 6749, The OAuth 2.0 Authorization Framework* (Oct. 2012), DOI: 10.17487/rfc6749, https://tools.ietf.org/html/rfc6749

IETF, *RFC 6819, OAuth 2.0 Threat Model and Security Considerations* (Jan. 2013), DOI: 10.17487/rfc6819, https://tools.ietf.org/pdf/rfc6819.pdf

IETF, *RFC 8446, The Transport Layer Security (TLS) Protocol Version 1.3* (Aug. 2018), https://tools.ietf.org/html/rfc8446

India, Gov't of, Ministry of Civil Aviation, *National Unmanned Aircraft System (UAS) Traffic Management Policy, Architecture, Concept of Operations and Deployment Plan for enabling UTM ecosystem in India, Discussion Draft,* Ver.1.0 (Nov. 30, 2020), https://www.civilaviation.gov.in/sites/default/files/National-UTM-Policy-Dicussion-Draft-30-Nov-2020.pdf ["India UTM Policy"]

India, Gov't of, Office of the Dir. Gen. of Civil Avi., Civil Aviation Requirements, Sect. 3-Air Transport, Series X Part 1, Issue I (Aug. 27, 2018, Effective Dec. 1, 2018), F. No. 05-13/2014-AED Vol. IV, http://bathindapolice.in/civil.pdf

ISO, *ISO/IEC 17000, Conformity assessment – Vocabulary and general principles* (2nd ed. 2020-05)

ISO, *ISO/IEC 17011:2017, Conformity assessment - Requirements for accreditation bodies accrediting conformity assessment bodies,* DOI: 10.3403/30312186, https://www.iso.org/standard/67198.html

ISO, *ISO/IEC 27007:2020, Information security, cybersecurity and privacy protection – Guidelines for information security management systems auditing,* https://www.iso.org/standard/77802.html?browse=tc

ITU-R, Recommendation ITU-R M.2083-0 (09/2015), *IMT Vision — Framework and overall objectives of the future development of IMT for 2020 and beyond*, https://www.itu.int/dms_pubrec/itu-r/rec/m/R-REC-M.2083-0-201509-I!!PDF-E.pdf ["ITU IMT 2020"]

Japan Ministry of Economy, Trade and Industry, *Roadmap for Application and Technology, Development of UAVs in Japan,* http://www.meti.go.jp/english/policy/mono_info_service/robot_industry/downloadfiles/uasroadmap.pdf

JARUS, *Guidelines on Specific Operations Risk Assessment* (SORA), JAR-DEL-WG6-D.04, Ed. 2.0 (Jan. 30, 2019), http://jarus-rpas.org/sites/jarus-rpas.org/files/jar_doc_06_jarus_sora_v2.0.pdf ["JARUS SORA"]

JARUS, *JAR doc 06 SORA (package),* http://jarus-rpas.org/content/jar-doc-06-sora-package

JARUS, JARUS guidelines on SORA, *Annex E, Integrity and assurance levels for the Operational Safety Objectives (OSO)*, Doc ID. JAR-DEL-WG6-D-0.4, Ed. No. 1 (Jan 25, 2019), http://jarus-rpas.org/sites/jarus-rpas.org/files/jar_doc_06_jarus_sora_annex_e_v1.0_.pdf

JARUS, Publications, http://jarus-rpas.org/publications

JARUS, *RPAS "Required C2 Performance" (RLP) concept* JAR_doc_13, v1.0 (Jan. 5, 2016), http://jarus-rpas.org/sites/jarus-rpas.org/files/storage/Library-Documents/jar_doc_13_rpl_concept_upgraded.pdf

JARUS, *Terms of Reference*, Ref. JARUS-ToR_v7.0.2020, Sect. 3.6, http://jarus-rpas.org/sites/jarus-rpas.org/files/imce/attachments/jarus_tor_v7.0.2020_and_annex_scb_tor_130818_released_20200304.pdf

JARUS, White Paper: Use of Mobile Networks to Support UAS Operations, JAR-DEL-WG5-D.05 (Jan. 21, 2021), http://jarus-rpas.org/

Jeppesen, *Jeppesen Distribution Manager and Jeppesen Services Update Manager End User License Agreement*, https://ww2.jeppesen.com/legal/jeppesen-distribution-manager-and-jeppesen-services-update-manager/

Jiang, Tao, et al., *Unmanned Aircraft System traffic management: Concept of operation and system architecture*, INT'L J. OF TRANSP. SCI. & TECH. (Vol 5, Issue 3, Oct. 2016), https://doi.org/10.1016/j.ijtst.2017.01.004

Johnson, Ronald D., UTM Project Mgr., NASA, *Unmanned Aircraft Traffic Management (UTM) Project* [slides] (April 10, 2018), https://ntrs.nasa.gov/archive/nasa/casi.ntrs.nasa.gov/20180002542.pdf

Jung, Jaewoo, et al., *Applying Required Navigation Performance Concept for Traffic Management of Small Unmanned Aircraft Systems*, 30th Cong. of the Int'l Council of the Aeronautical Sciences (Deejeon, Korea – Sept. 25-30, 2016), https://ntrs.nasa.gov/archive/nasa/casi.ntrs.nasa.gov/20160011496.pdf

Jung, Jaewoo, et al., NASA Ames Research Center, *Automated Management of Small Unmanned Aircraft System Communications and Navigation Contingency* (Jan. 2020), DOI: 10.2514/6.2020-2195, https://utm.arc.nasa.gov/docs/2020-Jung_SciTech_2020-2195.pdf ["Jung 2020"]

Jung, Jaewoo, et al., NASA Ames Research Center, *Initial Approach to Collect Small Unmanned Aircraft System Off-nominal Operational Situations Data* (Atlanta - June 25-29, 2018), DOI: 10.2514/6.2018-3030, https://utm.arc.nasa.gov/docs/2018-Jung-Aviation2018-Jun.pdf

Kopardekar, Parimal H., *Unmanned Aerial System (UAS) Traffic Management (UTM): Enabling Low-Altitude Airspace and UAS Operations*, NASA, NASA/TM-2014-21899 (April 2014), https://ntrs.nasa.gov/archive/nasa/casi.ntrs.nasa.gov/20140013436.pdf ["Kopardekar 2014"]

Kopardekar, Parimal, et al., *Unmanned Aircraft System Traffic Management (UTM) Concept of Operations*, 6th AIAA Aviation Technology, Integration, and Operations Conference (June 2016), https://utm.arc.nasa.gov/docs/Kopardekar_2016-3292_ATIO.pdf

Kopardekar, Parimal, Ph.D., Dir., NARI, NASA, *National Unmanned Aerial System Standardized Performance Testing and Rating (NUSTAR)* (2016), https://ntrs.nasa.gov/archive/nasa/casi.ntrs.nasa.gov/20160012460.pdf

Kopardekar, Parimal, Ph.D., Dir., NARI, NASA, *Unmanned Aerial System Traffic Management System*, Talks at Google (May 3, 2016), https://www.youtube.com/watch?v=iDjZzysw1MY

Kopardekar, Parimal, Ph.D., et al., NASA Ames, *Unmanned Traffic Management (UTM) Concept of Operations*, 16th AIAA Avi. Tech., Integration, and Ops. Conf. (June 2016), https://utm.arc.nasa.gov/docs/Kopardekar_2016-3292_ATIO.pdf ["Kopardekar 2016"]

Kopardekar, Parimal, Ph.D., Sr. Technologist, Air Transportation System, and Dir., NASA NARI, *Enabling Future Airspace Operations*, Vol. 62, No. 1 (Spring 2020), J. OF AIR TRAFFIC CONTROL

Larrow, Jarrett, UTM Prog. Mgr., FAA, and Marcus, Johnson, NASA, *Annex H— SORA & UTM* (2018), https://ntrs.nasa.gov/archive/nasa/casi.ntrs.nasa.gov/20180004530.pdf

Lascara, Brock, et al., *Urban Air Mobility Airspace Integration Concepts*, MITRE, 19-00667-9 (June 2019), https://www.mitre.org/publications/technical-papers/urban-air-mobility-airspace-integration-concepts ["MITRE 2019"]

Letter from Kratos SecureInfo to Amazon Web Services, *NIST Cybersecurity Framework (CSF)*, Amazon Web Services (Jan. 2019), *App. B — Third Party Assessor Validation* (Sept. 19, 2018), https://d1.awsstatic.com/whitepapers/compliance/NIST_Cybersecurity_Framework_CSF.pdf

Li, Sheng, and Egorov, Maxim, et al., *Optimizing Collision Avoidance in Dense Airspace Using Deep Reinforcement Learning*, Thirteenth USA/Europe ATM Research and Development Seminar (ATM2019), *available at* https://arxiv.org/abs/1912.10146 ["Sheng Li DRL 2019"]

Lin, Iuon-Chang, Asia Univ. & Nat'l Chung Hsing Univ., et al., *A Survey of Blockchain Security Issues and Challenges*, INT'L J OF NETWORK SECURITY, Vol. 19, No. 5 (Sept. 2017), *available at* https://pdfs.semanticscholar.org/f61e/db500c023c4c4ef665b-d7ed2423170773340.pdf

Liu, Yigang, et al., *SOA-Based Aeronautical Service Integration, INTECH* (2011), DOI: 10.5772/29307, https://cdn.intechopen.com/pdfs/20427/InTech-Soa_based_aeronautical_service_integration.pdf

Majority Staff of the Comm. on Transp. and Infrastructure, Final Comm. Rpt., *The Design, Development & Certification of the Boeing 737 MAX* (Sept. 20, 2020), https://transportation.house.gov/imo/media/doc/2020.09.15%20FINAL%20737%20MAX%20Report%20for%20Public%20Release.pdf

Marshall, Douglas M., *Drone versus Manned Aircraft: An Analysis of the Application of the Discretionary Function Exception to the Federal Tort Claims Act to Accidents Caused by a Collision Between a Drone and a Manned Aircraft*, ISSUES IN AVIATION LAW AND POLICY, Vol. 19:2 (Spring 2020), https://las.depaul.edu/centers-and-institutes/chaddick-institute-for-metropolitan-development/research-and-publications/Documents/IALPLatestIssueSpring2020b.pdf

Matthews, Melissa, SWIM Program Manager (A), Communications, Information and Network Programs, AJM-316, FAA, *Presentation on SWIM Cloud Distribution Service*, 2019 Air Transp. Info. Ex. Conf., McLean, Va. (Sept. 23, 2019), https://www.faa.gov/air_traffic/flight_info/aeronav/atiec/media/Presentations/Day%201%20PM%20009%20Melissa%20Matthews%20SCDS.pdf

McCarthy, Tim, et al., *Fundamental Elements of an Urban UTM*, Aerospace 2020 (June 27, 2020), DOI: 10.3390/aerospace7070085, https://res.mdpi.com/d_attachment/aerospace/aerospace-07-00085/article_deploy/aerospace-07-00085.pdf

Merkle, Jay, Exec. Dir., FAA, UAS Integration Office, Presentation at the Transportation Research Board (TRB) Annual Meeting, Orlando, Fl. (Jan. 13, 2020), https://www.c-span.org/video/?468065-1/automation-technology-transportation

Meyer, Andreas, *Integrated Risk Management: A Holistic Approach to Managing Aviation Risk,* UNITING AVIATION (FEB. 4, 2019), https://www.unitingaviation.com/strategic-objective/safety/integrated-risk-management/

Murakami, David, et al., *NASA Ames Research Center, Space Traffic Management with a NASA UAS Traffic Management (UTM) Inspired Architecture*, 10.22514 (June 2019), *available at* https://www.researchgate.net/publication/330206407_

Space_Traffic_Management_with_a_NASA_UAS_Traffic_Management_UTM_Inspired_Architecture, and http://sreejanag.com/Documents/murakami_scitech_final.pdf

Nakamoto, Satoshi *Bitcoin: A Peer-to-Peer Electronic Cash System* (Oct. 31, 2008), DOI: 10.2139/ssrn.3440802, https://bitcoin.org/bitcoin.pdf

NASA Aeronautics Research Mission Directorate: Team Seedling Solicitation, *Enabling Low Altitude Civilian Applications of Unmanned Aerial Systems by Unmanned Aerial System Traffic Management (UTM),* Submitted by Parimal Kopardekar, Ph.D., Principal Investigator (2013)

NASA Ames, *What is Unmanned Aircraft Systems Traffic Management?* (May 3, 2019), https://www.nasa.gov/ames/utm

NASA, *Advanced Air Mobility (AAM), Urban Air mobility (UAM) and Grand Challenge*, AIAA (July 24, 2020), https://ntrs.nasa.gov/archive/nasa/casi.ntrs.nasa.gov/20190026695.pdf

NASA, Aeronautics Research Mission Directorate, https://www.nasa.gov/uamgc

NASA, Aeronautics Research, *Advanced Air Mobility National Campaign, Overview* (Mar. 12, 2020), https://www.nasa.gov/aeroresearch/aam/description

NASA, Aweiss, Arwa S. et al., NASA Ames, *Unmanned Aircraft Systems (UAS) Traffic Management (UTM) National Campaign II* (2018), https://ntrs.nasa.gov/archive/nasa/casi.ntrs.nasa.gov/20180000682.pdf ["Aweiss 2018"]

NASA, Aweiss, Arwa, et al., NASA Ames, *Flight Demonstration of Unmanned Aircraft System (UAS) Traffic Management (UTM) at Technical Capability Level 3* (IEEE DASC, Sept. 8–12, 2019), https://utm.arc.nasa.gov/docs/2019_Aweiss_DASC_2019.pdf, and https://ntrs.nasa.gov/archive/nasa/casi.ntrs.nasa.gov/20190030743.pdf ["Aweiss 2019"]

NASA, *Grand Challenge Developmental Testing Partners* (March 3, 2020), https://www.nasa.gov/aeroresearch/grand-challenge-developmental-testing

NASA, *How to License NASA Technology*, https://technology.nasa.gov/license

NASA, Mueller, Eric & Kopardekar, Parimal, et. al., *Enabling Airspace Integration for High-Density On-Demand Mobility Operations* (2017), https://utm.arc.nasa.gov/docs/2017-Mueller_Aviation_ATIO.pdf

NASA, NASA Technology Transfer Program, *Unmanned Aerial Systems (UAS) Traffic Management*, https://technology.nasa.gov/patent/TOP2-237

NASA, Rios, Joseph L., et al., *UAS Service Supplier Framework for Authentication and Authorization*, NASA/TM–2019–220364 (Sept. 2019), https://utm.arc.nasa.gov/docs/2019-UTM_Framework-NASA-TM220364.pdf ["UFAA 2019"]

NASA, Rios, Joseph L., et al., *UAS Service Supplier Network Performance, Results and Analysis from Flight Testing Multiple USS Providers in NASA's TCL4 Demonstration*, NAS/TM–2020-220462 (Jan. 2020), https://utm.arc.nasa.gov/docs/2020-Rios_TM_220462-USS-Net-Perf.pdf ["Rios 2020"]

NASA, *UAM Vision Concept of Operations (ConOps), UAM Maturity Level (UML)4* (Dec. 2, 2020), Deloitte, et al., https://ntrs.nasa.gov/citations/20205011091 ["NASA UAM Vision ConOps"]

NASA, *UAS Integration in the NAS Project*, https://www.hq.nasa.gov/office/aero/iasp/uas/index.htm, and Small Business Innovation Research/Small Business Technology Transfer, NASA, *UAS Integration in the NAS* (Oct. 31, 2013), https://sbir.nasa.gov/content/uas-integration-nas

NASA, *UAS Traffic Management (UTM) Project* (Sept. 4, 2018), https://www.nasa.gov/aeroresearch/programs/aosp/utm-project-description/

NASA, *Upper E Traffic Management (ETM) Tabletop 2 Summary* (Feb. 20, 2020), https://utm.arc.nasa.gov/docs/2020-02-20-UTM-TableTop-2-Summary.pdf

NASA, *UTM Documents*, https://utm.arc.nasa.gov/documents.shtml

NASA, Wing, David J., et al., *New Flight Rules to Enable the Era of Aerial Mobility in the National Airspace System*, NASA/TM–20205008308 (Nov. 2020), https://www.researchgate.net/publication/346321764_New_Flight_Rules_to_Enable_the_Era_of_Aerial_Mobility_in_the_National_Airspace_System

NATA, *Urban Air Mobility: Considerations for Vertiport Operation* (June 2019), https://www.nata.aero/assets/Site_18/files/GIA/NATA%20UAM%20White%20Paper%20-%20FINAL%20cb.pdf ["NATA UAM"]

National Academies of Sciences, Engineering, and Medicine, Committee on Enhancing Air Mobility, *Advancing Aerial Mobility, A National Blueprint,* The Nat'l Academies Press (2020), https://doi.org/10.17226/25646 ["Nat'l Academies AAM"]

National Academies of Sciences, Engineering, and Medicine, *In-time Aviation Safety Management: Challenges and Research for an Evolving Aviation System,* The Nat'l Academies Press, 10.17226/24962 (2018), DOI: 10.17226/24962, https://www.nap.edu/catalog/24962/in-time-aviation-safety-management-challenges-and-research-for-an

National Conference of State Legislatures, *Current Unmanned Aircraft State Law Landscape* (Sept. 10, 2018), http://www.ncsl.org/research/transportation/current-unmanned-aircraft-state-law-landscape.aspx (presenting diverse approaches to state law addressing UAS)

National Cyber Security Center, UK, *Cyber Insurance Guidance* (Aug. 6, 2020), https://www.ncsc.gov.uk/guidance/cyber-insurance-guidance

National Technology Transfer and Advancement Act of 1995, Pub. L. No. 104-113 (Mar. 7, 1996), 110 Stat. 782, 15 U.S.C. 3701 et seq., *available at* https://www.law.cornell.edu/topn/national_technology_transfer_and_advancement_act_of_1995

NATO, *NATO Standardization Document Database*, https://nso.nato.int/nso/nsdd/listpromulg.html

NEXA Advisors and NBAA, *Business Aviation Embraces Electric Flight, How Urban Air Mobility Creates Enterprise Value* (Oct. 21, 2019), https://nbaa.org/wp-content/uploads/aircraft-operations/uas/NEXA-Study-2019-Business-Aviation-Embraces-Electric-Flight.pdf

NEXA UAM study sets 20-year market value of $318 billion, Air Traffic Management.net (Aug. 20, 2019), https://airtrafficmanagement.keypublishing.com/2019/08/20/nexa-uam-study-sets-20-year-market-value-of-318-billion-across-74-cities/?dm_i=4JU%2C6G8B6%2CPM6JE7%2CPKRIH%2C1

NextGen – SESAR, *State of Harmonisation*, 3rd ed., Report prepared by the Coordination Committee (CCOM) & Deployment Coordination Committee (DCOM) for the US–EU MoC Annex 1 Executive Committee (EXCOM) (Sept. 2018), DOI:10.2829/90536, https://www.faa.gov/nextgen/media/NextGen-SESAR_State_of_Harmonisation.pdf

NextGen – SESAR, *State of Harmonisation*, Report prepared by the Coordination Committee (CCOM) for the US–EU MoC Annex 1 High-Level Committee (2016), DOI:10.2829/979584, https://www.faa.gov/nextgen/media/nextgen_sesar_harmonisation.pdf

NextGen-SESAR Memorandum of Cooperation, NAT-I-9406, as amended (2014, 2017), Official J. of the EU, L 90/3 (April 6, 2018), https://eur-lex.europa.eu/legal-content/EN/TXT/PDF/?uri=CELEX:22018A0406(01)&from=EN

Nguyen, T. V., *Dynamic Delegated Corridors and 4D Required Navigation Performance for Urban Air Mobility (UAM) Airspace Integration*, J. of Avi./Aerospace Edu. & Research, 29(2) (2020), https://doi.org/10.15394/jaaer.2020.1828

NIS Cooperation Group, *EU coordinated risk assessment of the cybersecurity of 5G networks, Report* (Oct. 9, 2019), https://ec.europa.eu/newsroom/dae/document.cfm?doc_id=62132

NIST, *A Taxonomic Approach to Understanding Emerging Blockchain Identity Management Systems,* NIST Cybersecurity Whitepaper (Jan. 14, 2020), DOI: 10.6028/nist.cswp.01142020, https://nvlpubs.nist.gov/nistpubs/CSWP/NIST.CSWP.01142020.pdf

NIST, Computer Security Resource Center, *Publications,* https://csrc.nist.gov/publications/

NIST, *Enhanced Security Requirements for Protecting Controlled Unclassified Information: A Supplement to NIST Special Publication 800-172* (Feb. 2021), https://csrc.nist.gov/publications/detail/sp/800-172/final

NIST, FIPS Publication 200, *Minimum Security Requirements for Federal Information and Information Systems* (March 2006), DOI: 10.6028/nist.fips.200, https://csrc.nist.gov/publications/detail/fips/200/final, and https://csrc.nist.gov/csrc/media/publications/fips/200/final/documents/fips-200-final-march.pdf

NIST, *Framework for Improving Critical Infrastructure Cybersecurity*, Ver. 1.1 (April 16, 2018), https://nvlpubs.nist.gov/nistpubs/CSWP/NIST.CSWP.04162018.pdf

NIST, NISTIR 7628 Rev 1, *Guidelines for Smart Grid Cybersecurity for Smart Grid* (Sept. 2014), https://csrc.nist.gov/publications/detail/nistir/7628/rev-1/final

NIST, Special Publication 800-30, *Guide to Conducting Risk Assessments* (Sept. 2012), DOI: 10.6028/nist.sp.800-30r1, https://nvlpubs.nist.gov/nistpubs/Legacy/SP/nistspecialpublication800-30r1.pdf

NIST, Special Publication 800-37, Rev. 2, *Risk Management Framework for Information Systems and Organizations* (Dec. 2018), DOI: 10.6028/nist.sp.80037r2, https://nvlpubs.nist.gov/nistpubs/SpecialPublications/NIST.SP.800-37r2.pdf

NIST, Special Publication 800-52, Rev. 2, *Guidelines for the Selection, Configuration, and Use of Transport Layer Security (TLS) Implementations*, Kerry A. McKay and David A. Cooper, Computer Security Div. Info. Tech. Lab. (Aug. 2019), DOI: 10.6028/nist.sp.800-52r2, https://nvlpubs.nist.gov/nistpubs/SpecialPublications/NIST.SP.800-52r2.pdf

NIST, Special Publication 800-53, Rev. 5 DRAFT, *Security and Privacy Controls for Information Systems and Organizations* (Aug. 2017), DOI: 10.6028/nist.sp.800-53r5-draft, https://csrc.nist.gov/CSRC/media//Publications/sp/800-53/rev-5/draft/documents/sp800-53r5-draft.pdf [current version is Rev. 4] (April 2013), DOI: 10.6028/nist.sp.800-53r4, https://nvlpubs.nist.gov/nistpubs/SpecialPublications/NIST.SP.800-53r4.pdf ["NIST SP 800-53 Rev. 5"]

NIST, Special Publication 800-63-3, *Digital Identity Guidelines,* https://www.nist.gov/itl/tig/projects/special-publication-800-63 ["SP 800-63-3"]

NIST, Special Publication 800.122, *Guide to Protecting the Confidentiality of Personally Identifiable Information (PII),* DOI:10.6028/nist.sp.800-122, https://nvlpubs.nist.gov/nistpubs/Legacy/SP/nistspecialpublication800-122.pdf

NIST, Special Publication 800-160, Vol. 1, *System Security Engineering: Considerations for a Multidisciplinary Approach in the Engineering of Trustworthy Secure Systems* (Nov. 2016, updated Mar. 21, 2018), DOI: 10.6028/nist.sp.800-160v1, https://csrc.nist.gov/publications/detail/sp/800-160/vol-1/final

NIST, Special Publication 800-204, *Security Strategies for Microservices-Based Application Systems* (Aug. 2019), DOI: 10.6028/nist.sp.800-204-draft, https://csrc.nist.gov/publications/detail/sp/800-204/final

OpenSky Network, http://opensky-network.org/

P.R.C., State Post Bureau, *Specification for Express Delivery Service by Unmanned Aircraft,* YZ/T 0172—2020, National Postal Standardization Technical Committee (SAC/TC462) (Dec. 2020), http://www.spb.gov.cn/zc/ghjbz_1/201508/W020201204542195544172.pdf

Peisen, Deborah J., *Analysis of Vertiport Studies Funded by the Airport Improvement Program (AIP),* DOT/FAA/RD-93/37 (May 1994), https://apps.dtic.mil/dtic/tr/fulltext/u2/a283249.pdf

Porsche Consulting, *The Future of Vertical Mobility, Sizing the market for passenger, inspection, and goods services until 2035* (2019), https://www.porsche-consulting.com/fileadmin/docs/04_Medien/Publikationen/TT1371_The_Future_of_Vertical_Mobility/The_Future_of_Vertical_Mobility_A_Porsche_Consulting_study__C_2018.pdf

Prandini, Mari, et al., *Toward Air Traffic Complexity Assessment in New Generation Air Traffic Management Systems*, IEEE Trans. on Intelligent Transport. Sys., vol. 12:3 (Sept. 11, 2011), DOI 10.1109/tits.2011.2113175, https://ieeexplore.ieee.org/abstract/document/5723748

Prevot, Tom, Ph.D., Dir., Airspace Systems, Uber, Presentation at Uber Elevate Summit 2018, L.A., Cal. (2018), http://ushst.org/Portals/0/PDF/20180212_UberElevate2018Summit.pdf

Reisman, Ronald J., Aero Computer Eng'r, NASA Ames, *Air Traffic Management Blockchain Infrastructure for Security, Authentication, and Privacy* (2018), https://ntrs.nasa.gov/archive/nasa/casi.ntrs.nasa.gov/20190000022.pdf

Remotely piloted drones no longer accompanied by piloted planes flying into Hancock Field thanks to new radar system, Nextar Broadcasting, Inc. (Sept. 10, 2019), https://www.localsyr.com/news/local-news/remotely-piloted-drones-no-longer-accompanied-by-piloted-planes-flying-into-hancock-field-thanks-to-new-radar-system/

Ren, Liling, et al., *Integration and Flight Test of Small UAS Detect and Avoid on A Miniaturized Avionics Platform*, 10.1109/DASC43569.2019.9081780 (2019), DOI: 10.1109/DASC43569.2019.9081780, *available at* https://www.researchgate.net/publication/335960156_Integration_and_Flight_Test_of_Small_UAS_Detect_and_Avoid_on_A_Miniaturized_Avionics_Platform

Rios, Joseph L., et al., *Strategic Deconfliction Performance, Results and Analysis from the NASA UTM Technical Capability Level 4 Demonstration* (Aug. 2020), https://utm.arc.nasa.gov/docs/2020-Rios-NASA-TM-20205006337.pdf ["Rios Strategic Deconfliction Performance"]

Rios, Joseph L., et al., *UTM UAS Service Supplier Development, Sprint 1 Towards Technical Capability Level 4*, NASA Technical Memorandum, Moffett Field (Oct. 2018), https://utm.arc.nasa.gov/docs/UTM_UAS_TCL4_Sprint1_Report.pdf ["Rios TLC4 Sprint 1"]

Rios, Joseph L., et al., *UTM UAS Service Supplier Development, Sprint 2 Toward Technical Capability Level 4*, NASA/TM-2018-220050 (Dec. 2018), https://utm.arc.nasa.gov/docs/2018-UTM_UAS_TCL4_Sprint2_Report_v2.pdf ["Rios, TLC4 Sprint 2"]

Rios, Joseph, et al., *Flight Demonstration of Unmanned Aircraft System (UAS) Traffic Management (UTM) at Technical Capability Level 4*, AIAA Aviation Forum (June 25-19, 2020), https://utm.arc.nasa.gov/docs/2020-Rios-Aviation2020-TCL4.pdf ["Rios TCL4"]

Rios, Joseph, et al., *UAS Service Supplier Specification, Baseline requirements for providing USS services within the UAS Traffic Management System,* NASA/TM-2019-220376 (Oct. 2019), https://utm.arc.nasa.gov/docs/2019_Rios-TM-220376.pdf ["NASA USS Spec."]

Rios, Joseph, Ph.D., UTM Project Chief Eng'r, *UAS Traffic Management (UTM) Project Strategic Deconfliction: System Requirements Final Report*, UTM-SD.05 (July 2018), https://www.researchgate.net/publication/332107751_UAS_Traffic_Management_UTM_Project_Strategic_Deconfliction_System_Requirements_Final_Report/download, and (deck) https://utm.arc.nasa.gov/docs/2018-UTM-Strategic-Deconfliction-Final-Report.pdf ["Rios – Strategic Deconfliction"]

RTCA, Paper No. 099-18/PMC-1742, *Terms of Reference*, SC-228, *Minimum Performance Standards for Unmanned Aircraft Systems* (Rev 5) (Mar. 22, 2018), *available at* https://www.rtca.org/sc-228/ ["RTCA ToR SC-228"]

Sachs, Peter, *A Quantitative Framework for UAV Risk Assessment*, Vol. 1, Rpt. TR-008 (Sept. 13, 2018), https://storage.googleapis.com/blueprint/TR-008_Open_Risk_Framework_v1.0.pdf

Sachs, Peter, et al., *Evaluating Fairness in UTM Architecture and Operations,* Airbus UTM, Ver. 1.1, TR-010 (Feb. 2020), https://storage.googleapis.com/blueprint/UTM_Fairness_Tech_Report-v1.1.pdf ["Airbus UTM Fairness 2020"]

Sampigethaya, Krishna, and Kopardekar, Parimal, et al., *Cyber security of unmanned aircraft system traffic management (UTM),* ICNS (2018), https://ieeexplore.ieee.org/document/8384832 ["Sampigethaya, Kopardekar 2018"]

SESAR Joint Undertaking, CORUS, *Intermediate Concept of Operation for U-Space* (June 5, 2019), https://www.sesarju.eu/projects/corus and, https://www.sesarju.eu/projects/corus ["CORUS Intermediate ConOps"]

SESAR Joint Undertaking, CORUS, *U-Space Concept of Operation,* Vol. 1, *Enhanced Overview*, Ed. 01.01.03 (April 9, 2019), https://www.sesarju.eu/node/3411

SESAR Joint Undertaking, CORUS, *U-Space Concept of Operation,* Vol. 2, Ed. 03.00.02 (Oct. 25, 2019), https://www.sesarju.eu/node/3411 ["CORUS ConOps"]

SESAR Joint Undertaking, *Demonstrating RPAS Integration in the European Aviation System* (2016), https://www.sesarju.eu/sites/default/files/documents/reports/RPAS-demo-final.pdf

SESAR Joint Undertaking, *European ATM Master Plan, Digitalising Europe's Aviation Infrastructure* (2020), https://www.atmmasterplan.eu/exec/overview ["SESAR ATM Master Plan 2020"]

SESAR Joint Undertaking, *European ATM Master Plan: Roadmap for the safe integration of drones into all classes of airspace* (2017), http://www.sesarju.eu/sites/default/files/documents/reports/European%20ATM%20Master%20Plan%20Drone%20roadmap.pdf ["SESAR ATM Master Plan 2017"]

SESAR Joint Undertaking, *Exploring the boundaries of air traffic management, A summary of SESAR exploratory research results* 2016-2018 (2018), https://www.sesarju.eu/sites/default/files/documents/reports/ER_Results_2016_2018.pdf

SESAR Joint Undertaking, Media Release, *SESAR JU GOF U-space project: First demos successfully completed* (Feb. 2, 2019), https://www.frequentis.com/sites/default/files/pr/2019-07/PR_GOF_USPACE_072019.pdf

SESAR Joint Undertaking, *U-Space Blueprint* (2017), https://www.sesarju.eu/sites/default/files/documents/reports/U-space%20Blueprint%20brochure%20final.PDF ["SESAR U-space Blueprint"]

SESAR Joint Undertaking, *U-Space, Supporting Safe and Secure Drone Operations in Europe, A preliminary summary of SESAR U-space research and innovation results (2017-2019)* (2020), https://www.sesarju.eu/sites/default/files/documents/u-space/U-space%20Drone%20Operations%20Europe.pdf

SESAR, CORUS, *Intermediate U-space ConOps Annex E: List of Threats and Events*, Ed. (Mar. 19, 2019), https://tinyurl.com/U-space-Threats-Events

SESAR, *eATM PORTAL, European ATM Master Plan*, https://www.atmmasterplan.eu/, and *U-Space Implementation Map Tool,* https://www.atmmasterplan.eu/depl/u-space

SESAR, *Strategic Research and Innovation Agenda, Digital European Sky* (Sept. 2020), https://www.sesarju.eu/sites/default/files/documents/reports/SRIA%20Final.pdf ["SESAR SRIA"]

SESAR, U-space, *Supporting Safe and Secure Drone Operations in Europe, Consolidated Report on SESAR U-Space Research and Innovation Results* (Nov. 11, 2020), https://www.sesarju.eu/node/3691

Skygrid, *The Power of Blockchain in Unmanned Aviation* (June 2020), https://www.skygrid.com/whitepaper/blockchain-in-unmanned-aviation.pdf

Smith, Irene Skupniewicz, et al., NASA Research Center, *USS Service Supplier Checkout, How UTM Confirmed Readiness of Flight Tests with UAS Service Suppliers*, NASA/TM-2019-220456 (Dec. 1, 2019), https://ntrs.nasa.gov/archive/nasa/casi.ntrs.nasa.gov/20190034170.pdf ["UAS Service Supplier Checkout"]

Swiss FOCA, *Memorandum of Cooperation established between The Federal Office of Civil Aviation and skyguide* (Dec. 20, 2018), https://www.bazl.admin.ch/dam/bazl/en/dokumente/Gut_zu_wissen/Drohnen_und_Flugmodelle/moc_susi.pdf.download.pdf/Memorandum%20of%20Cooperation.pdf

Swiss FOCA, *Swiss U-Space, ConOps, Ver. 1.0* (April 1, 2019), https://www.bazl.admin.ch/dam/bazl/en/dokumente/Gut_zu_wissen/Drohnen_und_Flugmodelle/Swiss_U-space_Implementation.pdf.download.pdf/Swiss%20U-Space%20Implementation.pdf ["Swiss U-Space ConOps v1.0"]

Swiss FOCA, *Swiss U-Space, ConOps, Ver. 1.1* (2020), https://www.bazl.admin.ch/bazl/en/home/good-to-know/drones-and-aircraft-models/u-space.html ["Swiss U-Space ConOps v1.1"]

Taft, Jeffrey D., *Grid Architecture*, IEEE POWER AND ENERGY MAG. (SEPT./OCT. 2019), https://www.nxtbook.com/nxtbooks/pes/powerenergy_091019/index.php#/20

The Future of Aerospace: Interconnected from Surface to Space, MITRE (Jan. 15, 2020), https://www.mitre.org/news/in-the-news/the-future-of-aerospace-interconnected-from-surface-to-space

The White House, *National Strategy for Aviation Security of the United States of America* (Dec. 2018), https://www.aviationtoday.com/2019/10/08/homeland-security-dod-transportation-officials-focus-aviation-cyber-security/ ["Nat'l Strategy for Avi. Security 2018"]

Thurling, Andy, CTO, NUAIR Alliance, *The Drone Market*, ENTERPRISE TECH. REV. (AUG. 1, 2019), https://www.enterprisetechnologyreview.com/magazines/August2019/Display_Tech/#page=29

Transport Canada, *RPAS Traffic Management (RTM) Services Trials, CALL FOR PROPOSALS, Phase 1 – Round 1* (May 11, 2020), https://www.tc.gc.ca/en/services/aviation/drone-safety/drone-innovation-collaboration/remotely-piloted-aircraft-systems-rpas-traffic-management-services-testing-call-proposals.html ["Transport Canada RTM Services"]

Transport Canada, White Paper, *Drone Talks: Planning for Success* (May 29-30, 2019) Workshop #1: Airspace and RTM System, *available at* https://www.dropbox.com/s/uiozw2pwr8v4nku/Canada-RTM.doc?dl=0

TSCP, *Redacted X.509 CPS, v. 4 for the TSCP Bridge Certificate Authority (TBCA)* (Feb. 6, 2019), https://www.tscp.org/tscp-certification-practice-statement/

UK Civil Aviation Authority, CAP 1861, *Beyond Visual Line of Sight in Non-Segregated Airspace,* v2 (Oct. 8, 2020), https://publicapps.caa.co.uk/modalapplication.aspx?appid=11&mode=detail&id=9294 ["UK CAA CAP 1861"]

Unmanned Aircraft Safety Team, *Safety Enhancement No. 1, Airspace Awareness and Geofencing, Final Report, Out-of-the-Box Protection of High-Risk Airport Locations* (May 1, 2020), http://unmannedaircraftsafetyteam.org/safety-enhancement-no-1-%E2%80%A8airspace-awareness-and-geofencing/

UTM system to be implemented for the first time in Nordic region, Int'l Airport Rev. (Jan 16, 2020), https://www.internationalairportreview.com/news/110464/utm-system-implemented-nordic-region/

Vascik, Parker D., et al., *Assessment of air traffic control for urban air mobility and unmanned systems*, Report No. ICAT-2018-03 (June 2018), https://dspace.mit.edu/bitstream/handle/1721.1/117686/ICAT-2018-03_Vascik_2018a%20Vascik%20ICRAT%20UAM%20and%20UAS%20ATC.pdf?sequence=1

Vascik, Parker D., et al., *Constraint Identification in On-Demand Mobility for Aviation through an Exploratory Case Study of Los Angeles*, *17th* AIAA Aviation Technology, Integration, and Operations Conf., Denver, CO (June 2017), Report No. ICAT-2017-09, DOI:10.2514/6.2017-3083, https://dspace.mit.edu/bitstream/handle/1721.1/115341/2017c%20Vascik%20AIAA%20Constraint%20Evaluation%20ICAT%20Report_2017_9.pdf?sequence=1

Verizon, *The Eight Currencies*, https://www.verizonwireless.com/business/articles/business/5g-network-performance-attributes/

Virginia Tech Mid-Atlantic Aviation Partnership (MAAP), *MAAP UPP2 Final Report Attachment A, Security Considerations for Operationalization of UTM Architecture* (Jan. 12, 2021)

Volocopter, The roadmap to scalable urban air mobility, White paper 2.0 (Mar. 2021), https://press.volocopter.com/images/pdf/Volocopter-WhitePaper-2-0.pdf

Wee Keong Ng, Nanyang Technological Univ., et al., *A Study of Cyber Security Threats to Traffic Management of Unmanned Aircraft Systems* (Jan. 2007), *available at* https://www.researchgate.net/publication/313476896_A_Study_of_Cyber_Security_Threats_to_Traffic_Management_of_Unmanned_Aircraft_Systems

Whitney, Kaitlynn M., et al., *The Root Cause of Failure in Complex IT Projects: Complexity Itself*, Procedia Computer Science, Elsevier, V.20 (2013), pp. 325–330, https://doi.org/10.1016/j.procs.2013.09.280

Wing, David J., et al., Langley Research Center, *Autonomous Flight Rules—A Concept for Self-Separation in U.S. Domestic Airspace*, NASA/TP–2011-217174 (Nov. 2011), https://ntrs.nasa.gov/archive/nasa/casi.ntrs.nasa.gov/20110023668.pdf

Wing, *InterUSS Platform™—Data Node Stateless API*, http://app.swaggerhub.com/apis/InterUSS_Platform/data_node_api/

Wing, *InterUSS Platform™, Overview, Governance Requirements, Design and Implementation*, v. 1.0.1 (Aug. 28, 2018), GitHub, https://github.com/wing-aviation/InterUSS-Platform

Wolfe, Frank, *MITRE Aviation Lab Looking At What National Airspace System Will Look Like in 2035*, Aviation Today (Dec. 11, 2019), https://www.aviationtoday.com/2019/12/11/mitre-aviation-lab-looking-national-airspace-system-will-look-like-2035/

Woods, Daniel W., et al., *Does insurance have a future in governing cybersecurity*, IEEE Security & Privacy (2019), https://tylermoore.utulsa.edu/govins20.pdf ["Woods"]

Woods, Daniel W., et al., *Policy measures and cyber insurance: a framework*, *J. of Cyber Policy*, Vol. 2, Issue 2 (Aug. 2017), DOI: 10.1080/23738871.2017.1360927, *available at* https://www.tandfonline.com/doi/full/10.1080/23738871.2017.1360927

Wu, Stephen S., *Privacy and Security Challenges of Advanced Technologies*, PLI Current: The J. of PLI Press, vol. 3, no. 3 (Summer 2019), https://legacy.pli.edu/Content/Journal/PLI_Current_The_Journal_of_PLI_Press_Vol/_/N-bqZ1z0zkxf?ID=374249

Yang, Xuxi, and Egorov, Maxim, et al, *Stress Testing of Unmanned Traffic Management Decision Making Systems* (June 8, 2020), DOI: 10.2514/6.2020-2868, *available at* https://arc.aiaa.org/doi/abs/10.2514/6.2020-2868

Zhang, Jianping, *UOMS in China* (Shenzhen, June 6-8, 2018), https://rpas-regulations.com/wp-content/uploads/2018/06/1.2-Day1_0910-1010_CAAC-SRI_Zhang-Jianping_UOMS-_EN.pdf.

List of Abbreviated References

[5G GUIDE] GSMA, *THE 5G GUIDE, A Reference for Operators* (April 2019), https://www.gsma.com/wp-content/uploads/2019/04/The-5G-Guide_GSMA_2019_04_29_compressed.pdf

[AC 00-63A] FAA, *Advisory Circular AC 00-63A, CHG 1*, Subj: Use of Flight Deck Displays of Digital Weather and Aeronautical Information, App. 1 (Jan. 6, 2017), https://www.faa.gov/regulations_policies/advisory_circulars/index.cfm/go/document.information/documentID/1024126

[AC 20-153B] FAA, *Advisory Circular AC 20-153B*, Subj: Acceptance of Aeronautical Data Processes and Associated Databases (April 19, 2016), https://www.faa.gov/regulations_policies/advisory_circulars/index.cfm/go/document.information/documentID/1029446

[Airbus Blueprint] Airbus, *Blueprint For The Sky* (2018), https://storage.googleapis.com/blueprint/Airbus_UTM_Blueprint.pdf, and https://www.airbusutm.com/

[Airbus UTM Fairness 2020] Sachs, Peter, et al., *Evaluating Fairness in UTM Architecture and Operations,* Airbus UTM, Ver. 1.1, TR-010 (Feb. 2020), https://storage.googleapis.com/blueprint/UTM_Fairness_Tech_Report-v1.1.pdf

[Airspace Access Priorities ARC] FAA, *Aviation Rulemaking Committee Charter*, Subj: Airspace Access Priorities Aviation Rulemaking Committee (Effective, Nov. 20, 2017), https://www.faa.gov/regulations_policies/rulemaking/committees/documents/index.cfm/committee/browse/committeeID/677

[ANSI Roadmap] ANSI, *Standardization Roadmap for Unmanned Aircraft Systems*, Ver. 2.0 (June 2020), https://share.ansi.org/Shared%20Documents/Standards%20Activities/UASSC/ANSI_UASSC_Roadmap_V2_June_2020.pdf

[ASTM Autonomy Framework] ASTM International, Stephen Cook, Chair, ASTM AC377, et al., Technical Report, *Autonomy Design and Operations in Aviation: Terminology and Requirements Framework*, TR1-EB (July 2019), DOI: 10.1520/tr1-eb, https://www.astm.org/DIGITAL_LIBRARY/TECHNICAL_REPORTS/PAGES/1fe1b67c-5ff0-488e-ab61-cb0d329bc63c.htm, and https://www.astm.org/DIGITAL_LIBRARY/TECHNICAL_REPORTS/index.html

[ASTM F3442 M - 20 DAA Performance] ASTM International, F3442/F3442M-20, *Standard Specification for Detect and Avoid System Performance Requirements*, DOI: 10.1520/F3442_F3442M-20, https://www.astm.org/Standards/F3442.htm

[ASTM Remote ID] ASTM International, F3411-19, *Standard Specification for Remote ID and Tracking* (July 2, 2019), DOI: 10.1520/F3411-19, https://www.astm.org/Standards/F3411.htm

[ASTM Surveillance SDSP] ASTM International, F_ [WK69690], *New Specification for Surveillance UTM Supplemental Data Service Provider (SDSP) Performance*, https://www.astm.org/DATABASE.CART/WORKITEMS/WK69690.htm

[ASTM UTM Spec.] ASTM International, F_ [WK63418], *New Specification for UAS Traffic Management (UTM) UAS Service Supplier (USS) Interoperability*, https://www.astm.org/DATABASE.CART/WORKITEMS/WK63418.htm

[Aweiss 2018] NASA, Aweiss, Arwa S., et al., NASA Ames, *Unmanned Aircraft Systems (UAS) Traffic Management (UTM) National Campaign II* (2018), https://ntrs.nasa.gov/archive/nasa/casi.ntrs.nasa.gov/20180000682.pdf

[Aweiss 2019] NASA, Aweiss, Arwa, et al., NASA Ames, *Flight Demonstration of Unmanned Aircraft System (UAS) Traffic Management (UTM) at Technical Capability Level 3* (IEEE DASC, Sept. 8–12, 2019), https://utm.arc.nasa.gov/docs/2019_Aweiss_DASC_2019.pdf, and https://ntrs.nasa.gov/archive/nasa/casi.ntrs.nasa.gov/20190030743.pdf

[Baum 2019] Baum, Michael S., et al., Aviators Code Initiative, *Improving Cockpit Awareness of Unmanned Aircraft Systems Near Airports* (March 8, 2019), http://www.secureav.com/UAS-Awareness-Listings-Page.html

[Baum FCA Liability] Baum, Michael S., *Federal Certification Authority Liability and Policy*, NIST, by MITRE Corp. under Contract #50SBN1C6732 (1992), DOI: 10.1016/s0267-3649(00)80027-8, *available at* https://tinyurl.com/PKI-Baum

[Chicago Convention] ICAO, *Convention on International Civil Aviation*, Doc 7300/9 (9th Ed. 2006), https://www.icao.int/publications/Documents/7300_9ed.pdf

[Comm Delegated Reg. (EU) 2019/945 – UAS] European Commission, *Commission Delegated Regulation (EU) 2019/945 of 12 March 2019 on unmanned aircraft systems and on third-country operators of unmanned aircraft systems*, Annex 1, Part 2, Sect. 12, 2019 (O.J.) (L 152/1) (June 11, 2019), https://eur-lex.europa.eu/legal-content/EN/TXT/PDF/?uri=CELEX:32019R0945&from=EN

[Comm Implementing Reg. (EU) 2021 /664 - U-space] European *Commission, Commission Implementing Regulation (EU) 2021/664 of 22 April 2021 on a regulatory framework for the U-space*, https://eur-lex.europa.eu/legal-content/EN/TXT/?uri=CELEX%3A32021R0664

[Comm Implementing Reg. (EU) 2021/666 - U-space] *European Commission, Commission Implementing Regulation (EU) 2021/666 of 22 April 2021 amending Regulation (EU) No 923/2012 as regards requirements for manned aviation operating in U-space airspace*, https://eur-lex.europa.eu/legal-content/EN/TXT/?uri=CELEX%3A32021R0666

[Comm Implementing Reg. (EU) 2019/947 - UA] European Commission, *Commission Implementing Regulation (EU) 2019/947 of 24 May 2019 on the rules and procedures for the operation of unmanned aircraft* (O.J.) (L 152/1) (July 11, 2019), https://eur-lex.europa.eu/eli/reg_impl/2019/947/oj

[Connected Places Catapult] Connected Places Catapult, *Enabling UTM in the UK* (May 2020), https://s3-eu-west-1.amazonaws.com/media.cp.catapult/wp-content/uploads/2020/05/22110912/01296_Open-Access-UTM-Report-V4.pdf

[CORUS ConOps] SESAR Joint Undertaking, CORUS, *U-Space Concept of Operation*, Vol. 2, Ed. 03.00.02 (Oct. 25, 2019), https://www.sesarju.eu/node/3411

[CORUS Intermediate ConOps] SESAR Joint Undertaking, CORUS, *Intermediate Concept of Operation for U-Space* (June 5, 2019), https://www.sesarju.eu/projects/corus and, https://www.sesarju.eu/projects/corus

[Davis 2019] Davis, Mark, et al., *Aerial Cellular: What can Cellular do for UAVS with and without changes to present standards and regulations*, Presented at AUVSI Xponential, in Chicago, Ill. (May 2, 2019)

[DLR Blueprint] DLR, *DLR Blueprint, Concept for Urban Airspace Integration* (Dec. 2017), https://www.dlr.de/fl/Portaldata/14/Resources/dokumente/veroeffentlichungen/Concept_for_Urban_Airspace_Integration.pdf

[Drone Alliance Europe, U-Space Whitepaper 2019], Drone Alliance Europe, *U-space Whitepaper* (July 2019), www.sesarju.eu/sites/default/files/documents/reports/U-space%20Blueprint%20brochure%20final.PDF

[EASA Draft AMC & GM] *EASA, Draft acceptable means of compliance (AMC) and Guidance Material (GM) to Opinion No 01/2020 on a high-level regulatory framework for the U-space*, issue 1 (XX Month 2020), https://www.easa.europa.eu/sites/default/files/dfu/Draft%20AMC%20%26%20GM%20to%20the%20U-space%20Regulation%20—%20for%20info%20only.pdf

[EASA Easy Access 2019/945] EASA, *Easy Access Rules for Unmanned Aircraft Systems* (Regulation (EU) 2019/947 and Regulation (EU) 2019/945) (March 3, 2020), https://www.easa.europa.eu/document-library/general-publications/easy-access-rules-unmanned-aircraft-systems-regulation-eu

[EASA Opinion 01/2018] EASA, Opinion No 01/2018, *Unmanned aircraft system (UAS) operations in the 'open' and 'specific' categories*, https://www.easa.europa.eu/sites/default/files/dfu/Opinion%20No%2001-2018.pdf

[EASA Opinion 01/2020 - U-space] EASA, Opinion No 01/2020, *High-level regulatory framework for the U-space*, RMT.0230 (March 13, 2020), https://www.easa.europa.eu/document-library/opinions/opinion-012020

[EASA Reg. Framework U-space] EASA, Draft Opinion, *High-level regulatory framework for the U-space*, RMT.0230 (Oct. 2019), *available at* https://rpas-regulations.com/wp-content/uploads/2019/10/EASA_Draft-Opinion-on-U-space.pdf

[EASA SC Light UAS] EASA, *Proposed Special Condition for Light UAS*, Doc. No. SC Light-UAS 01, Issue: 1 (July 20, 2020), https://www.easa.europa.eu/document-library/product-certification-consultations/proposed-special-condition-light-uas

[Embraer UATM ConOps] Embraer^X and Airservices, *Urban Air Traffic Management, Concept of Operations,* Ver. 1 (Dec. 2020), https://embraerx.embraer.com/global/en/uatm

[Embraer^X 2019] Embraer^X, et al., *Flight Plan 2030, An Air Traffic Management Concept for Urban Air Mobility* (May 2019), https://daflwcl3bnxyt.cloudfront.net/m/4e5924f5de45fd3a/original/embraerx-whitepaper-flightplan2030.pdf

[Erotokritou 2019] Erotokritou, Chrystel, *The Legal Liability of Air Traffic Controllers*, INQUIRIES JOURNAL/STUDENT PULSE, Vol. 4(02) (2019), http://www.inquiriesjournal.com/a?id=613

[EUROCAE OSED-DAA] EUROCAE, *Operational Services and Environment Definition for Detect & Avoid in Very Low Level Operations*, Draft ED-269 for Open Consultation (June 2019), https://eurocae.net/news/posts/2019/june/eurocae-open-consultation-ed-267/

[EUROCAE Programme 2019] EUROCAE, *EUROCAE Technical Work Programme* (Public ver. 2019), https://eurocae.net/media/1567/eurocae-public-twp-2019.pdf

[EUROCONTROL UAS ATM Integration] EUROCONTROL and EASA, *UAS ATM Integration, Operational Concept*, Ed. 1 (Nov. 27, 2018), *available at* https://www.eurocontrol.int/sites/default/files/publication/files/uas-atm-integration-operational-concept-v1.0-release%2020181128.pdf

[FAA & NASA UTM RTT WG #3] FAA & NASA, Unmanned Aircraft System Traffic Management (UTM) Research Transition Team (RTT), Concept Working Group, *Concept & Use Cases Package #3: Technical Capability Level 4*, Ver. 1.0 (March 2019)

[FAA Aerospace Forecast] FAA, *FAA Aerospace Forecast Fiscal Years 2020–2040* (2020), https://www.faa.gov/data_research/aviation/aerospace_forecasts/media/FY2020-40_FAA_Aerospace_Forecast.pdf

[FAA ANPRM 2019] FAA, ANPRM, *Safe and Secure Operations of Small Unmanned Aircraft Systems*, 84 Fed. Reg. 3732-33 (2019) (to be codified at 14 C.F.R. pt. 107), https://www.federalregister.gov/documents/2019/02/13/2019-00758/safe-and-secure-operations-of-small-unmanned-aircraft-systems

[FAA ConOps UAM v1.0] FAA NextGen, *Concept of Operations for Urban Air Mobility (UAM)*, v1.0 (June 26, 2020), https://nari.arc.nasa.gov/sites/default/files/attachments/UAM_ConOps_v1.0.pdf

[FAA ConOps UTM v1.0] FAA NextGen, *Concept of Operations, Unmanned Aircraft System (UAS) Traffic Management* (UTM), v1.0 (May 2018), https://utm.arc.nasa.gov/docs/2018-UTM-ConOps-v1.0.pdf

[FAA ConOps UTM v2.0] FAA NextGen, *Concept of Operations, Unmanned Aircraft System (UAS) Traffic Management* (UTM), v2.0 (March 2, 2020), https://www.faa.gov/uas/research_development/traffic_management/media/UTM_ConOps_v2.pdf

[FAA High E ConOps v1.0] FAA NextGen, *Concept of Operations, Upper Class E Traffic Management*, v1.0 (April 22, 2020), https://nari.arc.nasa.gov/sites/default/files/attachments/ETM_ConOps_V1.0.pdf

[FAA Roadmap] FAA, *Integration of Civil Unmanned Aircraft Systems (UAS) in the National Airspace System (NAS) Roadmap*, Second Edition (July 2018), https://www.faa.gov/uas/resources/policy_library/media/Second_Edition_Integration_of_Civil_UAS_NAS_Roadmap_July%202018.pdf

[FAA SWIM Governance] FAA, *System Wide Information Management (SWIM), Governance Policies*, Ver. 3.1 (Feb. 6, 2020), https://www.faa.gov/air_traffic/technology/swim/governance/standards/media/SWIM%20Governance%20Policies%20v3.1_20200206_Final.pdf

[FAA UPP Phase 2 ConUse] FAA, *Unmanned Aircraft Systems (UAS) Traffic Management (UTM) Pilot Program Phase Two, Concept of Use V1.0* (Dec. 2019)

[FAA USS Performance Rules] FAA ATO, *Low Altitude Authorization and Notification Capability (LAANC), USS Performance Rules*, Ver. 4.1 (Apr. 17, 2020), https://www.faa.gov/uas/programs_partnerships/data_exchange/laanc_for_industry/media/LAANC_USS_Performance_Rules.pdf

[FAARA] *FAA Reauthorization Act of 2018*, Pub. L. No. 115-254, 132 Stat. 3186 (Oct. 5, 2018), *available at* https://www.congress.gov/115/bills/hr302/BILLS-115hr302enr.pdf

[FESSA] *FAA Extension, Safety, and Security Act of 2016*, Pub. L. No. 114-190, 130 Stat. 615 (2016), codified at 49 U.S.C. 40101 note, https://www.congress.gov/114/plaws/publ190/PLAW-114publ190.pdf

[GAO UAS] GAO, *Unmanned Aircraft Systems, FAA Could Strengthen Its Implementation of a Drone Traffic Management System by Improving, Communication and Measuring Performance*, GAO-21-165 (Jan. 2021), https://www.gao.gov/assets/720/712037.pdf

[GDPR] European Parliament, *Regulation (EU) 2016/679 of the European Parliament and of the Council of 27 April 2016 on the protection of natural persons with regard to the processing of personal data and on the free movement of such data, and repealing Directive 95/46/EC (General Data Protection Regulation)*, 2016 O.J. (L 119) 1 (April 27, 2016), https://eur-lex.europa.eu/legal-content/EN/TXT/HTML/?uri=CELEX:32016R0679&from=EN, and https://gdpr-info.eu/

[ICAO ICATF 2019] ICAO, *Working Outline for International Civil Aviation Trust Framework (ICATF)*, Draft Ver. 2 (Oct. 2019)

[ICAO Resilient ConOps] ICAO, *Global Resilient Aviation Network Concept of Operations, For a secure and trusted exchange of information* (2018), https://www4.icao.int/ganpportal/trustframework

[ICAO SWIM Concept] ICAO, *Manual on System-Wide Information Management (SWIM) Concept* (Doc 10039) AN/511 (20xx) (Adv. ed., unedited), https://www.icao.int/airnavigation/IMP/Documents/SWIM%20Concept%20V2%20Draft%20with%20DISCLAIMER.pdf

[ICAO UTM Framework Ed. 3] ICAO, *Unmanned Aircraft Systems Traffic Management (UTM) – A Common Framework with Core Principles for Global Harmonization*, Ed. 3 (Sept. 2020), https://www.icao.int/safety/UA/Pages/UTM-Guidance.aspx

[India UTM Policy] India, Gov't of, Ministry of Civil Aviation, *National Unmanned Aircraft System (UAS) Traffic Management Policy, Architecture, Concept of Operations and Deployment Plan for enabling UTM ecosystem in India, Discussion Draft*, Ver.1.0 (Nov. 30, 2020), https://www.civilaviation.gov.in/sites/default/files/National-UTM-Policy-Discussion-Draft-30-Nov-2020-updated.pdf

[ITU IMT 2020] ITU-R, Recommendation ITU-R M.2083-0 (09/2015), *IMT Vision — Framework and overall objectives of the future development of IMT for 2020 and beyond*, https://www.itu.int/dms_pubrec/itu-r/rec/m/R-REC-M.2083-0-201509-I!!PDF-E.pdf

[JARUS SORA] JARUS, *Guidelines on Specific Operations Risk Assessment* (SORA), JAR-DEL-WG6-D.04, Ed. 2.0 (Jan. 30, 2019), http://jarus-rpas.org/sites/jarus-rpas.org/files/jar_doc_06_jarus_sora_v2.0.pdf

[JO 7200.23A] FAA, *ORDER JO 7200.23A*, Subj: Unmanned Aircraft Systems (Aug. 1, 2017), https://www.faa.gov/documentlibrary/media/order/jo_7200.23a_unmanned_aircraft_systems_(uas).pdf

[Jung 2020] Jung, Jaewoo, et al., NASA Ames Research Center, *Automated Management of Small Unmanned Aircraft System Communications and Navigation Contingency* (Jan. 2020), DOI: 10.2514/6.2020-2195, https://utm.arc.nasa.gov/docs/2020-Jung_SciTech_2020-2195.pdf

[Kopardekar 2014] Kopardekar, Parimal H., *Unmanned Aerial System (UAS) Traffic Management (UTM): Enabling Low-Altitude Airspace and UAS Operations*, NASA, NASA/TM-2014-21899 (April 2014), https://ntrs.nasa.gov/archive/nasa/casi.ntrs.nasa.gov/20140013436.pdf

[Kopardekar 2016] Kopardekar, Parimal, Ph.D., et al., NASA Ames, *Unmanned Traffic Management (UTM) Concept of Operations*, 16th AIAA Avi. Tech., Integration, and Ops. Conf. (June 2016), https://utm.arc.nasa.gov/docs/Kopardekar_2016-3292_ATIO.pdf

[LAANC ConOps 2.1] FAA ATO, *Low Altitude Authorization and Notification Capability (LAANC) Concept of Operations*, Ver. 2.1 (March 20, 2020), https://www.faa.gov/uas/programs_partnerships/data_exchange/laanc_for_industry/media/FAA_LAANC_CONOPS.pdf

[LAANC MOA] FAA, *Memorandum of Agreement for Low Altitude Authorization and Notification Capability (LAANC) Between FEDERAL AVIATION ADMINISTRATION (FAA) And _______*, v2.3 (April, 2020), https://www.faa.gov/uas/programs_partnerships/data_exchange/laanc_for_industry/media/Memorandum_of_Agreement.pdf

[LAANC USS Onboarding] FAA, *USS Onboarding, Low Altitude Authorization and Notification Capability (LAANC)*, v1.0 (Feb. 2018), https://www.faa.gov/uas/programs_partnerships/data_exchange/laanc_for_industry/media/LAANC_UAS_Service_Supplier_onboarding_information.pdf

[LAANC USS Operating Rules] FAA ATO, *Low Altitude Authorization and Notification Capability (LAANC), USS Operating Rules*, Ver. 1.3 (Dec. 14, 2018), https://www.faa.gov/uas/programs_partnerships/data_exchange/laanc_for_industry/media/FAA_sUAS_LAANC_Ph1_USS_Rules.pdf

[LAANC USS Performance Rules], FAA ATO, *Low Altitude Authorization and Notification Capability (LAANC), USS Performance Rules*, Ver. 5.0 (Feb. 5, 2021), https://www.faa.gov/uas/programs_partnerships/data_exchange/laanc_for_industry/media/LAANC_USS_Performance_Rules.pdf

[LAANC-PHA] FAA, *Preliminary Hazard Analysis (PHA) for Low Altitude Authorization and Notification Capability (LAANC)* - Safety Risk Management Document, Ver. 1.0, p. 3 (Oct. 23, 2017)

[MITRE 2019] Lascara, Brock, et al., *Urban Air Mobility Airspace Integration Concepts*, MITRE, 19-00667-9 (June 2019), https://www.mitre.org/publications/technical-papers/urban-air-mobility-airspace-integration-concepts

[MNO-UTM Interface] ACJA, *Interface for Data Exchange between MNOs and the UTM Ecosystem, NetworkCoverage Service Definition*, v1.00 (Feb. 2021), https://gutma.org/acja/wp-content/uploads/sites/10/2021/02/ACJA-NetworkCoverage-Service-Definition-v1.00.pdf (A joint cooperation between GSMA and GUTMA)

[NASA UAM Vision ConOps] NASA, *UAM Vision Concept of Operations (ConOps), UAM Maturity Level (UML) 4*, Doc. ID 20205011091, Deloitte (Dec. 2, 2020), https://ntrs.nasa.gov/citations/20205011091

[NASA USS Spec.] Rios, Joseph, et al., *UAS Service Supplier Specification, Baseline requirements for providing USS services within the UAS Traffic Management System,* NASA/TM-2019-220376 (Oct. 2019), https://utm.arc.nasa.gov/docs/2019_Rios-TM-220376.pdf

[Nat'l Academies AAM] National Academies of Sciences, Engineering, and Medicine, Committee on Enhancing Air Mobility, *Advancing Aerial Mobility, A National Blueprint*, The Nat'l Academies Press (2020), https://doi.org/10.17226/25646

[Nat'l Strategy for Avi. Security 2018] The White House, *National Strategy for Aviation Security of the United States of America* (Dec. 2018), https://www.aviationtoday.com/2019/10/08/homeland-security-dod-transportation-officials-focus-aviation-cyber-security/

[NATA UAM] NATA, *Urban Air Mobility: Considerations for Vertiport Operation* (June 2019), https://www.nata.aero/assets/Site_18/files/GIA/NATA%20UAM%20White%20Paper%20-%20FINAL%20cb.pdf

[NIST SP 800-53 Rev. 5] NIST, Special Publication 800-53, Rev. 5 DRAFT, *Security and Privacy Controls for Information Systems and Organizations* (Aug. 2017), DOI: 10.6028/nist.sp.800-53r5-draft,https://csrc.nist.gov/CSRC/media//Publications/sp/800-53/rev-5/draft/documents/sp800-53r5-draft.pdf [current version is Rev. 4] (April 2013), DOI: 10.6028/nist.sp.800-53r4, https://nvlpubs.nist.gov/nistpubs/SpecialPublications/NIST.SP.800-53r4.pdf

[Remote ID Rule] FAA, *Remote Identification of Unmanned Aircraft,* Final rule (issued Dec. 28, 2020), 86 Fed. Reg. 4390 (Jan. 15, 2021), https://www.federalregister.gov/d/2020-28948

[RID NPRM 2019] FAA, NPRM, *Remote Identification of Unmanned Aircraft Systems*, 48 Fed. Reg. 72438-72524 (Dec. 31, 2019) (to be codified at 14 C.F.R. pts. 1, 47, 48, 89, 91, and 107), https://www.federalregister.gov/documents/2019/12/31/2019-28100/remote-identification-of-unmanned-aircraft-systems

[Rios 2020] NASA, Rios, Joseph L., et al., *UAS Service Supplier Network Performance, Results and Analysis from Flight Testing Multiple USS Providers in NASA's TCL4 Demonstration*, NAS/TM–2020-220462 (Jan. 2020), https://utm.arc.nasa.gov/docs/2020-Rios_TM_220462-USS-Net-Perf.pdf

[Rios – Strategic Deconfliction] Rios, Joseph, Ph.D., UTM Project Chief Eng'r, *UAS Traffic Management (UTM) Project Strategic Deconfliction: System Requirements Final Report*, UTM-SD.05 (July 2018), https://www.researchgate.net/publication/332107751_UAS_Traffic_Management_UTM_Project_Strategic_Deconfliction_System_Requirements_Final_Report/download, and (deck) https://utm.arc.nasa.gov/docs/2018-UTM-Strategic-Deconfliction-Final-Report.pdf

[Rios Strategic Deconfliction Performance] Rios, Joseph L., et al., *Strategic Deconfliction Performance, Results and Analysis from the NASA UTM Technical Capability Level 4 Demonstration* (Aug. 2020), https://utm.arc.nasa.gov/docs/2020-Rios-NASA-TM-20205006337.pdf

[Rios TCL4] Rios, Joseph, et al., *Flight Demonstration of Unmanned Aircraft System (UAS) Traffic Management (UTM) at Technical Capability Level 4*, AIAA Aviation Forum (June 25-19, 2020), https://utm.arc.nasa.gov/docs/2020-Rios-Aviation2020-TCL4.pdf

[Rios TLC4 Sprint 1] Rios, Joseph L., et al., *UTM UAS Service Supplier Development, Sprint 1 Towards Technical Capability Level 4*, NASA Technical Memorandum, Moffett Field (Oct. 2018), https://utm.arc.nasa.gov/docs/UTM_UAS_TCL4_Sprint1_Report.pdf

[Rios, TLC4 Sprint 2] Rios, Joseph L., et al., *UTM UAS Service Supplier Development, Sprint 2 Toward Technical Capability Level 4*, NASA/TM-2018-220050 (Dec. 2018), https://utm.arc.nasa.gov/docs/2018-UTM_UAS_TCL4_Sprint2_Report_v2.pdf

[RTCA ToR SC-228] RTCA, Paper No. 163-20/PMC-2034, *Terms of Reference*, SC-228, *Minimum Performance Standards for Unmanned Aircraft Systems* (June 11, 2020), https://www.rtca.org/wp-content/uploads/2020/08/sc-228_tor_rev_10_approved_06-11-2020.pdf

[Sampigethaya, Kopardekar 2018] Sampigethaya, Krishna, and Kopardekar, Parimal, et al., *Cyber security of unmanned aircraft system traffic management (UTM)*, ICNS (2018), https://ieeexplore.ieee.org/document/8384832

[SESAR ATM Master Plan 2017] SESAR Joint Undertaking, *European ATM Master Plan: Roadmap for the safe integration of drones into all classes of airspace* (2017), http://www.sesarju.eu/sites/default/files/documents/reports/European%20ATM%20Master%20Plan%20Drone%20roadmap.pdf

[SESAR ATM Master Plan 2020] SESAR Joint Undertaking, *European ATM Master Plan, Digitalising Europe's Aviation Infrastructure* (2020), https://www.atmmasterplan.eu/exec/overview

[SESAR SRIA] SESAR, *Strategic Research and Innovation Agenda, Digital European Sky* (Sept. 2020), https://www.sesarju.eu/sites/default/files/documents/reports/SRIA%20Final.pdf

[SESAR U-space Blueprint] SESAR Joint Undertaking, *U-Space Blueprint* (2017), https://www.sesarju.eu/sites/default/files/documents/reports/U-space%20Blueprint%20brochure%20final.PDF

[Sheng Li DRL 2019] Li, Sheng, and Egorov, Maxim, et al., *Optimizing Collision Avoidance in Dense Airspace Using Deep Reinforcement Learning*, Thirteenth USA/Europe ATM Research and Development Seminar (ATM2019), *available at* https://arxiv.org/abs/1912.10146

[SP 800-63-3] NIST, Special Publication 800-63-3, *Digital Identity Guidelines*, https://www.nist.gov/itl/tig/projects/special-publication-800-63

[Swiss U-Space ConOps v1.0] Swiss FOCA, *Swiss U-Space, ConOps, Ver. 1.0* (April 1, 2019), https://www.bazl.admin.ch/dam/bazl/en/dokumente/Gut_zu_wissen/Drohnen_und_Flugmodelle/Swiss_U-space_Implementation.pdf.download.pdf/Swiss%20U-Space%20Implementation.pdf

[Swiss U-Space ConOps v1.1] Swiss FOCA, *Swiss U-Space, ConOps, Ver. 1.1* (2020), https://www.bazl.admin.ch/bazl/en/home/good-to-know/drones-and-aircraft-models/u-space.html

[TCL 3] FAA & NASA, Unmanned Aircraft System Traffic Management (UTM) Research Transition Team (RTT), Concept Working Group, *Concept & Use Cases Package #2 Addendum: Technical Capability Level 3*, Ver. 1.0, Doc. No. 20180007223 (July 2018), https://ntrs.nasa.gov/search.jsp?R=20180007223

[Transport Canada RTM Services] Transport Canada, *RPAS Traffic Management (RTM) Services Trials, CALL FOR PROPOSALS, Phase 1 – Round 1* (May 11, 2020), https://www.tc.gc.ca/en/services/aviation/drone-safety/drone-innovation-collaboration/remotely-piloted-aircraft-systems-rpas-traffic-management-services-testing-call-proposals.html

[UAS Pilots Code] Baum, Michael S., et al., Aviators Code Initiative, *UAS Pilots Code* (2018), https://www.secureav.com/UAS-Listings-Page.html; *id.*, annotated ver., http://www.secureav.com/UASPC-annotated-v1.0.pdf

[UAS Service Supplier Checkout] Smith, Irene Skupniewicz, et al., NASA Research Center, *USS Service Supplier Checkout, How UTM Confirmed Readiness of Flight Tests with UAS Service Suppliers*, NASA/TM-2019-220456 (Dec. 1, 2019), https://ntrs.nasa.gov/archive/nasa/casi.ntrs.nasa.gov/20190034170.pdf

[UFAA 2019] NASA, Rios, Joseph L., et al., *UAS Service Supplier Framework for Authentication and Authorization, A federated approach to securing communications between service suppliers within the UAS Traffic Management system,* NASA/TM–2019–220364 (Sept. 2019), https://utm.arc.nasa.gov/docs/2019-UTM_Framework-NASA-TM220364.pdf

[UK CAA CAP 1861] UK Civil Aviation Authority, CAP 1861, *Beyond Visual Line of Sight in Non-Segregated Airspace*, v2 (Oct. 8, 2020), https://publicapps.caa.co.uk/modalapplication.aspx?appid=11&mode=detail&id=9294

[UPP Objectives] FAA, *Statement of Objectives for Test Site Selection to Participate in UTM Pilot Program Demo Program*, Rev#3 (July 20, 2018), https://www.faa.gov/uas/research_development/traffic_management/utm_pilot_program/media/UTM_Pilot_Program_Smart_Sheet.pdf

[UPP Phase 2] FAA & NASA, *Unmanned Aircraft Systems (UAS) Traffic Management (UTM) Pilot Program (UPP) Phase 2, Industry Workshop* (Dec. 2019), https://www.faa.gov/uas/research_development/traffic_management/utm_pilot_program/media/UPP2_Industry_Workshop_Briefing.pdf

[UPP Phase 2 Progress Report] FAA & NASA, *Unmanned Aircraft Systems (UAS) Traffic Management (UTM), UTM Pilot Program (UPP) Phase Two (2) Progress Report, V. 1.0* (March 2021), https://www.faa.gov/uas/research_development/traffic_management/utm_pilot_program/media/UTM_Pilot_Program_Phase_2_Progress_Report.pdf

[Volocopter Roadmap] Volocopter, *The roadmap to scalable urban air mobility, White paper 2.0* (Mar. 2021), https://press.volocopter.com/images/pdf/Volocopter-WhitePaper-2-0.pdf

[Woods] Woods, Daniel W., et al., *Does insurance have a future in governing cybersecurity,* IEEE Security & Privacy (2019), https://tylermoore.utulsa.edu/govins20.pdf

Index

B

C

D

E

F

G

H

I

J

K

L

M

N

O

P

T

V

W

Printed in the United States
by Baker & Taylor Publisher Services